MARINE ENVIRONMENTAL POLLUTION, 2

DUMPING AND MINING

TO: ~~Bob abel~~ — one of the
ablest and most outstanding
Pioneers in the field
of U.S. Oceanography, in
the broadest sense of the
Word; and in all of its
ramifications — and a
person who it has been
my Priviledge to call a
close friend for more
years than I would
care to admit —

Rith Geyer
3/27/83

FURTHER TITLES IN THIS SERIES

Elsevier Oceanography Series, 27B

MARINE ENVIRONMENTAL POLLUTION, 2

DUMPING AND MINING

Edited by

RICHARD A. GEYER

Department of Oceanography
Texas A & M University
College Station, TX 77843, U.S.A.

ELSEVIER SCIENTIFIC PUBLISHING COMPANY
Amsterdam—Oxford—New York   1981

ELSEVIER SCIENTIFIC PUBLISHING COMPANY
335 Jan van Galenstraat
P.O. Box 211, 1000 AE Amsterdam, The Netherlands

Distributors for the United States and Canada:

ELSEVIER/NORTH-HOLLAND INC.
52, Vanderbilt Avenue
New York, N.Y. 10017

(with 177 figures; 94 tables; and 806 references)

Library of Congress Cataloging in Publication Data (Revised
Main entry under title:

Marine environmental pollution.

    (Elsevier oceanography series ; v. 27)
    Bibliography: v. 1, p.   ; v. 2, p.
    Includes indexes.
    CONTENTS: v. 1. Hydrocarbons.--v. 2. Dumping and
mining.
    1. Marine pollution. I. Geyer, Richard A.
GC1085.M285   574.5'2632   80-14362
ISBN 0-444-41847-4 (Elsevier/North-Holland : v. 1)
ISBN 0-444-41855-5 (Elsevier/North-Holland : v. 2) :
fl 240.00 (v. 2 : est.)

Printed in The Netherlands

To My Children and Grandchildren

— with the hope that the problem of marine environmental pollution will be solved equitably in the time remaining.

PREFACE

The definition of pollution with its many connotations can be found readily in any dictionary. In addition, each person has his or her own interpretation of the meaning of this many-faceted word. But, in general, the majority of these individual interpretations is based primarily on subjective and emotional, rather than objective criteria and facts. In addition, the definitions generally carry negative connotations. However, there are certain environmental conditions which can be categorized as pollution in the strictest sense, but still the results, if pragmatic criteria are applied, can be demonstrated to be beneficial rather than deleterious. For example, cooling water necessary for the successful operation of a power plant is usually warmer than the ambient temperature of the body of water into which it is being discharged. In tropical areas and in the summer, in a portion of the temperate zones, this process might raise the temperature to the point where it exceeded optimum tolerance levels for a variety of marine organisms, indigenous to the area. On the other hand, the polar regions and during the winter seasons in the temperate zones it would still increase the temperature of the water body such as an estuary or a lagoon above ambient. But, this action would actually be beneficial by increasing the rate of growth of the marine organisms. This is especially important economically for those species having commercial value, or are an important link in the food chain.

Another difficulty with the use of the word pollution to describe a specific environmental situation is that the question of the degree of pollution is always not immediately evident or defined. If it were to be defined as, "an anomalous concentration of any constituent comprising a given environmental system", then the value of the ambient level must be known or agreed upon, before the term "anomalous" in the definition can have any significant meaning. To carry this concept one step further requires that valid baseline data for the concentration of each component must already be known and agreed upon, — "agreed upon", by whom? — and for how long a period of time must these data be obtained to be significant? Still another constraint may be found in reaching a concensus as to the accuracy and repeatability with which an individual parameter must be observed.

Further, there is still the important dimension of the gradient, or rate of change with time, in the concentration of a particular constituent. The actual quantitative difference between a "normal" and an "anomalous level" may not be as critical to the well-being of a marine organism, as the time required to get from an accepted baseline, or ambient level value, to one which can be agreed upon as being at a harmful level. Then again, — who and what are the qualifications of those who decide; and — whose decisions will, in turn, be accepted by whom? It is evident from this line of reasoning that many questions are raised and must be answered. Even more important, those answers must be generally accepted; and the problem involved in an actual or potential pollution situation be defined; and in turn must be solved to the satisfaction of a variety of persons and regulatory agencies.

It should be emphasized, however, that the difficulties described primarily involve the so-called "gray area" of the total pollution spectrum. These lie between conditions where most reasonably minded persons would agree that a state of pollution exists, in the generally accepted use of the term, in a given situation; to one where the concensus of the same group would be equally adamant that it does not exist. Again it is these gray areas in between, where subjectivity, rather than objectivity usually prevails in reaching a "valid" conclusion.

It is with these difficulties and constraints in mind that criteria were established to bring together yet another series of chapters on the broad and provocative subject of pollution. Heretofore, most books on the subject emphasize the effects of certain types of pollutants in a given geographic area, or comprise the proceedings of a particular workshop or seminar. The latter is generally concerned with a specific pollutant and sponsored by a pertinent investigative or governmental regulatory agency, or an academic institution specializing in research areas germane to the problem. Rarely is a book available in which the information includes the results of an industrial or group of industrial organizations. Therefore, it is the purpose of this book and its companion volume to bring together in a coordinated and thematic fashion the results of research of academic, governmental and industrial groups. The chapters in the two volumes describe research conducted over a broad and representative group of actual and potential pollutants which can or could have significant effects on life in the marine environment.

It is important to gain a more definitive and at the same time objective understanding of the effects of a given pollutant on a marine ecological system; and to be made aware of this problem from as many perspectives as possible. Hence, after selecting the diverse types of representative pollutants to be discussed, efforts were made to obtain contributions from each of the three groups describing the results of their research efforts. It was not possible in every case to achieve this objective. However, the response was so great that it soon became apparent that it would be necessary to add a second volume for two reasons; if the chapters were to be presented in one book, it would become rather unwieldy, but for an even more important reason. The interest in hydrocarbons is so great that there are sufficient chapters available to constitute a separate volume. It was, therefore, decided to devote Volume 1 to this subject, and this volume to other categories of major types of potential and actual pollutants. These include a wide variety of materials associated with such diverse processes and activities as heavy metals, sewage, particulate matter including dredge spoils, chlorinated hydrocarbons, thermal and other industrial effluents, as well as radioactivity and even cannery wastes.

When the terms pollutant or pollution are used, it almost always carries the connotation of some material or process caused by the activities of mankind, rather than by nature. This is indeed extremely unfortunate because the effects of pollution in the broadest sense caused by nature is generally many orders of magnitude greater. A very recent example of a major naturally occurring oil seep is described in *Science*, Vol. 205, 7 September 1979, pp. 999–1001 (G. R. Harvey et al.). It was discovered in the form of a weathered crude oil-rich layer of water about 1.5 km wide extending for hundreds of kilometers. It was conservatively estimated to have involved a total of more than one megaton of hydrocarbons. This material was found at a depth of about 100 m beneath the surface, with a thickness of over 100 m, starting in the eastern Caribbean, near Trinidad, and extending into the South Atlantic. It should be emphaized that this one event represents almost twice the value of $6 \cdot 10^5$ tons proposed by a Special Committee of the U.S. National Academy of Sciences for the total amount of naturally occurring petroleum hydrocarbons seeping into the world oceans each year. This is the reason for having two sections to Volume 1, one describing petroleum hydrocarbon pollution caused by nature, and the other by mankind. For example, the role of nature in polluting the ocean with petroleum hydrocarbons has been a most active one extending throughout much of geologic time; and yet the oceans literally teem with all types of botanical and zoological forms of life. These have obviously adapted successfully in a wide variety of ways to the presence of the hydrocarbons, as demonstrated in the chapters comprising Section A of Volume 1. Yet, the scientific literature on pollution is filled with terms such as "chronic," "sub-lethal" and "cumulative". These are used to justify the need for extensive long-term research to determine the effects of potential pollutants on various components of an ecological system. This is an example of the subjective and emotional reactions to the effects of possible pollution by petroleum hydrocarbons, and applies also to a variety of other substances.

The emotional reactions can be demonstrated by a similar demand to study in great detail over a long period of time the effects not only of mud used in drilling offshore oil wells, but of various constituents of these muds. The mineral barite is an example. It also is given internally in large quantities every day to countless humans, as a part of an accepted routine, medical, diagnostic method. In addition, the acceptable concentrations of barium (1 ppm) in drinking water approved by U.S. Public Health officials for human consumption is 30 times more than that found in drilling muds (0.03 ppm at 25°C)! Why is there not a comparable hue and cry by environmentalists to reduce the amount of barium in drinking water consumed by people to at least, if not less, than that which mollusks, corals, benthic and other forms of life in the ocean might be subjected to from drilling muds.

Similarly, there has been much concern about the effects on the ecology of the floor of the ocean from mining manganese modules. Yet, it can be readily demonstrated that the effect on the bottom sediments and benthic organisms from a single turbidity current active only for a few hours is equivalent to about one hundred years of manganese mining in a given area. Also, the particulate matter carried by rivers into the ocean is many orders of magnitude greater than that involved in ocean mining. For example, the Amazon each day deposits 3×10^6 tons of fine sediments into the Atlantic.

Extensive research at considerable expense also has been conducted of the effects of manganese mining on the possible generation of a "red-tide" and accompanying mass-mortality, because of possibly bringing to the surface excessive amount of nutrients. This remains to be demonstrated and again the volume of ocean water and sediments involved in maganese mining is miniscule in comparison. But, incidences of "red-tide" have resulted in the destruction of fish and other types of marine life in amounts representing several orders of magnitude greater than the varied types of pollution that can be attributed to man. Similarly, the solids, liquids and gases, of which some have high radioactive levels, emanating from volcanic eruptions in and beneath the sea contribute tremendous amounts of pollutants of all types to the marine environment.

These are but a few of the examples that could be cited to illustrate the fact that all too often reactions to pollution on the part of many have been one of subjectivity rather than objectivity; and emotionalism overcoming logic to severely cloud reasonable perspectives in evaluating a given pollution situation. It is the primary objective of this volume to put into better perspective the respective roles, not only of nature and man in using the oceans as a place to dump actual and potential pollutants, but the degree to which the coastal zones can be used for this purpose as compared to the remainder of the ocean. It is hoped that the companion Volume 1 will accomplish the same purpose for the types of hydrocarbons discussed therein.

RICHARD A. GEYER

GLOSSARY

List of units used in this volume (effort has been made to use S.I. units except in those cases where units are strongly associated with the particular field of science discipline; to assist, the relations for converting to S.I. units are listed here)

Angle	rad		$57.295778°$	(radian)
Area	acre		$4047\,\text{m}^2$	
Concentration	gAt.l^{-1}	(gram atom per liter)		
	M	(molarity)		
	N	(normality)		
	ppb	(parts per billion)		
	ppm	(parts per million)		
	vol.%	(per cent by volume)		
	wt.%	(per cent by weight)		
Electric current	A			(ampère)
Electric potential	V	(volt)	$1\,\text{kg m}^2\,\text{s}^{-3}\,\text{A}^{-1}$	
Energy	cal.	(calorie)	$4.184\,\text{J}$	(joule)
	erg		$10^{-7}\,\text{J}$	
	eV	(electron volt)	$1.602 \cdot 10^{-19}\,\text{J}$	
Flow rate	mgd	(million gallons per day)	$0.04375\,\text{m}^3\,\text{s}^{-1}$	
Length	in.	(inch)	$0.02452\,\text{m}$	(meter)
	ft.	(foot)	$0.3048\,\text{m}$	
	yd.	(yard)	$0.9144\,\text{m}$	
	mi.	(mile)	$1609\,\text{m}$	
	mil		$2.452 \cdot 10^{-5}\,\text{m}$	
	fathom	(nautical fathom)	$1.8288\,\text{m}$	
	n.m.	(nautical mile)	$1852\,\text{m}$	
Mass	mol	(mole)		
	lb.	(pound avoirdupois)	$453.6\,\text{g}$	(gram)
	t	(metric ton)	$10^6\,\text{g}$	
	ton	(long ton)	$1.016 \cdot 10^6\,\text{g}$	
Photoenergy	Ein	(einstein)	$0.11972\,\text{J m}^{-1}$	
Power	hp.	(horsepower)	$745.7\,\text{W}$	(watt)
Pressure	atm.	(atmosphere)	$1.013 \cdot 10^5\,\text{Pa}$	(pascal)
	bar		$10^5\,\text{Pa}$	
	lb. in.$^{-2}$	(pound-force per square inch)	$6.295 \cdot 10^3\,\text{Pa}$	
Radioactivity	Ci	(curie)		
	cpm	(counts per minute)		
	dpm	(disintegrations per minute)		

Temperature	°C	(degree Celsius)	K	(kelvin)
	°F	(degree Fahrenheit)	$T(°F) = 1.8T(°C) + 32$	
Time	min.	(minute)	60 s	(second)
	hr.	(hour)	3600 s	
	day		86,400 s	
	yr.	(year)	$3.16 \cdot 10^7$ s	
Velocity	kt.	(knot)	$0.5015 \, \text{m s}^{-1}$	
Volume	bbl	(barrel)	$0.1590 \, \text{m}^3$	
	gal.	(gallon)	$0.003785 \, \text{m}^3$	
	l	(liter)	$10^{-3} \, \text{m}^3$	

S.I. unit prefixes

	n	(nano)	10^{-9}
	μ	(micro)	10^{-6}
	m	(milli)	10^{-3}
	c	(centi)	10^{-2}
	d	(deci)	10^{-1}
	k	(kilo)	10^3
	M	(mega)	10^6
	G	(giga)	10^9

LIST OF CONTRIBUTORS TO THIS VOLUME

Dr. M. V. Angel
Institute of Ocean Science
Brook Road
Wormley Godalming, Surrey GU8 5UB
United Kingdom

Dr. Michael Bernhard
C.N.E.N.
Euratom 1 19030 Fiascherino
Environmental Protection Research
    Division
La Spezia, Italy

Dr. Roberto Blundo
Institute of Pharmacy and
    Toxicology
University of Naples
Naples, Italy

Dr. Robert E. Burns
N.O.A.A./Office of Marine Pollution
    Assessment
Deep-Ocean Mining Environmental
    Studies Project
7600 San Point Way Northeast
Seattle, WA 98115, U.S.A.

Dr. Michael Carnes
Department of Oceanography
Texas A & M University
College Station, TX 77843, U.S.A.

Dr. D. Eisma
Netherlands Institute for Sea Research
P.O. Box 59
1790 AB Den Burg, Texel
The Netherlands

Dr. M. J. R. Fasham
Institute of Ocean Science
Brook Road
Wormley Godalming, Surrey GU8 5UB
United Kingdom

Dr. Richard A. Geyer
Department of Oceanography
Texas A & M University
College Station, TX 77873, U.S.A.

Dr. Paul A. Gillespie
Cawthorn Institute
P.O. Box 175
Nelson, New Zealand

Dr. Charles G. Gunnerson
N.O.A.A.
Environmental Research Laboratories
Boulder, CO 80303, U.S.A.

Dr. John B. Herbich
Department of Civil Engineering
Texas A & M University
College Station, TX 77843, U.S.A.

Dr. Takashi Ichiye
Department of Oceanography
Texas A & M University
College Station, TX 77843, U.S.A.

Dr. J. W. Lavelle
N.O.A.A./Pacific Marine
    Environmental Laboratory
3711 15th Avenue Northeast
Seattle, WA 98105, U.S.A.

Present affiliation: National Oceanic and Atmospheric Administration, Office of Marine Pollution Assessment, Pacific Office, 7600 Sand Point Way NE, Seattle, WA 98115, U.S.A.

Dr. Alan J. Mearns*
Southern California Coastal Water
  Research Project
646 West Pacific Coast Highway
Long Beach, CA 90806, U.S.A.

Dr. Charles L. Morgan
Lockheed Ocean Laboratory
3380 N. Harbor Drive
San Diego, CA 92101, U.S.A.

Dr. Thomas P. O'Connor
N.O.A.A.
Ocean Dumping Program
Rockville, MA 20852, U.S.A.

Dr. Carl H. Oppenheimer*[1]
University of Texas
P.O. Box 368
Port Aransas, TX 78373, U.S.A.

Dr. Dorothy P. Oppenheimer*[1]
University of Texas
P.O. Box 368
Port Aransas, TX 78373, U.S.A.

Dr. E. Ozturgut*[2]
Department of Oceanography
University of Washington
Seattle, WA 98195, U.S.A.

Dr. P. Kilho Park
N.O.A.A.
Ocean Dumping Program
Rockville, MA 20852, U.S.A.

Dr. A. L. Rice
Institute of Ocean Science
Brook Road
Wormley Godalming, Surrey GU8 5UB
United Kingdom

Dr. William M. Sackett
Department of Marine Sciences
University of South Florida
Tampa, FL 33620, U.S.A.

Dr. Robert F. Shokes
Science Applications, Inc.
1200 Prospect Street
La Jolla, CA 92308, U.S.A.

Dr. Duane C. Simpson
N.O.A.A.
Ocean Dumping Program
Rockville, MA 20852, U.S.A.

Dr. Dorothy F. Soule
Harbors Environmental Projects
Allan Hancock Foundation 139
University of Southern California
Los Angeles, CA 90007, U.S.A.

Dr. John H. Trefry
Department of Oceanography and
  Ocean Engineering
Florida Institute of Technology
Melbourne, FL 32901, U.S.A.

Dr. Ralph F. Vaccaro
Woods Hole Oceanographic
  Institution
Woods Hole, MA 02543, U.S.A.

Present affiliation:
*[1] Department of Marine Studies, University of Texas, Austin, TX 78712, U.S.A.
*[2] Science Applications, Inc., 13400B Northrup Way, Suite 36, Bellevue, WA 98005,
U.S.A.

CONTENTS

INTRODUCTION

Richard A. Geyer

The reasons for the diversified and sometimes ambiguour definition of the word pollution appear in the Preface. These determine the primary criteria for selecting topics for this volume and Volume 1, as well as the need to use interdisciplinary and case history methods in the following chapters. These criteria also necessitate obtaining information from a wide variety of different ecological areas throughout the world. This requires contributions from a number of countries, as well as from industry, governmental agencies and from academe reporting on research programs conducted in many parts of the coastal zone and deep ocean. Therefore, an opportunity is provided to present a wider range of perspectives than is generally available in books on this subject.

The specific substances that could be considered as actual or possible pollutants described in this volume includes a broad spectrum ranging from liquid to solid phases. These include individual components, as well as total systems found in municipal sewage outfalls, industrial effluents, radioactive substances and solids associated with dredging and mining operations. All but the last are generally concentrated within the coastal zone. These except for placer mining are generally found in the ocean.

Emphasis in Section A is on appraisal of the coastal zone as an area which can successfully receive a wide variety of substances brought to it primarily by sewage outfalls, dredge spoils, industrial effluents, and/or river runoff. These activities markedly affect the composition and quality of this particular oceanic environment. The substances discussed in this section include heavy metals, PCB's, sewage, dredge spoils, radioactive and cannery wastes, as well as thermal effluents. The geographic distribution includes several areas on the east and west coasts of the U.S.A., the Gulf of Mexico including the Mississippi River delta, and various parts of the Mediterranean Sea, with emphasis on the Gulf of Naples, and the Adriatic Sea. These areas receive large quantities of a wide variety of materials, and therefore have been studied in considerable detail by various groups.

The presentation in Section B is dominated by the results of research on the effects of deep-ocean dumping, including radioactive materials, DDT and PCB's. It also includes the effects of such activities as deep-ocean mining of manganese nodules, and of dredging operations, including dredge spoils, on the environment. Again examples are presented from a wide geographic range, including the Gulf of Mexico, the North Atlantic and Mid-Pacific Oceans, as well as the east coast of the U.S.A.

It is the purpose of the following sections of this Introduction to summarize major opinions, reactions and generalizations of the authors, as they apply to the development of the basic objectives of this volume described in the Preface and discussed in the Introduction. This is done primarily by quoting or paraphrasing some of their most pertinent observations and con-

TABLE I

Examples of pollutants introduced by mankind into the oceans which are also present from natural causes

Class of pollutant found in the ocean	Originating from natural causes (sources)	Originating from mankind's activities (sources)
Petroleum hydrocarbons (a) biogenic (b) petrogenic	seeps (oil), (gas), (tars), rivers, runoff, volcanoes, gas hydrates, bacteria in water column, atmosphere	urban runoff (asphalt roads, crankcase oils, two-cycle engines), transportation, production, aerosols
Particulates	rivers, runoff, turbidity currents, nepheloid layers, high biological production/bioturbation, atmosphere	farming, fisheries (i.e. trawling), runoff, dredging (harbors, rivers, channels), industrial and municipal effluents, drilling muds/drilling operations
Heavy metals	volcanoes, rivers, runoff, fissures, faults, fractures, subduction zones, sediments, decomposition of organisms	industrial and municipal effluents
Radioactive materials	rivers, runoff, volcanoes, faults, fissures, fractures, placer deposits, subduction zones, atmosphere	industrial and municipal effluents, nuclear power plants, nuclear weapons testing
Nutrients	rivers, runoff, bottom sediments brought to the surface by upwelling, biological recycling, atmosphere	municipal effluents, agricultural fertilizers and slurry mixtures
Thermal effects	volcanoes, fissures, faults, fractures, subduction zones, supra-heated tropical lagoons/estuaries	cooling tower discharges, ocean thermal energy conversion
Brines	salt domes, faults, fissures, fractures, shallow lagoons rivers	industrial effluents including disposal of brine from salt dome storage cavities
Biological oxygen demand (BOD)	red-tide, eutrophication, decomposition	municipal and industrial effluents, cannery wastes

clusions. Details are found in the corresponding chapters as well as from the associated references. Additional supporting information is available in the appropriate chapters listed for those desiring a more complete development or elucidation of the critical common denominator questions to be discussed.

This method should facilitate reaching some objective assessments of the role of these substances in marine environmental pollution, especially the importance of those resulting from nature as compared with those from man's activities. It has been mentioned in the Preface that the effects of various types of pollutants from natural sources are frequently many orders of magnitude greater than those resulting from the activities of man. Table I, which also appeared in the Introduction of Volume 1, lists eight important actual and potential classes of pollutants found in the ocean, and the sources originating from the activities of man and nature. The role of hydrocarbons shown in this table has been presented in Volume 1.

COMMON DEMONINATOR QUESTIONS

(1) What are the Origins of the Actual and Potential Pollutants Found in the Coastal Zone and Deep Ocean?

(2) What Oceanographic and Biologic Processes, Including Those of Microbial Organisms, Could Cause the Ultimate Degradation of Pollutants?

(3) What Must Be Known to Be Able to Establish Valid Baselines to Monitor Anomalous Concentrations of Pollutants?

(4) What Constitutes Potentially Harmful Levels of Pollutants in the Marine Environment?

(5) What are the Relative Capacities of the Coastal Zone and the Deep Ocean for Disposing of Pollutants, As Well As Being a Potential Source of Living and Non-living Marine Resources?

(6) What Are Some of the Advantageous Effects of Disposing of Pollutants in the Marine Environment?

(1) *What are the Origins of the Actual and Potential Pollutants Found in the Coastal Zones and Deep Ocean?*

The term pollutant in its broadest useage includes a wide variety of compounds. For example, just among organic varieties, thousands of different compounds could be listed. Many of these are readily adsorbed to suspended particles in the ocean and eventually incorporated in bottom sediments. More specifically, and to put this problem of the number and diversity of potential pollutants in some perspective, a list of heavy metals and other chemical substances and the chapters in which they are discussed appear in Table II.

A tripartite definition may also be implied in using the word, "origin", in

TABLE II

Trace metals, heavy metals and other substances

	Chapter No.		Chapter No.
Mg	3, 9 and 14	Phosphates	1, 2, 5, 11 and 13
Ca	14	As	3, 4 and 9
Sr	3, 13 and 14	Sb	3 and 9
Ba	9	S	2
Ra	9	H_2S	1
Pb	1—4, 11 and 13	Sulfates	9
U	9	Se	11
Ti	3	Silicates	2, 5 and 9
TiO_2	12		
V	3	Chlorophyll a	5 and 11
Cr	1—4 and 11	Methanol	12
Mn	2—4, 9 and 14	Phenol	12
Fe	2—4, 9, 11, 12 and 14	Methyl sulfates	12
Co	3 and 9		
Ni	1—4, 11 and 13	DDT, DDE	3, 9 and 13
Zn	1—4, 9, 11 and 14	Polychlorinated	3, 9 and 13
Cd	1, 3, 4, 9, 11—14	biphenyls (PCB's)	
Hg	1—5, 9, 11 and 13	Aldrin	9
Cu	1—4, 9, 11—14	Endrin	9
Ag	3	Endosulfane	9
Al	9	Dieldrin	9
		Paraquet	9
N	9 and 11		
Ammonia	1, 5, 11	Dredge spills	6, 7, 11 and 12
Nitrates	1, 2, 5 and 11	Sewage	1, 2, 11 and 12

this question. One carries a geographic connotation, another differentiates between a source attributed to the activities of man, as compared with those of nature; and finally, one in which the emphasis is on the chemical, physical and biological aspects of the origin of these substances. Even in this last case a dinstinction is made for hydrocarbons between those of biogenic vs. those of petrogenic origin. Marine geochemists studying this problem have determined diagnostic chemical characteristics to classify samples into these two categories. Other characteristics of the potential geographic origin of these hydrocarbons are discussed in Volume 1.

However, emphasis in this volume is on the connotation of the word "origin", from the standpoint of the cause, namely, man or nature, as well as the geographic source, of pollutants, other than hydrocarbons found in the world oceans. The activities of mankind associated with the origin and subsequent dispersion of pollutants are even more widespread on land than in the ocean. However, in both of these regions the major rivers of the world and their tributaries that drain the continents, all join to transport pollutants to the ocean. These include such diverse major categories of substances as those listed in Table II. This table can also be used as in index to facilitate finding where each of these substances are discussed in some detail.

Historically, the properties, origin, presence and use of many of these pollutants in a wide variety of forms have been documented for periods varying from only decades to as much as hundreds of years. Results of specific studies as they apply to the solution of this question are presented in the following categories:

(a) Sewage — by Mearns (Chapter 1), Oppenheimer et al. (Chapter 2), Gunnerson (Chapter 11) and Simpson et al. (Chapter 12).

(b) Heavy metals — by Mearns (Chapter 1), Oppenheimer et al. (Chapter 2), Bernhard (Chapter 3), Trefry and Shokes (Chapter 4), Gillespie and Vaccaro (Chapter 5), Gunnerson (Chapter 11), Simpson et al. (Chapter 12) and Sackett (Chapter 13).

(c) PCB's, pesticides, herbicides and petroleum hydrocarbons — by Oppenheimer et al. (Chapter 2), Bernhard (Chapter 3), Gillespie and Vaccaro (Chapter 5) and Sackett (Chapter 13).

(d) Dredge spoils and general effects of particulate matter in the ocean — by Herbich (Chapters 6 and 7), Eisma (Chapter 9) and Gunnerson (Chapter 11).

(e) Potential pollution from deep-ocean mining activities — by Morgan (Chapter 14), Ozturgut et al. (Chapter 15) and Ichiye and Carnes (Chapter 16).

(f) Radioactivity — by Angel et al. (Chapter 10).

(g) Cannery wastes — by Soule (Chapter 8).

The above-mentioned question is also discussed in Volume 1 by Geyer and Giamonna (Chapter 2), Jeffrey (Chapter 5), Brown et al. (Chapter 6), Spies et al. (Chapter 8); and those with emphasis on gaseous hydrocarbons, by Bernard (Chapter 3) and Sauer and Sackett (Chapter 4).

(2) *What Oceanographic and Biologic Processes, Including Those of Microbial Organisms Could Cause the Ultimate Degradation of Pollutants?*

It is imperative to understand the role of oceanographic and biologic processes in degrading anomalous concentrations of pollutants originating from the activities of nature and man. Discussions of the basic principle together with examples are presented in some detail by Mearns (Chapter 1), Oppenheimer et al. (Chapter 2), Bernhard (Chapter 3), Trefry and Shokes (Chapter 4), Gillespie and Vaccaro (Chapter 5), Eisma (Chapter 9), Angel et al. (Chapter 10), Gunnerson (Chapter 11), Simpson et al. (Chapter 12) and Sackett (Chapter 13). The ultimate fate of pollutants found in the ocean depends on a wide variety of processes (Table III), as well as on their composition and the specific geographic location in the ocean where they occur. The manner and time when, how and where they are found is also critical (Table IV). Some processes are more important immediately after

TABLE III

Major oceanographic and biologic processes causing the ultimate degradation of pollutants

Surface drifting and spreading by winds and currents
Evaporation
Dissolution
Dilution—diffusion—dispersion
Emulsification
Photochemical oxidation
Auto-oxidation
Biodegradation
Adsorption on particles
Ingestion by zooplankton, and higher forms of marine organisms including crustacea and fish
Initial sedimentation
Subsequent resuspension of sediments into the water—sediment interface and in interstitial voids within sediments

TABLE IV

Oceanographic and operational factors affecting concentration of dumped liquids and solids

Chemical composition and/or particulate size if a solid
Rate of discharge
Duration of discharge
Quantity of water used to flush effluents down discharge pipe, which also can cause considerable pre-dilution
Speed and direction of water currents
Wave, current and wind regime
Eddies caused by water flow around type of ship or barge used
Depth of mixed layer (determined in general by thermocline, but could be controlled by any pycnocline)
Density of water beneath mixed layer (light particulates could float along on isopycnal surface and hence be concentrated and transported either beyond or to a target area)
Fractionation of different phases of fluids, i.e. light and heavy particulate fractions and dissolved fraction
Depth of discharge pipe
Water depth
Type of bottom including topography and composition

the substance enters the ocean (e.g., spreading); others may start several days thereafter (e.g., biodegradation especially by microbial action); and still others are important both immediately after and for several days or weeks later (e.g., drifting and weathering). Eventually, processes of ingestion by organisms, and incorporation within bottom sediments begin to dominate and continue for extended periods of time.

The rate of degradation of certain pollutants is closely related to microbial activity. For example, research reveals the ubiquitous nature of these organisms in the water and bottom sediments of the oceans. Although they occur worldwide, it has been observed they are more ambundant in areas of anomalous concentrations of petroleum hydrocarbons. If it were not for these organisms, substances such as petroleum residues, for example, would persist indefinitely in the marine environment. A detailed discussion of the degradation of petroleum hydrocarbons is presented in Volume 1 by Oppenheimer (Chapter 1), Bernard (Chapter 3), Jeffrey (Chapter 5), Brown et al. (Chapter 6), Spies et al. (Chapter 8), Oppenheimer et al. (Chapter 9), Gould and Koons (Chapter 11), Lee (Chapter 12), and Weller (Chapter 13). In open, exposed areas, with good circulation of water, most pollutants are more readily degraded, whereas, in protected shallow areas with poor circulation, those incorporated into fine sediments may remain relatively unaltered for several years. Photochemical oxidation, dissolution, emulsification, adsorption to particles, biodegradation and uptake by zooplankton, all interact to affect the fate of pollutants in the water.

Various types of sedimentation processes also are responsible for deposition of this material on the bottom. There it can be resuspended into the water, or penetrate deeper into the sediment, or be degraded. The sediment-rich community of microbes, meiofauna and macrofauna also plays an important role in the further degradation of pollutants in sediments.

Strong forces generated by major storms and hurricanes can mix near-surface bottom sediments at depths of as much as 80 m. This mobility of the bottom sediments is another factor aiding in the degradation of pollutants that eventually become interspersed with bottom sediments. Coarser bottoms (sand—gravel) result in greater penetration into the bottom. Although permitting greater penetration, these have higher biodegradation rates as compared with fine sediments. This is caused by greater aeration, and nutrient flow in the interstitial spaces. Also, areas of coarse sediments are generally found near exposed coasts, so that all of the weathering forces are more effective.

(3) *What Must Be Known to Be Able to Establish Valid Baselines to Monitor Anomalous Concentrations of Pollutants?*

A significant number of extensive research programs have been conducted in an attempt to establish valid baseline values for the critical marine environmental parameters; and much has appeared in the literature on this subject. This applies to the continental shelf areas of North America and Europe, as well as the Mediterranean Sea. Discussions of the various factors responsible for making it difficult to achieve these objectives is presented by Oppenheimer et al. (Chapter 2), Bernhard (Chapter 3), Trefry and Shokes (Chapter 4), Gillespie and Vaccaro (Chapter 5), Eisma (Chapter 9), Angel et al. (Chapter 10), Gunnerson (Chapter 11), Simpson et al. (Chapter 12), Sackett (Chapter 13), Morgan (Chapter 14) and Ozturgut et al. (Chapter 15).

One of the major problems is the extreme difficulty in measuring accurately the very low concentration of many of the pollutants found in the oceans. The problem is accentuated by the lack of agreement by knowledgable specialists as to what actually constitutes harmful levels for a wide variety of these substances found in the marine environment. Also, the major difficulties encountered in measuring accurately the amounts of pollutants found in the ocean affect the validity of the resulting values. Nevertheless, marine management decisions including the promulgation of governmental regulations affecting the dumping and mining of potential pollutants is determined to a large extent by the results of such measurements and studies.

These problems as they affect petroleum hydrocarbons are discussed in Volume 1 especially by Oppenheimer (Chapter 1), Bernard (Chapter 3), Sauer and Sackett (Chapter 4), Brown et al. (Chapter 6), Oppenheimer et al. (Chapter 9) and Lee (Chapter 12). The same problems again exist for other studies, because of the extreme accuracy required to obtain meaningful results for the various components that need to be studied.

The units generally used to measure the anomalous concentrations of pollutants in the ocean vary from a low of parts per billion (ppb) to parts per million (ppm). The former is more appropriate in attempting to establish baseline values in the open ocean far from contamination by substances brought to the ocean by rivers draining through densely populated areas into coastal zones.

It is evident from these values discussed in the previous paragraphs that the concentration of many potential pollutants found in the ocean is so small that valid results depend upon the use of extremely careful, as well as standardized laboratory methods. Otherwise, the actual values are not statistically significant, nor comparable for studies made at different locations and times. Only if these criteria are met, will temporal and spacial variations have any meaningful significance. Another fact not always recognized is that some values for concentrations measured represent only certain components in a given area. No attempt frequently has been made to differentiate between the components and the total system, or even components of certain specific substances.

Heavy metals

The rate of removal from solution into particulate form in estuarine and coastal waters varies for various types of pollutants. It may also for unknown reasons, vary from time to time for a specific pollutant. Further lowering of their concentration in solution in the sea may be caused by dilution, adsorption onto suspended particles, and uptake by organisms.

Pollutants associated with particulare matter will be deposited primarily in estuaries and in the near-shore sea. Fewer than 10% (and probably fewer than 5% of the continental supply of suspended particles will reach the deep ocean, diluted by a supply of material from sea-floor erosion. The fraction that reaches the deep ocean, however, may be relatively important for transport of pollutants, because of its small size and a relatively high content of adsorbed material. This applies also to PCB's, DDT and radioactive materials, as well as to heavy metals.

Near industrial centers in the coastal zone additional supply of pollutants from the atmosphere can be comparatively high. This supply to the ocean from these sources may, therefore, be more important than estimated, based on continental and near-shore supply (by runoff and direct discharges) alone. Another factor to be considered are the synergistic and antagonistic effects of potential pollutants. Most research emphasis is on short-term bioassays involving high concentration levels of a single material during an experiment. In addition, usually only the effects of dissolved pollutants and not those ingested in food are considered; and only one phase of a life cycle for a particular organism involved.

Thus, very little is known about the effects of the material in interactions between various components of a complex ecosystem. There is also the disadvantage in most laboratory experiments of reflecting only the unique behavior of a single or perhaps limited number of isolated species, whereas comprehensive response patterns dictated by entire faunal communities have been ignored.

Similarly, more effort needs to be directed toward understanding pollutant cycling in food webs, not just "the biota", and especially for trace elements. It is not at all clear that all of them build up through marine food webs. Also, in invertebrates, which can take up metals at the most contaminated sites, it is not all clear what effects to look for in an organism containing, for example, a 2-, 3-, 5-fold increase in Ag or Cr. Effects of such increases are rarely studied in laboratory experiments.

Recent findings indicate that Se acts antagonistically to Hg and that both in marine organisms and man high Hg concentrations are associated with high Se content. This is of particular importance because it has been observed that small amounts of selenite could definitely counteract the toxicity of $HgCl_2$. Results of other research in this area has shown, for example, the antagonistic effects of Se on inorganic and organic compounds as well. Some have been suggested that naturally occurring Se in tuna has a high protective action against methyl mercury toxicity, even in "high-Hg" tuna.

These factors are especially applicable to marine food obtained from open ocean where the Hg content is associated with the natural concentration level existing in these waters. However, fish caught in some coastal areas may have an unusually high concentration of Hg and some other heavy metals, because of the availability of these materials from anthropogenic, as well as natural sources. Also, the discharge of inorganic compounds which more closely resemble natural physical—chemical forms is not as hazardous as the addition of organic Hg compounds anthropogenic of origin. Table XXI in

Chapter 3 (Bernhard) (p. 189) summarizes hazardous minimal risk concentration for marine organisms, and the "normal concentration" of inorganic substances in seawater for the more common trace metals. It provides a quantitative base, as of the time of publication of these data, which can be used in further study of this most provocative and vital subject.

Observations of heterotrophic microbial activity can be extremely useful for evaluating changes in ecosystem stability following experimental perturbation marine systems. The ecosystem approach clearly demonstrates how interactions between various trophic components help determine the nature of subsequent biological response patterns. Indications are that the microbial population is particularly responsive to perturbations effecting all trophic levels. Enrichment with N, P and Si causes a stimulation in primary production by phytoplankton; and a concomitant increase in released organic substrates, which in turn affect heterotrophic activity. The microbial response to heavy metals also bears a striking relation to the above response to heavy metals for an initial but temporary reduction in heterotrophic activity. In the latter case a secondary buildup of readily utilizable organic substrates also occurs; and the heterotrophic population shows an adaptation to the experimental pollutant, which becomes increasingly obvious as organic availability increases.

Radioactivity

It only becomes possible to recognize significant changes in the marine environment in time to be able to take corrective action if data from a valid baseline study exists. This is especially important in the disposal of radioactive wastes. Similarly, absence of sufficient knowledge of rate processes in the deep ocean makes it extremely difficult if not impossible at this time to make accurate models of the deep-ocean ecosystem. However, what is known indicates that biological processes are not sufficiently intense to alter the transport rates of radionuclides appreciably by physical processes. This applies especially if the absence of accumulation of high-level radioactive isotopes up the trophic chain found in freshwater biota also applies to marine organisms. This also demonstrates the need for a sound taxonomic and zoogeographic foundation to facilitate comparisons of results between different research programs and to assess lateral spread of isotopes.

These are but some of the factors discussed in some detail by Angel et al. in Chapter 10 that make it extremely difficult to assess quantitatively the actual and potential effects on the biota of radioactive waste disposal in the ocean. The answer to this question is a basic requirement before an effective monitoring program can be designed. Another is that a control site where no dumping is occurring must be designated and observed. Observation at this site would permit any long-term fluctuations at time scales larger than those of the original baseline study in the dump-site area to be filtered out. Only recently the high level of long-term variability at scales of orders of

magnitude greater than 5—10 years was recognized in shallow-water populations. Therefore, it is now known how these changes relate to the seep-sea
community.

Ocean disposal

Experimentation at sea with ocean dumping parameters is difficult, and
the ability is needed to sample contaminated chemical and biological samples
for longer periods. Because the waste plumes are relatively small, they are
easily lost, and without concomitant scientific verification of the presence of
waste, no correlation can be made between observations of the biological
community and dumping events. The integrity and reliability of field experimentation must be improved.

There are numerous restraints on phytoplankton growth such as light,
temperature, nutrient concentrations and grazing pressure. These all bear
upon the size and structure of communities. It remains to be determined
definitely if waste dumping as presently practiced can exert a significant
effect as compared with natural variations. One result of a change in species
composition of phytoplankton is that certain organisms would come to
dominate, which are unacceptable as food to zooplankton. Concurrently,
the feeding rate of zooplankton in waste plumes may be decreased. This may
increase the possibility of waste affecting phytoplankton populations.

Extrapolation of laboratory results on waste effects is difficult because
the testing of all wastes against all possible planktonic species remains to
be accomplished. It has been demonstrated that effects upon growth of
phytoplankton vary with species and even among subgroups of the same
species. Therefore, it is possible that within waste plumes, species composition of the phytoplankton community is altered as growth among species
is affected differentially.

From this discussion it is evident that the critical component of the
marine ecosystem which could most likely suffer from dumping in the
deep ocean are those planktonic organisms which occupy surface layers.
Free swimming organisms can encounter plumes but unless they are
particularly attracted to it, their exposure should be brief.

(4) *What Constitutes Potentially Harmful Levels of Pollutants in the
Marine Environment?*

Diagnostic criteria, that are generally accepted, remain to be determined
for establishing such quantitative levels of concentration. The primary
reason is that a single value by itself is not universally valid. This applies
particularly to those that are based only on the results of tests with
organisms in a laboratory. The ocean environment is highly complex and
dynamic. It involves the interaction between its physical parameters on the
biological components and conversely. The value for each of these factors in

turn is affected by the added major dimension of time. This dimension has almost an infinite variety of scales ranging from annual and seasonal at one end of the ime spectrum, and extending to diurnal, hourly and finally to seconds on the other end.

In addition, it is not only the actual magnitude of a given change in the value of a specific parameter, but also the gradient or rate of change. The more rapid the gradient the greater the effect. Similarly, the role of both antagonistic and synergistic effects are important and may even be an overriding consideration in evaluating the presence of a given amount of potential pollutant on the marine environment. When studying longer-term effects, the rate at which a specific organism can successfully adapt to a changing environment must be considered.

These diverse factors are investigated extensively and reported on by Mearns (Chapter 1), Oppenheimer et al. (Chapter 2), Bernhard (Chapter 3), Trefry and Shokes (Chapter 4), Gillespie and Vaccaro (Chapter 5), Angel et al. (Chapter 10), Gunnerson (Chapter 11), Simpson et al. (Chapter 12), Sackett (Chapter 13), Morgan (Chapter 14) and Ozturgut et al. (Chapter 15). They are also addressed in Volume 1 by Oppenheimer (Chapter 1), Sauer and Sackett (Chapter 4), Jeffrey (Chapter 5), Giammona (Chapter 7), Spies et al. (Chapter 8), Oppenheimer et al. (Chapter 9), Lee (Chapter 12) and Soule (Chapter 19).

The major difficulties that occur in attempting to determine accurately quantitative levels of potential pollutants that might be harmful to life in the marine environment fall into three categories: (1) the lack of valid comparisons in general, of results obtained in the laboratory studies, as compared with those conducted in the field; (2) the difficulty of making accurate measurements of the parameters involved at the very low and often even minute concentrations of potential pollutants being studied; and (3) the ultimate tangible effect, if any, of a particular substance as organisms adapt to it over a given period of time.

As further examples, with emphasis on the question of adaptability of organisms to potentially toxic substances, interesting results were found in studies made of naturally occurring hydrocarbon seeps off eastern Mexico and southern California. In the first area a study was made of species diversity and equatability values. It was found that they did not vary from the control area sufficiently to be able to state that species composition and abundance were effected by exposure to the seep material (Spies et al., Volume 1, Chapter 8). This is important, because it did not mean that organisms were slowly dying in this area, and that the environment would deteriorate eventually because of some toxic effect of a substance such as the tar or petroleum.

(5) What Are the Relative Capacities of the Coastal Zone and the Deep Ocean for Disposing of These Pollutants, As Well As Being a Potential Source of Living and Non-living Marine Resources?

This question is addressed to varying degrees in every chapter. However, the reactions and conclusions of those contributors who discussed this question in some detail are summarized as follows:

Sewage

Results of studies in outfall areas off southern California demonstrate that for most areas, on the basis of over 2400 trawl samples, the abundance of fish, as well as the number of species and diversity increased toward the San Pedro area from San Diego in the south, or from Santa Barbara to the north. In addition, careful scrutiny of earlier data also indicates that previous catches made about a decade ago in Santa Monica Bay that were unusually low were caused by major differences in fishing gear that was used, as well as in the manner in which they were used.

In another part of the world, namely, in the Gulf of Naples in the Mediterranean Sea, studies were made to determine if this area could absorb without serious ecological damage multiple major sewage outfalls, if they were relocated in deeper waters. On the basis of existing hydrographic data it was concluded that the currents in the Gulf of Naples are continuous and flow at a relatively rapid rate, so that this water may pass through the Gulf in from 1—2 days accompanied, by upwelling and considerable mixing. These data also suggest that the amounts of nitrogen and phosphorus from the sewage, if properly distributed, may be accepted by these waters and would produce little or no adverse ecological effects. Data from the Zoological Institute on nutrients in the Gulf indicate the levels are quite low. This in turn suggests that there is little effect from shoreline pollution.

Finally, the fish catch from both the Mediterranean Sea and the Naples area itself indicates a continued increase, and no major decrease that could suggest adverse effects on pollution on productivity, even though public health problems are apparent. In contrast it is possible that this increase in nutrients from the sewage in the area may contribute to the increase in fish, because historically the Mediterranean has been significantly low in nutrients.

Heavy metals

It is unfortunate that much of the current popular literature concerned with environmental issues use the term, "heavy metals", or "trace metals" indiscriminately in referring to any and all of the approximately 65 transition elements, including the lanthanides and actinides. The use of this nomenclature effectively places into a single category a very heterogeneous group of elements. It can be most misleading because it places unusual toxins and inert materials in the same classification as harmful pollutants. Another difficulty experienced in attempting to study results of different research projects on potential toxic effects of heavy metals is the manner in which sampling and analyses were made. One of the major factors is the amount of contamination that occurs when obtaining samples, as well as significant differences in various steps and methods used by different investigators in a variety of analytical techniques.

Similarly much depends upon the specific ecological characteristics of a given area as to whether or not a particular heavy metal may be injurious. For example, recent studies in and around the Mississippi River delta show no well defined evidence of pollution for heavy metals, such as Cu, Zn, Ti, V, Cr, Mn, Fe, Co and Ni in this region, because of the high river particulate flux. This can dilute large anomalous inputs. Also these particulates which carry about 90% of the heavy metals load at this river precipitate rapidly entering the Gulf of Mexico.

It was concluded from another extensive study that heavy metals such as Hg and Pb do not seem to be serious contributors to pollution in the Gulf of Mexico, but may be important near some sewage outfalls and industrial waste sites. On the other hand, Pb and Cd pollution is observed in the same area, and is attributed to anthropogenic causes, because a significant increase began in the middle 1800's. This correlates with the advent of mining activities starting about the time in the watershed drained by the Mississippi and its tributaries. Similar marked increase in Hg occurs in many of the coastal zone areas around the Mediterranean Sea, because of the relatively widespread occurrence of mercury deposits and associated mining activities.

The previous history of a particular water mass with regard to heavy metals can affect tolerance levels of resident species. The resulting stressing of a community in turn can initiate a series of events which can significantly alter the behavior of bacteria. Fortunately, however, microbial recovery can be accomplished by a population exhibiting increased resistance to original stress. This development of tolerance emphasizes selective processes leading ultimately to dominance by fewer, but more resistant species.

A rigorous analysis and evaluation of extensive studies made on the effects of heavy metals in the Mediterranean has been made. This shows that in samples taken from the open-water areas concentrations of Hg, Cd, As and Zn are similar to those in other seas, such as the Irish and North Seas, as well as the North West Atlantic. However, this generalization does not apply to some heavy-metal concentrations in shallow coastal waters. For example, in the Adriatic Sea the concentrations of Zn, Co, Hg, Se and Sb were significantly higher. Zn samples were recorded as high as $365\,\mu g\,l^{-1}$. On the other hand, concentration of samples from the same region of Cs, Rb, Se and U were comparable to values observed in the open seas.

Another critical factor responsible for variations in concentrations is the grain size of the sediments in which heavy metals are found. In general, the coarser the sediment the less the concentration in a given area. For example, concentrations in coarse sands can be as low as $300\,\mu g\,Hg_t\,kg^{-1}$ DW and in adjoining finer sands can be as high as $3800\,\mu g\,Hg_t\,kg^{-1}$ DW.

Polychlorinated biphenyls (PCB's)

Of all the pollutants and potential pollutants discussed in this book the most definitive concensus exists for PCB's. The polychlorinated biphenyls are described by many investigators as:

"they continue to be the most serious pollutants of the Gulf of Mexico, primarily because of the magnification through the food chain,"

or

"they are the most persistent toxic substances and remain as contaminants for many years in high concentrations in offshore sediments."

Therefore, presumably effects of these kinds of toxins will be long lasting and continue even after discharges are terminated. There is evidence of chlorinated hydrocarbons having contaminated sediments to depths of as much as 400 m, with the more highly chlorinated found at increasing depths. In addition, higher concentrations have been found in the more industrialized portions of the Adriatic than to the south. For example, total DDT decreased from 1900 near outfall to less than $20 \mu g \, DDT_t \, kg^{-1}$ DW at a distance of only 5 km. Similar PCB's decreased from 775 to less than $20 \mu g \, kg^{-1}$ DW at about the same distance from the outfall.

The concentration of pollutants in bottom sediments reflects to a large extent the history of pollution in a given area. A decreasing content of pollutants including metals, PCB's and DDE, with increasing depth in the sediment, has been determined in a number of coastal zone areas near industrial centers, e.g., the Baltic and North Seas, and the coastal basins of California. These data permit determining the period and flux of contamination, as well as the degree of increase in concentration. But the interpretation depends on a reliable determination of the sedimentation rate during the past 200 years. This is complicated by reworking of the sediments by organisms.

Table XV in Chapter 3 (Bernhard) (pp. 172 and 173) can be used in further studies of "minimal" risk concentration of PCB's, which vary from 1 to 10 ng $PCB \, l^{-1}$ depending on the degree of chlorination. However, it should be emphasized that the values given in the tables should not be considered as final, because of the other factors discussed in the previous paragraphs. Advances in analytical instruments, techniques and an ever-increasing knowledge of the complexities and interactions existing in the solution of this problem will lead eventually to a more sophisticated research programs. This in turn will result inexorably in changes in these tabulated values.

Dredge spoils

Recent research results from major research efforts indicate that the original fears of water-quality degradation resulting from the resuspension of the dredged material during dredging operations are for the most part unfounded. Most of the short-term chemical and biological impacts of dredging and disposal have generally been minimal. No significant long-term increase in water-column contaminant concentrations have been observed at any aquatic disposal field sites. Most organisms studied were relatively insensitive to the effects of sediment suspensions, or turbidity. Release of heavy metals and their uptake into organism tissues have been rare. Similarly, the accumulation of oil and grease residues by organisms have been minimal.

Deep-ocean mining

Potential changes in the ecological characteristics in the water column of the open ocean as well as in the bottom are miniscule as compared with the change occurring daily throughout geologic time. For example, the rivers of the world discharge $\sim 10^7$ tons per day of dissolved metals into the world's oceans. Assuming that all the discharge solids $(500\,\text{t}\,\text{day}^{-1})$ can be put into solution, it would take 2000 mining operations to put an equivalent amount of metals into the sea. The clearest water in the Atlantic Ocean is found $\sim 1000\,\text{km}$ seaward from the mouth of the Amazon River, but visibly turbid water can be found $\sim 80\,\text{km}$ from shore. This river carries a load of fine material averaging $\sim 3 \cdot 10^6\,\text{t}\,\text{day}^{-1}$. By analogy, a mining system discharging $5000\,\text{t}\,\text{day}^{-1}$ into the sea would be undetectable by any means, 1.5 km from the discharge, and would be invisible 175 m from the discharge. Additional analogies involving the activities of nature vs. man including specific reject of manganese mining are found in the chapters discussing mining and dredging (Chapters 6, 7 and 14—16).

(6) *What Are Some of the Advantageous Effects of Disposing of Pollutants in the Marine Environment?*

It is a pertinent question to ask because it can serve to focus on the discussions presented in this volume on this subject. This is necessary because so much is readily available in the literature in the opposite vein. These discussions are emphasized in chapters by Mearns (Chapter 1), Oppenheimer et al. (Chapter 2), Bernhard (Chapter 3), and Morgan (Chapter 14), and in detail by Ozturgut et al. (Chapter 15). Similar effects of petroleum hydrocarbons have been described in Volume 1 by Oppenheimer (Chapter 1), Spies et al. (Chapter 8), Oppenheimer et al. (Chapter 9) and Gould and Koons (Chapter 11).

Petroleum hydrocarbons

Micro-organisms are found universally in the water column and bottom sediments of the ocean. However, it has been observed that they are more abundant in areas of anomalous concentrations of hydrocarbons. These organisms constitute an important link in the food chain and in the productivity cycle. Because natural microflora are actively oxidizing hydrocarbons in the ocean, petroleum may, therefore, be considered as another source of carbon for use in this cycle.

Nutrients cycled by bacteria may also improve growth of phytoplankton. Therefore, the presence of hydrocarbons could be considered to be a positive factor in these processes. For example, a creel census in the Corpus Christi Bay area showed that the region around oil platforms yielded the highest catch per unit effort for sports fisherman, as did similar studies on the Louisiana continental shelf. In the latter area, it was found that the fish yield over an extended period of time was directly proportional to the production of petroleum. This has been attributed to the results in the field of experiments with mirco-organism concentrations that suggest petroleum is broken down to its first degradation product, namely, fatty acids. These in turn enter directly into the food chain.

Most infaunal species have planktonic larvae, and it therefore seems possible that an altered microbial community or the petroleum itself, could be encouraging metamorphosis and juvenile growth. It is also known that metamorphosing larvae select sediments having specific organic or microbial contents.

Adult lobsters (*Homarus*) have positive feeding responses to substances soaked in the cyclic and branched-aliphatic fractions of petroleum. Such compounds are especially prevalent in petroleum from natural seeps, and these may possibly influence larval recruitment. One investigator studying adult mussels taken from Coal Oil Point, California, and exposed to oil in aquaria, claims increased survival rates over those experienced by the controls.

Deep-ocean mining

The importance of nutrients to the food cycle has already been discussed in the previous section on petroleum. The major source from natural processes in the coastal zones of the ocean is from upwelling. But this does not occur effectively in the deep ocean. However, during the course of deep-ocean manganese-nodules mining some bottom sediments are also brought to the surface, where they are separated from the nodules and returned to the ocean. Therefore, a source of nutrients becomes available to organisms living in these areas which they would not otherwise have.

Still another source of nutrients becomes available to organisms in the surface waters of the coastal zone other than from the natural process of upwelling. It comes from fertilizers used in agriculture on land. Some of the fertilizers are carried into the rivers and eventually reach the coastal waters where the rivers enter the ocean. These fertilizers qualify as pollutants in the strict definition of the word, yet they actually are beneficial to the marine environment in the same way as the rejects from ocean mining.

Sewage

The principal responses of the biota (benthos, fish and phytoplankton), to the discharges from outfalls are mainly biostimulatory excluding the effects of chlorinated hydrocarbons. The enrichment generally is not sufficient to alter significantly the dissolved oxygen content or pH of the water. But it does change both feeding and structure and species composition of the benthos in a manner which appears to be reversible. More research is needed to evaluate any additional beneficial or detrimental aspects of these changes, in processes involving the feeding ecology of fishes and larger invertebrates, and also on processes that affect colonization and growth. This includes the relative reproductive potential of affected populations and influences of discharges on dispersal and relative survival of eggs and larvae. It has been suggested that the increase in nutrients from the sewage in the Gulf of Naples area may possibly contribute to the increase in fish catch, because the Mediterranean Sea historically has been known to be an area that is poor in nutrients.

Thermal effluents

Cooling water is necessary to successfully operate a power plant, but it is usually warmer than the ambient temperature of the body of water into which it is being discharged. In tropical areas and in the summer, in a portion of the temperature zones, this process might raise the temperature to the point where it exceeded optimum tolerance levels for a variety of marine organisms indigenous to the area. On the other hand, in polar regions and during the winter in the temperate zones it would still increase the temperature of the water body such as an estuary or a lagoon above ambient. But, this action would actually be beneficial by helping to increase the rate of growth of the marine organisms. This is especially important economically for those species having commercial value, or are an important link in the food chain.

Warm water also facilitates hatching, and an aquaculture study in San Diego, California, is counting on a rapid-growth, nonseasonal production schedule which would be managed by "fooling" the shellfish with an "always-summer" flow of warm water from an electric power plant.

CONCLUSIONS

Of the examples given of the advantages of using the ocean to dispose of certain selected wastes, one of the most obvious and important reasons is generally ignored. It is, if they are not disposed of in the ocean, where will they be? The obvious answer is on land, because it can be demonstrated readily that it is not economic to send it in capsules into outer space. Similarly, using economics as a critical factor land in general is more valuable for uses other than waste disposal, as compared with the most of the ocean.

This applies particularly to land adjacent to the coastlines of the continents.

The greatest population densities of the world are found in this region, which is also the closest to the ocean. Thus, a minimum of transportation is required between the maximum source of the waste material and a potential disposal site. This generalization is not to be considered as a recommendation for the unequivocal disposal of all the wastes of the activities of our society into the coastal zone or deep ocean. There are materials that could cause damage to the ecology of the ocean that should not be disposed of in this manner. But similarly, there are waste products that the ocean is capable of receiving with no danger, and in fact some are discussed in this volume that would be beneficial to the marine environment.

The point to be made, therefore, is which wastes fall into either of these two basic categories, and what criteria should be used to decide? For some substances these are not available to make such a decision, but for many they exist already. Therefore, it is necessary to determine the necessary guidelines to be followed in making these extremely important decisions. Some of the more critical ones are listed as follows:

(1) The disposal of substances that can definitely be proven to be injurious to the marine environment should not be permitted.

(2) Those wastes materials that are not injurious should be disposed of in the ocean, if it can be proven to have advantages over dumping them on land.

(3) Objective criteria should be established to determine in which category, (1) or (2), a substance may be classified for the purpose of where and how it should be dumped. These include:

(a) Giving high priority whenever possible to avoid areas that are unusually biologically active, or those which are used for recreation or esthetic purposes.

(b) Obtaining definite evidence to determine the degree of irreversibility of the impact to the marine environment of dumping a certain substance.

(c) Determining the volume and in what concentration the substance is to be disposed of in a given location.

(d) Considering the exact location of a specific disposal site, with respect to any others in the same area, and the possible interaction effects between them.

In short, the ocean is large enough to support the diversified needs of all biological, economic and sociological activities of mankind. This statement assumes judicious use is made of available, valid, scientific and objective criteria to differentiate between what can and cannot be disposed of in the ocean and still preserve a viable marine environment.

An era of tremendous and widespread concern began in the late 1960's, as to the ability of the ocean to cope successfully with the diverse substances that were being dumped into it in ever-increasing amounts from anthropogenic sources. This culminated in almost a decade of large-scale interdisciplinary scientific investigations conducted at an accelerated pace on a global scale. Research programs were designed to obtain more information to improve our knowledge of existing and potential pollutants and their effect on the marine environment, with emphasis on living organisms.

This activity was accomplished in part by the availability of increased funding, as well as by significant advances in marine chemistry, including more sophisticated analytical techniques. The latter provided a means to accurately measure increasingly smaller concentrations of pollutants with methods that decreased the contribution from contamination during the sampling phase. For example, concentrations of Pb were known to exist in the oceans at ppb levels. It became evident with these improvements in sampling techniques that Pb in uncontaminated seawater sometimes exists in concentrations less than one part per trillion. These results necessitated a revision in thinking as to what quantitatively constituted a pollution level for a specific substance.

This also became an era in which laboratory studies involving the effect of a pollutant on a single species was supplanted by field experiments in which pelagic communities were studied in a natural environment, but confined in large plastic bags. These communities were then subjected to physical and chemical stresses to determine how the community would react under such conditions. As a result a useful index of stress was established and found to be the ratio of flagellates to diatoms, with flagellates dominating the phytoplanktons when the system was under stress. The results of such studies in which heterotrophic microbial activity is described is presented in Chapter 5, Gillespie and Vaccaro.

A complete interpretation, including all the nuances of the results of this new method of studying effects or pollution in the marine environment, remains to be accomplished. However, it has demonstrated that a variety of additional factors regarding a given organism must be considered. These include specificity, biological variability, rate of detoxification and adaptation as well as the environmental history of the organism being studied.

Same significant results have been obtained from this era of new experimental techniques and improved analytical methods. These have been somewhat at variance with some of the preconceived and subjective opinions based more on a combination of emotion and incomplete facts than on an objective and factual basis. Instead of the frail envisioned by many for a number of years prior to this era, it now appears that the ocean enjoys a resilience and a recuperative power greater than most people would have imagined at that time. It now appears from the results of research conducted under the auspices of the I.D.O.E.[*] that the marine environment can recover from concentrations of Hg and Cu in the 1—20 ppb range, and as oil in the

---

[*] International Decade of Ocean Exploration.

the ppm range. With the advent of additional data in the future it is conceivable that these concentrations could increase because of the adaptive resilience of many organisms.

Much, however, remains to be done before these generalizations can be applied to some highly-stressed coastal zone areas. It is also to be hoped that these results will be taken into consideration favorably by those in regulatory agencies responsible for the promulgation of realistic rules for equitably using the ocean for the benefit of all concerned.

Emphasis in this Introduction has been on discussing a half a dozen major questions, which when answered should do much to lead to a better basic understanding of and to the placing in better perspective, the problem of pollution in the marine environment. The data and information presented in the following chapters can be applied toward achieving these objectives. In this regard, attention should be called to the voluminous number of references appearing in the back of this volume totaling over eight hundred. These can be used as another avenue to explore further in greater depth not only additional details, but also various philosophies espoused by different investigators that are pertinent to the solution of these major questions. It also indicates the tremendous body of literature available on this controversial subject with almost eight hundred, for example, in Volume 1. This in turn reflects the extreme preoccupation of, at least, one segment of the broad scientific community with the solution of these questions.

Perhaps the only major point of concensus that may be reached by those who actively persue the subject of marine pollution is that it is a controversial one. But it has both qualitative and quantitative aspects. The former has been explored to a much greater degree by both scientists and layman. However, before some of the many questions and doubts can be satisfactorily answered and allayed, it is the quantitative aspects that must be addressed to a much greater extent than heretofore. Not until then can the correct answers to these questions be obtained to the satisfaction of those concerned. This also is a formidable task as indicated by six of the basic common demoninator questions discussed in the Introduction, for which more definitive quantitative answers are needed. These are not meant to comprise a complete list, nor are they listed in order of priority.

Nevertheless, this type of question must be answered, before the highly emotionally charged atmosphere now dominating many discussions of marine environmental pollution, can yield to more serious objective evaluation of this problem. It is undeniably a problem of far reaching ecologic, economic and sociologic importance. But, if it is not solved soon in a manner equitable to the overall needs of society as it exists today, only the continuing alternative of increasing divisiveness will remain, leading in turn to anarchy and chaos.

Section A — Coastal Zone

CHAPTER 1

EFFECTS OF MUNICIPAL DISCHARGES ON OPEN COASTAL ECOSYSTEMS*

Alan J. Mearns

INTRODUCTION

Many coastal communities discharge their sewage wastes directly into the ocean. In North America municipal wastes from some large coastal cities are partially treated and discharged by barges (New York and Philadelphia) or through shallow-water outfalls (Vancouver and Oakland). However, in southern California, Hawaii, and in Puget Sound, Washington, municipal wastes are discharged through large outfalls and into relatively deep water several kilometers from shore.

These wastes cause measurable changes in the abundance, diversity and health of adjacent marine plant and animal communities. This chapter summarizes the quantity and composition of these wastes, coastal conditions which influence their distribution, and their effects on coastal marine life. Effects are examined first in terms of field observations on populations and communities and then in terms of biological processes (growth, food web bioaccumulation of pollutants, response and recovery of affected ecosystems). Emphasis is focused on similarities and differences among five sites in southern California and one in Puget Sound, Washington.

Present deep-ocean outfalls and their onshore treatment plants were designed and constructed within the past several decades to alleviate ecological and public health hazards which were becoming acute in estuaries, harbors, lakes, and at public beaches. The net response by various municipalities was to consolidate numerous small, uncontrolled, near-shore domestic and industrial discharges into larger municipal systems, remove coarse and settleable solids and floating material (primary treatment), and then discharge the wastes into deep water offshore. In most cases these events achieved their goals. However, new kinds of environmental changes now occur offshore. Sediments at depths from 30 to 200 m contain increased concentrations of organic matter, trace metals and synthetic organic chemicals. There are changes, in some cases reversible, in the structure, abundance and diversity of benthic marine communities. Fish near one site are

* Contribution No. 142 of Southern California Coastal Water Research Project, 646 W. Pacific Coast Highway, Long Beach, CA 90806, U.S.A.

26

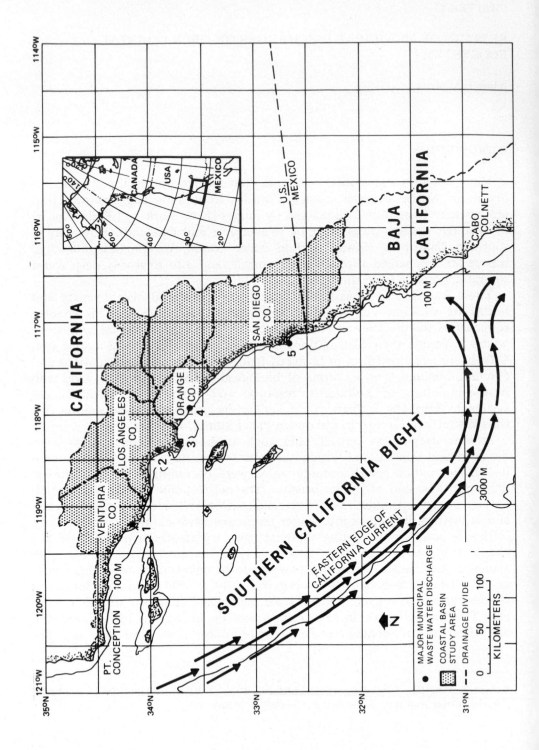

contaminated by chlorinated hydrocarbons but not trace metals; in contrast, invertebrates accumulate slight, but significant amounts of trace metals at several sites. Biostimulatory responses are readily observed in benthic, near-shore and pelagic ecosystems at some sites while responses attributable to toxic materials are less apparent at most sites. Finally, it appears that pollutant mass emission rates, rather than effluent concentrations, are more directly related to the magnitude of observed ecological changes.

This chapter is largely a synthesis of the biological and chemical studies carried out by the U.S. Coastal Water Research Project (S.C.C.W.R.P.) during the past several years. It also includes observations from the *Puget Sound Interim Studies* sponsored by the Municipality of Metropolitan Seattle. Much of the work from both programs is still in preparation or press. It is nonetheless relevant and is documented in the reports and papers cited.

Study areas

The five major southern California ocean outfalls are located on a relatively narrow (5—15 km) mainland shelf adjacent to major urban areas (Fig. 1). The shelf drops steeply into basins which reach a depth of 1000 m. The near-shore waters are stratified most of the year with a seasonal mixed layer extending to 20 m. Most of the year, surface waters move downcoast (SE-ward), at a velocity averaging $25 \, \text{cm s}^{-1}$. In contrast, subsurface waters (below the mixed layer) generally move alongshore and upcoast at an average velocity in the range of $3—5 \, \text{cm s}^{-1}$ but with velocities as high as $50 \, \text{cm s}^{-1}$ during some periods (Hendricks, 1977a).

The mainland shelf forms the coastal eastern edge of the 10^5-km^2 Southern California Bight, a 400 km long eastward indentation of the west coast of North America (Fig. 1). The outer edge of the Bight is bounded by the California Current, a meandering, southward flowing water mass (S.C.C.W.R.P., 1973).

Outfalls in Puget Sound, Washington, discharge into an even more active 10^4-km^2 (300 m deep) fjord-like saltwater system (Fig. 2) driven by freshwater runoff and with a 120—140-day water mass residence time (Friebertshauser and Duxburry, 1972). Average surface water flow is north-ward and out the Strait of Juan de Fuca to the ocean. This is compensated by a net subsurface flow of $59 \cdot 10^3 \, \text{m}^3 \, \text{s}^{-1}$ of seawater through the Strait. Tidal driven bottom currents off West Point in the central basic of Puget Sound are high, reaching velocities of $\sim 100 \, \text{cm s}^{-1}$ (2 kt.).

Fig. 1. The Southern California Bight and the adjacent coastal basin. Major municipal wastewater dischargers are: *1* = City of Oxnard; *2* = City of Los Angeles (Hyperion Treatment Plant) in Santa Monica Bay; *3* = Joint Water Pollution Control Plant (J.W.P.C.P.) off Palos Verdes; *4* = Orange County Sanitation Districts in southern San Pedro Bay; and *5* = City of San Diego, Point Loma Treatment Plant. Note 100-m isobath occurs within 5—10 km from shore.

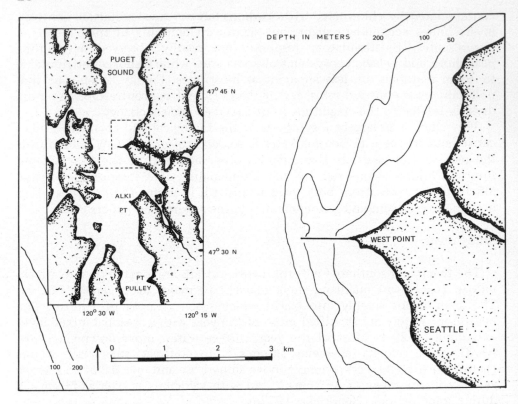

Fig. 2. Central basin of Puget Sound, Washington, showing narrow shelf, discharge sites of major METRO treatment plants (West Point and Alki Point) and control site (Point Pulley).

Both regions are characterized by well-oxygenated surface waters and poorly-oxygenated deep waters (3—5 mg l^{-1} at 60 m). Coastal shelves in both areas also are occupied by soft sediments ranging from clay to silt, and silty-sand occasionally mixed with coarse sand, gravel and shell debris penetrated by rock outcroppings. In both areas the bottom fauna is composed of a number of taxonomically similar benthic invertebrates and bottom fishes. The most significant faunal differences occur in surface waters; southern California is dominated by subtropical fish, macroinvertebrate and algae, while Puget Sound is dominated by a subarctic fauna and flora.

Outfall locations and characteristics

The five largest southern California discharges account for over 90% of the $1.1 \cdot 10^9$ gal. ($4.2 \cdot 10^9$ m^3) of municipal sewage discharged daily into the Southern California Bight. The wastes are discharged through outfalls which range in depth from 20 to 80 m and terminate 4—10 km from shore. The largest are those of Los Angeles County (Joint Water Pollution Control Plant), and the City of Los Angeles (Hyperion Treatment Plant), which each discharge approximately 350 million gallons per day (mgd). Approximately one-third of the Hyperion flow (100 mgd) is given secondary treatment (biological oxidation and sludge removal) but the resulting sludge is discharged via a second outfall located 11.2 km from shore and 3—4 km north of the effluent outfall. The Orange County outfall (160 mgd primary and 20 mgd secondary) discharges at a depth of ~ 60 m and about 8 km from shore in southern San Pedro Bay. The City of San Diego's Point Loma plant discharges ~ 117 mgd of primary treated effluent through a 3.2 km long outfall located at a depth of 60 m. The next largest outfall, from the City of Oxnard, discharges about 12 mgd (1976) of primary treated effluent through a 2 km long outfall at a depth of 20 m. Finally, there are 17 other outfalls in southern California which discharge a total of 63 mgd through outfalls ranging in depth from 0 to 30 m. Most of the small plants chlorinate regularly but the four major plants rarely chlorinate.

At present there are 28 municipal sewers discharging directly or indirectly into central Puget Sound (total, 230 mgd; Collias and Lincoln, 1977). The largest of these is the West Point system; in 1974, 118 mgd of chlorinated primary treated wastes were discharged through the 1.2 km West Point outfall at a depth of 75 m (Table I). A discussion of the problem with sewer outfalls in the Gulf of Naples, and in the New York Bight area is also presented by Oppenheimer et al. (Chapter 2 of this volume) and Gunnerson (Chapter 11 of this volume), respectively.

Concentrations and mass emission rates

Concentrations and mass emission rates of a large variety of chemicals from southern California discharges are compiled annually by S.C.C.W.R.P. (Schafer, 1977). The highest concentrations of most constituents (including suspended solids, oil and grease, trace metals, total DDT's and total PCB's) occur in the Hyperion sludge (Table I); however; because of order-of-magnitude differences in flow rates, the Joint Water Pollution Control Plant (J.W.P.C.P.) outfall is responsible for the largest mass emission rates of these constituents (Table II). There are other competitive sources of these pollutants into the area including harbors, runoff and aerial fallout (Young et al., 1978). In addition, these wastes contain a diversity of petroleum hydrocarbons (R. Eganhouse, pers. commun., 1978) and other volatile chemicals and solvents at concentrations ranging from < 0.01— 4.2 mg l^{-1} (Young, 1979).

Concentrations and mass emission rates from Seattle, Washington, METRO's West Point outfall are most comparable to those at Point Loma, California (Tables I and II). However, unlike the Southern California

TABLE I

Concentrations of some general constituents, trace metals and organic chemicals in final effluents of six west coast municipal discharges

Discharger	Oxnard	Point Loma	West Point	Orange County	Hyperion 5-mi.	Hyperion 7-mi.	J.W.P.C.P.
Year Region	1976[1] Ventura County	1976[1] San Diego County	1974[2] Seattle	1976[1] San Pedro Bay	1976[1] Santa Monica Bay	1976[1] Santa Monica Bay	1976[1] Palos Verdes
Flow							
$(10^6\ 1\,yr.^{-1})$	16.0	162	163	251	496	5.7	488
(mgd)	11.6	117	118	182	359	4.1	353
General constituents $(mg\,l^{-1})$							
total suspended solids	91	128	36–96	128	77	9,900	284
5-day BOD	225	189	49–103	199	125	n.d.	231
ammonia nitrogen	19.7	24.1	6.9–13.5	32	14.6	n.d.	36.3
Trace metals $(mg\,l^{-1})$							
chromium	0.028	0.104	0.01–0.11	0.190	0.190	11.7	0.75
copper	0.086	0.216	0.12–0.18	0.360	0.17	16.8	0.41
lead	0.054	0.184	0.02–0.17	0.11	0.03	1.65	0.220
zinc	0.167	0.311	0.25–1.14	0.540	0.22	21.1	1.32
Organic chemicals[3]							
phenols $(mg\,l^{-1})$	0.193	0.299	n.d.	0.240	0.050	0.610	3.48
chlorinated benzenes $(\mu g\,l^{-1})$	16.85	5.75	n.d.	9.3	5.9	150	21.0
total DDT's $(\mu g\,l^{-1})$	0.10	0.06	n.d.	0.12	0.5	3.2	2.6
total PCB's $(\mu g\,l^{-1})$	0.31	1.6	0.14	3.9	0.23	21	2.7

n.d. = no data.

[1] From Schafer (1977); Hyperion effluent and sludge separated.

[2] From R. Bain (pers. commun., 1978).

[3] Chlorinated benzenes modified from Young and Heesen (1978). See Table II, footnote [5] for description of compounds.

TABLE II

Mass emission rates of some general constituents, trace metals, and organic chemicals in final effluents of municipal waste dischargers

Discharger Year Region	Oxnard 1976[1] Ventura County	Point Loma 1976[1] San Diego County	West Point 1974[2] Seattle	Orange County 1976[1] San Pedro Bay	Hyperion 1976[1] Santa Monica Bay	J.W.P.C.P. 1976[1] Palos Verdes
Flow						
$(10^9 \ l \ yr.^{-1})$	16.0	162	163	251	502	488
(mgd)	11.6	117	118	182	361	353
General constituents $(t \ yr.^{-1})$						
total suspended solids	1,465	20,700	11,315	32,300	94,490	138,400
5-day BOD	3,620	30,600	13,140	50,200	>62,200[4]	112,800
ammonia nitrogen	320	3,900	1,662[3]	8,070	260[4]	17,800
Trace metals $(t \ yr.^{-1})$						
chromium	0.45	16.9	11.0	47.9	161	367
copper	1.38	35.0	32.9	90.8	180	200
lead	0.87	29.8	18.3	27.7	24.3	108.0
zinc	2.68	50.4	7.3	136	229	646
Organic chemicals $(t \ yr.^{-1})$						
phenols	3.1	40.3	n.d.	60.5	28.4	1,700
chlorinated benzenes[5]	0.27	0.92	n.d.	2.33	3.78	10.2
total DDT's	0.002	0.010	n.d.	0.031	0.043	1.24
total PCB's	0.005	0.267	0.022	0.991	0.228	1.32

n.d. = no data.
[1] From Schafer (1977); for Hyperion, effluent and sludge discharges are combined.
[2] From S.C.C.W.R.P. (1975), with exceptions noted.
[3] From R. Bain (pers. commun., 1978).
[4] Effluent from 5-mi. outfall only; not measured in sludge.
[5] Modified from Young and Heesen (1978); p-DCB, para-dichlorobenzene, o-DCB, ortho-dichlorobenzene; 1,2,4-TCB, 1,2,4-trichlorobenzene; "others" include 1,3,5-trichlorobenzene and hexachlorobenzene.

Bight, sources other than the METRO outfall account for most of the metal mass emission rates into Puget Sound; the most prominent source is river runoff (Schell and Nevissi, 1977).

Dilution and subsequent fate of pollutants

The ocean outfalls through which the wastes are discharged are fitted with long (1—2 km) multiport diffusers which achieve the initial dilutions in the range of 1:100—1:500 (Young and Jan, 1975; Hendricks, 1977b). During most of the year, when a thermocline is present, plumes remain submerged, reaching an equilibrium depth just below the thermocline (Fig. 3). Dilutions at Seattle METRO's outfall are on the order of 1:220 (Domenowske and Matsuda, 1969) but surface filaments and patches with dilutions as low as 1:10 also occur (Ebbesmeyer and Helseth, 1975).

Following initial dilution, the southern California sewage plumes tend to move along-shore and upcoast (to the northwest), along the isobath of discharge (generally 60 m) and usually opposite to the surface currents. During transit, components of the diluted effluent begin to separate; ~ 10—20% of the heavier particulates fall out of the plumes and onto the bottom near the discharge (Herring and Abati, 1979). The fate of the remaining mass of fine particulates is largely unknown but a portion of it appears to remain suspended in the discharge area (L. L. Peterson, 1974) and perhaps many kilometers beyond. The fate of dissolved constituents is largely unknown, but ammonia is taken up by phytoplankton (MacIsaac et al., 1979).

Since most of the potentially toxic pollutants are attached to sewage particulates (Young et al., 1978), bottom sediments in the fallout area around the discharges contain above-normal concentrations of trace metals and chlorinated hydrocarbons. The sediments also contain elevated levels of organic matter [as measured by nitrogen, carbon, 5-day BOD, COD and volatile solids and total sulfides (R. W. Smith and Greene, 1976; Word and Mearns, 1979)]. As a result of extensive decomposition, hydrogen sulfide gas is produced in sediments at one site [Palos Verdes (R. W. Smith and Greene, 1976)]. Table III shows that all these materials occur in measurable concentrations in sediments at control sites and that the concentrations increase at discharge sites roughly in proportion to the size of the discharges as measured by mass emissions (see Table I). Contamination factors (outfall to control) generally range from 2- to 5-fold except at Palos Verdes.

Unlike the sediments, the waste discharges cause few enhancements of *dissolved* constituents in the water column near the discharge sites. For example, while ammonia concentration in the outfall plume off Palos Verdes averaged 17 times higher than control seawater (0.25 vs. 0.015 mg l^{-1}, respectively), dissolved copper and nickel were increased only 3 to 5 times (0.45 vs. 0.1 and 0.97 vs. 0.29 μg l^{-1}, respectively); and, there were no significant increases in dissolved cadmium or chromium (0.07 vs. 0.07 and

33

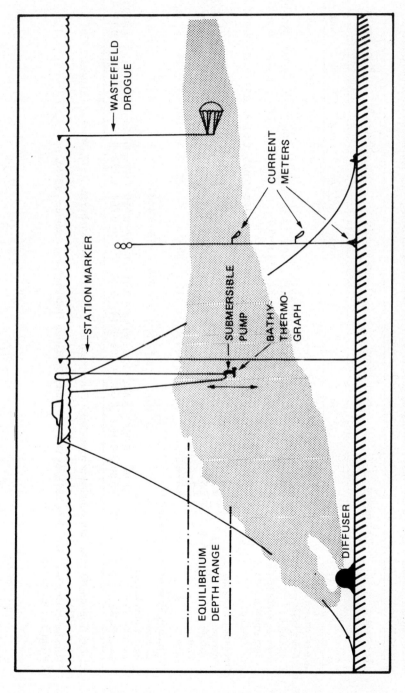

Fig. 3. Section through water column and an outfall diffuser showing configuration of a diluted (greater than 100:1) sewage field rising to the thermocline (equilibrium depth range) and drifting along the depth of discharge. Methods for sampling the plume and adjacent waters are depicted (from Hendricks, 1977b).

TABLE III

Concentrations (medians and ranges) based on dry weights, of selected general constituents, trace metals and chlorinated hydrocarbons in sediments from 29 "control" sites and four outfall sites on the southern California mainland shelf

Location	Control sites 29		Point Loma 10		Orange County 10		Santa Monica Bay 9		Palos Verdes 7	
Number of stations	$\bar{m}$	range	$\bar{m}$	range	$\bar{m}$	range	$\bar{m}$	range	$\bar{m}$	range
General constituents										
5-day BOD ($mg\,kg^{-1}$)	636	266–1,017	754	407–2,175	927	499–1,931	817	486–1,850	7,781	943–25,048
volatile solids (%)	2.8	1.8–3.8	2.8	1.9–4.7	2.5	1.9–3.0	3.2	2.4–3.9	17.3	5.0–27.0
acid solids sulfides ($mg\,kg^{-1}$)	0.01	<0.007–0.07	0.03	0.005–0.32	0.039	0.011–0.109	0.035	0.011–0.203	0.67	0.026–1.6
Trace metals ($mg\,kg^{-1}$)										
chromium	22.0	6.5–43.0	23	4.4–28	37	33–63	74	54–146	680	86–1,317
copper	8.3	2.8–31.0	10	2.8–62.0	27	18–56	22	17–192	289	33–782
lead	6.1	2.7–12.0	6.2	3.1–42.7	22	15–36	16	11–40	162	25–537
zinc	42.5	9.8–62.0	41.2	9.8–155	47	39–88	68	58–143	526	57–2,096
Chlorinated hydrocarbons ($mg\,kg^{-1}$)										
phenols[1]	<0.007		n.d.	n.d.	0.076	0.015–0.440	n.d.	n.d.	n.d.	
chlorinated benzenes	n.d.		n.d.	n.d.	n.d.	n.d.	n.d.	n.d.	0.36	n.d.
total DDT's	0.007	<0.001–0.09	0.001	0.001–0.08	0.018	0.003–0.019	0.173	0.154–0.499	25	1.3–175
total PCB's	<0.002	<0.002–0.04	0.037	0.004–0.14	0.063	0.020–0.254	0.122	0.060–0.513	2.1	0.11–10.9

All data from the 60-m isobath sampled between Point Conception and the U.S.A.—Mexico border, 1977 (Word and Mearns, 1979)

n.d. = no data.

[1] Data taken October 1974–June 1975, by Orange County Sanitation Districts (pers. commun., 1978).

[2] From Young and Heesen (1978).

0.30 vs. $0.23 \mu g l^{-1}$, respectively) (Young and Jan, 1975). In contrast, suspended particulate concentrations of these materials were one to two orders of magnitude above control seawater levels (Young and Jan, 1975). Thus even in the water column, particulates remain the most important reservoir of trace contaminants. In Puget Sound, Schell and Nevissi (1977) found that the concentrations of several trace metals in surface sediments taken near the West Point outfall ranged from 1 to 3.6 times greater than sediments deposited over 50 years ago at this site and, except for lead and cadmium, the concentrations of metals in Puget Sound water are similar to reported values for open-ocean water (Schell and Nevissi, 1977). These observations suggest that bottom contamination near the West Point outfall is not unlike the condition at Point Loma (Table III). A slight but significant increase in sediment COD was observed in deep water during the first year of discharge (Domenowske and Matsuda, 1969).

BIOLOGICAL EFFECTS

During the past two decades the coastal discharge sites cited in this chapter have been surveyed to document the abundance, diversity and health of marine life. Much of the work has been done by the dischargers themselves or through private and university contracts. Independent researchers, including the U.S.W.R.P., have also conducted a number of regional surveys of benthic and water column life and contrasted conditions with those at distant coastal sites.

The benthic environment

Visual conditions around the discharge sites. Using submersibles, remote television, and 35-mm cameras the S.C.C.W.R.P. found that the Hyperion outfalls in Santa Monica Bay were colonized by a variety of large attached invertebrates including several species of gorgonians inshore of 30 m and large anemones (*Metridium senile*) and colonial anemones (*Corynactis* sp.) in deeper water (Allen et al., 1976). Large schools of a variety of water-column feeding rock-fish were common over both outfalls (including the diffusers) in deeper water, and the rock ballast appeared to provide refuge for other fishes and invertebrates such as crabs and lobsters (Allen et al., 1976). Similar results have been obtained from photographic and television surveys at outfalls at Point Loma, Orange County, and Palos Verdes (S.C.C.W.R.P., unpublished data, 1975—1977) and at the West Point outfall in Puget Sound (R. Matsuda, pers. commun., 1979). In most areas, the bottom adjacent to the pipes had a normal color and was occupied by fishes common in bottom trawls. However, television surveys near the Hyperion sludge outfall in the fall of 1976 revealed a bottom containing loosely aggregated sludge-like material with large numbers of white croaker (*Genyonemus lineatus*), northern anchovy (*Engraulis mordax*), and numerous large Pacific electric rays (*Torpedo californicus*). A grab survey revealed the area of bottom containing the loose sludge-like material was $\sim 2\,km^2$ in water $\sim 100\,m$ deep at the head of a submarine canyon (Schafer and Bascom, 1976).

Benthic infauna. Despite the relative simplicity of the soft-bottom environment the benthic macrofauna that normally colonizes the sediments is one of the most diverse communities in the coastal zone. For example, on the southern California mainland shelf at a depth of 60 m, a 0.1-m^2 grab captures an average of 71 species and 423 individuals with a Shannon–Weaver diversity, H', of 3.05, an Infaunal Trophic Index, ITI [a measure of dominance by suspension feeding biota (Word, 1979)] of 93.5 and a wet weight of 7.05 g (Word and Mearns, 1979). This fauna occupies sediments with the physical and chemical characteristics shown in Table III (5-day BOD, 636 mg kg^{-1}, dry weight; chromium 22 mg kg^{-1}, dry weight; etc.).

Overall, 4500 species of benthic invertebrates are known to occur in southern California. However, only $\sim 1\%$ are frequent and abundant enough to contribute to the bulk of biomass and abundance. Echinoderms (particularly the brittle star, *Amphiodia urtica*) and polychaetes dominate abundance, but polychaetes, arthropods and molluscs dominate species richness (Word and Mearns, 1979). Surveys of Puget Sound reveal a similar richness (Lie, 1967; Thom et al., 1979).

As one approaches the epicenter of a deep-water (60 m) outfall impact zone, sediment contaminant concentrations infaunal abundance and biomass increase while the number of species and diversity generally decrease (Fig. 4a–f). Specific changes include a major reduction in the brittle star, *A. urtica*, increases, and then decreases, in several polychaetes (*Tharyx* sp. and *Mediomastus californiensis*) and ostracods (*Euphilomedes* sp.) and increases in several pelycypods (notably *Parvilucina tenuisculpta*); in the areas of severe impacts (mainly at the two largest outfalls in Santa Monica Bay and off Palos Verdes) there are increases in another pelycypod (*Solemya panamensis*) and three polychaetes [*Shistomeringus longicornis, Capitella capitata* and *Armadia bioculata*; (Word et al., 1977)]. During this general shift, total abundance and biomass of the community can increase by as much as an order of magnitude to over 30,000 animals, weighing over 100 g per 0.1 m^2 (at the Palos Verdes site). Increased biomass is partly caused by increased size of some organisms such as *Parvilucina* (Word et al., 1977). The total number of species drops from about 70 to less than 50 in a 0.1-m^2 sample.

Previous surveys by dischargers [summarized in Mearns and Young (1978)] indicated substantial differences among the discharge sites in terms of general characteristics of the benthic infauna. As indicated in Table IV there was a 75-fold difference in benthic biomass between Oxnard (12 g m^{-2})

and Palos Verdes ($918 \, g \, m^{-2}$) at the 60-m isobath. In addition, low numbers of species per sample were reported (ranging from 21 per sample at Palos Verdes to 64 at Orange County). Recent data taken by uniform sampling procedures along the 60-m isobath reveal similar trends but somewhat smaller differences between the discharge sites (Table V). The differences are due to several factors including past variations in sampling gear (Word, 1978; Table IV) and past variations in sample processing and taxonomic expertise. The 1977 samples were all taken by uniform procedures and processed and analyzed by one team, thus minimizing these difficulties. In addition, however, the differences in biomass at some discharge sites are due to a 1973—1976 invasion by a large burrowing echiuroid, *Listriolobus peloides*. This organism is once again rare (Word, 1979) so that overall biomass at most sites was lower in 1977 than in 1974—1975. Finally, the new surveys placed more emphasis on distant control sites and their variability; in past discharger surveys, less than 10% of the survey effort was focused on "control" stations, and some of these fell within the influence of outfalls.

Studies to determine the effect of Seattle's METRO outfalls on benthic communities are in progress. However, several preliminary studies suggest there are some effects. Earlier, Harman et al. (1977) found it extremely difficult to assess outfall-related patterns in the benthos due to considerable small-scale variation in substrate type (ranging from sand and gravel to wood and organic debris from runoff) which in turn affected sample sizes. However, they did find evidence for increased abundance of several aerenaceous foraminifera near the diffuser as well as an accumulation of seeds centered about the diffuser and covering an elliptical area of $2-3 \, km^2$. During the first two years of operation, Domenowske and Matsuda (1969) found significant variations in abundance of prominent benthic organisms but none that could be attributed solely to the discharge.

The area of the coastal shelf where benthic infaunal communities are changed by the discharges range from less than $1 \, km^2$ to $\sim 100 \, km^2$. For example, using earlier data (1974—1975) from areal surveys of the five southern California outfalls, bottom areas occupied by above normal benthic biomass values were calculated. The affected bottom areas were less than $1 \, km^2$ at the Oxnard outfall, $\sim 6 \, km^2$ at Point Loma, $10 \, km^2$ at Orange County, $26 \, km^2$ around the Santa Monica Bay outfalls and at least $45 \, km^2$ associated with the Palos Verdes outfall. Using the Infaunal Trophic Index data from more recent aerial surveys, Bascom et al. (1979) found that bottom areas dominated by deposit-feeding benthic communities were $\sim 4 \, km^2$ at Point Loma, $11 \, km^2$ at Orange County, at least $60 \, km^2$ in Santa Monica Bay and over $94 \, km^2$ associated with the Palos Verdes outfall; there was no area around the Oxnard outfall dominated by a deposit-feeding infauna. In total, then, as much as $170 \, km^2$ or 4.7% of the 3640-km^2 southern California mainland shelf (between the 20- and 200-m isobaths) is

38

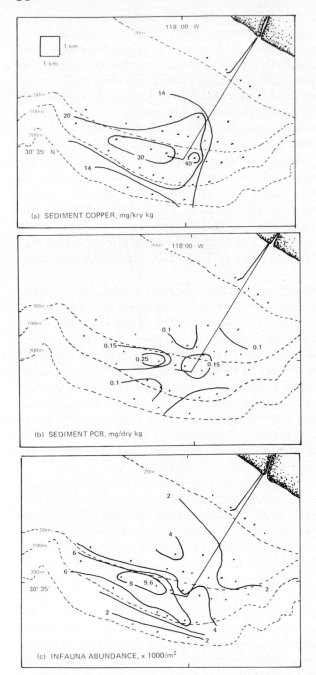

(a) SEDIMENT COPPER, mg/kry kg

(b) SEDIMENT PCB, mg/dry kg

(c) INFAUNA ABUNDANCE, x 1000/m²

Fig. 4. Contours of sediment chemistry and benthic infaunal community characteristics surrounding the Orange County Sanitation Districts deep-water outfall, July 1975: (a) copper, mg kg^{-1}, dry weight; (b) PCB's mg kg^{-1}, dry weight; (c) infaunal abundance,

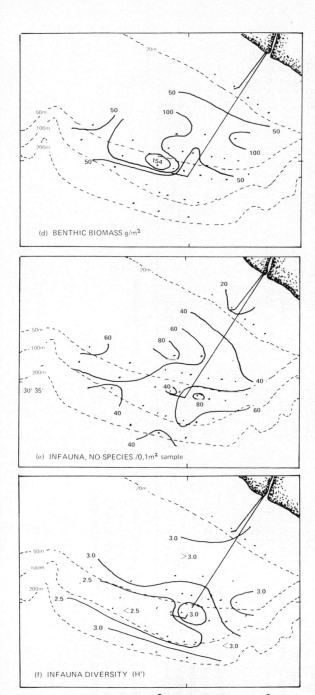

(d) BENTHIC BIOMASS g/m²

(e) INFAUNA, NO-SPECIES /0.1m² sample

(f) INFAUNA DIVERSITY (H')

number of animals per m²; (d) biomass, g m⁻²; (e) number of species per 0.1-m² sample; and (f) Shannon—Weaver diversity. Major area of change is a 2—4-km² elliptical zone to the west of the diffuser. After data in Greene (1976a).

TABLE IV

Summary of sampling statistics (mean ± SE) from surveys of benthic stations located between the 55- and 60-m isobaths at five southern California discharge sites; summers 1974 and 1975 (modified from Mearns and Greene, 1976)

Region	Oxnard*1	Point Loma*2	Orange County*3	Santa Monica Bay*1	Palos Verdes*1
Biomass (g m^{-2})	12 ± 4.0	28 ± 2.9	69 ± 15	153 ± 42	918 ± 213
Abundance (10^3 m^{-2})	3.1 ± 0.5	1.5 ± 0.1	5.7 ± 0.7	4.0 ± 0.6	6.6 ± 1.2
Species (number per sample)	48 ± 4.3	42 ± 1.8	64 ± 3.7	24 ± 1.8	21 ± 1.3
Shannon—Weiner diversity, H'	3.4 ± 0.1	3.0 ± 0.2	2.7 ± 0.2	2.2 ± 0.1	1.8 ± 0.1
Richness, D	23 ± 1.4	19 ± 0.9	23 ± 1.5	11 ± 0.9	8.4 ± 0.5
	9.5	349	341	175	109
Discharge depth (m)	16	61	56	60 and 100	60 and 65
1975 flow (mgd)	9.5	109	175	349	341
mass emission rate of suspended solids (t yr.$^{-1}$)	2,181	18,725	33,396	110,180	130,966
concentration (mg l^{-1})	166	125	138	229*4	278

*1 Surveyed using 0.04-m^2 Shipek grab; 0.5-mm sieve used at Oxnard, 1.0-mm elsewhere.
*2 Surveyed using 0.16-m^2 Petersen grab; and 0.1-mm sieve.
*3 Surveyed using 0.10-m^2 Van Veen grab and 0.1-mm sieve.
*4 Flow-weighted average for effluent and sludge.

TABLE V

Summary of conditions of the benthic infauna at a series of stations (N) along the 60-m isobath at four major discharge sites (no stations are taken at Oxnard discharge zone) and at 29 coastal control sites, all sampled by a 0.1-m^2 Van Veen grab during 1977 (values are medians and, in parentheses, ranges)

	Control sites	Point Loma	Orange County	Santa Monica	Palos Verdes
Number of samples	29	10	9	9	7
Abundance (m^{-2})	4,167	3,550	5,530	3,670	8,560
	(910–12,130)	(1,860–8,330)	(3,960–10,780)	(1,980–12,590)	(2,810–30,570)
Biomass (g m^{-2})	70.5	52	89	87	207
	(28–112)	(43–94)	(76–218)	(15–135)	(77–707)
Species (number per 0.1 m^2)	73	67	80	59	47
	(32–167)	(47–125)	(65–96)	(38–90)	(36–78)
Infaunal trophic index*	93	77	54	63	54.8
	(83–98)	(69–94)	(44–79)	(48–99)	(21–64.4)
Shannon–Weiner diversity, H'	3.17	3.23	3.16	2.93	2.09
	(2.19–4.16)	(3.09–3.94)	(2.52–4.16)	(1.54–3.31)	(1.34–3.30)
5-day BOD (mg kg^{-1}, wet wt.)	632	754	927	817	7,781
	(266–1,017)	(407–2,175)	(499–1,931)	(486–1,850)	(943–25,048)

*Values > 60 dominated by suspension feeding fauna; < 60, > 30 by surface deposit feeding fauna; and < 30 by subsurface deposit feeding fauna such as *Capitella capitata* (see Word, 1979).

experiencing measurable change in benthic communities as a result of deep-water waste discharge. As suggested by Bascom et al. (1979), about $12\,km^2$ (or 0.3%), principally off Palos Verdes, can be considered "degraded", i.e. very dense benthic communities dominated by sub-surface deposit feeding organisms such as the indicator polychaete, *Capitella capitata*.

Bottom fish and larger invertebrate populations. There is a great deal of dis-agreement about whether ocean outfalls enhance or degrade fish populations. With enough sampling sufficient to account for seasonal, year-to-year and regional variability, it is possible to find increased abundance of some species at small and medium-sized discharge sites, and at the periphery of very large discharge sites, and to find significant depressions in abundance of some species in the main impact areas of the largest outfalls.

During 1957–1963, the California Department of Fish and Game con-ducted a six-year series of quarterly trawls at nearly 30 stations surrounding the sewage outfalls in Santa Monica Bay; over 104 species and 100,000 fish were captured, but Carlisle (1969) found the overall catch statistics to be too variable to identify major changes due to the discharges. However, sev-eral species of bottom fish (e.g., speckled sanddab, *Citharichthys stigmaeus*) occurred in significantly higher numbers near the discharges while several others (e.g., California tongue fish, *Symphurus atricauda*) appeared to avoid the site.

Since 1969 all the large discharge sites have been subjected to otter trawl surveys by the dischargers, their consultants, and in southern California, by the S.C.C.W.R.P. In a summary of data from over 2,400 trawl samples taken through 1975, Allen and Voglin (1976) concluded that fish abundance, number of species and diversity increased as one approached the San Pedro Bay area from San Diego in the south or from Santa Barbara in the north. Catches off Palos Verdes and in San Pedro Bay usually exceeded the coastal averages of 175 fish of 11 species weighing 7.1 kg and with a Brillouin diversity of 1.28 ($H' = 1.36$). These data suggest that the coastal area off Los Angeles and Orange Counties was supporting the most abundant and diverse bottom fish fauna along the southern California coast. Careful inspection of earlier data also indicated that past catches in Santa Monica Bay (e.g., Carlisle, 1969) were unusually low due to significant gear and gear use differences (Mearns and Greene, 1974).

A synoptic trawl survey of the three largest areas of sewage discharge using uniform sampling procedures revealed that catches in Santa Monica Bay and off Palos Verdes were higher in San Pedro Bay while those in San Pedro Bay were more diverse; also these areas exceeded the overall coastal average in terms of average catch and number of species per haul (Mearns and Green, 1974).

Despite this trend, past surveys still suffered from inadequate control sampling and, in some cases, inadequate stratification by depth. These

TABLE VI

Summary of catch statistics for fishes and invertebrates collected in 10-min. on-bottom time otter trawls along the 60-m isobath; 28 coastal control sites and four waste discharge sites on the southern California mainland shelf, spring–summer, 1977 (see Word and Mearns, 1979)

Region Number of samples, N	Control sites[1] 28	Point Loma[2] 5	Orange County[2] 3	Santa Monica[2] 7	Palos Verdes[2] 7
Fish:					
Abundance (number per tow)					
all fish	378 (186–623)	232 (114–593)	523 (236–917)	151[4] (17–393)	56[4] (26–852)
less recruits[3]	188 (109–291)	190 (114–370)	172 (104–757)	100 (17–340)	34[4] (25–145)
Species (number per tow)	14.5 (12–16)	17 (14–21)	16 (14–16)	16 (4–23)	10[4] (7–15)
Biomass (kg per tow)	4.7 (3.2–7.9)	6.3 (4–16.5)	6.2 (2.5–14.8)	3.6 (1.5–14.8)	2.5[4] (0.8–4.9)
Shannon–Weiner diversity, H'	1.38 (1.22–1.57)	1.73[5] (1.63–2.45)	1.87[5] (1.02–1.94)	1.83[5] (1.21–2.08)	1.66 (0.88–1.96)
Invertebrates:					
Abundance (number per tow)	181 (90–351)	313 (110–3640)	167 (112–190)	488[4] (225–638)	286 (25–843)
Species (number per tow)	10.5 (7–14)	11 (9–16)	22[5] (22–26)	15[5] (8–20)	11 (9–32)
Biomass (kg per tow)	7.4 (2.5–12.5)	1.17 (0.8–6.0)	7.4 (2.1–16.7)	7.9[5] (3.1–23.1)	18.9[5] (4.9–21.3)

[1] Median and, in parentheses, 95% confidence limits.
[2] Medians and ranges.
[3] Recently settled post-larval young of the year occurred in patches along the coast; this statistic excludes these.
[4] Value significantly lower than control.
[5] Value significantly higher than control.

problems were overcome during the 1977 Coastal Water Research Project 60-m control survey which effectively reversed the ratio of "background" (i.e., $N = 28$) to "outfall" stations ($N = 3-7$) (Word and Mearns, 1979). As shown in Table VI, the 28 coastal control sites produced an average of 378 fish (188 of which were considered one year and older) of 14.5 species weighing 4.7 kg with $H' = 1.38$ and 181 large invertebrates (prawns, crabs, urchins, starfish, etc.) of 10.5 species weighing 7.4 kg. Two of the discharge sites (Point Loma and Orange County) produced catches that were statistically indistinguishable from background in terms of abundance but which were significantly higher in terms of diversity (Shannon—Weiner, H').

The Orange County site also had significantly higher numbers of invertebrate species (22) than the control regions (10.5; Table VI). In Santa Monica Bay, total catch was significantly lower than in control areas, but catch of fish one-year and older was not. Also fish diversity and invertebrate abundance, number of species and biomass was significantly higher in Santa Monica Bay than at control sites. From these statistics it is difficult to conclude that the fish and larger invertebrate fauna were degraded at this depth (60 m) in these outfall regions during the late spring through early summer of 1977. However, it is also not at all clear that diversity indices are a proper way to assess pollution effects on fish communities; Bechtel and Copeland (1970) concluded that low diversity indices were useful indicators of response of fish to pollution in Galveston Bay, Texas; by this criteria we could conclude that fish communities around these outfalls were improved over background!

The fourth site, Palos Verdes, produced fish catches that were significantly lower than the control site in terms of abundance (56 fish per haul), number of species (10) and biomass in spite of an apparently normal Shannon—Weiner diversity ($H' = 1.66$) and normal or above normal catches of invertebrates (Table VI). Clearly the fish fauna at this site was less abundant than background at this depth and time.

Closer inspection of data from the chemically most contaminated site at each outfall reveals no substantial departures from the trends noted above. As shown in Table VII, catches at the most contaminated sites along the 60-m contour at Point Loma, Orange County and in Santa Monica Bay were generally as productive and more diverse than the control sites even though sediment BOD reached concentrations as high as three times background, chromium over six times and PCB's over 100 times (see Table X, last rows); only at the most contaminated site at Palos Verdes, where BOD was 40 times higher than background (Cr 60 times and PCB's over 5000 times), did fish catches show significant depression in abundance (80% below background for fish older than about one year) and Shannon—Weiner diversity ($H' = 0.88$ or 40% below a background of 1.30; Table VII).

Monthly trawls taken at Seattle's METRO West Point outfall and a control site during 1975—1976 indicated that the total abundance of fish caught was

TABLE VII

Summary of trawl catch statistics for single 60-m deep outfall trawl station, each representing the chemically most contaminated location at each of four major outfall sites

	Control sites[1]	Point Loma	Orange County	Santa Monica	Palos Verdes	$\bar{m}$
Number of samples, N	28	1	1	1	1	
Fish:						
Abundance (number per tow)						
all fish	378 (186–623)	245	236	247	852[2]	246
less recruits	188 (109–291)	188	172	242	32[3]	180
Species (number per tow)	14.5 (12–16)	19[2]	14	17[2]	12	15.5
Biomass (kg per tow)	4.7 (3.2–7.9)	9.7[2]	6.2[2]	5.0	n.a.	6.2
Shannon–Weiner diversity, H'	1.38 (1.22–1.57)	2.02[2]	1.94[2]	1.92[2]	0.88[3]	1.93
Invertebrates:						
Abundance (number per tow)	181 (90–351)	191	190	580[2]	286	238
Species (number per tow)	10.5 (7–14)	12	26[2]	19[2]	10	15.5
Biomass (kg per tow)	7.4 (2.5–12.5)	1.3[3]	2.1	17.3[2]	4.9	3.5
Sediment chemistry (mg kg^{-1}, dry wt.)						
BOD	636 (266–1017)	888	1,931	1,390	25,048	
PCB's	<0.002 (<0.002–0.04)	0.037	0.254	0.288	10.90	
Chromium	22 (6.5–43.0)	28.0	63.0	146.0	1,317.0	

n.a. = not analysed.
[1] Median and, in parentheses, 95% confidence limits.
[2] Value significantly higher than 95% CL.
[3] Value significantly lower than 95% CL.

similar at each site at depths between 5 and 45 m (B. S. Miller et al., 1977). However, in nearly every season, total abundance was much higher at the West Point site at depths of 70 and 90 m due principally to high catches of ratfish (*Hydrolagus collei*) at 70 m (the depth of discharge); in addition, Shannon—Weaver diversity was lower at these depths than at the control site.

Recurrent group analysis has been used to examine community structure of bottom fish populations near and away from waste discharge sites in both southern California (Mearns, 1974) and in Puget Sound (B. S. Miller et al., 1977). In both areas, at least three depth-related species groups have been described. Allen (1975) used the results to aid in identifying feeding roles and then noted that the variety of feeding roles in deep water appeared to be lower at Palos Verdes than in San Pedro Bay. The analyses also pointed out that certain species, such as yellowchin sculpin (*Icelinus quadriceriatus*) and California tongue fish (*Symphurus atricauda*) occurred in unusually low abundance at this largest discharge site.

The 1977 60-m survey (Word and Mearns, 1979) offered a new opportunity to identify possible anomalies in the abundance of bottom fish species which Allen (see Mearns, 1974) deemed important members of the community. As shown in the upper portion of Table VIII, there were a number of significant differences in abundances of some of these species. For example, Pacific sanddabs (*Citharichthys sordidus*) were captured in unusually low abundance at Point Loma, Santa Monica Bay, and especially Palos Verdes; similar differences were noted for stripetail rockfish (*Sebastes saxicola*). Inspection of the data revealed the differences were due primarily to lower numbers of young of the year at these outfall sites and to a few patchy, but high catches of young of the year juveniles at several control sites off Santa Barbara. Further examination of Table VIII indicates that pink sea perch (*Zalembius rosaceus*) and English sole (*Parophrys vetulus*) occurred in unusually high abundance in the Point Loma area, yellowchin sculpin, California tonguefish and speckled sanddabs (*Citharichthys stigmaeus*) in the Orange County area and speckled sanddabs in the Santa Monica Bay area. At Palos Verdes, only the Dover sole (*Microstomus pacificus*) was apparently enhanced at this depth and season.

Examination of invertebrates revealed a similar mix of possible enhancements and depressions. For example, the urchin, *Lytechinus anamesus*, appeared to be unusually abundant in the Point Loma trawl area but unusually reduced in Santa Monica Bay and off Palos Verdes. In contrast, the ridgeback prawn (*Sycionia ingentis*) appeared to occur in normal abundance at all sites except Palos Verdes were it was nearly an order of magnitude more abundant than in the control areas.

An approach to examining the net result of these possible changes was to summarize the number of "enhancements" and "depressions". As shown at the bottom of Table VIII for the 1977 data, there were twice as many enhancements (4) as depressions (2) at Point Loma (the smallest discharge),

TABLE VIII

Summary of median catch per 10-min. otter trawl tow for 18 species of bottom fish and invertebrates at control sites ($N = 28$) and four discharge sites along the 60-m isobath of the southern California mainland shelf, 1977 (see Word and Mearns, 1979)

	Control sites 28	Point Loma 5	Orange County 6	Santa Monica 7	Palos Verdes 7
Number of samples, N					
Fishes:					
Pacific sanddab	124 (80—209)	57*1	89	36*1	1*1
Stripetail rockfish	27 (10—177)	1*2	6*2	20	7*2
Plainfin midshipman	21 (2—40)	3	18	7	3*2
Pink seaperch	9.5 (5—15)	21*2	7.5	1*1	0*1
Dover sole	4.5 (2—7)	8	10.5	7	13*2
Bigmouth sole	1 (0—3)	3	2	0*2	0*2
Yellowchin sculpin	5.5 (3—15)	1	65*1	2	0*1
Longspine combfish	2 (0—12)	0	7.5	3	0*1
California tonguefish	2 (1—4)	5	8*2	<1*2	1*1
English sole	2 (1—5)	12*3	8.5	1	0
Speckled sanddab	<1 (0—1)	0	130*1	19*1	0
Invertebrates:					
Lytechinus anamesus	9.5 (3—162)	218*2	26	0*2	0*1
Parastichopus californicus	6 (2—15)	2	22*2	12	0*1
Astropectin verrilli	<1 (0—9)	2	1	27*1	0
Luidia foliata	3 (1—6)	1*2	9	1*2	0*1
Sycionia ingentis	24 (16—98)	15	32	149	228*2
Mursia quadichaudi	<1 (0—<1)	0	2.5*1	0	1*2
Pleurobrachaea californica	<1 (0—1)	0	1.5*2	0	3*1
Number of enhancements		4	5	2	4
Number of depressions		2	1	6	11
Enhancement/depression ratio		2.0	5.0	0.3	0.4

*1 $p < 0.05$;   *2 $0.1 > p > 0.05$;   *3 $0.2 > p > 0.1$.

48

TABLE IX

Summary of possible enhancements and depressions in abundances of 16 trawl caught fish at five depths near the West Point outfall, Puget Sound, 1975—1976 (adapted from B. S. Miller et al., 1977)

Species	Depth (m)				
	5	25	45	70	95
Speckled sanddab	0.8				
C-O sole	0.9				
Striped perch	>1.3				
Padded sculpin	>3.7				
Rock sole	1.1	1.1	0.9		
Shiner perch	<0.6	>1.5	>1.6	>1.3	
English sole	0.4	1.6	2.6	1.6	0.6
Roughback sculpin		0.7	0.5		
Tomcod		1.9	1.7	1.3	
Pacific sanddab			0.9		
Ratfish		<1.5	<1.0	6.2	2.5
Bluespotted poacher				0.9	
Dover sole				3.0	1.9
Rex sole				2.0	0.6
Slender sole				<0.5	<0.5
Quillback rockfish				<1.3	>0.6
Possible enhancements (≥ 1.6)	1	2	3	4	2
Possible depressions (≤ 0.5)	0	0	1	1	1

Each number is average catch per unit effort at West Point divided by average catch per unit effort at Point Pulley, a control site.

five times as many at Orange County (5 and 1, respectively) but only about one-third as many at Santa Monica Bay and Palos Verdes (the largest discharge).

B. S. Miller et al. (1977) also found evidence for possible enhancements (ratfish, English sole, and rex sole, *Glyptocephalus zachirus*) and depressions (e.g., slender sole, *Lyopsetta exilis*) near the West Point outfall. Using their data, outfall/control ratios were calculated for average catch per tow for 16 fishes; the possible alterations in these species are apparent in Table IX. In addition, the data in this table suggests that the enhancements most outweigh depressions at 70 m, the depth of discharge.

Diseased fish populations. While abundance and diversity of bottom fish appear to be only moderately changed at major discharge sites, more profound changes occur in the health of some fishes at a few sites. For example, a fin erosion disease occurs or has occurred at three of the sites but is absent at the remaining three and at control sites. In addition, bottom fish living at some of the sites have enlarged and somewhat altered livers.

The fin erosion disease was discovered in Dover sole from the Palos Verdes peninsula when trawls were taken there for the first time in 1969. Since then over 30 species of fish-bearing eroded fins have been reported from Palos Verdes (Sherwood and Mearns, 1977). As described by Sherwood (1979) and Sindermann (1979) a similar disease has been observed in bottom fish from dumping sites in the New York Bight (see also Chapter 12 of this volume), an industrial waterway in Seattle, off Orange County, in deep water in Santa Monica Bay, and in Boston Harbor, Massachusetts. Additional trawl surveys confirmed this disease was absent at Point Loma, Oxnard, and at control sites in southern California (Sherwood and Mearns, 1977; Table X) and at West Point and control sites in Puget Sound (B. S. Miller et al., 1977).

The diseases do not appear to be caused by infectious micro-organisms but may be related to high chlorinated hydrocarbon levels (especially PCB's) in tissues of affected fishes and to sediments highly contaminated with trace metals and chlorinated hydrocarbons (Sherwood and Mearns, 1977; Sindermann, 1979). The disease was experimentally induced in previously healthy Dover sole exposed to sediments from Palos Verdes (Sherwood and Mearns, 1977). It also affects post-larval fish within one to two months after they settle out of the plankton into the Palos Verdes discharge area (Sherwood, 1979). Finally, another study (McDermott-Ehrlich et al., 1977) suggested that affected Dover sole at the Orange County outfall were in fact migrants from Palos Verdes and that the disease was probably affecting Dover sole in Santa Monica submarine canyon within a year after discharge of sludge was initiated in 1957 (Sherwood, 1979). Together, these studies suggest that of the six discharge sites, the disease is clearly initiated at one (Palos Verdes), possibly at another (the Hyperion sludge discharge site) but not at any other site.

Liver size, color and structure of flatfish also seem to respond to some of the waste discharges. For example, livers in Dover sole from Palos Verdes are over twice the size of control fish; fish from Orange County, Santa Monica Bay, and at a natural oil seep have livers of intermediate size (Sherwood, 1979). Several histological studies indicate livers of some fish have increased lipid vacuolation and compensatory changes in structure and organization of the tissue (K. V. Pierce et al., 1977; Sherwood and Mearns, 1977) and laboratory experiments indicated that both Palos Verdes sediments and sediments contaminated only by PCB's caused increased liver size in Dover sole (Sherwood, 1979). The significance of this change to the fish, or its reversibility, is unknown.

Skin tumors occur in pleuronectid flatfish throughout the North Pacific but have also been implicated as responses to waste discharges [reviewed in Sindermann (1979)]. However, all studies in Puget Sound and southern

TABLE X

Prevalence of fin erosion and skin tumors in Dover sole collected in trawls taken along the 60-m isobath between Point Conception and the U.S.A.–Mexico border, summer, 1977 (samples grouped to increase regional sample sizes)

	Pt. Conception to Pt. Dume	Santa Monica Bay	Palos Verdes	Southern San Pedro Bay	Laguna Beach to Pt. La Jolla	Point Loma
Number of trawls	(13)	(8)	(7)	(3)	(12)	(3)
Dover sole						
total number	120	53	201	36	66	81
catch/haul	9.2	6.6	28.7	12.0	5.5	27.0
number and percentage (in parentheses) with fin erosion	0	0	63 (31%)	0	0	0
number ≤ 120 mm SL*	90	16	147	15	18	58
number and percentage (in parentheses) with tumors	3 (3.3%)	1 (6.2%)	3 (2.0%)	1 (6.7%)	0	1 (1.7%)

* Disease is initiated in young of the year and decreases in prevalence with age; size limitation (≤ 120 mm SL) includes mainly 1976–77 year class.

California have failed to show that tumor prevalence near discharge sites is any higher than at control sites (Mearns and Sherwood, 1977; Table X).

In contrast to skin tumors, McCain et al. (1977) have recently found that nearly half the English sole in the contaminated Duwamish Waterway in Seattle are afflicted with hepatomas (liver tumors). To date, hepatomas have not been observed in English sole from two Puget Sound outfall discharge sites. Work is in progress to determine whether or not English sole at southern California discharge sites have hepatomas.

The inshore environment

Despite the fact that these discharges are located in deep water offshore, inshore waters and subtidal habitats are not totally immune to their effects. Forests of giant kelp (*Macrocystis pyrifera*) are among the most prominent and ecologically important features of southern California rocky coastal areas, such as at Point Loma and Palos Verdes. During three decades from the mid-1940's to the early 1970's the 5-km^2 surface canopy of kelp beds off Palos Verdes completely disappeared. In addition, the diversity and abundance of the entire rocky subtidal fauna and flora was greatly reduced (Grigg and Kiwala, 1970). The situation was attributed, in part, to the Los Angeles County's sewage discharge and, in part, to overgrazing by urchins and abnormally warm sea temperatures in 1957 and 1958. Suspected outfall-related factors included reduced light penetration from very fine, suspended material (L. L. Peterson, 1974) and buildup of a flock-like material on rock surfaces to which plants and a variety of animals were attached (Grigg and Kiwala, 1970). In the early 1970's, the California Department of Fish and Game began a restoration program using adult transplants.

Although there were no major changes in sewage treatment other than source control of DDT's in 1971, reproductive kelp beds recently returned to much of the Palos Verdes peninsula (Mearns et al., 1977); by fall, 1978, surface canopy of *Macrocystis* covered over 1.2 km^2 with plants occurring to depths of 12 m (K. Wilson, pers. commun., 1977). In addition, there were dramatic increases in the abundance and diversity of seaweeds, attached invertebrates, crustaceans and fish at 6 m throughout the peninsula and at 15 m at all sites except near the outfall itself (Grigg, 1979).

At Point Loma and at La Jolla even larger kelp beds suffered a decline during 1940—1970. In 1964, Dr. Wheeler North (California Institute of Technology, Pasadena) began controlling sea urchins, which are major grazers of kelp. During the same year, the City of San Diego diverted sewage from San Diego Harbor to the new deep-water outfall just offshore of the Point Loma kelp bed site. By 1969, the Point Loma beds returned to their 1930's abundance, covering over 10 km^2; this occurred despite several episodes of sludge discharge; partial recovery has occurred at La Jolla, but

52

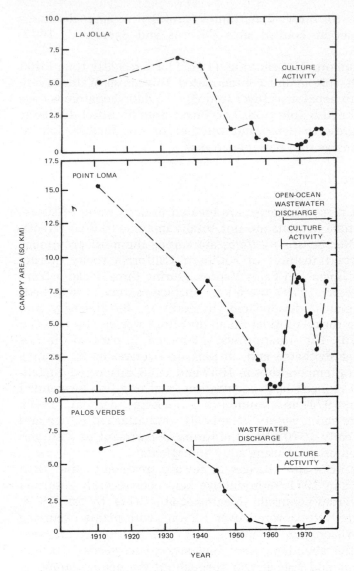

Fig. 5. Summary of historical changes in kelp canopy size at three coastal sites [see Mearns et al. (1977) for data sources]. During the period 1935 to about 1965 there were simultaneous decreases in kelp canopies at the San Diego sites (Point Loma and La Jolla) and off Palos Verdes. Periods of sewage discharge and kelp restoration programs are indicated. Most pronounced recovery occurred at Point Loma. Canopy size increases from less than 0.1 km² in 1970—1972 to over 1.2 km² in 1978 at Palos Verdes.

the beds were still well below their historical sizes by the mid-1970's (Mearns et al., 1977; Fig. 5).

Seaweeds and algae communities at the West Point discharge in Seattle also showed possible evidence of response to that discharge. Compared to four other Seattle area beaches, at West Point, there appeared to be a slight reduction in number of seaweed species, changes in community structure, increased growth rate of two brown algae (*Fucus* and *Laminaria*), increased reproductive activity in one of them (*Fucus*) and decreased diversity of benthic diatoms (Chew, 1977); in contrast, there were apparently no outfall related differences in the intertidal and shallow subtidal macrofaunal communities. Finally, although beach seine studies revealed that West Point shallow-water fish communities differed from a control site, B. S. Miller et al. (1977) felt the differences were due principally to between site habitat differences.

Sampling and examination of trace metals and chlorinated hydrocarbons in intertidal animals confirms that the sewage waste field from Palos Verdes occasionally enters some of these shallow subtidal sites (Young et al., 1978; Grigg, 1979).

Taken together, the ecological surveys suggest that growth of some seaweeds such as *Laminaria* and *Fucus*, and possibly *Macrocystis*, may be enhanced by several waste discharges (Seattle, Point Loma). However, the cause of disappearance and resurgence of kelp and subtidal seaweed and macrofaunal communities off Palos Verdes is an obvious exception. A specific toxin, DDT, was controlled at this site, suggesting the possibility that it may have once played a role in damaging adjacent near-shore resources. Supporting this possibility is the recent observation that PCB's, a group of chlorinated hydrocarbons related to the DDT's, are capable of being rapidly absorbed by diatoms from PCB-contaminated secondary wastewater particles; intracellular levels can reach concentrations that inhibit photosynthesis (Harding and Phillips, 1978).

The pelagic environment

The occurrence of DDE, originating from the Palos Verdes discharge, in intertidal biota (Young et al., 1978) and at a distant pelican breeding site (D. V. Anderson et al., 1977) confirms that materials from deep-water discharge can enter the upper pelagic zone. However, other than DDE, the effects of potentially toxic materials on euphotic-zone organisms is largely unknown.

Several past studies confirm that the plankton community has responded to some of the larger discharges. S.C.C.W.R.P. (1973) reported that in Santa Monica Bay over a 15-yr. period from 1957 to 1970, annual variations in net-haul plankton volumes varied according to changes in the plankton abundance in the entire California Current region. However, the study also found evidence that total plankton volume in the bay was higher when the Hyperion Treatment Plant was operating a shallow-water discharge than after discharge was diverted in the deep-water sites offshore. Presumably, in the past years, much more of the wastes were entering the mixed layer and contributing to plankton growth. However, recent evidence for stimulation of phytoplankton by nutrients (especially ammonia) from the discharges has been gathered by several workers. Throughout several seasons, Eppley et al. (1978) found consistently higher phytoplankton standing crops and productivity in Santa Monica Bay than at other sites of equivalent depth and distance from shore. In an earlier study, MacIsaac et al. (1979) found that ammonia levels in the mixed layer near both the Santa Monica Bay and Palos Verdes discharges were sufficient to enhance phytoplankton growth and yet inhibit phytoplankton nitrate uptake.

Attempts to determine responses of fishery populations have so far been inconclusive. Using catch and catch per effort data compiled by the California Department of Fish and Game from 16 km x 16 km statistical blocks, Allen and Voglin (1977) and Mearns (1977a) were unable to identify geographical trends that could be conclusively attributed to proximity to the discharges. For example, in 1973 in 37 statistical areas ranging from Point Conception to the U.S.A.—Mexico border, the average commercial party boat coastal catch was 1.47 fish per angler hour with an accumulative average of 24.3 species per area. Seven statistical areas occupied by the five major discharge sites produced an average of 1.67 fish per angler hour (ranging from 1.25 off Point Loma to 3.84 in Santa Monica Bay) and a cumulative average of 34 species per block (Mearns, 1977a). Lack of an observable adverse response may simply be due to the large geographical areas (10 mi. x 10 mi.) over which catch and effort data are summarized. However, Allen et al. (1975), using standardized fishing procedures with set lines and rod and reel, produced a slightly higher catch per unit effort and variety near the Hyperion sludge outfall than at more remote sites in Santa Monica Bay. For example, they caught 8.5 bottom fish per angler hour (6 species total) at the Hyperion sludge outfall compared to 5.2 fish/hour (5 species) at a station 5 km NW of the outfall and 5.6 fish/hour (4 species) at a third station 10 km NW of the outfall (all stations 91 m deep). Standardized set lines produced a slightly higher catch at the distant site (7 fish/hour) than at the 4-km and sludge outfall site (6 and 7 fish/hour, respectively).

The commercial landings data did produce one important observation; seventeen statistical blocks within a 50-km radius of the three largest outfalls produced 73% of the 1973 southern California catch ($17.0 \cdot 10^4$ t; Allen and Voglin, 1977). Thus, most of the commercial fish landed are taken from those waters closest to the urban areas and outfalls.

Pollutant residues in the coastal biota

Young et al. (1978) have recently reviewed the cycling of trace metals and chlorinated hydrocarbons in the southern California ecosystem. In summary, trace elements and synthetic hydrocarbons, such as PCB's, are taken up by marine organisms living near major discharge sites. In most cases, the increases are small and the organisms meet all existing U.S. Federal guidelines for safe human consumption. However, the authors have repeatedly advised Federal and State agencies that fish from the Palos Verdes area approach and in some cases exceed the U.S. Food and Drug Administration (F.D.A.) limit of 5 ppm DDT; fish captured from the area have been responsible for deaths of captive birds in the Los Angeles Zoo (Young et al., 1979).

In contrast to the chlorinated hydrocarbons, marine fish do not accumulate heavy metals in their flesh despite their proximity to the discharge sites and metal-contaminated sediment. For example, Dover sole living in direct contact with Palos Verdes sediments, which are highly contaminated with trace metals, had no liver trace-metal elevations above control animals (deGoeij et al., 1974); six other popular seafood fishes from Palos Verdes also had no increases in metals in their muscle tissue [see review and references in Young et al. (1978)]. Moreover, deGoeij et al. (1974) found evidence that semi-volatile trace elements (Hg, Se and Sb) were actually depressed in the livers of Dover sole from the most contaminated sites.

Outfall-related enhancements of trace metals occur more frequently in lower animals and plants than in fishes. For example, digestive glands of intertidal mussels (*Mytilus californianus*) near almost all discharge sites contain increased levels of non-volatile and volatile trace elements (Eganhouse and Young, 1976; Schell and Nevissi, 1977; Young et al., 1978); maximum enhancements were on the order of 2- to 8-fold, but were lower than maximum enhancements due to harbor pollution [see review in Mearns and Young (1978)]. Moreover, edible tissue from each of six popular shellfish from the rocky subtidal zone of Palos Verdes each contained enhanced concentrations of at least one of the eight trace metals analyzed; most enhancements were about 2- to 3-fold, but chromium was nearly 10 times above normal in abalone and scallops [see review and references in Young et al. (1978)]. Despite large emissions of grease and oils (Schafer, 1977), which could contain a variety of petroleum hydrocarbons, Dunn and Young (1976) were unable to detect increased levels of benzo(a)pyrene in mussels (*Mytilus californiensis*) that could be related to the offshore discharges. Chlorinated benzenes were detected in sediments and bottom fish; however, due to their volatility, it appears that the potential for bioconcentration of chlorinated benzenes is weak compared to DDT's and PCB's (Young and Heesen, 1978).

PROCESSES AND SOME RELATIONS TO POLLUTANTS

The previous section focused mainly on recent conditions of marine communities as observed from coastal surveys. Such surveys represent "snapshots" of conditions at a few points in time but do not by themselves identify biological and chemical processes which cause changes in marine populations. Although few studies have been specifically directed at understanding causes of change, there is some information which helps to identify the importance of processes such as feeding, growth, reproduction, recruitment, mortality, pollutant uptake and the reversibility of changed communities.

Biostimulatory responses

It seems clear from data on the infauna and possibly on benthic fish and plankton that the wastes discharges are increasing biomass. For the infauna, this occurs, in part, at the expense of diversity. Presumably production is also increased, but at these discharge sites, this has been demonstrated only for phytoplankton and only at a few sites (MacIsaac et al., 1979).

Increased biomass of the infauna is due both to increased abundance of deposit feeding animals and to increased size of some of them. For example, in control areas, the clam, *Parvilucina tenuisculpta*, is not abundant and rarely reaches 4 mm in length or shell weight of 5 mg; however, both abundance and size increase with proximity to discharges so that at the most contaminated sites, there can be as many as 3000 per m^2 with many 11 mm long with a shell weight of 150 mg (Word et al., 1977). Differences in growth and age have not been determined yet. In related studies, Mearns and Harris (1975) found evidence for major differences in growth rate among various local populations of southern California Dover sole. Fish from coastal localities appeared to grow faster than those from an offshore study area (Santa Catalina Island). Along the coast, the fastest growing fish occurred in Santa Monica and San Pedro Bay discharge areas followed by a coastal control site (Dana Point). The slowest growing fish occurred at Palos Verdes. No extensive feeding habits studies have yet been done at these discharge sites. However, occasional inspection of stomachs indicates that several species of flatfish (including the Dover sole) are consuming the abundant and pollution-tolerant infaunal organisms including *Parvilucina tenuisculpta*.

Recently, S.C.C.W.R.P. was asked to forecast ecological changes that might occur as a result of reductions in discharge of wastewater solids (U.S. Congress, 1978). As part of the forecast the group identified a possible quantitative relationship between waste solids emission and biomass. Assumptions were made that the overall "excess" standing crop of infauna at each discharge site was produced by the organic matter associated with suspended solids and that standing crop was in equilibrium with total solids emission

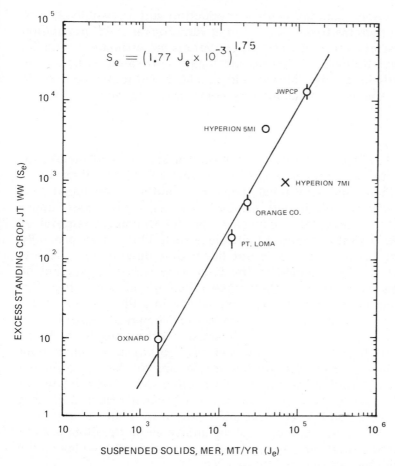

$$S_{\varrho} = \left(1.77 \; J_{\varrho} \times 10^{-3}\right)^{1.75}$$

Fig. 6. Relation between mass emission rates of suspended solids, J_e (t yr.$^{-1}$), and estimates of average excess standing crop of benthic infauna (metric tons, wet weight) at each of six discharge sites in southern California. *Bars* indicate range of estimates from calculations with several data sets, 1974—1976.

rates at the time of sampling. To estimate the total "excess" standing crop at each outfall, biomass (g m^{-2}) was plotted at each station from one or more areal surveys at each site, the area exceeding background biomass at each depth was contoured, background values subtracted, and the total excess calculated in metric tons. As shown in Fig. 6, there was in fact a direct relationship between solids mass emission rates and the estimates of excess infaunal standing crop.

While such a relationship may represent a gross simplification of benthic community responses to waste discharge, it does suggest an approach toward developing a unified theory to define the magnitude of biostimulatory responses for emissions of organic wastes into coastal waters. It also suggests a need to more closely examine benthic production and a possible need to redirect benthic programs toward monitoring variations in *total* population sizes (much the same way that fishery populations are estimated). Finally, since diversity and number of species are highly correlated with benthic biomass at these discharge sites (Mearns and Young, 1978) it may be possible to forecast these parameters of the infaunal communities as well.

Toxic responses

Stober et al. (1977) used a variety of local marine fish and shellfish in bioassays to determine lethal and chronic effects of diluted West Point effluent. Using these and additional data, these authors concluded that 0.3% (333:1 dilution) chlorinated effluent was an approximate upper limit for a maximum acceptable concentration for discharge; removal of chlorine decreased toxicity by about half. Since such dilutions are generally approached at the present sites, it is not likely that directly toxic effects from the effluent can be detected in the field. However, this may not be true for organisms exposed to contaminated sediments at some sites. For example, hydrogen sulfide, a very toxic gas, occurs in sediment pore water at the Palos Verdes site. Using a combination of numerical classification, principal coordinate analyses and multiple regression techniques, R. W. Smith and Greene (1976) found that the presence of H_2S gas in sediment pore water was the most significant outfall-related abiotic factor, affecting infaunal community structure during a 1973 benthic survey of this site. An earlier study [cited in R. W. Smith and Greene (1976)] also found, that hydrogen sulfide, DDT, and percent organic nitrogen were the most important chemical factors explaining biological variability along the 60-m isobath at Palos Verdes while total mercury and redox potential (Eh) explained the least amount of the biological variability.

In yet another study of the Orange County discharge area, Greene (1976a) found that physical/chemical factors fell into two distinct groups; (*1*) sediment coarseness, total volatile solids and total DDT; and (*2*) trace metals (Cu, Cd, Cr) and total PCB's and acid volatile sulfides. The altered faunal composition and enhanced abundance observed in the impact area was found to be most related to the group-*2* chemicals, especially Cu and PCB's. However, it is highly unlikely that enhanced concentration of ·these chemicals would cause enhanced abundance of the two organisms most closely correlated with them — the clam, *Parvilucina tenuisculpta*, and the ostracod, *Euphilomedes charcharadon*. Thus, these investigations concluded that factors having greatest effect on these animals were probably not measured or could not be determined from the analyses.

The decreased diversity of infauna in the epicenter of impacted areas is due in part to a loss of suspension feeding animals such as the brittle star (*Amphiodia urtica*), and a number of crustaceans and polychaetes. Herring and Abati (1979) estimated that in these areas, wastewater solids may be falling out at rates exceeding 10^{-2} cm s^{-1}, and possibly well above the natural fallout rate. Thus it is possible some of these organisms are unable to cope with this amount of fallout.

Overall then, it is likely that hydrogen sulfide and possibly DDT may be acting as toxicants at Palos Verdes. It is not yet clear, however, that metals in sediments are contributing to toxic responses.

As indicated earlier, DDT's have occurred at toxic concentrations in sea birds and may have been a factor in the past destruction of kelp beds and other plant and animal communities at Palos Verdes. Also, as indicated above, chlorinated hydrocarbons, especially PCB's, may be implicated in fin erosion disease and liver abnormalities in benthic fish at Palos Verdes. A similar discussion on this subject for the Mediterranean area is presented by Bernhard (Chapter 3 of this volume).

Food web accumulation of pollutants

Data cited above indicate that, while chlorinated hydrocarbons from ocean outfall accumulated in tissues of benthic, near-shore and pelagic organisms, it is not clear that trace metals are doing so. Moreover, it is not at all clear that pollutants are accumulating in organisms at concentrations proportional to their apparent trophic position. For example, benthic infauna (ostracods, polychaetes and clams) from the Orange County discharge area had higher overall concentrations of five trace elements than sediments (suggesting some bioconcentration), while their predators (sand-dabs and English sole) had tissue concentrations that were generally ten

TABLE XI

Summary of median concentration (mg kg^{-1}, dry wt.) of five trace metals in fish, benthic infauna and sediments in the Orange County discharge area, 1974—1977

	Ag	Cd	Cu	Cr	Zn
Predatory fish (secondary carnivores):					
California halibut	0.03	0.04	0.18	4.4	15
Scorpionfish	0.05	0.31	0.14	3.3	19
Benthic feeding fish (primary carnivores):					
English sole	0.03	0.27	0.26	2.5	25
Speckled sanddab	0.113	0.33	0.43	6.5	29
Benthic infauna:					
Polychaeta	1.25	12.0	4.1	79	96
Ostracoda	4.9	8.3	5.3	121	206
Clams	14.0	2.2	6.2	184	507
Sediments	0.41	1.0	33.9	30.8	60.0

TABLE XII

Summary of median concentrations of cesium (Cs), the Cs/K ratio, and median concentrations of seven trace metals, total DDT and total PCB in fishes and invertebrates from the Palos Verdes coastal shelf, samples collected 1975—1976

Common name	Trophic assignment	Cs ($\mu g\,kg^{-1}$)	Cs/K ($\times 10^6$)	Median concentration ($\mu g\,kg^{-1}$)					Median concentration ($mg\,kg^{-1}$)			
				Ag	Cd	Cr	Cu	Ni	Hg	Zn	DDT	1254 PCB
Boccacio	predatory fish	77.4	16.6	8	2	<10	150	58	0.14	4.7	0.61	0.072
Scorpionfish	predatory fish	54.4	13.6	22	4	36	150	154	0.38	3.9	3.5	0.39
Sanddab	epifaunal feeder	48.1	12.1	5	3	32	190	56	0.081	3.2	6.1	0.38
Yellowcrab	infaunal feeder	21.1	6.5	95	4	80	7,840	261*2	0.064	25	1.5	0.19
Prawn	infaunal feeder	37.3	11.2	<4	32	<19	2,000	<30	0.080	9.8	0.15	0.058
Scallop	phyto-planktivore	21.4	5.4	<3*1	803*1	250*3	240*1	46*1	0.056*1	19.8	0.16	0.012
Abalone	grazing herbivore	24.3	7.6	28*1	41	950*3	3,350	680*1	0.010	6.1	0.001	0.006

All concentrations based on wet weights (not comparable to data in Table XI). All DDT and 1254 PCB concentrations are higher than distant controls.

The complete study (Jan et al., 1977) indicates these values are: (*1) 2—3 times non-outfall controls; (*2) 6 times non-outfall controls; and (*3) 10 times non-outfall controls.

times lower; the largest carnivore in the series, the California halibut (*Paralichthys californicus*) had the lowest concentrations (except for Cr, Table XI).

The alkali trace element, cesium, and the Cs/K ratio are known to increase in concentration 2- to 3-fold in tissues of predators relative to their known prey and has been used to identify trophic structure in linear marine food chains (Isaacs, 1972). Recent measurements of Cs/K in organisms from coastal sites including the Palos Verdes outfall area confirm that some trophic structure is present and that mercury and chlorinated hydrocarbons in part follow that structure. However, other trace metals did not; as in the case at Orange County, there was evidence of decreasing concentration of trace metals with increase in trophic position (Young and Mearns, 1979), moreover, outfall enhancements were limited to lower trophic-level organisms such as scallops and abalone (Table XII). Thus, the assumption that all heavy metals from ocean waste discharges concentrate through marine food webs needs to be re-examined and that concern about uptake should be directed mainly to lower trophic levels organisms such as filter feeding mollusks.

Rates of recovery and impact

Changes in the benthic fauna caused by coastal sewage outfall discharges are not necessarily permanent. During the period August 1970—August 1972, G. B. Smith (1974) recorded both biological and chemical changes accompanying termination of a 15-yr. long discharge of primary treated effluent from an inshore 20 m deep outfall off Orange County. Flow (at the time, 130 mgd) was diverted to the deep (60 m) outfall in March, 1971. G. B. Smith reported a rapid recovery, with organic carbon and acid volatile (total) sulfide concentrations reaching background within three months and a return to normal diversity, abundance and community structure occurring within the first year of recovery. Further reduction in sediment chemical concentrations, benthic biomass and abundance of fish and invertebrates also took place between 1972 and 1975 (Greene, 1976a, b; Fig. 7). Enhanced concentrations of metals also returned to background during the first year (S.C.C.W.R.P., 1973).

Changes were much slower and less conspicuous at the newer deep-water site. For example, fish catches increased from 1969 through 1976, then decreased at the deep site, with catches generally higher following initiation of discharge (Fig. 8). The abundance and number of species of benthic infauna showed a similar trend, but an increase in screen size (from 0.5 mm in 1970—1972 to 1.0 mm from 1976 onward) and other changes in sampling techniques prevented a firm conclusion about overall changes in the benthic infauna at G. B. Smith's (1974) deep outfall site. However, individual species showed dramatic responses: *Capitella capitata* densities increased

62

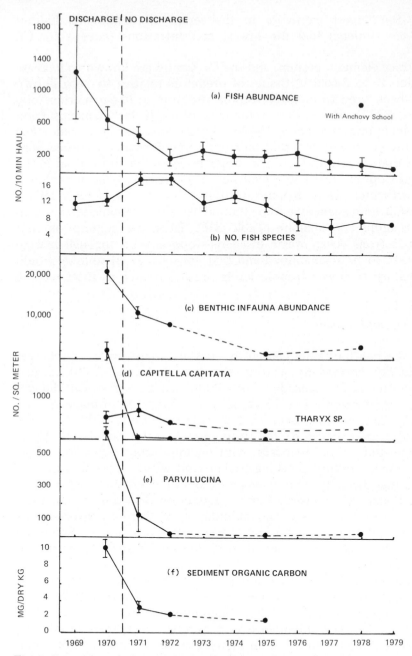

Fig. 7. Summary of changes in fish abundance and number of fish species per 10-min. trawl, sediment chemistry (organic carbon and total sulfides), benthic infaunal abundance and number of species and abundance of four species in infaunal organisms before and after termination of sewage discharge from the shallow (20 m) Orange County outfall. Data from G. B. Smith (1974), Greene (1976b) and Mearns (1977b).

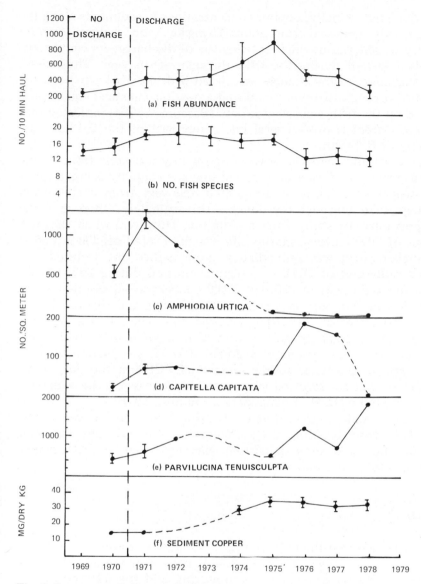

Fig. 8. Summary of changes in fish abundance and number of species per 10-min. trawl infaunal abundance and number of species, abundances of four selected species and sediment copper (mg kg^{-1}, dry weight) before and after initiation of sewage discharge at the deep water (60 m) Orange County outfall, 1969—1978.

between the fourth and fifth year after initiation of discharge while the brittle star (*Amphiodia urtica*) appeared to increase in abundance within the first two years and then disappear from the site. *Parvilucina tenuisculpta* and *Tharyx* sp. also appeared to increase in the fourth and fifth year after initiation of discharge. Finally, copper concentrations in sediments along the 60-m isobath slowly increased from about $15\,mg\,kg^{-1}$, dry weight, to about $31\,mg\,kg^{-1}$, dry weight, during the first five years of discharge; concentrations were relatively constant during the fifth through eighth year. Thus, these data suggest that most of the changes associated with the discharge occurred during the third through fifth years and that a possible chemical equilibrium was established by the fifth year. Newer data (Bascom, 1979a) indicated that the epicenter of effect is now several kilometers east of the $0.4\text{-}km^2$ site first sampled by G. B. Smith.

Presumably the rapid recovery was aided by physical conditions at the site, including currents and wave action, and that such activity is much lower at the new, deeper sites. Thus, if terminated, recovery from deeper-water discharges may take considerably longer. Unfortunately, there has yet been no opportunity to study such a situation. However, when the high concentrations of DDT in local marine life was discovered off Palos Verdes in 1970, source control was immediately implemented and resulted in a nearly parallel reduction of DDT in intertidal mussels during 1971—1973. However, further reduction in effluent DDT concentration resulted in no further reduction in the mussels through 1975 (Young and Heesen, 1978). Subsequently, Young et al. (1978) concluded that DDT in the offshore sediments was slowly leaching and is now the major source to the ecosystem. The half-life of the sediment residues of DDT is estimated to be at least seven years. In contrast, source control of PCB's at the deep-water Orange County outfall has resulted in measurable decreases in the sediments and in the benthic biota (D. R. Young, pers. commun., 1978).

Overall, available data suggest that the benthic biota can recover rapidly (months to years) from biostimulatory effects of sewage discharged through coastal outfalls, but that some persistent chemicals such as DDT remain serious pollutants long after controls are implemented.

CONCLUSIONS

It has been claimed that the large sewage discharges in southern California are creating a sea of sludge and that coastal areas surrounding them are biological deserts. Nearly a decade of monitoring and special surveys confirms that there are indeed ecological problems associated with some discharges and ecological changes associated with all of them. However, it is also apparent that the discharge sites have not created large sludge fields and biological deserts. Taken together the most severe local and long-range ecological problems are associated with the Palos Verdes discharge site where bottom fish are afflicted with a fin erosion disease and where the near-shore ecosystem has been reversibly damaged. The wastewater-related factors that seem to distinguish this site from all others are the high mass emission rates of suspended solids and the unusually high past emissions of chlorinated hydrocarbons (especially DDT's) which still reside in high concentrations in offshore sediments. Presumably effects of these kinds of toxins will be long lasting and continue even if discharges were terminated. The lesson is clear; persistent toxic halogenated hydrocarbons remain contaminants for years, and should not be discharged into the near-shore marine environment.

Excluding this site and the possible effects of chlorinated hydrocarbons, the principal responses of the biota (benthos, fish and phytoplankton) to the remaining discharges are mainly biostimulatory in nature. The enrichment is generally not sufficient to significantly alter dissolved oxygen or pH of the water (S.C.C.W.R.P., 1973), but does change both feeding structure and species composition of the benthos in a manner which appears to be predictable (Word, 1979) and reversible (G. B. Smith 1974; and this chapter). To evaluate any beneficial or detrimental aspects of these changes, more research is needed in areas such as the feeding ecology of fishes and larger invertebrates and on processes that affect colonization and growth including the relative reproductive potential of affected populations and influences of the discharges on dispersal and relative survival of eggs and larvae.

More effort needs to be directed toward understanding pollutant cycling in food webs, (not just chemical analyses of available biota), especially for trace elements. It is now clear that not all of them build up through marine food webs as a result of domestic discharges. Also, in invertebrates, which can take up metals at the most contaminated sites, it is not at all clear what effects to look for in an organism containing a 2-, 3-, or 5-fold increase in silver or chromium. Effects of such slight increases are rarely studied in laboratory experiments although enzymes studies offer a fruitful direction (Brown and Parsons, 1978).

Finally, there is an immediate need to develop a more direct link between either the quality or magnitude of the discharges and the magnitude of total change experienced by affected populations and communities. New, simplified, non-traditional approaches, such as the Infaunal Trophic Index, liver-weight/body-weight ratios, and calculation of total population size can help speed such assessment, while maintaining sensitivity in measuring effects. Mass emission rates, rather than effluent concentrations, are also much more relevant in making this link. This kind of information needs to be developed further so that dischargers and regulatory agencies can implement treatment strategies that will result in real ecological improvements and that future monitoring and research programs will be directed toward testing the effectiveness of these programs.

ACKNOWLEDGEMENTS

The recent 60-m coastal survey under the direction of Mr. Jack Word, Southern California Coastal Water Research Project, provided important and timely data base for this report. I thank him for allowing me to publish some of the data in advance of a more complete report and appreciate the efforts of the Project's benthic ecology team. I also thank Marjorie Sherwood (Coastal Water Research Project) for reviewing the manuscript, Bonnie Mearns and Karla Crum for typing and editorial assistance, and those investigators and coworkers who provided the personal communications cited. This work was supported, in part, by Grant R801152 from the U.S. Environmental Protection Agency, Grant ENV 77-15376 from the U.S. National Science Foundation, and a contract from the Los Angeles—Orange County Metropolitan Area Regional Wastewater Solids Management Program.

Marine Environmental Pollution, 2. Dumping and Mining by R. A. Geyer (Editor)
© 1981 Elsevier Scientific Publishing Company, Amsterdam — Printed in The Netherlands

CHAPTER 2

AN ECOLOGICAL SURVEY OF THE GULF OF NAPLES AREA, CONDUCTED DURING SEPTEMBER 15 TO OCTOBER 15, 1976, FOR THE REGION OF CAMPANIA

Carl H. Oppenheimer, Dorothy P. Oppenheimer and Roberto Blundo

INTRODUCTION

The coastal environments of the world are one of man's greatest assets. Coastal zones are not only esthetically pleasing as used for various forms of recreation, such as swimming, diving, and boating, they also can be evaluated in economic terms, e.g., transportation, resources of water and minerals, industry, fisheries. Throughout the history of mankind, the oceans have played a vital role as a main means of transportation and communication. Most of the large cities of the world originating as centers of trade were established on or near the coastline or on rivers where they were easily accessible by ships. However, modern-day man is placing a much heavier burden on the coastlines of the world. At the same time that man is beginning to realize the finite value of his coastal environment, both esthetically and economically, he is altering the coastal environment at an ever-increasing rate, often without looking ahead to future consequences.

The purpose of this study is to apply methodologies developed for an environmental evaluation of the Texas coastal environment to the Bay of Naples area. In the short time allowed, we have tried to evaluate the present status of the Bay of Naples through a search of the literature and personal observations and to show man's use of the environment, as well as its economic value. We have also tried to show where new information will be required to complete a thorough evaluation of the environment in future years.

Table I illustrates the associated effects of man's activities on his environment, generally applied to the region of the Gulf of Naples and surrounding land. The symbol x indicates an impact of the action to a specific aspect of the ecology of the area. The following sections address some of the most pertinent effects as related to man's activities.

Finally, we have developed an intermediate concept of ocean disposal for sewage and industrial wastes. This intermediate concept is only viable if the continual development of secondary sewage and industrial waste treatment is followed. However, it is possible that future research may show that the ocean outfall concept may provide adequate disposal criteria for many years to come.

TABLE I

Ecological effects of man's activities on his environment

Man's actions	Ecological effects																							
	land				water																air			
	land drainage and leaching	fertilizers	river runoff	street runoff	photo synthesis	photo synthesis	BOD	oxygen	nutrients	salinity	toxic chemicals	pesticides	floating debris	color	suspended solids	temperature	pH	odors	hydrocarbons	coliforms	sulfur	N-ox	organics	hydrocarbons
Urban development:																								
Sewage	×		×		×	×	×	×	×		×	×	×	×	×		×	×	×	×	×			
Streets	×			×									×	×	×			×	×		×			
Autos				×	×																×			×
Industry:																								
Chemical waste	×	×	×	×	×	×	×	×	×		×	×	×	×	×		×							
Water use		×	×		×					×														
Water release	×	×	×		×		×			×				×		×								
Resource exploitation:																								
Seawater	×				×	×				×														
Freshwater		×	×							×			×	×										
Fishing						×							×		×									
Transportation:																								
Ship										×	×		×	×				×	×	×	×	×	×	×
Land	×			×	×												×				×	×	×	×
Recreation:																								
Swimming																				×				
Fishing and boating															×				×	×		×	×	×
Esthetics	×	×	×			×	×	×	×	×	×	×	×	×	×	×	×	×	×	×	×	×	×	×

Fig. 1. Map of the Italian coast adjac

DESCRIPTION OF THE AREA

Coastal environments throughout the world are complicated, balanced systems influenced by time, hydrography, biology, chemistry, meteorology, coastal morphology and man's activities. Each portion of the coast is further influenced by the surrounding major geographic features. The Bay of Naples, while representing a finite portion of the Italian coast (Fig. 1), must also be considered as a part of the Mediterranean—Black Sea system and the extensive weather system of southern Europe. Throughout the years volcanic activity has had a major impact on the area. The Naples area with its very mild climate has attracted people from all over the world. The Mediterranean coast of Italy is well-known for its beauty. Archeological findings and remains are common and provide an historical insight into the growth of the area and mankind, as well as attracting tourists and scientists from all over the globe. The land is fertile and productive due to the long growing season, mild winters and abundant rainfall. The coastline is picturesque; its warm, clear waters are inviting to the swimmer and diver. Boating is both a commercial business and widely enjoyed for recreation. Tourism is a major industry of the area; it has many inducements to offer for short and lengthy tourism visits.

Naples originated as the town of Parthenope, approximately 700 B.C. This tranquil bay, enclosed by the Islands of Capri and Ischia, provided a haven for seamen and a safe port for central Italy. At this early date, the original plan of the city was laid out much as we see it today. The many caves in the limestone cliffs provided shelter for the early inhabitants. Hovering in the distance, the great Vesuvius Mountain, with its fertile lava slopes, must have seemed to reign supreme to these early settlers even though it violently affected the area at times. Naples has been a major new- and old-world port almost since its settlement. Because of its port facility, it was one of the first Italian towns to become industrialized. This industry is significant since the city has grown around many of the major industrial sections.

The following ecological description of the Campania region, Italy, and specifically the Bay of Naples, has been organized to show how the regional approach to an environmental assessment of a coastal area may be used to provide data and information for management purposes. During the short period of the study, from September 15 to October 15, 1976, *it was not possible* to obtain all pertinent information for the area. However, sufficient information was obtained to permit an evaluation of the regional approach and to show that information can be derived or planned in the future that may allow responsible maintenance of the environment.

Population

In 1961, the population of Campania was 4,738,000 (Table II), and Mendia et al. (1976) estimates a total population of almost 7 million by

TABLE II

The population of the Campania region for 1961 and extrapolated to 2015 (Mendia et al., 1976), and sewage outfall to the Gulf of Naples by percent habitation

Year	Number of people		Percentage to the Gulf of Naples
	total	population discharging sewage to the Gulf of Naples	
1961	4,738,000	2,204,361	46.5
2015	6,999,750	3,270,780	46.7
Increase (%)	47.7	48.3	

2015. Fig. 2 shows the population of the Campania region relative to runoff into the Bay of Naples. Table III shows estimated population for the years 1986 and 2016 for the study area of the Cassa per il Mezzogiorno. Fig. 3 shows the growth of principle sites of Campania including Naples from 1871

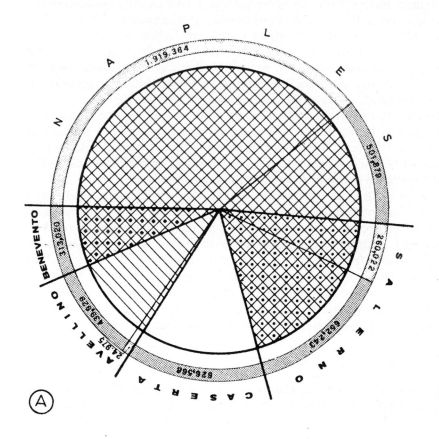

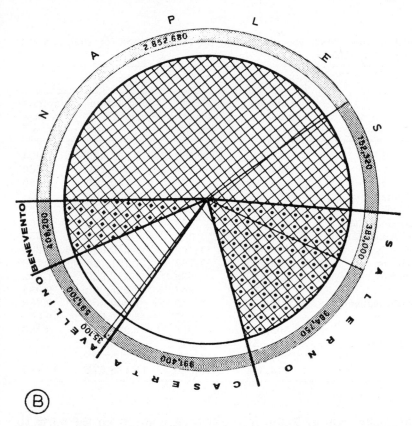

Fig. 2. Distribution of population in the Campania region for: (A) 1961; and (B) 2015 (estimated) (Mendia et al., 1976).
Total population 1961: 4,738,000, estimated total population 2015: 6,999,750. *Light stippled*: impact on the ecology of the Bay of Naples; *dash stippled*: no impact on the ecology of the Bay of Naples.

to 1971. This has been one of the most populated towns in Italy since the days of man's first written accounts. It was one of the first major sea ports for the Mediterranean and was the first area to establish modern industry.

Physiographic features

Physiographically, the coastal area is one of the alternating alluvial plains, carbonate-bluff hills, and the Vesuvius Mountain, with its slopes of lava and volcanic ash. Considerable volcanic activity has been present both in the surrounding upland areas and in subterranean zones near the coast (Fig. 4). The temperature is mild with adequate rainfall, as shown by typical precipitation and rainfall records in Figs. 5 and 6 and Table IV. Wind speed and direction are shown in Figs. 7 and 8 and Table V.

TABLE III

Estimated population and average waste water for the Naples region for 1986 and 2016 (Cassa per il Mezzogiorno, 1975)

No.	Area	1986		2016	
		population number	average sewage $(m^3 s^{-1})$	population number	average sewage $(m^3 s^{-1})$
1	Naples West	1,488,870	4,182	1,837,450	7,004
2	Naples East	1,961,015	5,376	2,218,220	6,869
3	Ischia—Procida	143,660	258	190,590	420
4	Alveo Camaldoli	195,203	433	356,401	903
5	Sarno River	905,850	2,222	1,127,262	3,121
6	S. Sorrentina—Capri	218,140	423	284,230	668
7	Costiera Amalfitana	159,525	363	210,395	559
8	Salerno	700,081	1,799	1,240,000	3,282
9	Lower Sarno	750,565	1,762	1,191,780	3,168
10	Upper Sarno	350,575	538	682,620	1,072
11	Nola	435,865	947	925,065	2,034
12	Acerra—Pomigliano	828,000	1,823	1,062,000	2,788
13	Naples North	841,000	2,005	1,555,630	6,665
14	Caseria	805,110	1,868	1,216,870	3,360
15	Regi Lagni River	632,855	1,345	1,090,060	2,797
Total		10,414,314	25,346	15,188,573	42,710

At Capri, the higher winds, found in the winter, come from the north to the east. The data for Naples indicate coastal influence and a seasonal change in the wind direction. The predominant winter winds at Capri may be correlated with the surface currents in the Bay of Naples (Fig. 9), for January of 1973. In the spring of 1966, the current direction at the surface is shown to be northwest (Fig. 10). The currents at different depths indicated by numbers are shown by the arrow lines representing velocity. The surface current drift is shown by the solid lines between dots, representing observation time of a surface drogue. The depth is given in meters. The general current circulation for the coast of Italy is presented in Fig. 11. It is obvious that the Bay of Naples is influenced both by the offshore currents and the local wind structure. In general, the waters in the Bay of Naples are in continuous motion and are constantly being replenished through the Straits of Capri and Ischia.

Upwelling occurs at certain times, as indicated in Fig. 12.

The bottom contours, shown in the hydrographic chart (Figs. 13 and 14) which shows the two submarine canyon-type structures, allow a tongue of deeper water to move toward the coast. These data are also supported by more recent data collected by the Zoological Institute of Naples. The hydro-

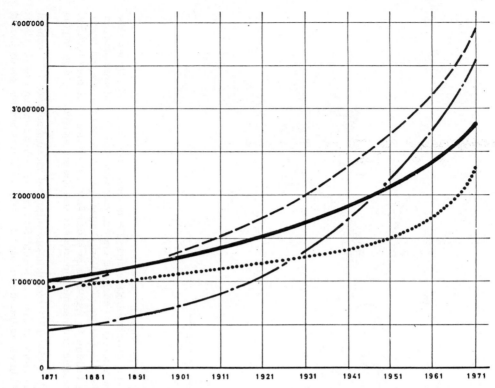

Fig. 3. A comparison of the growth of principle regions of Italy since 1871 (Eurostaff S.p.A., 1973, Vol. II).
- - - = Milan; — · — · — = Rome; ——— = Naples; . . . . = Turin.

graphic data from the Bay of Naples by Hapgood (1960) in 1957—1958 indicate a surface warming trend in the spring with a thermocline at between 50 and 100 m. However, little mixing of the deep waters appears, as the water below 100 m remains remarkably stable throughout the year. This coastal water is relatively high in phosphate, which may indicate the presence of upwelling, as the near-shore waters are almost free of phosphate. More recent data collected from the various research projects may clarify the circulation in the bay, and could in turn be used to determine mixing processes of offshore and coastal waters.

Nutrients, oxygen, salinity, temperature and pH

A study of the nutrients in the Bay of Naples has been made by the Zoological Institute (Cassa per il Mezzogiorno, 1975). Only one month of data were available: March, 1975. These data show a deficiency in the ratio of nitrogen to phosphate, which would indicate, as is seen in many inshore

Fig. 4. Volcanic structures in the Naples area (Eurostaff S.p.A., 1973, Vol. I).

Phlegraean—Ischia regions: 1 = dome extrusion of trachyte lava in pericentric structures; 2 = domes of lava buried slightly less than 100 m average; 3 = volcanic structure, recent crater formation, craters; 4 = enclosed craters, reconstructed; 5 = volcanic structure either buried or removed; 6 = lateral explosions, calderas; 7 = old centers of explosive openings; 8 = cone of scoria and lapilli; 9 = lava extrusions of the Phlegraean district; 10 = folds in relation to fractures in subsurface; 11 = volcano-tectonic faulting, fractures due to concentric collapse; 12 = dikes; 13 = shore margin and marine terraces exposed and submerged in average depths in meters.

Vesuvius region: 14 = area of openings of the flow of 1751—1906; 15 = eccentric openings and radial fractures; 16 = conical pericentric fractures; 17 = domes of lateral lava diffusion; 18 = zone of dikes of Mount Somma; 19 = Vesuvian lava 1831—1944; 20 = buried lava at shallow depths; 21 = possible position of buried central eruptions; 22 = edge of Mount Somma, partly buried portion.

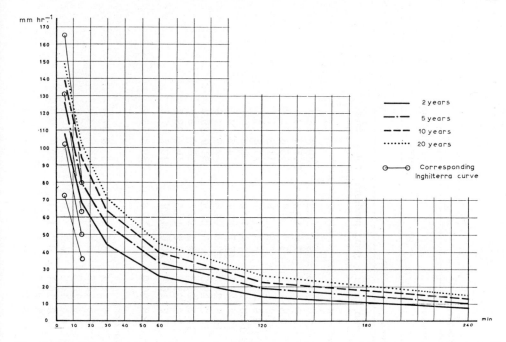

Fig. 5. Rainfall intensity curves of the Naples area for 2- to 20-yr. averages (Eurostaff S.p.A., 1973, Vol. II).

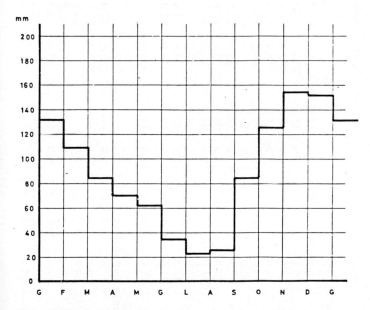

Fig. 6. Monthly rainfall averages for 47-yr. period (Eurostaff S.p.A., 1973, Vol. II) average observations for 27 stations.

TABLE IV

Temperatures and monthly precipitation for Naples area (hydrological tables)

Naples Hydrographic Service:

Month	Temperature (°C) extremes		average			Precipitation (mm)	
	max.	min.	max.	min.	medium	total	max.
Jan.	19.2	5.6	15.0	9.6	12.3	70.2	40.4
Feb.	16.6	5.0	13.4	8.6	11.0	227.2	94.4
Mar.	16.8	−2.0	12.5	6.7	9.6	190.0	41.8
Apr.	19.0	1.1	14.6	8.3	11.5	65.4	25.6
May	24.5	6.5	16.5	10.5	13.5	58.8	19.2
Jun.	34.0	10.8	25.2	15.0	20.1	—	—
Jul.	33.4	17.0	26.1	20.2	23.2	25.4	23.4
Aug.	35.0	13.8	29.4	17.9	23.7	33.8	29.6
Sep.	34.0	15.4	29.8	18.1	24.0	58.8	27.4
Oct.	33.0	15.0	27.5	20.5	24.0	143.2	...
Nov.	27.2	10.5	21.8	15.2	18.5	54.2	20.4
Dec.	21.0	0.0	16.3	11.1	13.7	32.8	29.0
Year	35.0	−2.0	20.7	13.5	17.1	959.8	...

Naples Capodichino Airport:

Month	Temperature (°C) extremes		average			Precipitation (mm)	
	max.	min.	max.	min.	medium	total	max.
Jan.	17.4	0.5	14.2	6.6	10.4	30.8	11.0
Feb.	15.4	0.0	12.7	5.1	8.9	216.8	73.8
Mar.	16.6	−3.0	11.7	3.6	7.7	170.6	34.4
Apr.	18.4	−2.8	13.7	4.5	9.1	70.8	20.8
May	24.4	2.6	16.1	6.0	11.1	57.4	10.4
Jun.	33.2	5.0	24.6	11.8	18.2	—	—
Jul.	33.4	12.6	28.0	16.1	22.1	29.4	12.4

Capri Semaforo (aeronautical station):

Month	Temperature (°C) extremes		average			Precipitation (mm)	
	max.	min.	max.	min.	medium	total	max.
Jan.	18.4	6.6	14.6	9.7	12.2	39.8	18.8
Feb.	15.1	5.0	12.6	7.9	10.3	165.0	53.4
Mar.	15.4	0.8	11.3	5.8	8.6	153.6	22.8
Apr.	18.4	1.2	13.3	6.6	10.0	41.8	7.6
May	21.4	3.9	15.0	8.3	11.7	24.8	7.6
Jun.	29.8	10.3	23.5	14.4	19.0	0.6	0.4
Jul.	32.2	15.2	26.5	17.7	22.1	4.2	3.2
Aug.	35.2	16.6	28.8	20.3	24.6	18.6	14.0
Sep.	33.6	15.2	29.2	20.6	24.9	51.0	29.2
Oct.	30.2	15.2	26.9	19.2	23.1	18.6	9.0
Nov.	28.2	9.6	21.8	15.2	18.5	48.6	28.8
Dec.	19.8	2.2	16.8	10.6	13.7	23.2	7.2
Year	35.2	0.8	20.0	13.0	16.6	589.8	53.4

Ischia Port:

Month	Temperature (°C) extremes		average			Precipitation (mm)	
	max.	min.	max.	min.	medium	total	max.
Jan.	17.5	2.5	14.2	6.9	10.6	133.8	87.0
Feb.	16.5	0.5	12.6	5.8	9.2	225.6	73.0
Mar.	16.5	0.0	12.2	4.6	8.4	192.0	45.0
Apr.	19.0	0.0	14.2	4.9	9.6	123.6	37.8
May	22.0	3.0	16.1	6.7	11.4	22.0	9.0
Jun.	32.0	8.5	24.5	12.4	18.5	—	—
Jul.	32.0	13.5	28.2	16.5	22.4	11.5	7.0

Aug.	35.2	13.8	29.4	18.1	23.8	8.0	6.4	34.5	15.0	30.1	18.9	24.5	—
Sep.	34.0	15.4	29.8	18.3	24.1	156.6	92.6	...	...	...	...	...	...
Oct.	32.6	12.6	27.4	16.0	21.7	141.4	71.0	32.0	14.5	28.5	16.5	22.5	21.4
Nov.	28.0	5.4	20.6	9.3	15.0	55.0	25.0	...	...	...	...	...	...
Dec.	20.2	−3.5	16.7	6.7	11.7	37.2	27.0	...	...	...	...	...	7.2
Year	35.2	−3.5	20.4	10.2	15.3	974.0	92.6	...	...	...	...	...	...

Salerno:

Jan.	16.9	6.2	14.5	9.6	12.1	62.4	27.6
Feb.	16.0	5.7	12.9	8.4	10.7	261.2	90.2
Mar.	16.2	0.8	12.5	6.6	9.6	211.8	37.2
Apr.	20.2	2.1	13.9	7.6	10.8	103.2	39.6
May	21.8	5.0	16.4	9.2	12.8	92.0	24.8
Jun.	32.0	10.2	24.9	15.5	20.2	2.2	1.6
Jul.	35.0	15.8	27.7	19.0	23.4	19.6	9.8
Aug.	35.5	18.1	30.2	21.6	25.9	24.8	9.8
Sep.	34.0	15.2	30.3	21.5	25.9	89.4	41.2
Oct.	33.1	14.9	28.1	19.4	23.8	103.2	27.0
Nov.	30.2	11.2	22.9	15.9	19.4	100.2	52.4
Dec.	21.7	1.8	17.1	11.0	14.1	72.8	46.8
Year	35.5	0.8	21.0	13.8	17.4	1,142.8	90.2

— = no record; ... = not known.

TABLE V

Frequency (F) and velocity (V, km hr.⁻¹) of winds for Capri and Naples (hydrological tables)

	N		NE		E		SE		S		SW		W		NW		Variable	Calm	Monthly average	
	F	V	F	V	F	V	F	V	F	V	F	V	F	V	F	V	F	F	direction	velocity
Capri Semaforo (aeronautical station):																				
Jan.	7	8	10	8	28	6	11	4	4	10	—	—	—	—	2	7	2	29	var.	41
Feb.	14	8	12	9	19	6	8	5	6	5	7	8	5	9	6	11	1	15	N	50
Mar.	9	10	16	11	12	6	5	7	4	5	9	9	8	8	5	13	—	16	NE	63
Apr.	18	9	15	11	12	6	4	6	2	5	1	9	4	9	7	10	1	29	SW	46
May	13	11	7	9	5	6	7	4	3	3	2	6	25	14	12	15	—	16	NW	56
Jun.	20	10	3	10	13	5	3	7	2	6	4	3	5	3	6	7	—	37	SE	34
Jul.	14	10	1	16	14	3	4	3	3	4	2	3	8	8	9	10	—	35	W	46
Aug.	11	9	5	8	8	4	1	3	—	—	6	4	15	7	13	11	—	35	NW	50
Sep.	18	10	2	2	6	4	—	—	1	5	1	2	2	4	3	7	—	55	N	47
Oct.	13	10	3	9	6	6	3	8	4	7	1	4	4	10	9	9	—	47	NW	46
Nov.	15	7	10	16	10	6	2	4	1	3	5	9	7	12	4	11	1	38	NE	54
Dec.	9	7	10	15	2	4	4	1	2	3	4	13	12	7	11	10	—	40	NW	62
Year	161	9	94	11	134	6	49	5	31	6	42	7	100	9	87	11	5	392	NE	63
Naples Capodichino Airport:																				
Jan.	7	3	39	9	8	2	—	—	2	11	2	3	—	—	3	2	—	38	E	44
Feb.	13	7	27	9	10	9	2	10	6	6	6	6	3	7	5	4	—	21	NE	43
Mar.	8	5	21	11	9	6	1	4	7	8	9	8	6	9	5	7	—	18	NE	52
Apr.	22	5	28	9	8	12	1	5	7	7	4	5	1	4	6	8	—	16	NE	38
May	8	7	6	13	1	4	1	7	12	7	18	10	20	13	8	4	—	16	W	55
Jun.	5	3	3	8	6	1	1	1	9	5	23	8	4	9	9	6	—	34	SW	32
Jul.	6	4	7	8	1	5	2	4	6	5	25	9	10	9	11	6	—	22	NW	42
Aug.	7	5	9	6	1	7	1	—	5	8	25	8	11	9	11	7	—	24	W	42
Sep.	8	5	6	5	3	1	6	6	11	6	16	8	4	12	17	6	—	29	NW	62
Oct.	8	4	4	4	4	9	11	11	8	7	19	6	5	5	5	3	—	35	NW	68
Nov.	12	6	20	11	—	—	1	2	5	10	12	10	5	11	16	5	—	22	NE	58
Dec.	6	5	15	8	6	6	—	—	1	14	18	9	5	7	6	5	—	33	W	45
Year	110	5	185	9	48	8	11	6	79	7	177	8	74	10	103	6	—	308	NW	68

— = not determined; var. = various.

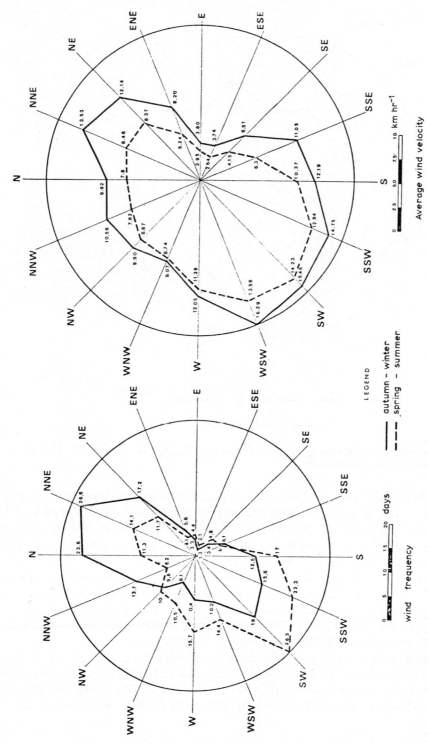

Fig. 7. Windrose — frequency and velocity of winds for the Naples area (Eurostaff S.p.A., 1973, Vol. II).

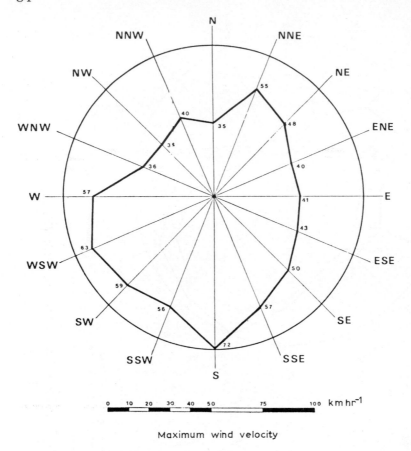

Maximum wind velocity

Fig. 8. Windrose of maximum velocity (Eurostaff S.p.A., 1973, Vol. II). *Heavy line* is annual maximum wind velocity.

areas, that nitrogen is the limiting nutrient for photosynthesis. Surface phosphate values offshore from the Port of Naples had 0.0—0.26 μgAt. 1^{-1} P and 0.31—0.10 μgAt. 1^{-1} N. About 3.5 km off the coast, the phosphorus was zero and the nitrogen was 0.16 μgAt. 1^{-1}. In Pozzuoli Harbor the phosphorus was 0.32 μgAt. 1^{-1}, while nitrogen was 0.16 μgAt. 1^{-1}. In the center of the Bay of Naples, the values were between the values listed above and zero. In all areas, the nutrients increased or decreased with depth to a maximum value of phosphorus 0.5 μgAt. 1^{-1}, and nitrogen 0.6 μgAt. 1^{-1}. There was no noticeable general trend with depth and the average values at depth were about half of the maximum values.

Oxygen values were generally near saturation, even in the inshore areas. This would indicate that eutrophication was not present and is supported by the relatively low values of nutrients found in the water column. The review of the nutrients reported by Hapgood (1960) and summarized by

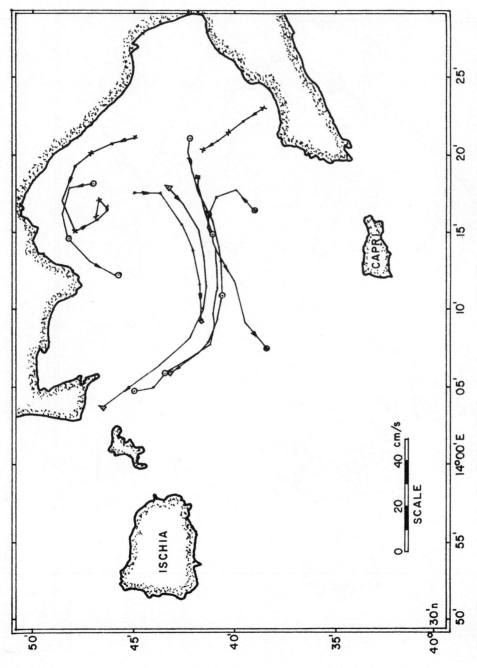

Fig. 9. Water movement in the Gulf of Naples for January, 1973 (University of Naples, Institute of Hydrography, unpublished data).

86

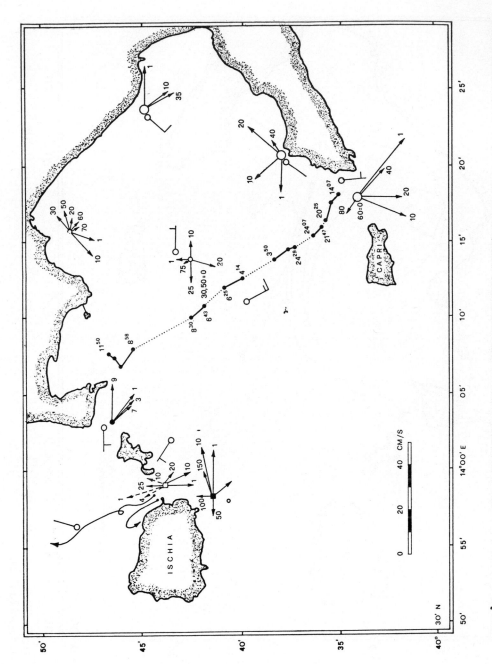

Fig. 10. Water movement in the Gulf of Naples (De Maio and Moretti, 1973).

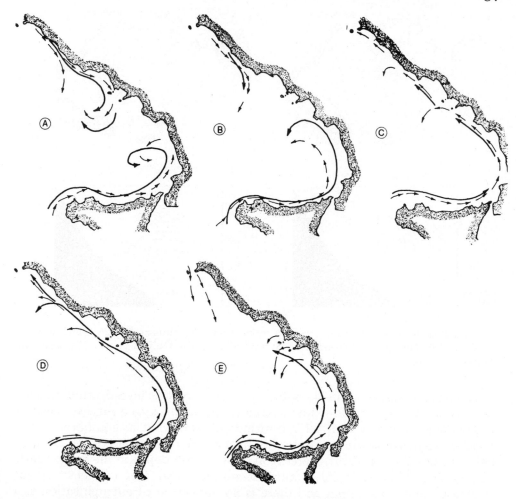

Fig. 11. General circulation of the coastal waters of the southwest Italian coast (De Maio and Moretti, 1973) for
A: July 1—10, 1957; B: September 17—25, 1957; C: February 26, 1958; D: March 21—22, 1958; and E: September 28—October 2, 1958. *Large arrow* = circulation from salinity measurement; *small arrows*: circulation from temperature measurement.

Carrada and Rigillo-Troncone (1973) does not indicate any significant changes with time.

The salinity of the Bay of Naples is typical of the higher-salinity waters of the Mediterranean. Fig. 15 shows the average salinity profile which shows a slightly lower value along the shores. This latter suggests that the high-salinity Gulf water mixes rapidly with the runoff and sewage from land. Fig. 16 illustrates the effect of upwelling which occurs as the more dense water is transported up the canyons of the central part of the bay.

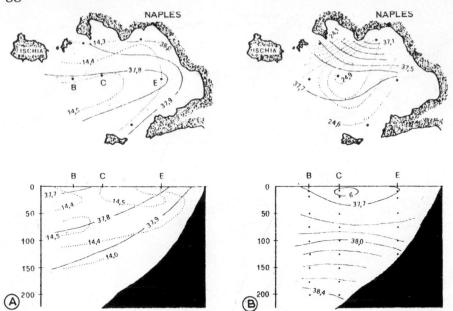

Fig. 12. Distribution of temperature and salinity on: (A) January 24—25, 1957; and (B) July 29—30, 1957, showing evidence of upwelling in the Gulf of Naples (De Maio and Moretti, 1973).

All oxygen values available within the scope of this investigation indicate saturation. The presence of low nutrients and high oxygen suggests that the indigenous populations of all organisms are somewhat at equilibrium with oxygen exchange at the air—water boundary. The recent data of the Zoological Institute show a slight decrease in oxygen with depth in the deeper parts of the Gulf. This is typical of most deep-water areas. The oxygen does not decrease more than 1 ppm and there is no indication of eutrophication as a result of the sewage and freshwater input. The data of Carrada et al. (1974) (Fig. 17), Paoletti (1975a, b) (see also section on sewage) and others indicate by oxygen and pH data, that the effluent along the shores is diluted or transported to near background within 2—4 km from the coast or origin of the effluent. Even in the most concentrated area of effluent, the oxygen is near saturation and the salinities are only a few parts per thousand less than background.

The hydrogen ion concentration (Fig. 17) shows the same effect as salinity and oyxgen. There is no appreciable ecological difference as related to the sewage impact.

Unfortunately, we found no definitive data on the Naples Harbor area. Carrada et al. (1974) measured hydrographic features outside of the harbor but did not report inside the harbor.

Generally, the ecological effect of the various effluents is restricted to a

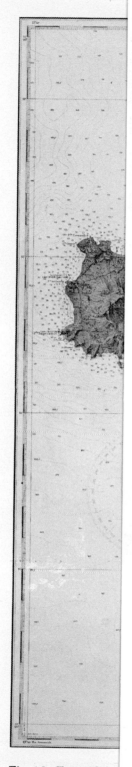

Fig. 13. Chart of the

Fig. 53. Bottom of the Gulf of Naples.

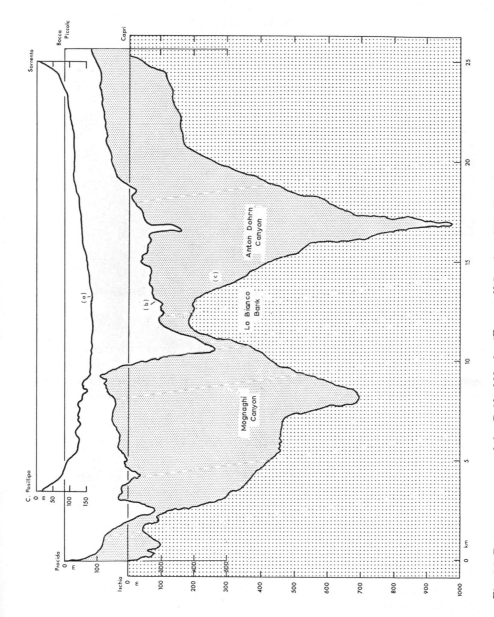

Fig. 14. Bottom contours of the Gulf of Naples (Eurostaff S.p.A., 1973, Vol. I).

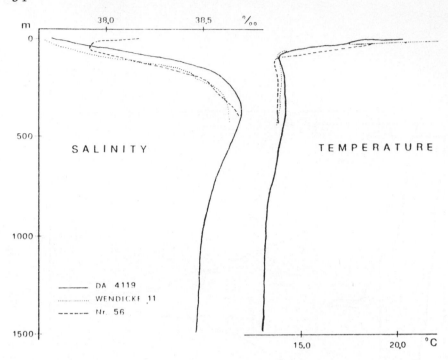

Fig. 15. Salinity—temperature profile in the Gulf of Naples (Carrada and Rigillo-Troncone, 1973).

zone adjacent to the shoreline. Wind mixing and current transport are quite large, which restricts the zone of significance. There is no evidence that these effects are noted in the main body of the Gulf or outside a zone of ~ 4 km from the sewage effluent.

URBAN DEVELOPMENT AND IMPACT ON THE GULF OF NAPLES

The population increase of Naples with time (Fig. 3) is indicative of the rate of urban growth of the city. Presently, the population of the region indicates a megalopolis that extends along most of the shoreline of the Gulf of Naples. The Port and its industrial center and the industry at Coroglio has been surrounded by urban development. Thus the industrial air-borne wastes affect the surrounding areas and after a rain may be significant to the local runoff pollution. The air pollution, which is present from the Port industrial area (Fig. 18) is significant, and must affect the health of the surrounding inhabitants.

The concentration of people within the area (Fig. 18) has increased solid-waste loading, much of which is carried away, but a small percentage persists

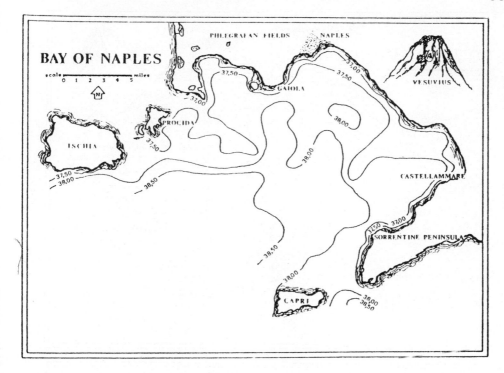

Fig. 16. Salinity contours (‰) in the Gulf of Naples (Carrada and Rigillo-Troncone, 1973).

along the streets. The leaching of metals of copper and zinc, asphalts from roofs, metals from paints, erosion of cars on the streets, grinding of solid materials by tires of autos, excrements of dogs, cats and occasionally man, and the droppings of birds, etc., all contribute to the runoff pollution of the area. It is possible that much of the coliform contamination of the area results from street runoff after a rain. The same material will be air-borne as dust during dry periods to increase human contact and produce allergic responses in man.

Another aspect of contamination is the wind disbursal of aerosols from coastal waters. High winds will remove surface waters from waves along the bay which acts as an aerosol to be blown inland. Pollution of the water surface can thus be taken back to land.

Capri

The authors spent two weekends at the Island of Capri as the guests of the mayors of Capri and Anacapri. Unfortunately, the notes of this part of the program were lost at the Rome Airport and therefore the detailed coverage of the total ecology of man's use of the Island of Capri cannot be provided

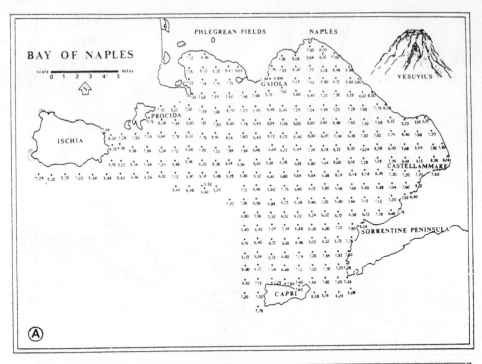

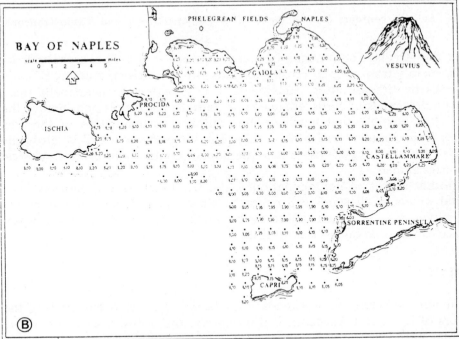

Fig. 17. Oxygen (A, in milligram per liter) and pH (B) values in the Gulf of Naples (Carrada and Rigillo-Troncone, 1973).

Fig. 18. Photographs of Naples industrial area.

TABLE VI

Boating and tourists at Capri [data from Azienda Autonoma di Cura Soggiorno e Turismo Isola di Capri (Tourist Office of Capri)]

Month (1975)	Total		Overnight visitors					
	harbor traffic		arrivals			staying		
	passengers	ships	Italians	foreigners	total	Italians	foreigners	total
Jan.	42,743	1,201	177	236	413	747	597	1,344
Feb.	52,059	1,114	190	229	419	412	520	932
Mar.	108,802	1,736	1,348	1,164	2,512	3,672	3,755	7,427
Apr.	180,644	2,474	1,923	1,925	3,848	5,288	7,725	13,013
May	213,172	2,702	3,373	4,106	7,479	8,593	19,383	27,976
Jun.	238,977	3,600	2,427	4,992	7,419	10,731	26,008	36,739
Jul.	302,435	4,018	4,185	5,508	9,693	29,463	29,706	59,169
Aug.	709,931	5,914	5,414	4,376	9,790	53,157	27,559	80,716
Sep.	394,179	4,226	4,186	5,825	10,011	17,180	28,526	45,776
Oct.	169,849	2,382	496	2,404	2,900	1,802	12,749	14,551
Nov.	66,659	1,436	245	328	573	768	1,519	2,287
Dec.	53,244	1,386	516	331	847	1,157	763	1,920
Total	2,532,694	32,189	24,480	31,424	54,904	132,970	158,880	291,850

Visitations for 1973:

Month (1973)	Gr. Azzurra (Grotto Blue)	Villa Jovis	San Michele
Jan.	994	350	1,414
Feb.	799	174	1,252
Mar.	8,638	535	3,048
Apr.	26,965	3,947	14,936
May	55,351	2,254	18,888
Jun.	62,456	2,490	24,362
Jul.	67,157	2,137	23,380
Aug.	108,157	2,175	26,270
Sep.	33,405	1,079	11,279
Oct.	12,703	659	6,893
Nov.	7,674	340	—
Dec.	935	176	—
Total	374,234	16,316	131,722

at this time. However, there are aspects of the Island that can be used to illustrate examples of man's impact.

Table VI provides an outline of the tourists's daily and overnight use of the Island of Capri. The resident population is ~ 12,000 and therefore the visitor population has a very great impact on the Island. There are three

Fig. 19. Photographs of solid and liquid wastes at Capri.

sewage treatment plants on the Island. All these plants dispose effluent into the surrounding waters. The one plant on the Marina Grande side sends its effluent into the harbor. The Anacapri effluent runs over the cliff a short distance from the famous Blue Grotto (Grotto Azzura) (Fig. 19F). Table VI shows how many people pass by this effluent each month as they visit the

Blue Grotto. The number of private boats using the waters of the Island are significant.

Thus the Island has a considerable amount of sewage entering the waters. Most of this sewage is treated only by primary treatment or Imhoff chamber. During periods of large tourist activity all the treatment plants overreach their capacity and the treatment plant is bypassed. This is also true during heavy rainfall as the plants have no capacity to take the larger runoff.

Solid wastes are a problem that all small overpopulated islands must overcome. One of the problems on the Island is related to the use of plastics. Plastic bags are not biodegradable and therefore if they are disposed into the waters around the Island they will accumulate. This is demonstrated by the photographs in Fig. 19A, C and E).

Capri has other problems related to the availability of fresh-water, the exhaust fumes of ships and vehicles, the by-products of the large influx of tourists in the summer, the sewage treatment plants, energy, etc. The Island could be used as a good model of an impacted environment and, thus, be used to develop the criteria for environmental evaluation and control. It is a beautiful Island steeped in history and worthy of concern.

STREET RUNOFF

The data at hand will not permit a complete analysis of the effects of rainfall and its contribution of street runoff to the coastal environments. It would be necessary to determine the major areas of runoff along the coast and to estimate the numbers of street kilometers. The impact of man's activities on contamination of the streets is considerable and arises from leaching of metals of roofs and cars such as Cu and Cr; hydrocarbons from roofs, streets and cars (Fig. 20); washing of air-borne pollutants from the air by the rain; the activities of man when he uses the streets. Animals also have a great impact. It has been estimated in the U.S.A. that for each person (in an average city) ~ 31 kg of feces will be deposited by dogs on the streets per year.

A publication by Sartor and Boyd (1972) provides values for pollutants as averaged from seven cities in the U.S.A. The values are related to a rainfall frequency of 1.3 cm of rain at intervals of 5 days, and in which the average street is cleaned once a week with a mechanical street cleaner. Table VII shows the amounts of pollutants washed from the average size streets in kilograms per curb kilometer.

These contamination values can be multiplied by the number of rains of 1.3 cm or better per year and by the number of kilometers of streets. The numbers would be appreciably greater if no mechanical street cleaning is performed on a routine basis. The reference is included as a guideline to show the magnitude of street runoff effect on adjacent bodies of water. For example, if there are 5000 linear kilometers of streets draining into the Gulf

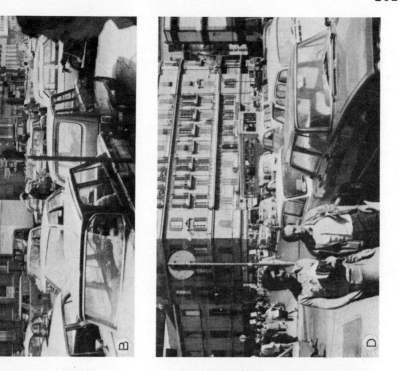

Fig. 20. Photographs of people's activities.

TABLE VII

Street source of pollutants (Sartor and Boyd, 1972)

Measured constituents	Weighted means for all samples (kg/curb km)
Total solids	1,017.0
Oxygen demand	
BOD$_5$	9.8
COD	69.0
volatile solids	72.6
Algal nutrients	
phosphates	0.7
nitrates	0.068
Kjeldahl nitrogen	1.5
Bacteriological	
total coliforms	$158.4 \cdot 10^9$
fecal coliforms	$9.0 \cdot 10^9$
Heavy metals	
zinc	0.47
copper	0.14
lead	0.41
nickel	0.002
mercury	0.053
chromium	0.01
Pesticides	
p,p-DDD	$48.7 \cdot 10^{-6}$
p,p-DDT	$44.3 \cdot 10^{-6}$
dieldrin	$17.4 \cdot 10^{-6}$
polychlorinated biphenyls (PBC's)	$799.0 \cdot 10^{-6}$

of Naples waters, a 1.3-cm rain would wash 700 kg Cu, 2100 kg Pb, $8 \cdot 10^{14}$ total coliforms, etc., off the roads. These data could then be compared with the amounts entering from the sewage, if the two could be separated.

SOLID WASTE

It is difficult to determine the environmental impact of solid waste. One could start with the average figure used in the U.S.A., that each person generates ~ 3.2 kg of solid waste per day. This would produce $\sim 3.2 \cdot 10^6$ kg per day per each million of population. Figs. 19—21 illustrate the solid wastes in the area of the Gulf of Naples. In general, such solid wastes except for excreta, only affect the esthetic or cosmetic aspects of our environment. The plastics that accumulate along the shore of the Island of Capri or along the beaches of the Gulf of Naples really do not harm the total ecology but

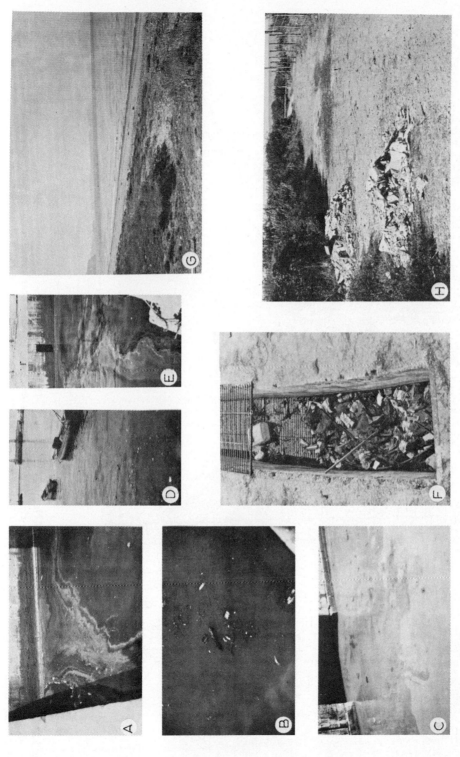

Fig. 21. Photographs of solid wastes.

Fig. 22. Site of sewage effluents in the Gulf of Naples (Mendia et al., 1976).
1 = sewage outfall to the sea; *2* = street runoff outfall; *3* = rainwater discharge via gutter network for normal rain and storm rain;
4 = critical discharge via elevated station; *5* = discharge via sewers merged to water treatment plants; *6* = mixed discharge; *7* =
artificial sewer network for effluents not originating from the community of Naples.

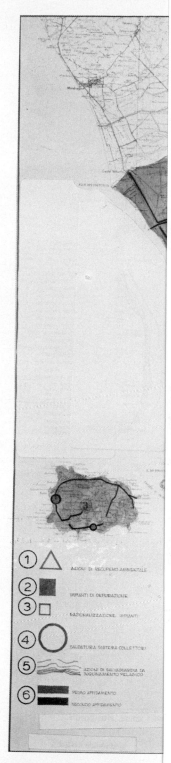

Fig. 23. Sewage treatment e:
1975).
1 = environmental treatment
4 = merging of sewer network
treatment area; *7* = secondary

Fig. 38. Sewerage installations, existing and planned for Naples (Casa per il Mezzogiorno,
.......................) .... plan; 2 = purification installation; 3 = planned installation;
... = ... network; 5 = protection of cleanliness of beaches; 6 = primary
.... 7 = ... for treatment area.

do hurt the esthetic feeling of the inhabitants and, especially, the visitors. There is direct evidence that the plastics are ingested by some fish, turtles, and other organisms and may cause death as they impair the normal metabolic functions. However, the general aspect is one of cosmetic appearance. Education of the people, enforcement of existing laws, and the discouragement of the use of plastics, the most general offender, will alleviate much of the cosmetic effect of solid waste.

The other consideration must be made in relation to solid waste. The solid by-product of sewage and industrial waste must be disposed of by methods which will not affect the groundwaters or land use. Also, solid wastes must not be placed in the marine waters without evaluation of the environmental and cosmetic effects. This is especially true of those solid-waste materials that provide floating residues or are not effectively bio-degraded by micro-organisms. Sanitary landfills must be situated so that the leaching of the wastes does not affect surrounding waters.

SEWAGE EFFLUENT

Fig. 22 identifies the numerous sewage effluents that enter directly along the shore into the Gulf of Naples. Such diffuse type of sewage disposal has grown throughout the centuries with the city. There is a considerable amount of information available on the direct impact of the sewage outfalls to the beaches.

The Cassa per il Mezzogiorno produced Fig. 23, which outlines the sewage system and types of disposal for the present and planned for future. The Eurostaff S.p.A. (1973) Report fully covers representative portions of the effluents, as shown in Table VIII that cover a large part of the shoreline of the Gulf. Paoletti et al. (1973), and Paoletti (1975a, b) Figs. 24 and 25, and Carrada et al. (1974), Fig. 26, reported on the offshore distribution of sewage adjacent to outfalls. The Eurostaff report (Fig. 27) shows the disbursion of sewage relative to wind and currents.

These data can be used to determine the impact of the sewage on both the coastal environment and the Gulf of Naples. While it is not possible to cover all the data in this chapter, the material was analyzed relative to ocean outfalls in the section entitled "Waste treatment alternatives".

The sewage outfalls in Naples Harbor are shown in Fig. 22.

Other discussions on the effect of sewage outfalls are presented by Mearns (Chapter 1), Gunnerson (Chapter 11) and Simpson et al. (Chapter 12) in this volume.

TABLE VIII

Water-quality data of the coastal plume of the River Sarno near Naples, 1972 (Eurostaff S.p.A., 1973, Vol. II)

Date (1972)	September 22			September 29			October 2			October 6			October 9		
Sample No.*	1S	2C	3D	4S	5C	6D	7S	8C	9D	10S	11C	12D	13S	14C	15D
Air temp. ($^\circ$C)	25	25	25	21	21	21	21	21	21	22	22	22	20	20	20
Water temp. ($^\circ$C)	15.80	15.80	15.80	14	14	15	13.5	13.5	15	13.70	13.20	15	13.40	13.40	14.50
pH	6.70	6.50	6.30	6.15	6.45	7.20	6.20	6.95	6.85	6.50	6.45	7.00	6.20	6.90	7.00
BOD$_5$ (mgl^{-1})	81.80	81.80	101	8.9	70.2	83.6	70	71.2	112	70	70	85	70	112	112
COD (mgl^{-1})	100	100	200	85	85	100	100	200	260	80	80	100	90	200	280
DO (mgl^{-1})	3.92	3.72	2.00	5.22	5.12	3.75	7.80	4.00	4.00	4.20	4.20	3.70	7.08	4.00	4.00
NH$_4^+$ (mgl^{-1})	0.60	0.50	0.25	0.50	0.50	0.40	0.75	0.75	0.50	0.60	0.65	0.20	0.80	0.80	0.60
NO$_2^-$ (mgl^{-1})	0.10	0.05	0.2	0.05	0.05	0.10	0.05	0.05	0.05	0.05	0.05	0.20	0.05	0.06	0.06
NO$_3^-$ (mgl^{-1})	1.50	2.00	3.00	1.30	1.30	1.70	1.50	1.50	1.50	1.50	1.50	1.80	1.50	1.60	1.50
Ferrous ion (mgl^{-1})	0.002	0.002	0.002	0.002	0.002	0.002	0.002	0.002	0.002	0.002	0.002	0.002	0.002	0.002	0.002
Methylene blue test		aerobic			aerobic			aerobic			aerobic			aerobic	

Date (1972)	October 13			October 16			October 20			October 27			December 1		
Sample No.	16S	17C	18D	1S	2C	3D	4S	5C	6D	7S	8C	9D	10S	11C	12D
Air temp. ($^\circ$C)	21	21	21	20	20	20	18	18	18	17.20	17.20	17.20	18	18	18
Water temp. ($^\circ$C)	13.10	13.10	14	13.10	13.10	14	13	13	15	13	13	15	14	14	15
pH	6.20	6.20	7.00	6.80	6.80	7.12	6.8	6.8	6.10	6.80	6.80	6.30	6.20	6.50	6.55
BOD$_5$	81	81	110	88	88	101	80	80	88	80	80	88	112.4	106	111.4
COD	110	110	200	90	90	166	92	92	140	95	95	172	170	160	168
DO	4.20	4.00	3.00	4.00	4.00	2.95	4.00	4.00	3.50	4.00	4.00	3.28	1.20	1.20	1.00
NH$_4^+$ (mgl^{-1})	0.50	0.50	0.75	0.75	0.75	1.60	0.75	0.75	2.00	0.75	0.75	2.00	1.50	1.00	0.75
NO$_2^-$ (mgl^{-1})	0.06	0.06	0.10	0.01	0.01	0.03	0.01	0.01	0.05	0.01	0.01	0.05	0.20	0.15	0.10
NO$_3^-$ (mgl^{-1})	1.50	1.50	1.50	1.50	1.50	2.00	1.00	1.00	1.00	1.50	1.50	1.50	2.00	1.80	2.00
Ferrous ion (mgl^{-1})	0.002	0.002	0.002	0.002	0.002	0.002	0.002	0.002	0.002	0.002	0.002	0.002	0.002	0.002	0.002
Methylene blue test		aerobic			aerobic			aerobic			aerobic			aerobic	

S = left of river; D = right of river; C = center.

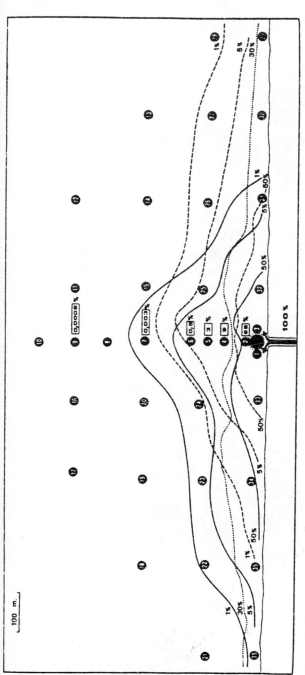

Fig. 24. Distribution of silicate and coliforms given in percent of river concentration in the River Cuma coastal plume. *Solid line represents a south wind; dashed line north wind. Stippled line off-coast wind; values in full dots sampling point number; values in boxes values of coliforms.* Values are average of eight days of data (Paoletti et al., 1973).

Water-quality parameters	Sewer decharging on the River Cuma	Sampling point No.						
		1, 2, 3	4	5	6	7	8, 9, 10	
Silicates (μ mol l^{-1})	141	89.5	47.5	34.5	21.4	3.9	1.8	
Coliforms (number in 100 cm^3)	$1.8 \cdot 10^9$	$5.1 \cdot 10^8$	$17.2 \cdot 10^7$	$6.1 \cdot 10^7$	$9 \cdot 10^6$	$5.4 \cdot 10^4$	$3.6 \cdot 10^3$	
Silicates (%)	100	63	33	23	14	2	1.1	
Coliforms (%)	100	28	9	3	0.5	$3 \cdot 10^{-3}$	$2 \cdot 10^{-4}$	

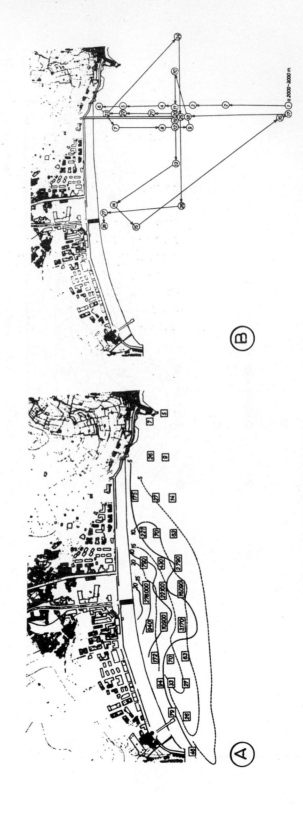

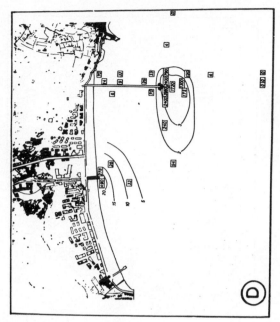

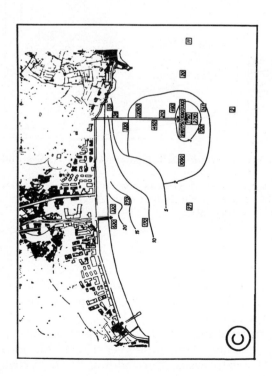

Fig. 25. Distribution of silicate and coliforms at the mouth of the Regenna River (Paoletti et al., 1973).
A and B. *Full lines* and *broken lines* indicate silicates in micromoles per liter when sampled in the morning and afternoon, respectively; *figures in boxes* coliforms per 100 cm^3. C and D. *Full lines* indicate silicates in micromoles per liter; *figures in boxes* coliforms per liter; *figures in boxes* coliforms per liter; *figures in boxes* coliforms per 100 cm^3.

Fig. 26. Effect of sewage effluents on several water-quality parameters at Naples and Pozzuoli (Carrada et al., 1974).
A. Nitrites. B. Nitrates. C. Phosphates. D. Silicates. E. Organic matter. F. BOD$_5$. Values in A—D in micromoles per liter; in E and F in parts per million. *Campi Flegrei* = Phlegrean Fields; *Napoli* = Naples; *Data prelievo* = sampling dates.

INDUSTRIAL DEVELOPMENT

Time did not permit an analysis of all the industrial effects of pollution per se. However, observations were made that can be used to outline the extent of industrial effects on the environment.

The first obvious effect of industrial impact is to increase the urban development for factory employee housing. The employees of the industrial area generally live in the adjacent urban areas, thus adding to the urban

115

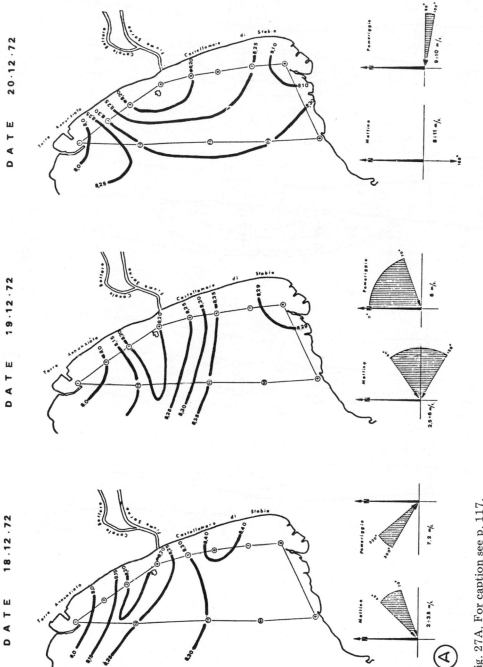

Fig. 27A. For caption see p. 117.

116

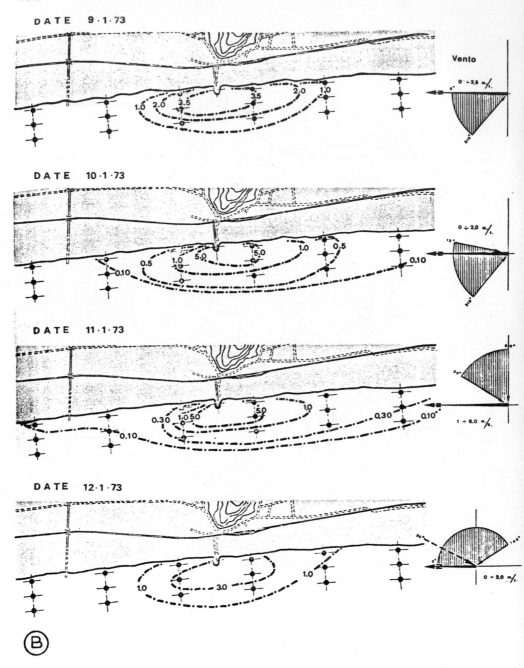

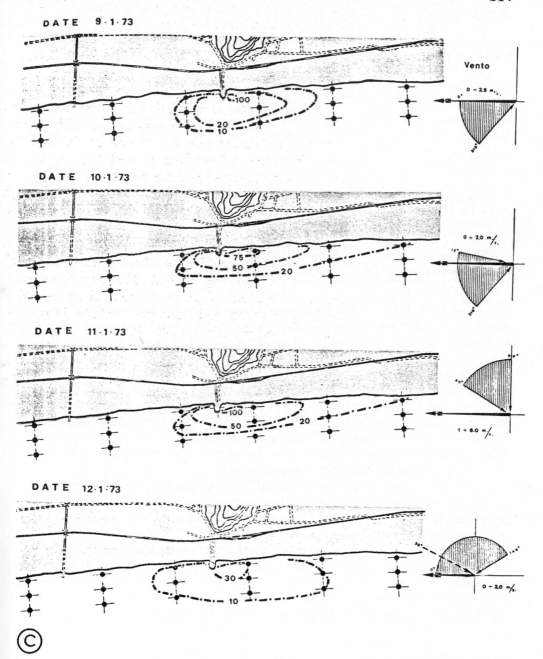

DATE 9·1·73

Vento

0 ÷ 2.5 m/s

100

20
10

DATE 10·1·73

0 ÷ 2.0 m/s

75
50
20

DATE 11·1·73

100
50
20

1 ÷ 6.0 m/s

DATE 12·1·73

30
10

0 ÷ 2.0 m/s

Ⓒ

Fig. 27. Sewage and wind effects at the River Sarno and at Cuma (Eurostaff S.p.A., 1973, Vol. I).
A. pH and wind effects at the mouth of the River Sarno.
B. Phosphate and wind effects at Cuma.
C. Turbidity and wind effects at Cuma.

TABLE IX

Domestic and industrial BOD$_5$ in metric tons per day in the Sarno River near Naples, Salerno and Avellino (Eurostaff S.p.A., 1973, Vol. II)

	Sewage BOD$_5$ (t day^{-1})		
	Naples	Salerno	Avellino
Domestic	2,050	12,500	1,180
Industrial			
tannery	2,000	1,300	72,800
food	0.976	14,374	0.366
mechanical	0.004	0.049	—
chemical	2,843	0.123	—
various	1,211	17,225	—
Total	9,084	45,571	74,346
Total of three cities			129,001

pollution load. Most of the industries release untreated wastes into the local rivers or sewage lines. Those industries with gas wastes, pass untreated vapors and solids into the atmosphere. The gas materials, depending on the wind, may either return directly to the surrounding area or be carried inland or seaward for many kilometers. During certain meteorological conditions, the industrial area is remarkably clear in comparison to other times when the materials remain in the vicinity. Local rainfall will wash these materials from the air and add them to the street runoff. Most of the industrial wastes are added to either the Harbor of Naples or the area north of Monti di Procida. Both areas are highly polluted. The latter area was previously a favorite swimming spot before the industries were built.

Table IX, taken from the Eurostaff S.p.A. report indicates the BOD pollution for three areas entering the Sarno River. Tables X and XI taken from Insola et al. (1973) and the Eurostaff report show a summary of pollutants in effluents from industries of the area.

Before one can assess the effects of industrial contaminants, it would be necessary to determine the amounts and types of air pollution per day for each industry and to determine the amount of effluent and amounts of contaminants in the effluents. These data can be used to determine the loading effect on the sewage and adjacent waters where the material is being disposed.

HYDROCARBONS

No data on hydrocarbon concentration of the Bay of Naples waters were found during the preliminary review of the literature. Some information is

TABLE X

Pollutants in industrial effluents (Insola et al., 1973)

	Desali-nation	Distil-lation* (1 atm.)	Vacuum distillation	Thermic and catalytic cracking	Neutralisation of basic phenols
pH	7—9	4—7	6—7	—	—
NH_4^+ (ppm)	2	10—70	—	80—7,000	—
Oil (ppm)	20—500	7—10	—	30—100	—
Phenols (ppm)	10—20	5—15	10—30	55—1,400	5,000—10,000
Sulfur (ppm)	1—10	15—40	10—30	40—9,000	3,000
Sewage (kg per ton load)	30—100	50	50—250 (stripped) 50—250 (ejected)	50—200	—

*These data are from only one Italian oil refinery.

available on the concentrations of hydrocarbons in industrial effluents and could be related to volumes of water used if these data are available. Tables X and XII, taken from Insola et al. (1973), indicates the various pollutants, including hydrocarbons, related to industrial wastes. Normally, values of hydrocarbons in excess of 50 ppm in effluents are considered possible pollutants. The amounts of hydrocarbons indicated in the tables as being introduced into the beach waters via the sewage could provide a negative impact. As much of the refining capacity is adjacent to the harbor (Fig. 28), most of the hydrocarbons will undoubtedly be added to the harbor waters through the six sewage disposal inlets (Fig. 29).

One source of major concern is the hydrocarbons found in the harbor waters. The photographs in Fig. 30 illustrate the amounts of oil on the surface of the water. The area devoted to oil shipping was noticeably contaminated. The bottom sediments, as shown in Fig. 30, were so enriched with hydrocarbons that the sediment could be burned as a fuel. The surface of the water was covered with methane bubbles, indicating microbial action. The area was anaerobic, as evidenced by the hydrogen sulfide odor associated with the oily sediments.

Oil was present on most waterline surfaces of piers, bulkheads, vessels, etc., in the harbor. There was a distinct gradient, however, that could be detected as a function of distance from the oil docks. The oil in the inner harbor was more abundant than that found in the outer harbor. It is obvious that the oil was being spilled during loading operations. However, other oil was being added from the bilges and from the sewage outfalls. In the photographs of Fig. 29, one can see the obvious oil on the water adjacent to the sewage outfalls.

Some information has been recently published for hydrocarbons in the Mediterranean. Fig. 31 shows the relative abundance of hydrocarbons and

TABLE XI

Water-quality parameters and industrial pollutants (concentrations in mg l^{-1}) entering the Gulf of Naples at sources shown in Figs. 22–26 (Eurostaff S.p.A., 1973, Vol. II, Appendix)

Analysis	Distribution site No. (see Fig. 22)																						
	F1c	10b	26	27	28	29	29b	29d	30	120	12g	129d	129e	129f	130	132	135A	136	139	142	142a	169a	170
BOD$_5$	78	44	222	109	107	40	38	41	230	373	177	64	197	80	136	370	513	238	100	424	189	669	285
COD	286	181	803	410	353	157	160	166	606	633	333	106	473	160	405	586	1,920	546	260	880	850	963	1,616
pH	6.9	7.1	7.3	7.60	7.9	7.8	8.05	7.43	7.35	7.39	7.28	7.95	7.76	7.1	7.78	7.21	7.73	7.68	10	8.03	4.6	7.55	9.06
TDS	7,741	28,838	1,013	613	1,223	1,480	1,521	2,151	1,010	4,365	2,389	14,956	19,264	2,424	44,025	1,850	4,069	1,290	20,078	1,468	13,192	2,760	5,568
Blue methylene test	<5	5	1	>1	5	5	5	<5	1	1	2	<5	3	>1	<5	>1	<1		<5		<5		5
NH$_4^+$	28.2	0.9	60.0	35.7	4.12	14.1	12.2	9.3	12.7	38.3	7.5	0.82	5.2	15	0.89	12.86	59.7	43.5	1.61	50.3	3.12	44.74	17.3
NO$_2^-$	73	9.4	62.2	27.7	59.5	12.7	19.3	26.1	68	12.0	9.7	5.53	10.4	20	7.68	35.6	289	19.7	3.45	38		50.6	188
NO$_3^-$	2.4	2.8	11.7	10.5	0.9	4.2	2.40	2.30	5.4	5.34	3.23	0.18	2.4	5.0	0.15	6.02	4.68	4.73	0.15	10.2	4.25	7.86	20.4
PO$_4^{3-}$	1.40	3.5	2.13	0.9	4.16	1.80	1.73	4.66	2.33	4.1	5.4	0.29	0.04	0.11	0.90	0.11	1.26	1.55	0.95	0.07	0.93	0.09	×
Total Cr	0.06	0.03	0.03	1.2	0.08	0.08	0.03	0.12	×	0.23	0.1	0.01	1.55	3.0	0.12	2.21	0.15	0.11	0.20	1.64		1.21	0.56
Iron	37.7	0.21	0.17	0.36	0.18		0.04	0.05	0.08	2.40	1.85	×	0.53	×	0.08	×	0.30	0.08	0.05	0.16	8.40	×	×
Manganese				0.06	0.08		0.05			0.53	0.68	×	0.07	×		0.12				0.07	0.35	0.19	×
Zinc										×	0.07	×	×	×	×	0.27	×	×	×	×	×	0.12	
Copper			×							×	×		×		×				×	×	×		
Lead										×			×										
Silica										×			×								5.17		
Cyanides										×	×		×							0.02	×	×	×
Phenols	0.07	0.11	0.24	0.17	0.08	0.09	0.11	0.11	0.06	0.06	0.07	0.07	0.04	0.08	0.10	0.07	0.12	0.07	0.08	0.03	0.03	0.07	×
Extractives and solvents	20	66	6.66	6.0	6.0	9.0	2.6	9.0	2.0	6.2	120.2	×	2.00	×	×	×	205	4.00	3.3	8.6	3.0	6.00	×
Cl$_2$		×	×	×	×	×	×	×	×	×	×	×	×	×	×	×	×	×	×	0.6	×	×	×
Sulfur									0.3	0.29	0.47	0.16	0.07	1.10		0.80	×	0.44	0.20	0.41	0.38	0.65	1.4
Detergents (ABS)	0.15		0.62						0.7	0.71	0.17	0.15	0.30	0.36	0.35	0.24	0.10	1.83	0.06	0.8	0.13	1.09	0.01

Analysis	Distribution site No. (see Fig. 22)																					
	171	172	173	206	219	220	221	223	227	228	229	230	231	232	235a	236	237	248	251	252	253	254
BOD$_5$	155	525	317	61	82	73	127	62	10	11	30	26	35	278	94	155	194	4	21	49	17	10
COD	550	3,175	1,225	110	441	416	500	333	25	25	70	100	100	450	250	640	410	53	183	223	42	72
pH	8.25	8.27	7.52	7.57	6.73	7.55	7.25	8.20	7.67	7.35	7.62	7.56	6.9	7.35	7.63	7.68	7.4	6.5	6.77	6.91	6.43	6.36
TDS	1,468	13,552	1,784	885	1,820	1,551	1,530	576	225	256	909	780	218	1,192	1,244	840	896	1,170	30,606	25,962	2,162	17,789
Blue methylene test	5	<5	>1	<5	<5	<5	<5	<5	>5	>5	>5	>5	3	>1	5.0	5	5	>5	4	2	>5	>5
NH$_4^+$	18.9	27.3	43.7	4.7	0.99	0.60	0.77	0.61	0.06	1.08	0.21	0.23	4.82	14.6	0.07	4.43	4.12	1.71	1.88	3.17	2.52	1.10
NO$_2^-$	111.5	209	82	6.23	214	76.6	47.6	22.3	3.51	3.8	61.0	67.7	150		11.80	16.7	9.00	7.27	7.88	5.6	4.66	6.3
NO$_3^-$	2.4	1.57	9.5	5.2	0.64	0.35	0.20	1.44	0.09	0.75	0.80	0.08	0.37		0.12	0.03	0.13	0.71	1.18	3.82	2.0	0.44
PO$_4^{3-}$	0.32	5.3	0.04	0.8	13.9	0.54	18.1	1.65	1.00	0.40	5.28	15.4	2.30	1.10	0.45	1.71	0.31	0.38	0.25	0.45	1.75	0.30
Total Cr	×		0.30									×						×	×	×	×	
Iron	×																					
Manganese																						
Zinc																						
Copper			×	×	×	×		×	×			×						×	×	×	×	
Lead																						
Silica																						
Cyanides		×		×	×	×	×	×	×			×						×	×	×	×	×
Phenols	×	×	×	×	×	×	×	×	×			×						×	×	×	×	×
Extractives and solvents	×	22.0	×		4.0		20.0	×														
Cl$_2$	×	×	×	×	×	×	×	×	×									×	×	×	×	×
Sulfur	0.06	5.55	1.05		×	0.01	0.03	0.33									0.80					
Detergents (ABS)	0.15	0.05	0.62		0.04	0.03	0.03	0.01		0.7	0.71	0.17	0.15	0.30	0.36	0.35	0.24	0.10	1.83	0.06	0.8	0.13

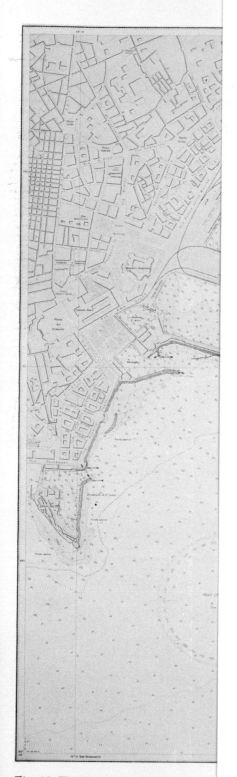

Fig. 28. The Harbor of Naples.

TABLE XI (*continued*)

Area	Shoreline (km)	Pollution load	
		BOD$_5$ (kg day^{-1})	toxicity
Industrial pollution loading to the shoreline of the Gulf of Naples:			
0	5.75	18,494	considerable
F	0.25	37	moderate
10	1.20	65.5	notable
11	3.40	7.6	present
12	8.40	12,584.6	notable
14	3.20	166,894	considerable
15	1.95	50,371	present
16	0.45	22,307	present
19	1.20	14,325	present
28	5.00	2,808	present
29	1.40	1,788	present
29A	0.50	113,270	present
31	2.00	9.0	present
Total		402,960	

The latter table confirms the extensive development of industries at the Gulf of Pozzuoli; the Naples Harbor area and at the mouth of the River Sarno.

sampling sites in various parts of the world. As this figure indicates, the Mediterranean, in spite of its enclosed geography, major shipping, and other oil-related activities, is not significantly greater in hydrocarbon concentration than other areas cited. These data may be related to the fishery data which indicate that the concentrations of hydrocarbons have not significantly altered the fishing in the Mediterranean.

NAPLES' PORT AND BOATING ACTIVITIES

Naples Harbor

The Harbor of Naples (Fig. 28) is a major shipping area with an annual metric tonnage of $45 \cdot 10^6$ t, of which $9 \cdot 10^6$ t are hydrocarbons. In addition, six sewage outfalls enter its waters with untreated municipal and industrial waste and street runoff. The amounts of sewage added to the Harbor in one year is large: it will carry 10—20 ppm N and 10—20 ppm P. Figs. 29 and 32 show the sewage outfalls of Naples Harbor and other areas, people fishing adjacent to outfalls, picking up trash from the bay waters and the average amount of vessels in an average day. During the period of our observations, the Harbor waters were grossly contaminated. However, fish were seen everywhere except in the oil dock areas. The sewage outfalls

TABLE XII

Hydrocarbons in refinery effluents (Insola et al., 1973)

source	boiling point (°C)	phenols (ppm)	sulfur (ppm)	oil (ppm)	NH_4^+ (ppm)
		Aquatic phase			
Gasoline (FCC)	38—221	300—600	3,000	—	2,200
Gasoline (FCC)	—	55	—	30	—
Gasoline (FCC)	—	70—260	40—305	—	85—440
Gasoline (FCC and topping)	—	200—300	1,000—5,000	—	1,000—2,000
Gasoline (cracking)	—	700—1,400	6,000—9,000	—	4,740—7,000
Gasoline (catalytic cracking)	—	340	1,640	—	—
Gasoline (TCC)	38—217	600	2,960	—	—
Gasoline (catalytic cracking)	—	600	—	—	—
Gasoline (catalytic cracking)	—	290	1,500	—	—
Gasoline (FCC)	C_3—199	300	1,000	100	—
Gasoline (FCC)	—	550	3,200	—	2,500
Gasoline (FCC)	—	—	—	30	—
Gasoline (catalytic cracking)	38—200	235	1,300	—	—

FCC = fluid catalytic cracking; TCC = temperature catalytic cracking.

introduced both dissolved and particulate materials, the latter remain floating on the surface. Hydrocarbons were present in most parts of the harbor.

The pollution is a public health hazard as the wind action and other activities allows the bay waters to contaminate the wharves and downwind areas. The odors are considerable and may cause allergic reaction in the dock workers, ships company and visitors.

Many parts of the bay indicated anaerobic conditions. This is caused by the high nutrient and BOD loading from the sewage, ships and oil. The poor circulation with the bay, a result of the breakwater configuration, allows the material to concentrate in the harbor waters.

Boating

The number of boats per capita is increasing throughout the world. In Texas there is approximately one boat per family. This increase in boating has required more boat facilities and access to them. This is quite apparent in the Bay of Naples area. Small boats for commercial fishing and recreation

Fig. 29. Photographs of sewage effluent in the Harbor of Naples.

are abundant. This activity adds to the pollution load of the environment through the release of hydrocarbons from the bilges, two-cycle engines, and spillage during fueling. Sewage from these boats is introduced into the water. Trash left over from eating or other activities is discarded into the water. The total effects of these activities can be estimated. However, time did not permit a good data-base to be accumulated. Therefore, it is only possible to indicate the types of pollution that may occur.

128

Fig. 30. Photographs of various pollution in the Harbor of Naples.
A—C. Oil docks. D. Oil released from bilge in Marina Grande. E. Diesel exhaust in Hydro-foil Harbor. F and G. Smoke and solid waste at Anacapri incinerator.

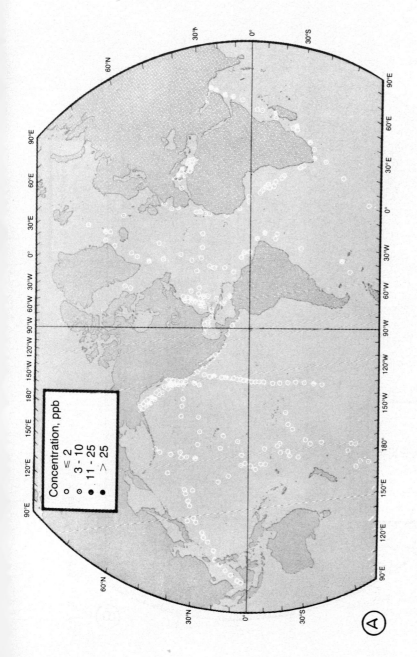

Fig. 31A. For caption see p. 132.

130

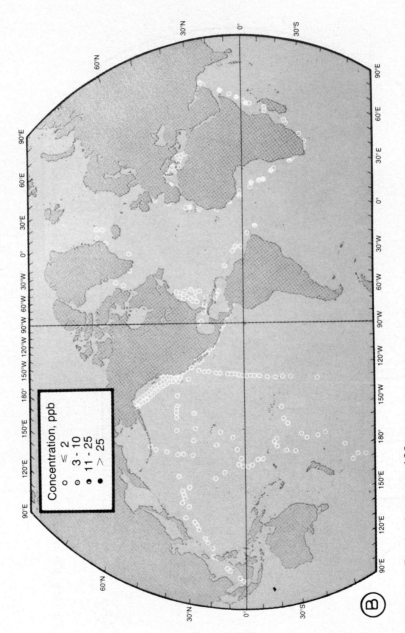

Fig. 31B. For caption see p. 132.

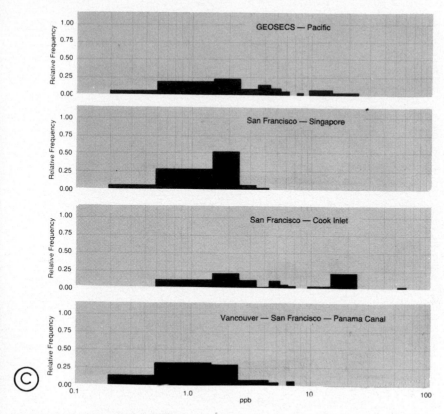

Fig. 31C. For caption see p. 132.

FISHING

The commercial fishing yield must be considered in a regional evaluation of an environment in the Mediterranean such as the Gulf of Naples. Fig. 33 shows the relationships between the Mediterranean—Black Sea area and other fishing areas of the world. These data, taken from F.A.O. (1938—1975), have been summarized over the years for Italy and the Mediterranean—Black Sea areas. Table XIII lists the fish catch in these two regions from 1938 to present. No abnormal perturbations can be seen and the fish catch has continually increased through the reported years. The average catch reported for Italy in percent of the Mediterranean—Black Sea catch is 34% of the average catch reported for Italy for the last five years recorded. This shows the significance of the Italian fishery in the area and the relative value of the fishery.

The relatively stable fish catch indicates that the area is perhaps at the highest sustaining yield per unit area. If the environment were overfished and the standing stock decreased, a general continuing decline would be noted.

132

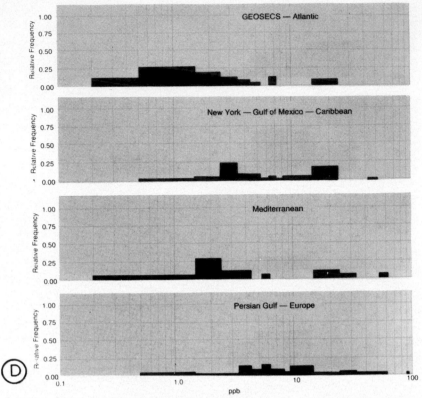

Fig. 31. Non-volatile hydrocarbons in the world's oceans (Myers and Gunnerson, 1976).
A. Concentration in surface waters.
B. Concentration in subsurface waters.
C and D. Frequency histograms from measurements in surface waters of some heavy-traffic trajects.

These data also suggest that pollution from coastal areas, such as the Gulf of Naples, has not materially affected the fish yield.

The Eurostaff S.p.A. (1973) report provided a graph of fishboat tonnage from 1959 to 1971 (Fig. 34). Other data (Fig. 35) show the catch taken in the Naples area. These figures indicate a continual increased yield of the area. This is remarkable for an active fishery area since the time of the Romans. One would expect that such a continual increase in fish yield indicates that it is in equilibrium with fishing effort. The area of the Gulf of Naples is $\sim 960 \, \text{km}^2$. If the total fish catch is $\sim 7 \cdot 10^6$ kg, the yield of the Bay would be $\sim 7200 \, \text{kg km}^{-2}$. Even though much of the field yield is shell-fish, the productivity of the area is considerable.

Fig. 32. Photographs of various harbors: A. Marina Grande, Capri. B. Naples Harbor. C. Hydrofoil Harbor, Naples. D. Naples Harbor. E. Marina Piccola, Capri.

DISCUSSION OF WASTE TREATMENT ALTERNATIVES

It became obvious during the course of literature review that the problem of street runoff, sewage and industrial waste disposal has been accumulating throughout the years. As the city grew, it continually compounded the problem of waste disposal. The older sewage lines were covered over by the growing city and new lines were added. The flow of sewage followed the contours of the terrain and the easy way to cope with the engineering was to provide a multitude of small outfalls directly into the waters edge of the

134

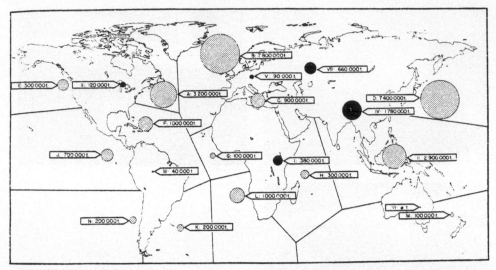

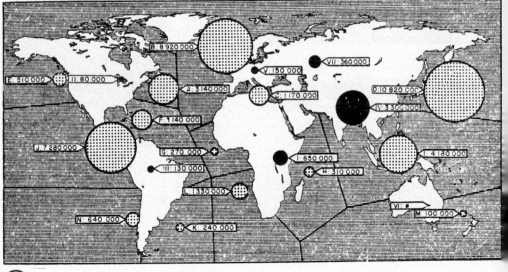

Fig. 33. Fish catch statistics (live weight, in million metric tons) of the world (F.A.O., 1938—1975).
A. 1956 fish catch of major fishing areas, *stippled circles* indicate marine areas; *full circles* freshwater areas.
B. 1962 fish catch of major fishing areas, *stippled circles* indicate marine areas; *full circles* freshwater areas.
C. 1964 fish catch of major fishing areas, *stippled circles* indicate marine areas; *full circles* freshwater areas.
D. 1967 fish catch of major fishing areas, *"open" fishes* indicate fish catch in marine areas; *"full" fishes* in freshwater areas.

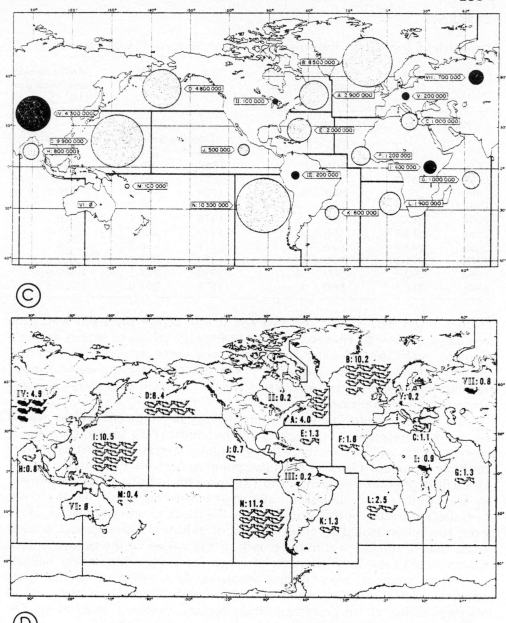

(C)

(D)

Maritime areas: $A =$ Northwestern Atlantic; $B =$ Northeastern Atlantic; $C =$ **Mediterranean**
and Black Sea; $D =$ Northwestern Pacific; $E =$ Northeastern Pacific; $F =$ Western-Central
Atlantic; $G =$ Eastern-Central Atlantic; $H =$ Western Indian Ocean; $I =$ Indo-Pacific area;
$J =$ Eastern-Central Pacific; $K =$ Southwestern Pacific; $L =$ Southeastern Atlantic; $M =$
Southwestern Pacific; and $N =$ Southeastern Pacific.
Freshwater areas: $I =$ Africa; $II =$ North America; $III =$ South America; $IV =$ Asia;
$V =$ Europe; $VI =$ Oceania; and $VII =$ U.S.S.R.

TABLE XIII

Fish catch for Italy and the Mediterranean and Black Sea areas (F.A.O., 1938—1975)

Year	Italy $(10^3$ t)	Mediterranean and Black Sea $(10^3$ t)	Year	Italy $(10^3$ t)	Mediterranean and Black Sea $(10^3$ t)
Total fish catch for Italy (all areas):			*Fish catch for Italy (Mediterranean and Black Sea only):*		
1938	181.2	700	1965	277.7	990
1948	182.8	600	1966	285.5	1,030
1955	257.7	700	1967	284.3	1,110
1956	258.9	800	1968	281.0	1,030
1957	247.7	800	1969	288.0	940
1958	245.7	800	1970	320.3	1,070
1959	253.5	800	1971	321.1	1,115.5
1960	249.5	800	1972	348.7	1,136.7
1962	271.3	800	1973	334.2	1,150.5
1963	290.4	900	1974	344.1	1,367.0
1964	266.3	960	1975	351.8	1,322.5

surrounding coast. At first the area was able to accept the input with little ecological problems. However, as our technology advanced wastes accumulated to the ~ 757 l of water and 3 kg of solid waste per person in each day. As the industrial growth increased, it was simple to provide waste disposal through the sewage lines. Also street runoff for much of the area also was routed to the sewage lines.

As a result considerable expense and facility development will be needed to provide treatment: (1), there is no convenient place to construct major sewage plants within the city; (2) the irregular terrain would either require multiple treatment plants or extensive pumping facilities to route the sewage through large collecting plants; and (3), space for industrial waste treatment at the plants is quite limited. In all instances the cost would be very great.

Industrial waste control would reduce the required sewage capacity. However, this action would involve recycling of industrial effluents and in most cases require oxidation ponds because of the nature of the industry. Such action would take large amounts of space for the ponds and very delicate control to eliminate air pollution problems. As a visitor, it became quite apparent that the region could not afford and did not have the time to alleviate all the above problems within present economic constraints. The excessive cost would prohibit a quick solution to the problem.

Therefore as the Eurostaff report had already accomplished an excellent job identifying the problems, it seemed obvious that some alternate approach than building secondary sewage and industrial waste plants would be required for immediate action. One alternative was to see if the Bay of Naples could absorb, without ecological damage, the contents of multiple major sewage outfalls if they could be relocated into deeper water as has been advocated

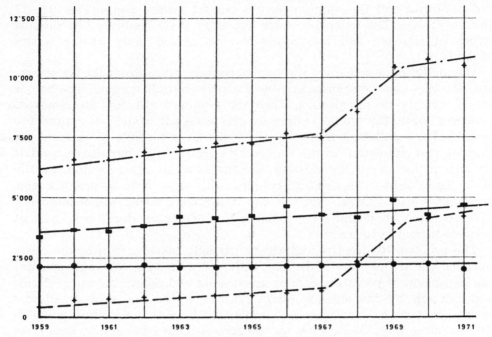

Fig. 34. Fishboat tonnage for the Gulf of Naples (Eurostaff S.p.A., 1973, Vol. II).
— — = motorboat; ——— = rowboat; - - - - — trawler; — · — · — = total.

by several Italian scientists. It would be within reason to collect several of
the existing outfalls into single offshore outfalls thus eliminating the problem of merging sewage pipes in the city.

The question then would be whether the Bay of Naples could absorb the
sewage combined wastes of the area. While there is not a large amount of
hydrographic information available for the bay, enough information was
available to make a preliminary judgement and to show the need for more
information.

The fish catch for both the Mediterranean and the Naples area (Figs.
33—35) indicates a continual increase and no major decrease that could
suggest adverse effects of pollution on productivity, even though public
health problems are apparent. In contrast, it is possible that the increase
in nutrients from the sewage of the area may contribute to the increase
in fish, as historically the Mediterranean has been notably poor in nutrients.

The current data available for the bay suggest that considerable circulation
takes place with associated replenishment from the surrounding bodies of
water. The upwelling that occurs from the circulation and deep canyons of
the center of the bay is shown by the rise in salinity near the far end of the
channels (Fig. 12). This causes the somewhat circular gyres of the water
found in the bay (Figs. 9 and 10). The water movement and circulation
data are supported by the absence offshore of high numbers of nutrients

138

and coliforms and the dilution seen in coastal sewage zones (Figs. 24—27
and Table XI). The extensive water exchange is also indicated by the low
values of nitrogen and phosphorus in the central body of the waters
of the Gulf.

The pollution problem of the Naples Harbor arising from the six sewage
outfalls may also be approached by the open outfall method. The harbor
would quickly be self-cleaning when the pollutant effluents are removed.
However, while this would be true for the sewage it would not control the
hydrocarbons polluting from the oil docks. It is possible to increase the
flushing and circulation of the Harbor by constructing a circulation conduit
through a portion of the existing breakwater at an inner location on the
Molo San Vicenzo. This action, coupled with strict laws to prohibit bilge
pumping, waste disposal over the side of ships or docks, the maintenance
of clean dock areas and strict maintenance of the oil docks would be of
immediate benefit to the area.

The problem of street runoff will be partially taken care of by the ocean
outfalls. The architecture of the city, with so few green areas, would prohibit
the expansion of parks or roadside green areas and relieve the use of streets
and sidewals by the animals. Street runoff may be best approached as a
social problem that could be best confronted by a citizens action group to
devise better ways of cleaning up the streets. One approach to determine
the capability of the Gulf of Naples to absorb the sewage effluent is to
construct mathematically a mass balance of the pollutants. This generally
requires considerable information and data about the system. However,
preliminary data are available that will allow a partial demonstration of a

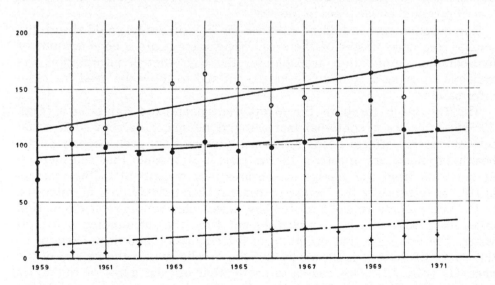

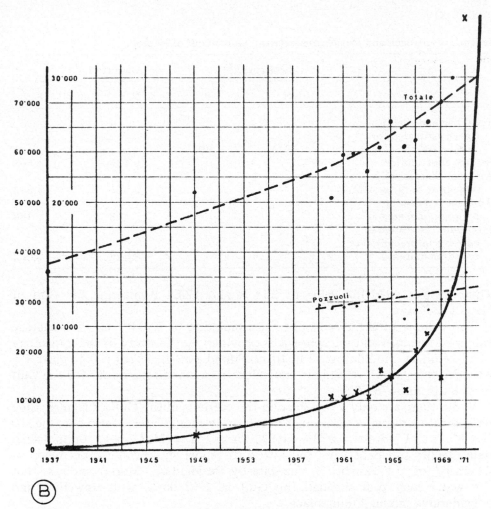

Fig. 35. Fish catch statistics for the Gulf of Naples (Eurostaff S.p.A., 1973, Vol. II).
A. Catch (in metric tons) of crustaceans in the Pozzuoli and Procida areas.
– – – – = Pozzuoli; – · – · – = Procida; —— = total.
B. Total fish catch and catch of edible mollusks (in hundreds of kilos).
– – – – = total fish catch (outside ordinate scale); —— = mollusks (inside ordinate scale).

mass balance for certain materials from the sewage as they relate to the Gulf system.

Paoletti (1975a) indicated that approximately 2.2 million people lived in the drainage area of the Gulf of Naples. If these 2.2 million people produced $0.55\,m^3$ of sewage each day, including runoff and industrial effluent (this is an estimate used in the U.S.A.), $1.25 \cdot 10^6\,m^3$ of sewage would be produced per day. The average content of the sewage according to an abstract of data from the Eurostaff report is; nitrogen 20 ppm, phosphorus 10 ppm.

TABLE XIV

Balance of nitrogen and phosphorus nutrients in the Gulf of Naples

	Loss (t yr.$^{-1}$)		Gain (t yr.$^{-1}$)	
	N	P	N	P
Total fish and shellfish (catch estimated for 1975: 10^4 t)	300	20		
Sewage contribution			9,125	4,563
Ambient theoretically remaining in the system			8,825	4,543
Ambient reported in hydrographic data			403	480
Lost from the system by currents or precipitation or sedimentation			8,422	4,063

These figures yield the value of 25 t N and 12.5 t P entering the Gulf of Naples each day.

The volume of the Bay has been estimated to be $2 \cdot 10^{11}$ m^3. The average values of nitrogen and phosphorus as provided by the study of the hydrology of the Gulf by the Zoological Institute (Mendia et al., 1976) is 0.15 μgAt. l^{-1} N and 0.08 μgAt. l^{-1} P, this is equivalent to a standing amount for the Gulf of 403 t N and 476 t P.

While there are only a few data on the currents in the Gulf of Naples, they do indicate a rather continuous and relatively rapid water movement. Fig. 10 (De Maio and Moretti, 1973), data given in Fig. 9 (M. Monetti, unpublished data, 1976) and data from the Eurostaff report show water movements of 51 cm s^{-1} or ~ 1.6 km hr.$^{-1}$. The data by De Maio and Moretti indicate that the water may pass through the Gulf in 1—2 days, with upwelling and considerable mixing taking place.

These data also suggest that the amounts of nitrogen and phosphorus entering from the sewage may be accepted by the waters if properly distributed and thus would produce little or no ecological effect. This latter is supported by the data in various reports cited in this chapter, which indicate that the shoreline contamination from the various sewage outfalls disappears only a short distance from the outfall. Data from the Zoological Institute (Cassa per il Mezzogiarno, 1975) on nutrients in the Gulf are quite low which suggests that there is little effect from the shoreline pollution.

A balance of nutrients for the Gulf of Naples is shown in Table XIV. Nitrogen and phosphorus are continually being removed from the system by the fish catch. The ambient nitrogen and phosphorus calculated from the available hydrographic data, indicate a much lower standing amount than indicated by the sewage input. This excess nitrogen and phosphorus must be removed by the currents passing through the Gulf or from chemical precipitation or sedimentation. No data were available for nitrogen or phosphorus contents of the sediments. However, the data for currents in the Gulf support a conclusion that much of the sewage nitrogen and phosphorus are moved from the system to the Mediterranean Sea.

It must be understood that the above values are only preliminary, taken from a very inadequate data-base, and should be used only to demonstrate a method for sewage impact measurement. It does, however, suggest that the Gulf of Naples can absorb the sewage without ecological impact.

Coliforms can also be treated the same way. Approximately $2.2 \cdot 10^{13}$ coliforms enter the Gulf each day from sewage. There are $2 \cdot 10^{11}$ m^3 of water in the Gulf and therefore the coliforms undergo mixing to a non-public health status. It is known (Paoletti, 1970), that the survival of coliforms in seawater is between two weeks and one month. This suggests that the dilution, natural death and the water exchange rate in the Gulf would reduce any public health problem if the sewage were properly distributed.

For the purposes of discussion we have assumed a well-mixed Gulf. One must also realize that the circulation in the waters may decrease or change directions and thus allow higher concentrations to occur. However, the differences between materials introduced and their concentration, assuming a well-mixed system, are large and thus allow a concentration effect to build up without producing an environmental effect.

During some conversations, it was brought out that there is a thermocline in the Gulf during certain periods of the year. If sewage were disbursed below the thermocline it may become entrapped at the interface and thus be concentrated. However, visual observations of ocean outfalls in the U.S.A. indicate that at single-source outfalls the freshwater will rise to the surface and that the amount of material released at any one point can be regulated to avoid either the large surface plume of the rising freshwater or entrapment in a thermocline.

ACKNOWLEDGEMENTS

It is a pleasure to present this manuscript to Assessore Dr. Pavia, Region of Campania, Naples, Italy. Appreciation is also expressed to the Italian National Biological Association who provided the opportunity to visit other environmental areas of Italy; and who introduced us to other ecologists and scientists interested in systems ecology. We should like to thank the many colleagues who are listed in the references of the manuscript and other individuals too numerous to list who made our visit complete.

CHAPTER 3

HEAVY METALS AND CHLORINATED HYDROCARBONS IN THE MEDITERRANEAN

Michael Bernhard

INTRODUCTION

Major concern about environmental contamination of the sea by heavy metals and chlorinated hydrocarbons has stimulated research into the distribution of these substances in Mediterranean marine systems, on their effects on biota and on possible routes of transfer to man. In this chapter the main emphasis is on the level of heavy metals and chlorinated hydrocarbons in seawater, sediments and biota; for the vulnerability of the Mediterranean ecosystems to pollutants the reader is referred to Pérès (1978). The consequences of these levels, i.e. the toxic effects, on biota will only be discussed shortly. The different areas of the Mediterranean are discussed, starting from the western part and proceeding eastwards.

CONCENTRATION OF HEAVY METALS IN THE MARINE ENVIRONMENT

Many toxic or potentially toxic heavy metals are released into the oceans, especially into the coastal zones. Iron, copper, zinc, cobalt, manganese, chromium, molybdenum, vanadium, nickel, zinc and selenium are known to be essential for living organisms, but even these essential metals become toxic if present in excessive quantities.

Apart from waste discharges, the oceans receive heavy metals (together with other materials) through the atmosphere, land runoff and rivers. Natural processes such as degassing, weathering and erosion mobilize these metals. Heavy metals mobilized by man enter into the natural biogeochemical cycles at rates comparable to or even exceeding the natural rates. The prediction of the rate and the pathway of a heavy metal mobilized by man is further complicated by the observation that the pathways and rates of mobilization depend on the physico-chemical state of the element. Often, however, anthropogenic mobilization produces physico-chemical forms which are different from the natural ones. As a consequence, an artificially-mobilized heavy metal can appear in a component of the marine system in which it was not anticipated and cause unexpected damage to marine biota and create hazards for human health.

Heavy metals in seawater

Owing to the difficulties in determining heavy metals in seawater (low concentration, contamination of the sample, etc.) relatively few measurements have been made in the Mediterranean Sea. The comparison between values given by different authors is further complicated by the use of different methods which often determine different physico-chemical forms and fractions of the total amount present. Some methods such as instrumental neutron activation (INAA) or chelating resin extraction will yield "total concentrations", while extracting seawater with organic solvents at pH 4—6 will only determine fractions of the total amount present. In most cases it is not even easy to decide which physico-chemical form or fraction of the total amount has been determined by a certain method. On the other hand, the distinction between different physico-chemical forms is very important because different physico-chemical forms will participate in different ways in the biogeochemical cycles. They also have different effects on marine organisms. For example, ionic forms are preferentially accumulated by some marine organisms and particulate forms by others [for review see, e.g., Bernhard and Zattera (1979)]. Organo-mercury compounds are more toxic than inorganic mercury compounds. On the other hand, most heavy metals are more toxic in ionic than in complexed form. Also heavy metal—sediment interactions depend very much on the physico-chemical form of the metal and on the mineralogical composition of the sediment. Only in the mid-1970's has the chemical speciation of seawater received the attention it deserves (see Kester et al., 1975).

Since influences from land will obviously affect the concentration of heavy metals in near-shore regions, a distinction is made between determinations of "open-sea samples" and samples taken in the vicinity of the coast. The former will supply an indication of the general level of heavy-metal concentration in the Mediterranean while the latter will reveal mainly local influences. At the same time, an attempt is made to indicate the most likely physico-chemical forms determined by each method. It is realized that, owing to the scarce knowledge of the actual forms determined, the indication of the physico-chemical forms is very approximate and its main purpose is to draw attention to the chemical speciation problem and stimulate further research on speciation. The role of particulates in the ocean as it might apply to heavy metals is presented by Eisma in Chapter 9 of this volume.

Heavy metals in "open-sea samples"

Robertson et al. (1972) were probably the first to compare the Hg concentration in open-sea water samples from the Mediterranean with samples from other areas. A summary of published data, together with a "typical" concentration, i.e. a concentration considered average for unpolluted water, is given in Table I. Concentrations of Hg in samples from the tropical North-west Atlantic, Irish Sea, North Sea and English Channel were not significantly different from those of the Mediterranean. The "typical value" for total mercury (Hg_t) is $0.02\,\mu g\,Hg_t\,l^{-1}$ with a range of $0.01{-}0.02$. Determining hydrolysable soluble ($< 0.45\,\mu m$) mercury, Fukai and Huynh-Ngoc (1976a) found similar values in several areas of the Mediterranean. According to Robertson et al. (1972), the particulate fraction of typical open seawater constitutes maximally only 2.5 wt.% of the total Hg (Hg_t). Therefore, it is unlikely that the particulate fraction of Hg_t in open-sea seawater is considerable. The differences between values obtained by the two groups are probably caused by the use of different methods. Robertson et al. (1972) like Kosta et al. (1978), have determined total mercury content with neutron activation analysis (NAA) in the Adriatic Sea. Their range of values is the same as that of Robertson et al. (1972). Fukai and Huynh-Ngoc (1976b) also analysed for Cd, Cu and Zn. Their values can be compared with those of Bubic et al. (1979), for Cu and Zn with those of Fonselius and Koroleff (1963) and Fonselius (1970), and for Zn with data of Bernhard et al. (1975). Frache et al. (1976) determined Cu, Co, Ni and Cd concentrations in surface waters of the Ligurian Sea. In comparing the data, one has had to keep in mind that, for example, the range of a normally distributed population of two data estimates 1.25σ, a range from four data estimates 2σ, and a range of nine data 3σ (Snedecor, 1962). Since the ranges from the different regions overlap (with sample number n between 3 and 13), no significant difference was observed. Bubic et al. (1979) observed the highest Cd values, but their sample stations are located on the fringe of the Gulf of Kvarner (Rijeka).

The Cu concentrations vary considerably between authors and areas. Contamination during sampling may be one reason. Also the Zn concentrations observed show differences which are not easy to explain. Fukai and Huynh-Ngoc (1976a, b) and Bubic et al. (1979) have determined the ionic plus labile-complex fraction, with anodic stripping voltametry (ASV); Fonselius and Koroleff (1963) and Fonselius (1970), using a method based on dithizone, probably determined particulate Zn and the Zn which is bound to complexes which have a smaller stability constant than dithizone, while Bernhard et al. (1975) have determined hydrolysable Zn with ASV at pH 2. This method includes all Zn which can be liberated from complexes at pH 2. According to the last-named authors, in Ligurian seawater the ionic fraction is about $10{-}15\%$, the particulate $20{-}35\%$ and the complexed Zn $45{-}50\%$ of the total hydrolysable. Chelex extraction yielded about 10% higher Zn concentrations than those determined with ASV at pH 2.

Although these data are by no means conclusive, and additional data, intercalibration exercises and new analytical methods are badly needed, it seems that the concentration of Hg, Cd, Cu and Zn in Mediterranean open-seawater samples is similar to that of other oceans and seas.

TABLE I

Concentrations ($\mu g\,l^{-1}$) of mercury, cadmium, copper and zinc in the open Mediterranean

Sampling depth (m)	n	Hg	Species	Cd	Species	n	Cu	Speci
15—300	3	0.11M (0.062—0.11)	T	—	—		—	—
0	13	—		<0.1M (<0.05—0.5)	I + L MF		<0.2M (0.1—4.9)	I + L MF
0—5	5	0.024 M (0.016—0.03)	TH MF	0.09M (<0.05—0.51)	I + L MF	10	<0.06M (<0.06—0.62)	I + L MF
0—5	4	0.021M (0.017—0.03)	TH MF	0.12M (0.05—0.43)	I + L MF		0.07M (0.06—0.93)	I + L MF
0	11	—	—	—	—		3.7M (1.0—12.7)	I + L + P
20	11	—	—	—	—		0.72M (0.15—1.16)	I + L + P
10—2,250	15	—	—	—	—		0.23 (<0.03—3.32)	I + L + P
0	7—9	—	—	<0.05M (<0.005—0.2)	T?		1.3M (0.72—2.26)	T?
0—175	121	—	—	—	—		—	—
5	4	0.025M (0.02—0.03)	TH MF	0.1M (0.08—0.33)	I + L MF		<0.06M	I + L MF
0—500	69	—	—	—	—		—	—
0	3	—	—	0.06M (0.06—0.12)	I + L MF		0.33M (0.08—0.74)	I +
0—1100	20	0.07M (0.01—0.21)	T	—	—		—	—
?	>454	—	—	$0.1\bar{x}$ (0.05—0.75)	I + L		$0.5\bar{x}$ (0.1—127)	I +
0.5	40	—	—	$0.14\bar{x}$ (<0.05—0.60)	I + L MF		$1.12\bar{x}$ (0.1—1.6)	I +
0—495	10	—	—	—	—		0.57M (<0.03—2.86)	MF I +
0	3	—	—	0.07M (<0.05—0.12)	I + L MF		0.36M (0.08—0.74)	+ I +
15—300	2	0.12 (0.09—0.14)	T	—	—		—	MF —
Typical concentration range		0.01—0.2	T	0.06—0.07			0.5—2	

$\bar{x}$ = mean; M = median; (...) before $\bar{x}$ or M number of samples analysed; concentration range in (...) below the $\bar{x}$ or M; I = ionic species; P = particulate species; L = labile complexes with regard to anodic stripping voltammetry; D = labile complexes with regard to dithizone; H = hydrolysed by acid treatment; MF = filtrate through 0.45-μm Millipore filter; T = total; and TH = total, hydrolysed by acid treatment.

n	Zn	Species	Sampling area and date	Reference
	—	—	Str. Gibraltar (Jul. 1968—Aug. 1968)	Robertson et al. (1972)
	2.1M (0.8—7.7)	I + L MF	NW Mediterranean (Jun. 1973—Jun. 1974)	Fukai and Huynh-Ngoc (1976a)
10	1.0M (0.1—2.6)	I + L MF	SW Mediterranean (May 1975—Sep. 1975)	
	1.3M (0.5—4.8)	I + L MF	W. Ligurian Sea (Jul. 1978—Oct. 1975)	
	30.0M (9.6—86.0)	I + D	Ligurian Sea (May 1962)	Fonselius and Koroleff (1963)
	5.4M (3.9—10.6)	I + D + P	Ligurian Sea (Sep. 1962)	
	4.5M (3.3—16.4)	I + D + P	40 km S-Imperia (Apr. 1963)	
	—	—	Imperia—Livorno transect (Nov. 1974)	Frache et al. (1976)
	7.6$\bar{x}$ (0.84—12.5)	HT	20 km SW La Spezia (Feb. 1967—Feb. 1969)	Bernhard et al. (1975)
	0.7M (0.2—2.3)	I + L MF	Tyrrhenian Sea (Sep. 1975)	Fukai and Huynh-Ngoc (1976a)
	10.5$\bar{x}$ (5.8—28.4)	HT	Taranto Gulf (Apr. 1969—Nov. 1970)	Bernhard et al. (1975)
	1.1M (0.7—2.4)	I + L MF	Ionian Sea (May 1975)	Fukai and Huynh-Ngoc (1976a)
	—	—	Adriatic (1976)	Kosta et al. (1978)
	4.5$\bar{x}$ (1.1—314)	I + L	Adriatic (May 1973—Apr. 1974)	Branica et al. (1976)
	5.35$\bar{x}$ (0.8—23.6)	I + L MF	G. Kvarner, northern Adriatic (Aug. 1973—Feb. 1976)	Bubic et al. (1979)
	18.0M (0.1(?)—29.4)	I + D + P	off Crete (Jul. 1963)	Fonselius (1970)
	4.7M (1.5—5.8)	I + L	Aegean + Cretan S. (May 1975)	Fukai and Huynh-Ngoc (1976a)
	—	—	off Cyprus (Apr. 1968)	Robertson et al. (1972)
	0.5—10		world oceans	Robertson and Carpenter (1976)

Heavy metals in coastal waters

The concentrations in coastal seawater are markedly influenced by anthropogenic sources, river inputs and land runoff. Establier (1969a) observed seasonal fluctuations of Cu and high Cu concentrations in the coastal waters near Cadiz (Table II). In the near hinterland of Cadiz are the Huelva copper mines, so anthropogenic Cu pollution is superimposed on a geochemical Cu anomaly. This may also be reflected in the Cu concentration of the marine sediments of the Bay of Cadiz. Fukai and Huyhn-Ngoc (1976b) investigated the concentration of Cu, Zn and Cd in the coastal and offshore waters from Sète to Genoa. They observed no significant differences between concentrations in coastal and offshore water except for areas with significant anthropogenic units, e.g., in March high Cu concentration ($6 \mu g l^{-1}$) near Marseilles and between Cap de l'Aigle and Cap Camarat ($20 \mu g l^{-1}$). Their values were obtained by ASV at pH 8 in Millipore-filtered seawater and hence they determined the ionic and labile complexes with regard to the electrochemical method, i.e. the electroreducible forms of Cu, Zn and Cd. Capelli et al. (1976) collected water samples in ports and harbours from San Remo to Viareggio at the same sites from which they also sampled *Mytilus* (p. 163). Their values should be considered "total" concentrations since they extracted the trace elements with a chelating resin (Chelex-100). Total hydrolysable Cd, Pb and Cu were determined by ASV at pH 2 near the coast from Ventimiglia to Anzio (Rome) in 225 sampling stations in May—June 1976 (Nürnberg et al., 1977; Mart et al., 1978). Bernhard et al. (1975) determined total hydrolysable, ionic and particulate Zn in the harbor of La Spezia and its immediate surroundings. They found that total hydrolysable Zn increased towards the harbour, while the percentage of the ionic and particulate Zn remained practically constant until the outskirts of the harbour were reached. In the outer and inner harbour the ratios between particulate and ionic vary considerably, probably due to anthropogenic inputs from the ship-building industry and town of La Spezia.

The influence of an anthropogenic Hg source on the Hg concentration in seawater and biota was studied south of Livorno by monitoring the Hg discharges of a chloralkali plant (Solvay, Rosignano) along the coast of Tuscany (Renzoni et al., 1973; Table III). Near the outfall (1 km south) they observed relatively high values of $\sim 0.2 \mu g l^{-1}$, but at ~ 10 km south and north of the outfall the Hg concentration was already below its detection limit of $0.02 \mu g l^{-1}$.

Heavy-metal concentrations in the Adriatic received attention from several workers. Using NAA, Grancini et al. (1975) analysed Millipore-filtered seawater from the northern Italian Adriatic coast from Triete to Ravenna. Here, Cs, Rb, Se and U mean concentrations in the area were comparable to those of the open sea, while the means of Ag, Co, Sc, Sb and Zn were significantly higher. The authors did not exclude the possibility of sample

Cd, Cu, Zn, Pb, Ni and Ag concentrations ($\mu g \, l^{-1}$) in the coastal waters of the Mediterranean Sea

Sampling sites	Cd	Cu	Zn	Pb*	Ni	Ag	Reference
Gulf of Cadiz	—	(40) 5.2; I + P (2.6–8.6)	—	—	—	—	Establier (1969a)
Sète to Genoa	(14)?; I (DL = 8.0)	(14)?; I (DL = 22.4)	(14)?; I (DL = 11.2)	—	—	—	Fukai and Huynh-Ngoc (1976b)
San-Remo to Livorno (mainly in ports)	(7) 0.1; T (0.1–0.28)	(14) 2.4; T (0.8–4.0)	—	—	(12) 1.7; T (0.6–5.7)	—	Capelli et al. (1976)
Imperia to Livorno (4—20) km offshore	(28) 0.07M; T? (<0.05–0.59)	(30) 1.43; T? (0.71–2.43)	—	—	(30) 0.99; T? (0.35–2.93)	—	Frache et al. (1976)
Gulf of La Spezia	—	—	(40) 10.5; T (6.5–15)	—	—	—	Bernhard et al. (1975)
Ventimiglia to Anzio	(222) 0.15; H (0.005–0.45)	(221) 3.5; H (0.13–3.6)	—	(227) 0.15; H (0.018–2.42)	—	—	Nürnberg et al. (1977); Mart et al. (1978)
Trieste to Ravenna	—	—	(21) 13.6; T (1.4–36.4)	—	—	(22) 0.83; T (0.26–1.9)	Grancini et al. (1975)
Plonim to Nin (Adriatic surface)	(85) 0.12; I (<0.05–6.9)	(85) 1.2; I (0.2–3.6)	(85) 13.8; I (1.5–365)	(85) 1.2; I (0.02–2.8)	—	—	Bubic et al. (1979)
10—<85 m	(255) 0.9; I (<0.05–2.25)	(225) 1.1; I (0.2–3.5)	(225) 6.2; I (1.5–63.6)	(225) 0.55; I (0.1–3.3)	—	—	Bubic et al. (1979)
Saronikos Gulf (Athens)	—	(1?) 3.0; ?	(1?) 20; ?	—	—	—	Papadopoulou (1972)
Rhodes Island	—	(1?) 8.0; ?	(1?) 25; ?	—	—	—	Papadopoulou (1972)
Rosh Haniqra to Haifa	(60) 0.94; TF (0.6–2.9)	(60) 3.7; TF (0.8–31.2)	(60) 38.3; TF (1.0–256.9)	(59) 6.4; TF (2.1–11.4)	(60) 3.3; TF (2.0–5.4)	—	Roth and Hornung (1977)
	(53) 0.15; TP (DL–0.8)	(53) 0.53 TP (DL–2.1)	(51) 5.7; TP (10–24.2)	(52) 2.4; TP (0.6–13.3)	(53) 0.48; TP (DL–2.4)	—	Roth and Hornung (1977)
Hazardous concentration	10	50	100	50	100	5	N.A.S. (1973)
Minimal risk concentration	0.2	10	20	10	2	1	

For explanation of symbols see footnote of Table I. (DL = detection limit.)

* Lead may be overestimated (Patterson and Settle, 1976).

TABLE III

Hg, Cr, Co, As, Ag and Mn concentrations (μg l^{-1}) in coastal waters of the Mediterranean and minimal risk concentration

Mercury, Hg:		
Tuscan coast		
10 km N. Solvay plant	<0.02	Renzoni et al. (1973)
near Solvay plant	0.180	
10 km S. Solvay plant	<0.02	
Isonzo river mouth (Trieste)	0.16—0.23	Kosta et al. (1978)
Gulf of Trieste	0.025—0.1	Majori et al. (1976a)
Limski Canal and Banjole (Istria)	0.04	Strohal and Dzajo (1975)
Roch Haniqra to Haifa	0.06 (0.01—0.18)	Roth and Hornung (1977)
"Minimal risk concentration"	0.1	N.A.S. (1973)
Chromium, Cr:		
Monte Carlo	0.1—0.57	Fukai and Vas (1967)
Trieste to Ravenna	0.6—50(?)	Grancini et al. (1975)
Rosh Haniqra to Haifa	0.4 (< 0.1—5.8)	Roth and Hornung (1977)
"Minimal risk concentration"	10.0	N.A.S. (1973)
Cobalt, Co:		
Imperia to Livorno	<0.05—0.81	Frache et al. (1976)
Ventimiglia to San Remo	0.1—1.5	Capelli et al. (1976)
Trieste to Ravenna	0.2—0.76	Grancini et al. (1975)
"Minimal risk concentration"	2	N.A.S. (1973)
Arsenic, As:		
Rhodes Island	2	Papadopoulou (1972)
Saronikos Gulf	3.5	
"Minimal risk concentration"	10.0	N.A.S. (1973)
Manganese, Mn:		
Rhodes Island	1.6	Papadopoulou (1972)
Saronikos Gulf	2.0	
"Minimal risk concentration"	20.0	N.A.S. (1973)

(?) High concentration probably due to sample contamination.

contamination as an explanation for their high Hg and Cr values. Branica et al. (1976) determined ionic Zn, Cd, Pb and Cu in samples from the open Adriatic, and Bubic et al. (1979) in the Kvarner region (Rijeka) with ASV at pH 8 (four times from November 1973 to August 1974). They observed a seasonal, horizontal and vertical variation of ionic Zn, but not of ionic Cd, Pb and Cu. The average Zn concentration was much higher (365 μg l^{-1}) near anthropogenic inputs (Rijeka) than at the more "open-sea" stations. At different depths at the same stations they often found greatly varying concentrations, an indication of either very heterogeneous distribution of the heavy metals or possible sample contamination by the sampling gear or the ship. Strohal and Dzajo (1975) determined Hg$_t$ concentrations near Banjole and Limski Canal on the Istrian coast with NAA. They found values near the "typical" open-sea concentrations.

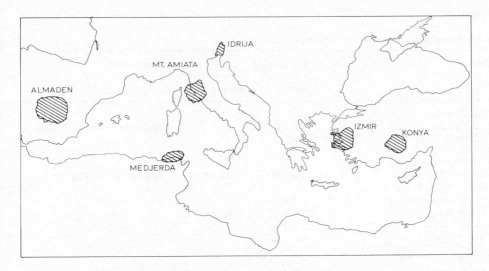

Fig. 1. Commercially exploited mercury anomalies in the Mediterranean Basin.

Of special interest are the values reported by Kosta et al. (1978). At the mount of the River Isonzo (Soča), which originates in the Hg anomaly of the Idrija mining district (Fig. 1), they found much higher concentrations than in open-sea samples. Also the Hg concentrations in the sediments were much higher there than in the open-sea sediments (Table IV). Similar results were obtained by Majori et al. (1976a).

Papadopoulou (1972) compared the concentrations of several elements in samples from the "clean waters" of the Island of Rhodes with those from the polluted Saronikos Gulf (Athens). In the elements examined no direct pollution was evident. But the collateral data on the sediment concentration in heavy metals indicate pollution by heavy metals in the Saronikos Gulf. Also the As concentrations in fish from the Saronikos Gulf were higher than from other areas. Using a chelating resin, Roth and Hornung (1977) investigated the distribution of soluble ($<0.45\,\mu$m) total Hg, Cd, Cu, Pb and Ni and particulate ($>0.45\,\mu$m) Cd, Cu, Pb, Ni and Cr mainly in surface waters along the Israel coastline from Rosh Haniqra to Haifa. In general, they found concentrations near the "typical" values except for high concentrations near densely populated and industrialized areas. High values could also be attributed to the runoff of the River Qishon, which is heavily polluted by industrial discharges, and the River Na'amon polluted by agricultural and domestic wastes.

A summary of available data for Hg, Cd, Cu, Zn, Pb, Ni, Ag, together with concentrations considered hazardous and a minimal risk, are listed in Tables II and III. A detailed discussion of the effects of heavy metals in the Gulf of Mexico is presented by Trefry and Shokes (Chapter 4 of this volume)

TABLE IV

Hg in marine sediments from the Mediterranean

Region	n	Hg (mg kg^{-1} DW)		Reference
		mean	range	
Alboran Sea				
(2,720 m)	1	0.26		Robertson et al. (1973)
Off Monte Carlo	2	0.3		
Tyrrhenian Sea				
(bottom samples)				
(93—1,715 m)	9	0.1M	0.05—0.24	Selli et al. (1973)
5 km WSW River				
Mignone (68 m)	1	1.57		
(core samples)	4	0.1M	0.045—0.16	
(390—3,520 m)				
(0—700 cm)				
9.5 km SW River Fiora				
(108 m)				
(0—1 cm)	1	0.58		
(60—61 cm)	1	0.13		
Tuscany coast				
Solvay plant	2		1.1—1.3	Renzoni et al. (1973)
2.5 km S and N	18		0.1—0.8	
∼ 6 km S and N	6		0.04—0.1	
(conc. proportional to sand grain sizes)				
Adriatic Sea				
bottom samples	20	0.11M	0.07—0.97	Selli et al. (1973)
(5—1,195 m)				
core samples	2	0.08M	0.05—0.1	
(64 + 888 m)				
(0—720 cm)				
(15—1200 m)	38	0.05	0.01—0.16	Kosta et al. (1978)
(0—35 cm)				
Isonzo river mouth	3	35.1M	24.5—56.8	
(1.5 m)				
(0—15 cm)				
1.5 km S River Isonzo	3	14.4M	13.1—15.1	
(0—15 cm)				
20 km SW, Isonzo	2		1.25—1.65	
transect Isonzo	18		0.2—47.0	Majori et al. (1976a)
Trieste				
Off Trieste	1	3.0		Robertson et al. (1972)
Po delta	23	0.42	0.07—0.97	Viviani et al. (1973b)
Po to Rimini	?	0.3	0.06—0.97	Selli et al. (1977)
S off Crete	1	<0.03		Robertson et al. (1972)
(2,360 m)				
Saronikos Gulf		0.5		Grimanis et al. (1977)
background				
Elefsis Bay			0.05—1	
Athens outfall			0.5—3	
Piraeus harbour		10		
Haniqra to Haifa	60	0.13	0.01—0.57	Roth and Hornung (1977)

For explanations see Table I.

and Sackett (Chapter 13 of this volume) as well as off the west coast of the U.S.A. by Gillespie and Vaccaro (Chapter 5 of this volume).

Heavy metals in sediments

The great importance of sediments in the distribution of heavy metals becomes immediately evident when one takes into consideration that for the majority of the elements and substances sediments are the ultimate "sink" or deposit. In general, a heavy metal, introduced into the sea from a "natural" or an anthropogenic source, stays only briefly in the receiving water.

Natural sediments are mixtures of sands, clays and organic substances. The relative abundance of these components various considerably with different types of sediments. Since each component has its particular physico-chemical characteristics, the interaction of heavy metals with natural sediments (sorption, ion exchange, complexation and chelation with inorganic and organic substances) depends on the composition of the sediments. Higher heavy-metal concentrations are generally found in mud, silt, muddy sand and mixtures of them, rather than in pure sand. The interpretation of the values observed is in most cases very difficult, if not impossible, if the mineralogical and sedimentological characteristics of the sediment samples are not reported. In addition, the analytical methods for sediment analysis vary widely, and results of intercalibration exercises conducted in the Mediterranean area have not yet been published. It is therefore very difficult to compare data obtained by different authors.

Despite the great public concern about Hg, only few and sporadic data on Hg in sediments have been published. It is astonishing that the extensive work on Hg distribution on land around mercury mines, especially in the Monte Amiata and Idrija regions (Dall'Aglio, 1968; Benvegnú et al., 1974; Kosta et al., 1975; Cigna-Rossi et al., 1976), and the possibility of an environmental impact from these natural sources, has only stimulated investigations into the Hg concentration in marine sediments near the Idrija area and not in the other areas of known terrestrial geochemical Hg anomalies (Fig. 1). These and other such areas should be common in many parts of the Mediterranean.

The "natural" concentration of Hg in sediments ranges from 0.05 to $\sim 4.7 \cdot 10^4$ $\mu g\,kg^{-1}$ DW (dry weight) (Table IV). The high values are found in apparently unpolluted sediments in the coastal zone west of the Monte Amiata mining area and very high concentrations around the mouth of the River Isonzo which drains the Idrija region. Concentrations as high as $3.8 \cdot 10^3$ $\mu g\,Hg_t\,kg^{-1}$ DW have been observed in the mud of the River Fiora which comes from the Monte Amiata area (Bernhard and Renzoni, 1977). In the coarse sand of the river mouth only 300 $\mu g\,Hg_t\,kg^{-1}$ DW was found, showing the importance of the mineralogical composition for the interpretation of heavy-metal pollution in sediments. Mud-feeding fish (*Mullus barbatus*) collected in this region have a higher Hg concentration than specimens sampled in other areas (see Table XI). Recently, data have also been published on the heavy-metal concentrations from another Hg anomaly (Idrija), Kosta et al. (1978) investigated the heavy-metal concentrations in the Adriatic, and Majori et al. (1976a) especially in the Gulf of Trieste. Both found that the Hg concentrations in the sediments show a gradient from the mouth of the River Isonzo (highest value ~ 50,000 $\mu g\,Hg_t\,kg^{-1}\,DW$) towards the open Adriatic, where concentrations are around 40 $\mu g\,Hg_t\,kg^{-1}$ DW. The high concentration of Hg in the sediments around the mouth of the Isonzo, are explained by Kosta et al. (1978) as the contribution of the fine-particulate river sediments containing spent mercury ore processed in Idrija. Viviani et al. (1973b) studied the Hg concentration off the Po delta. They observed a significant correlation between Hg concentration and relative amount of clay.

Majori et al. (1976b) also found that the Zn, Cu and Pb concentrations in sediments decreased from the town of Trieste and its industrial zone towards the south of the Gulf of Trieste, which received less waste matter. Cd, Co, Cr, Cu, Mn, Ni, Pb, Ti, V and Zn were determined in sediment samples taken in the seas around Italy by Selli et al. (1977). So far only mean values and ranges have been published. J. Paul and Meischner (1976) and Kosta et al. (1978) examined the concentration of As, Cd, Co, Cr, Cu, Fe, Hg, Sb, Ni, Pb and Zn in sediments from a sampling grid covering the entire Adriatic. With the exception of Cr the values in the open-sea sediments did not exceed "typical" concentrations and hence did not indicate pollution. Cr concentrations were higher in the SE Adriatic. This was explained by natural river inputs.

Roth and Hornung (1977) investigated the heavy-metal concentrations in sediments from the northern shoreline of Israel. Comparison of their data with those of Amiel and Navrot (1978) for the southern shoreline shows that the concentrations of Cu, Zn and Cr are higher along the northern shoreline, while the Ni and Hg concentrations are about the same in both areas. As expected, element concentrations are higher in estuarine sediments. The highest values were found near Akhziv, Nahariyya and Akko:

$(\mu g\,kg^{-1}\,DW)$		$(\mu g\,kg^{-1}\,DW)$	
Cd	2,200	Zn	18,200
Pb	19,700	Ni	9,300
Cu	2,800	Cr	12,400

The sediments near Fuarom contained relatively high Hg concentrations: 210—570 $\mu g\,kg^{-1}$ DW. Besides the distribution pattern of heavy metals in the sediments of the Gulf of Trieste, the anthropogenic influence has been investigated in two other areas. The dispersion of mercury from the chlor-alkali plant situated on the Tuscan coast (Solvay, Rosignano) and the dispersion of seven trace elements through the Athens outfall showed clearly the limited influence of anthropogenic sources. For example, at both sides the Hg concentrations decreased to background values within 10 km of the discharge point (Tables IV and V).

A summary of the published concentrations of heavy metals in marine sediments from the Mediterranean is given in Tables IV and V, together with average heavy-metal concentrations in near-shore clay-shales and in carbonate rocks.

Heavy metals in marine organisms

The concern about heavy metals in food items has also especially stimulated the analysis of Hg in fisheries products: edible crustaceans, mollusks and, above all, fish. In order to obtain an idea of the general level of concentration of heavy metals, the concentrations in pelagic organisms are discussed separately from those in coastal and benthic organisms, since the latter are more subject to local influence. As shown below, different species collected at the same site have shown great differences in their heavy-metal concentrations. Consequently, body contents can only be compared between specimens of the same species. In order to avoid confusion, body contents of species listed only with their local or common name were not considered when these names left doubt as to their identification. For many pollutants positive or negative correlations between body content and body length are observed. If marine organisms are used as pollution indicators, specimens of the same length have to be compared.

Pelagic organisms. Only a few authors (Vucetic et al., 1974; Fowler et al., 1976b; Kosta et al., 1978) have analysed mixed plankton samples. Since plankton samples vary widely in species composition and contain phyto- and zooplankton, and because predators feed selectively on different plankton species, it is very difficult to draw conclusions about heavy-metal concentration in the marine environment from these data. More interesting are analyses of single zooplankton species (e.g., Vucetic et al., 1974; Belloni et al., 1976; Fowler et al., 1976b; Zafiropoulos and Grimanis, 1977). However, here, too, only very few data exist.

Heavy-metal concentration data from pelagic fishes are more numerous. Of special interest are surveys in which both Mediterranean and Atlantic specimens of the same species have been analysed by the same author by the same method. This will facilitate the comparison between areas, since possible systematic errors can be neglected. In Table VI Hg concentrations are listed from authors who have analysed anchovy, European hake, sardine, mackerel and bluefin tuna, both from the Mediterranean and the Atlantic. Since the Hg concentration is expected to increase with age, the length of

TABLE V

Heavy metals (mg kg^{-1} DW) in marine sediments from the Mediterranean Sea

Region	Type of sediment	n	Cd	Zn	Pb[*]
Nice to Monte Carlo (50—100 m) (0—6 cm)	?	18	1.7 (0.1—2.3)	70.6 (45—114)	57.6 (35—95)
Adriatic (3—7 m) (0—4 cm)	sand	4	—	35M (31—50)	14.5M (8—24)
	silty sand	5—6	—	62.5M (40—76)	12M (9—21)
	muddy sand	1	—	85.5M	11.5M (11—12)
	mud	10—18	—	117M (105—131)	12M (7—17)
	?	47	0.14 (0.02—0.95)	65 (24—123)	—
Gulf of Trieste	?	13	1.0M (0.6—1.8)	97M (24—147)	36M (2.8—67)
Po to Rimini		<60	0.71 (0.44—1.52)	89 (66—113)	32 (10—92)
Saronikos Gulf (0—5)(cm) background	?	?	—	50	—
Elefsis Bay	?	16	—	(<100—>500)	—
Athens outfall	?	>40	—	(50—1,800)	—
Piraeus harbour	?	?	—	2,500	—
Rosh Haniqra to Haifa	?	40	0.72 (0.3—2.2)	7.0 (2.1—18.2)	8.4 (4.3—19.7)
. . . average near-shore clay-shales			—	95	20
. . . average carbonate rocks			—	23	9

n = number of samples; the concentrations are given in mean or median (M) and the range beneath in brackets.
[*] Lead may be overestimated (Patterson and Settle, 1976).

the fishes, where available, is also reported. It is realized that length is only a very approximate estimator for age, but, as yet, age has not been reported for the species analysed. In comparing the data one has to bear in mind what has been aid about sigma estimation from ranges (p. 145). Allowing for the fact that the data are not normally distributed and consequently the range will estimate smaller sigmas, the differences are still significant for anchovy, sardine, mackerel and tuna ($n = 130$; 5.2σ). It is very likely that significant differences will be revealed for hake, if a comparison with more specimens

i	Ag	Mn	n	Cr	Reference
3.7	2.5	227		18.8	Renfro and Oregioni
8—33)	(1.9—3.6)	(95—320)		(12—30)	(1974)
20M	—	290M		14M	J. Paul and Meischner
		(250—305)		(10—18)	(1976)
20M	—	525M		75M	
		(450—1275)		(17—83)	
20M	—	890M		75.5M	
		(780—1,000)		(72—79)	
20M	—	1250M		112M	
20—35)		(680—1,770)		(35—222)	
	—	1100	7	125	Kosta et al. (1978)
		(300—4,250)		(94—185)	
)		463		—	Majori et al. (1976c)
2.7—59)	—	(338—557)			
)	—	585		45	Selli et al. (1977)
5—64)		(410—695)		(24—86.5)	
	0.04	—		80	Papakostidis et al.
	?	—		(80—>150)	(1975); Grimanis
	(<0.25—2)	—		(80—1,100)	et al. (1977)
	2.7	—		250	
8	—	—		4.2	Roth and Hornung
3—9.3)	—	—		(1.7—12.4)	(1977)
	—	850		90	Turekian and Wedepohl
					(1961)
	—	700		11	Wedepohl (1970a, b)

were carried out. The concentrations determined by the authors who investigated only Mediterranean fishes (Table VII) support, with few exceptions, the hypothesis that Mediterranean fishes of the same species have higher Hg concentrations than fishes from the Atlantic. These data also show the great importance of single-specimen analysis for a statistical treatment of the results, and that intercalibration is absolutely necessary for a comparison of data. No data on other elements which would allow a similar comparison have come to the present author's attention.

158

TABLE VI

Comparison of mercury concentrations ($\mu g\ Hg_t\ kg^{-1}$ FW) in pelagic fishes from the Mediterranean Sea and Atlantic Ocean

Species	n	Atlantic		Mediterranean		Length (cm)		Sampling area and date	Reference
		mean	range	mean	range	mean	range		
Engraulis encrasicholus	5H, 5–6n	50	15–80			13	11.5–14	SW French Atlantic coast off Marseilles	Thibaud (1971)
	1H, 5–6n			165		?			
	5	70	50–94			13.2	12.5–13.8	Str. Gibraltar (Jul. 1976)	Stoeppler et al. (1979)
	17			380	210–590	16.3	16.0–17.5	N. of Sardinia (Jun. 1976)	
	?	60				?		Atlantic Ocean (?)	G. Cumont (pers. commun., 1977)
	2			230	210–250	15.5	15–16	off Corsica (?)	
	37			250	40–1,040	?		Mediterranean Sea (?)	
	10	70	50–90			12.6	11.5–14	off Ceuta (Jul. 1977)	Bernhard and Renzoni (1977)
	10			160	70–330	14.6	13.5–16	off SW Sardinia (Jun. 1977)	
Merluccius merluccius	1H, 5–6n	50				?		Charente coast (Jul. 1971)	Thibaud (1971)
	1H, 5–6n			125		?		off Sète (Mar. 1971)	
Sardina pilchardus	2H, 5–6n	65	60–70			15	12–18	Gulf of Gascony (Mar. 1971)	Thibaud (1971)
	1H, 5–6n			235		?		off Sète (Mar. 1971)	
	1H, 5–6n			205		?		off Marseilles (Apr. 1971)	

Species	n	mean	range	mean	range	mean	range	Location (date)	Reference
	28	28	5—52			13.7	12—14.7	Str. Gibraltar (Jul. 1976)	Stoeppler et al. (1979)
	26	26		220	120—310	15.4	14—16.5	off La Spezia (Jul. 1976)	
	19	19		300	160—475	16	14.5—16.5	N. of Sardinia (Jun. 1976)	
	?					?		Atlantic Ocean (?)	Cumont et al. (1972)
	24	26	20—40	150	110—250	16—17		off Corsica (Apr. 1975)	G. Cumont (pers. commun., 1977)
	10	54	47—70			15.8	15.5—16.5	Str. Gibraltar (Jul. 1977)	Bernhard and Renzoni (1977)
	10	65	35—110			13.4	9.0—16.0	off Monte Carlo (Jul. 1977)	
	10	150	100—210			14.0	13.0—14.5	off SW Sardinia (Jun. 1977)	
Scomber japonicus	9	76	43—96			19.5	17.3—24.2	Str. Gibraltar (Jul. 1976)	Stoeppler et al. (1979)
S. scombrus	15			340	125—510	19.9	14.7—21.5	off La Spezia (Sep. 1976)	
S. scombrus	? 11	70	15—160	230	100—580	? ?		Atlantic Ocean (?) Mediterranean Sea (?)	Cumont et al. (1972) G. Cumont (pers. comm, 1977)
Thunnus thynnus	285 132	485	200—760	1,145	200—2,460	? ?		Atlantic Ocean (?) Mediterranean Sea (?)	Cumont et al. (1972)

H = composite sample from *n* specimens.

TABLE VII

Heavy metal (μg kg^{-1} FW) in pelagic fishes and mammals from the Mediterranean Sea

Species	n	Hg	Cu	Zn
FISHES:				
Engraulis	1H	110	—	—
encrasicholus	1H	60	—	—
	4H	400*M*	. 1,250*M*	11,350*M*
		(130—660)	(500—2750)	(4,850—18,20
	2H	(160—300)	(1,150—1,180)	(17,600—20,6
	6	(310*M*)	—	—
		(100—400)		
	1H	240	1,100	15,250
	9H	140	700	—
		(70—215)	(600—800)	
Merluccius	18	240*M*	—	—
merluccius		(120—870)		
	2H	(60—120)	—	—
	8H	215*M*	185*M*	3,000*M*
		(130—530)	(50—1,750)	(2,300—3,75C
	4	60*M*	—	—
		(10—140)		
	9	280*M*	880*M*	3,300*M*
		(110—780)	(160—320)	(2,800—4,30(
	4	180*M*	730*M*	2,200*M*
		(140—230)	(600—1,100)	(600—5,600)
Sardina	2H	(50—160)	—	—
pilchardus				
	2H	(55—70)	—	—
	13	170*M*	—	—
		(20—760)		
	4	175*M*	—	—
	19	160*M*	—	—
		(35—360)		
	10H	430*M*	880*M*	5,950*M*
		(200—870)	(320—1,600)	(830—13,25C
Sardinella	5—7H	80*M*	1,350*M*	21,700*M*
aurita		(30—120)	(770—16,650)	(11,100—23,
Scomber	1H	80	—	—
colias				
S. scomber	4	360*M*	—	—
		(100—500)		
	3H	580*M*	1,000*M*	4,600
MARINE MAMMALS:				
Globicephale	1	fat 1,300	—	—
melaena 3.9 kg		musc. 16,000		
Grampus griseus	1	fat 1,700	—	—
3.0 m		musc. 16,000		
Physeter	1	fat 2,150	—	—
catodon 8.0 m		musc. 4,000		
Stenella coeru-	2	fat (1,300—5,500)	—	—
leoalba		musc. (1,950—23,800)		
1.7 + 2.1 m				
Tursiops trun-	1	musc. 41,000	—	—
catus 140 kg				

For explanation of symbols see Tables I and VI; musc. = muscle tissue.
* Pb values are doubtful (Patterson and Settle, 1976).

Cd	Pb*	Ni	Region	Reference
—	—	—	NW African coast	Establier (1972b)
		—	Str. Gibraltar	
50 (50—50)	11,000 (1,000—1,250)	—	Italian Tyrrhenian coast	Ciusa et al. (1973)
(340—390)	(1,400—1,450)	(850—870)	Gulf of Trieste	Majori et al. (1976c)
—	—	—	off Biscéglie S. Adriatic	Perna et al. (1972)
—	940	—	S. Benedetto S. Adriatic	Ciusa and Giaccio (1972)
<100	<100	300 (200—400)	Rovinj, Istria	Gilmartin and Relevante (1975)
—	—	—	Spanish coast	Ballister et al. (1978)
<50	—	—	French coast	Thibaud (1971)
	500M (120—1,250)	—	Italian Adriatic coast	Ciusa et al. (1973)
—	—	—	Italian Adriatic coast	Perna et al. (1972)
	750M (500—1,100)		Italian Adriatic coast	Ciusa and Giaccio (1972)
>0 (40—60)	750 (550—930)	30 (<20—200)	Coast of Israel	Roth and Hornung (1977)
	—	—	NW African coast / Gulf of Cadiz	Establier (1972b, 1973)
	—	—	Spanish coast	Ballister et al. (1978)
	—	—	off Pescara	Caracciolo et al. (1972)
	—	—	Fano, Bescéglie, Morfetta (S. Adri.)	Perna et al. (1972)
	1,100M (800—1,500)	—	Italian Adriatic coast	Ciusa and Giaccio (1972)
70M (40—170)	140M (80—170)	635M (450—950)	coast of Israel	Roth and Hornung (1977)
	—	—	NW African coast	Establier (1973)
	—	—	Spanish coast	Ballister et al. (1978)
0	750	—	Tyrrhenian coast	Ciusa et al. (1973)
	—	—	French Medit. coast	Thibaud and Duguy (1973)
	—	—		
	—	—		
	—	—		
	—	—	off Pescara, Ital. S. Adriatic	Caracciolo et al. (1972)

Other heavy metals in pelagic fishes have received much less attention. Selected data are listed in Table VII.

The Hg levels in tuna, swordfish and pelagic marine mammals deserve special consideration. Cumont et al. (1972, 1975) analysed a large number of tuna from the Mediterranean and the Atlantic (Table VI). The only swordfish data so far published were those from Establier (1972b) and Riolfatti (1977), unfortunately without length and weight measurements. The Hg concentrations range from 860 to 3000 μg Hg$_t$ kg^{-1} FW. The Hg concentrations in tuna are the highest reported for teleost fish from the Mediterranean, but are by no means uncommon in tuna (C. L. Peterson et al., 1973) and in other large teleost fish from other oceans. Merlin caught off NE Australia, where specimens reach 2—3.5 m, had Hg body contents of between 500 and 16,500 μg Hg$_t$ kg^{-1} FW (Mackay et al., 1975; Schultz et al., 1976). Only in marine mammals have similar and higher concentrations been observed. In the blue dolphin (*Stenella coeruleoalba*) from the Mediterranean, the Hg$_t$ concentration in the muscle tissue reached; (Doi and Ui, 1975) 23,800 μg kg^{-1} FW (fresh weight), in the liver 344,000 μg Hg$_t$ kg^{-1} FW, and in the spleen as much as 530,000 μg Hg$_t$ kg^{-1} FW (Thibaud and Duguy, 1973). Other whales have similarly high Hg concentrations (Table VI).

Marine mammals with high Hg$_t$ concentrations have relatively low methyl mercury concentrations (1—10%). In teleost fish with low Hg$_t$ concentrations, the methyl mercury comprises between 75 and 95% (Cumont et al., 1972; G. Cumont, pers. commun., 1977). In the merlin and in marine mammals and birds a positive correlation between Hg and Se concentration has been observed (Koeman et al., 1975; Mackay et al., 1975). This correlation is of very great practical importance, since an antagonism exists between Hg and Se reducing the toxicity of both metals. Kosta et al. (1975) suggest that the natural level of Se in man is raised (co-accumulation) when exposed to high Hg levels. This co-accumulation of Se and Hg observed in man seems to occur also in marine mammals with high Hg concentration. It appears from the few data so far available that in organisms with concentrations higher than 1000 μg Hg$_t$ kg^{-1} FW (> 1 ppm FW) also the Se body burden is raised to a Hg/Se molar ratio of ~ 1.

Coastal and benthic organisms. Very few data are available on macroscopic algae and marine plants (Fukai and Broquet, 1965; Strohal and Dzajo, 1975; Roth and Hornung, 1977). The values for Hg are low (10—100 μg Hg$_t$ kg^{-1} FW) and the concentrations of other metals are similar to those of the higher organisms investigated. The data available on crustaceans are also scarce.

Mytilus galloprovincialis is probably the marine organism which has received the widest attention in the Mediterranean area (Tables VIII and IX) followed by the oyster, especially in the area of Cadiz. These sessile organisms show wide variations in heavy-metal concentration, indicating local conditions. Establier (1969a, b; 1972a, b, 1973) studied heavy metals in oysters in the Gulf of Cadiz. He observed very high Cu content in oysters from the Huelva River (Gulf of Cadiz) with more than $1.5 \cdot 10^6$ μg Cu kg^{-1} FW (= 0.15%) (see also p. 148). The Cu and Zn concentrations in oysters from the Gulf of Cadiz are high in comparison with oysters from non-Mediterranean areas (Bernhard and Zattera, 1975). On the other hand, Hg concentrations of various mollusks from the Gulf of Cadiz were similar to those of non-Mediterranean areas. Thibaud (1973) and Alzieu et al. (1976) compared Hg concentrations in *Mytilus* collected on the French Atlantic and Mediterranean coasts. The overall average concentrations in the Mediterranean tend to be higher than those from the Atlantic, although the ranges overlap. The wide ranges are caused by contamination from local sources, especially noticeable in the Seine estuary, the Morlaix and Brest roads and the mouths of the rivers Charente and Adour. In some sampling sites *Crassostrea gigas*, *Patella vulgata*, *Pecten maximum* and *Murex trunculus*, were also collected and analysed. The Hg levels in the two bivalves were similar, but the Hg concentrations in the gastropods were markedly lower. Fowler and Oregioni (1976) determined the Cd, Cu, Ni, Pb, Zn, Cr, Ag, Co, Fe and Mn content in mussels from Sète to Genoa. They found that near ports and river discharges the heavy-metal content of the mussels increased. Seasonal variation in the concentrations was evident and may, in part, be due to the changes in biomass during the life of the mussel. In some cases the seasonal variation observed in the mussels could be related to high metal concentrations in the ambient water collected simultaneously (Fukai and Huynh-Ngoc, 1976b). However, Fowler and Oregioni (1976) point out that the filtered sea water analysed by Fukai and Huynh-Ngoc (1976b) does not contain the particulate metal fraction available to the filter feeding mussels. They also draw attention to the fact that water samples represent short-time situations in heavy-metal concentrations, while mussels will integrate water concentrations over a certain time interval. Comparing their values with those obtained by other authors reveals that the Mediterranean values do not differ significantly from those of other areas.

For a period of two years (1963—1964) Macchi (1966) examined monthly the Cd, Cu and Pb concentrations in *Mytilus galloprovincialis* collected from a mussel bed situated near Palmaria Island (La Spezia). Capelli et al. (1976) investigated the concentrations of Cd, Cu, Ni, Pb, Cr, Co and Mn in *Mytilus* collected in ports and harbours from San Remo to Viareggio. In La Spezia one of their collection sites (Palmaria Island) is practically at the same location as the sampling station of Macchi. Contrary to Fowler and Oregioni (1976), Macchi (1966) did not observe a significant seasonal variation, probably because his sampling sites was less exposed to the influence of runoff and anthropogenic contamination sources.

TABLE VIII

Cd, Cu, Ni, Pb, Zn, Cr and Ag concentrations ($\mu g\ kg^{-1}$ DW) in *Mytilus*

	Cd	Cu	Ni
Ligurian Sea	400—5,900	2,400—154,000	900—14,100
Ligurian—Tuscan coast	200—4,700	8,000—25,200	2,600—11,500
La Spezia, Medit.	560—2,600	32,000—190,000	—
Gulf of Trieste	70—165[*1]	3,900—12,500[*1]	2,900—6,350[*1]
Saronikos G., Medit.	—	(4,500)	39,000

Values in brackets are calculated from ash weight/dry weight ratios (C. Papadopoulou, pers. commun., 1976).
[*1] Fresh weight times 5.
[*2] Lead levels may be overestimated (Patterson and Settle, 1976).

The influence of the River Isonzo, which carries Hg from the Idrija area, is clearly reflected in the high Hg mercury body content of mussels collected near the river mouth:

	n	($\mu g\ Hg_t\ kg^{-1}$ FW)	Reference
Venice (lagoons)	2	30—50	Kosta et al. (1978)
Tagliamento to Isonzo River mouth	4	140—280	
Trieste to Pula	7	10—70	
East coast Gulf of Trieste	58	20—70	Majori et al. (1976c)
Grado to Isonzo River mouth	10	95—240	

Of considerable interest is the multi-element analysis carried out on seven mollusks from the coastal waters of the Saronikos Gulf (Table X). The concentrations of all elements varied remarkably between the different species collected in the same area. Similar results were obtained by Stancher and Chimenti (1970) who analysed four major (Na, K, Ca, Mg) and ten minor elements in 13 mollusks from the Gulf of Trieste. These data supply Mediterranean examples of similar observations made earlier in the English Channel by Segar et al. (1971) who analysed 17 elements in seven Lamellibranchia and four Gasteropoda. For example, in the English channel the Cu concentrations ranged from $360\ \mu g\ kg^{-1}$ FW for *Anodonia* sp. to $40,500\ \mu g\ kg^{-1}$ FW in *Crepidula fornicata*. The Ni concentration showed an even wider range for these two species ($40—127,500\ \mu g\ kg^{-1}$ FW). This shows that the concentration of heavy metals from different species cannot be easily compared.

Pb[*2]	Zn	Cr	Ag	Reference
2,700—117,000	97,000—644,000	500—28,800	100—18,900	Fowler and Oregoni (1976)
14,000—32,000	—	2,000—24,300	—	Capelli et al. (1976)
145,000—310,000	100,000—400,000	—	—	Macchi (1966)
1,850—68,000[*1]	50,000—190,000[*1]	—	—	Majori et al. (1976c)
—	(87,000)	7,800	—	Papadopoulou and Kanias (1976)

TABLE IX

Concentration range of total mercury in *Mytilus*

	Hg_t ($\mu g\ kg^{-1}$ FW)	Reference
French Atlantic	10—200	Thibaud (1973)
French Mediterranean	23—310	Thibaud (1973)
French Atlantic	(10—110)[*1]	Alzieu et al. (1976)
French Mediterranean	(65—165)[*1]	Alzieu et al. (1976)
Vigo, Atlantic	60—140	Establier (1973)
Spanish Mediterranean	70—160	Ballister et al. (1978)
Pontevedra, Atlantic	10—200	Torre and Masso (1975)
Ligurian Sea	30—160	Stoeppler et al. (1979)
Pescara, river mouth	1,500	Caracciolo et al. (1972)
North Adriatic	65	Strohal and Dzajo (1975)
North Adriatic	74	Lulic and Strohal (1974)
Saronikos G., Mediterranean	(220)[*2]	Papadopoulou and Kanias (1976)

[*1] Calculated from fresh weight/dry weight ratio.
[*2] Calculated from ash weight/dry weight ratio and freshweight/dry weight ratios (C. Papadopoulou, pers. commun., 1976).

The few data available on other mollusks show relatively high Hg concentrations for cephalopods (*Octopus* and *Sepia*). Both are of a greater age and higher position in the food chain compared to mussels, which may be the explanation. A large number of different fish species has been analysed for Hg in various parts of the Mediterranean. Establier (1972b, 1973) has compared the Hg concentrations of fishes from the NW African coast with those from the Gulf of Cadiz. Ballister et al. (1978) investigated species

TABLE X

Multi-element analysis (μg kg^{-1} DW) of the soft part of mollusks from the coastal waters of the western part of Saronikos Gulf between Megara and Slamis Islands (from Papadopoulou and Kanias, 1976)

	Hg	As	Sb	Cr	Ni	Co
Mytilus galloprovincialis	210	25,500	140	7,800	39,000	19,400
Venus verrucosa	22	15,000	40	4,700	—	1,700
Glycymeris glycymeris	15	7,300	100	4,900	14,000	520
Ensis ensis	2,350	15,000	60	12,500	—	510
Meretrix chionae	70	59,000	370	4,100	26,000	14,000
Ostrea edulis	320	83,000	500	12,500	15,000	730
Tapes decussatus	290	56,000	100	18,000	45,000	22,500

Samples were collected between 0 and 12 m depth during October 1969.
Coefficient of variance $\sim$ 10%.
Values in brackets are calculated from ash weight/dry weight ratio (C. Papadopoulou, pers. commun., 1976).

from the northern Spanish coastline; Thibaud (1971), Cumont et al. (1972, 1975) and Aubert (1975) those from the French Mediterranean coast; Ciusa et al. (1973), Renzoni et al. (1973), Renzoni and Baldi (1975), and Stoeppler et al. (1979) examined species from the Italian west coast. Of special interest for a correct interpretation of Hg body contents are the very recent findings of Bernhard and Renzoni (1977) who observed that *Mullus barbatus*, from the area along the Tuscan coast bordering the Monte Amiata, a geochemical Hg anomaly, had very high Hg body contents (up to 3700 μg Hg kg^{-1} FW) far away from possible anthropogenic Hg sources. From Table XI, it can be seen that *M. barbatus* caught south of the Isle of Elba and north of the Isle of Giglio have considerably higher Hg concentrations than the *M. barbatus* from other regions. The previously reported concentrations from Piombino and Orbetella by Ciusa et al. (1973) confirm these observations. The often-reported Hg concentration vs. length (weight) correlation was much better in *M. barbatus* specimens which had relatively low concentrations. In all samples with high Hg the data scattered widely, an observation made also by other authors (e.g., C. L. Peterson et al., 1973). At present no reason can be given to explain this phenomenon.

In this respect it is interesting that the Hg body contents of six out of eight marine organisms from the Isonzo area, although not very high, were significantly higher than those of the same species from the eastern Gulf of Trieste (Majori et al., 1976c). Unfortunately, *M. barbatus* was not examined. The influence of the Hg discharge from the Solvay plant on the concentrations in marine organisms was also investigated (Renzoni et al., 1973; Renzoni, 1977). Similar to the Hg distribution in water and sediments (pp. 148 and 154), the highest Hg concentrations were observed near the outfall.

Mg $(\times 10^6)$	Cu	Zn	Mn	Sr	V
5.5	(4,500)	(87,000)	(13,500)	(26,000)	(230)
7.2	(5,200)	(17,500)	(2,100)	(24,300)	(100)
6.0	(28,000)	(75,000)	(3,500)	(135,000)	(750)
0.6	(9,400)	(24,000)	(13,000)	(15,000)	(700)
6.7	(4,100)	(33,000)	(26,000)	(38,000)	(450)
4.5	(100,000)	(685,000)	(2,000)	(24,000)	(20)
4.3	(10,500)	(83,000)	(650)	(32,000)	(320)

At about 10 km north and south of this point the concentrations were only slightly above the "typical" concentrations. Some 28 months after the Solvay plant had to reduce its Hg release, the concentrations in the limpet *Patella coerulea* and the two fishes *Serranus scriba* and *Scorpaena porcus* decreased by 20—30%, and in the crab *Pachygrapsus mormoratus* by as much as by 80%.

Hg concentrations in coastal organisms are listed in Tables VIII—XII. Data on other heavy metals are less abundant (Tables VIII, X, XI). Ciusa and Giaccio (1972) and Ciusa et al. (1973) determined Cu, Pb, Cd and Zn in organisms collected along the Italian coast. Papadopoulou et al. (1973a, b) analysed *Pagellus erythrinus* for Ni, Ca, Mg, As, Pb, Co, Sr, Zn, Fe and Mn and found considerable concentrations of As in muscle tissue. Comparison of *P. erythrinus* from unpolluted waters (Rhodes Island) with specimens from the polluted Saronikos Gulf showed that the As concentration in fish from polluted waters was 2—2.5 times higher than in fish from the unpolluted waters. No difference could be established for the Hg concentrations. Finally, Roth and Hornung (1977) analysed fishes from the Israel coast, determining Cd, Pb, Cu, Zn, Ni and Cr, in addition to Hg. Unfortunately, despite the many samples analysed, only very few species were analysed by several authors, making impossible a comparison of the data from all the regions mentioned. A wide variation of concentrations has been observed. At present it is not possible to say whether these observed differences in the concentration are actually due to anthropogenic sources, as is undoubtedly the case for the fish from the Saronikos Gulf, or if the differences are caused by Hg anomalies or special enrichment factors inherent in the species. Finally, since no intercalibration exercise data are as yet available, methodological difficulties may also be the cause of high and low Hg concentrations.

TABLE XI

Mercury levels ($\mu g\,kg^{-1}$ FW) in *Mullus barbatus* and *M. surmuletus* from the Mediterranean Sea

Region	M. barbatus			M. surmuletus			Reference
	n	$\bar{x}$	range	n	$\bar{x}$	range	
Gulf of Cadiz	—			2	280	80—80	Establier (1973)
Str. Gibraltar	20	265	50—615	4	180	190—390	Stoeppler et al. (1979)
Ebro to Blanes	18H	190M	110—3,450	3H	—	160—500	Ballister et al. (1978)
La Spezia to Carrara	66	130	20—760				Stoeppler et al. (1979)
Off River Arno	51	220	60—900		—		Bernhard and Renzoni (1977)
	1F	1,500					
Secca di Vada (Livorno)	—			6	630 ± 600		Renzoni et al. (1973)
Piombino	1H	3,000		1H	360		Ciusa et al. (1973)
Orbetello	1H	1,300					
Isle Monte Cristo	22	500	180—1,750		—		Bernhard and Renzoni (1977)
Talamone coast	19	200	55—335		—		
South, Isle Giglio	61	775	100—2,500				
North Sardinia	5	200	165—245				Stoeppler et al. (1979)
Civitavecchia to Reggio Calabria	6H	310M	120—680		—		Ciusa et al. (1973)
Trapani	7H	250M	140—1,050	6	90	70—110	Stoeppler et al. (1979)
Trieste—Pescara	2	250M	55—145				Ciusa and Giaccio (1972)
Off Pescara					—		Caracciolo et al. (1972)
Coast of Israel	3H	220M	50—290		—		Roth and Hornung (1977)
Isle Pilau, Tunis	10	250	90—560		—		Stoeppler et al. (1979)

For explanation of symbols see Table I. F = female.

Some typical concentrations of the more important heavy metals have been summarized only for health-protection purposes (Table XIII). These data are still very approximate and they can only be used for a first estimation of the heavy-metal intake by humans from fishery products. The main limitations are lack of representative data from different areas of the Mediterranean and accuracy of analytical results. Accuracy merits attention, especially in the light of recent findings of Patterson and Settle (1976) that the concentration of Pb in the marine environment has been overestimated by several orders of magnitude owing to contamination during sampling, sample preparation and analysis.

The data-base for the heavy-metal levels in the Mediterranean Sea is unevenly distributed. Most data come from the NW and central Mediterranean; few are available from the eastern and practically none from the southern Mediterranean. Comparing the data from the pelagic environment with those from other oceans shows that the levels are of the same order of magnitude. The data from coastal areas reflect local pollution conditions in the same way as has been observed in other non-Mediterranean areas, with the exception of the Hg levels. The data available clearly demonstrate statistically higher levels in migratory pelagic fishes which cannot be attributed to anthropogenic sources. These higher natural levels are caused by the higher average Hg concentration in the Mediterranean Basin (Bernhard and Renzoni, 1977).

A comprehensive survey of geochemical anomalies in the Mediterranean does not exist, but the commercially exploitable resources can give an idea of the Hg anomaly of the Mediterranean Basin. Comparing the Hg resources of the Mediterranean (Fig. 1 and Table XIV) with world resources, shows that the Mediterranean, with about 65% of the world resources, occupies only 1% of the Earth's surface. Also, the Hg production illustrates this Hg anomaly of the Mediterranean. Further support is given by the higher than average Hg concentration in sediments and Hg concentrations in the benthic fish *Mullus barbatus* from the Hg anomaly of the Monte Amiata region and similar observations for the northern Adriatic (Idrija anomaly). These high concentrations are natural and not of anthropogenic origin, although the contribution of spent mercury ore residue from the River Isonzo superimposes anthropogenic mobilization on natural sources. A detailed survey would show the actual mechanisms involved. Higher than typical Hg seawater concentrations were not observed in the open-sea water samples from the Mediterranean.

CONCENTRATION OF CHLORINATED HYDROCARBONS IN THE MARINE ENVIRONMENT

Unlike heavy metals, all chlorinated hydrocarbons are anthropogenic in origin. Non-persistent pesticides are very unlikely to contaminate the

TABLE XII

Heavy metals (μg kg^{-1} FW) in selected coastal fishes from the Mediterranean Sea

Species	n	Hg	Cu	Zn
Mugil auratus	12H	260M (80—1,000)	650M (270—1,250)	450M (50—19,000)
	2	(200—350)	—	—
M. cephalus	4H	220M (100—250)	250M (120—370)	5,000M (4,450—6,100)
	11	390 (± 270)	—	—
Pagellus erythrinus	2	(250—290)	—	—
	9	120M (60—260)	—	—
	4H	1,550M (400—3,300)	200M (<120—370)	2,750M (2,200—3,500)
Solea solea	30	45 (15—115)	—	—
	3	60 (50—70)	—	—
	5H	250 (180—330)	250 (<120—500)	4,800M (2,500—5,400)
	2	(185—345)	—	—
	1H	40	1,400	22,100
	9H	280M (140—860)	500M (100—1,420)	5,750M (580—9,450)

* Pb values are doubtful (Patterson and Settle, 1976).

marine environment significantly, since they will be degraded before reaching it (Duursma and Marchand, 1974). In fact, only the PCB's and pesticides such as dieldrin, endrine, DDT and its metabolites, heptachlor, aldrin, lindane, chlordane, toxaphene, hexachlorocyclohexane, endosulphan, methoxychlor and wastes from PVC production (aliphatic chlorinated hydrocarbons) have been detected in marine organisms (N.A.S., 1973). The difficulties in identification and quantification of chlorinated hydrocarbons are numerous (S. Jensen et al., 1973; Hattula, 1974; Chau and Sampson, 1975; Bernhard, 1976) and the comparison of data from different authors is not easy, since different pretreatments and quantification techniques have often been used. In the present state of knowledge, probably only data from the same author can be used for the detection of differences between samples taken at different sites and/or different dates. Again, as in the discussion of heavy-metal concentrations, a distinction is made between open-ocean (pelagic) and near-shore or coastal samples.

Cd	Pb*	Ni	Region	Reference
—	950M (50—2,100)	—	Adriatic coast	Ciusa and Giaccio (1972)
—	—	—	Chioggia, Venice	Perna et al. (1972)
50M (<50—50)	3,600M (870—6,200)	— —	Tyrrhenian coast	Ciusa et al. (1973)
—	—	—	Secca di Vada (Livorno)	Renzoni et al. (1973)
—	—	—	Gulf of Cadiz	Establier (1973)
—	—	—	Corsica	G. Cumont (pers. commun., (1977)
50M (<50—100)	500M (250—500)	—	Tyrrhenian coast	Ciusa et al. (1973)
—	—	—	NW African coast	Stoeppler et al. (1979)
—	—	—	Gulf of Cadiz	Establier (1973)
50M (<50—100)	370M (<250—870)	—	Tyrrhenian coast	Ciusa et al. (1973)
—	—	—	Chioggia, Venice	Perna et al. (1972)
200	300	1,100	coast of Israel	Roth and Hornung (1977)
	880M (380—1,380)	—	Adriatic coast	Ciusa and Giaccio (1972)

TABLE XIII

Tentative "typical" heavy metal concentrations ($\mu g\,kg^{-1}$ FW) in selected pelagic and benthic marine organisms of economic importance

	Hg	Cd	Pb*	Cr	Cu	Zn	Ni	As
Mytilus galloprovincialis	50	350	(25)	1,000	1,500	30,000	500	5,000
Sardina pilchardus	200	20	(1)	40	1,000	10,000	300	—
Merluccius merluccius	100	50	(1)	—	1,000	3,000	—	—
Mullus barbatus	300	20	(1)	—	500	3,000	300	—
Pagellus erythrinus	200	30	(1)	—	500	5,000	200	4,500

* Taking Barnett and Patterson's (1978) results into consideration.

TABLE XIV

Reasonably assured mercury reserves and production of mercury for the year 1972 (after Brinck and Van Wambeke, 1974; *Mining Journal*, 1974)

	Reserves (t)	Average ore grade (%)	Production (76-lb. flasks)
Mediterranean[1] :			
Spain	87,000	1—2	60,000
Italy	21,000	0.5—0.3	42,120
Yugoslavia	20,000	0.16—0.9	16,419
Turkey	11,000	0.3—0.4	7,722
Algeria	?	?	>10,000
Tunisia	?	?	?
	>140,000[2]		>136,000[3]
Non-Mediterranean:			
U.S.S.R.	30,000	0.1—0.6	?
Mexico	10,000	0.25	22,510
Canada	10,000	0.23	14,600
China	10,000	?	?
U.S.A.	6,000	0.25	7,286
Non-Med. socialist countries	?	?	81,800
World	>215,000		>280,000

[1] Mediterranean area is ~ 1% of the Earth's surface.
[2] Mediterranean reserves are 65% of the whole-world reserves.
[3] Mediterranean production is ~ 50% of the whole-world production.

TABLE XV

PCB concentration ($\mu g\,l^{-1}$) in seawater

		n
1975	Mediterranean Sea, open sea	80
1972	North Atlantic	52
1973—1975	Atlantic, north of $30°$N	26
	Atlantic ($30°$N—$30°$S)	18
	Atlantic, south of $30°$S	3
1973	Sargasso Sea	8
1974	southern California coast	14

Chlorinated hydrocarbons in seawater

Few data are available on chlorinated hydrocarbons in seawater of the Mediterranean area, either from open-sea or from coastal regions.

Chlorinated hydrocarbons in "open-sea samples"

Elder and Villeneuve (1977) determined the PCB's in various parts of the open Mediterranean (37 stations, including six east of Sicily). At most stations only surface samples were taken, but from nine stations subsurface water was also sampled. Grouping the stations into various areas (Aegean Sea, Ionian Sea, Strait of Sicily and Malta, Tyrrhenian—Ligurian Sea, and the western Mediterranean), the PCB concentration tends to increase from east to west. The ranges overlap and the differences are not significant. The PCB concentration does not decrease with depth. For comparison, Harvey et al. (1973) and Harvey and Steinhauer (1976) (with whom D. G. Elder and G. P. Villeneuve had intercalibrated seawater samples) observed ranges between 0.0003 and $0.008\,\mu g\,PCB\,l^{-1}$ in the surface of the North and South Atlantic. These and results from other non-Mediterranean regions are given in Table XV. Open-sea concentrations of other chlorinated hydrocarbons have not come to the present author's attention.

Chlorinated hydrocarbons in coastal waters

During 1975, Elder et al. (1976) investigated the dissolved PCB's at eleven locations off the French Mediterranean coast. From the lowest values $0.0015\,\mu g\,l^{-1}$), 3—5 km off Monaco, the concentrations increase towards Marseilles ($0.038\,\mu g\,l^{-1}$, 500 m from the mouth of the Rhône). The data can be summarized as follows:

	n	PCB ($\mu g\,l^{-1}$)	
		mean	range
French Mediterranean coast	11	0.013	0.0015—0.038

$\bar{x}$	Range	Reference
0.002	0.0002—0.0086	Elder and Villeneuve (1977)
0.035	0.001—0.150	Harvey et al. (1973)
0.002	0.0004—0.0071	Harvey and Steinhauer (1976)
0.003	0.004—0.008	Harvey and Steinhauer (1976)
0.0017	0.0003—0.0037	Harvey and Steinhauer (1976)
0.0011	0.0009—0.0036	Bidleman and Olney (1974)
0.0092	0.0023—0.0356	Scura and McClure (1975)

TABLE XVI

Pesticides in the surface layer and in seawater 40 cm below the surface, sampled in the outflow of the River Huveaune (Marseilles) (Raybaud, 1972)

Seawater	Concentration ($\mu g \, l^{-1}$)	
	$\bar{x}$	range
Surface film, dissolved:		
p,p-DDT + *p,p*-DDD	81	55—117
p,p-DDE	70	17—97
Lindane	25	12—45
40 cm below surface, dissolved:		
p,p-DDT + *p,p*-DDD	0.11	0.07—0.18
p,p-DDE	0.08	0.03—0.11
Lindane	0.07	0.04—0.09
Surface film, bounded to particles:		
p,p-DDT + *p,p*-DDD	0.0021	(particulate matter:
p,p-DDE	0.0029	1,700—3,000)
Lindane	0.00066	

To our knowledge, pesticides (DDT's and metabolites, lindane) have only been determined in Mediterranean coastal seawater by Raybaud (1972) (Table XVI). Near the mouth of the River Huveaune, near Marseille, he sampled an outflow of sewage and industrial wastes, and determined 45—121 $\mu g \, PCB \, l^{-1}$ in the surface film, and, in the surface water 40 cm below, only 0.1—0.21 $\mu g \, PCB \, l^{-1}$. Owing to the contamination of the river outflow, his values are expected to be higher than even high values of Elder (1976) off the mouth of the River Rhône.

Chlorinated hydrocarbons in sediments

Open-sea sediments. Elder et al. (1976) studied sediments from the southern Mediterranean at eight stations, west of the Strait of Gibraltar to the Ionian Sea PCB's were detected only in the top 1 cm. According to the authors, PCB's are probably only present in the top 1 mm. The concentrations ranged from 0.8 to 9.0 $\mu g \, kg^{-1}$ DW. It is remarkable that the PCB's have contaminated sediments at a depth of over 4 km. More highly chlorinated hydrocarbons were found in the deeper sediments, which Elder et al. explained by assuming that the highly-chlorinated hydrocarbons are more readily adsorbed on sedimenting particles and the less-chlorinated hydrocarbons are more easily decomposed.

Coastal sediments. Duursma et al. (1974) and Marchand et al. (1975) determined DDT's and their metabolites and PCB's in coastal sediments along the French—Italian coast from Sète to Genoa. The PCB's ranged from 0.8 to about $30 \,\mu g\, kg^{-1}$ DW. Exceptionally high values were found at 100 m depth off Nice ($80-1200 \,\mu g\, PCB\, kg^{-1}$ DW). Total concentration of DDT's ranged from 0.4 to $30 \,\mu g\, kg^{-1}$ (100 m depth off Nice). Mestres et al. (1975) investigated the chlorinated hydrocarbons in sediments of the Gulf of Lions without discovering a distribution pattern. Stirn et al. (1974) found much higher concentrations in the more industrialized northern Adriatic than in the southern Adriatic. The distribution pattern around the Athens outfall in the Saronikos Gulf (Dexter and Pavlou, 1973) is very marked. Total DDT's decreased from a maximum of $1900 \,\mu g\, DDT_t\, kg^{-1}$ DW near the outfall to values lower than $20 \,\mu g\, kg^{-1}$ DW at about 5-km distance. Similarly, the PCB's decreased from $775 \,\mu g\, kg^{-1}$ DW to below $20 \,\mu g\, kg^{-1}$ DW at about the same distance from the outfall.

Chlorinated hydrocarbons in marine organisms

Pelagic organisms. Very few data are available on pelagic mollusks. Relevante and Gilmartin (1975) analysed one specimen each of *Sepia officinalis* and *Loligo vulgaris*, both from Rovinj (North Adriatic). DDT_t concentration was between 50 and $100 \,\mu g\, kg^{-1}$ FW and PCB's varied from trace amounts to $5 \,\mu g\, kg^{-1}$ FW.

Anchovy, sardine and dogfish (*Squalus acanthias*) have been studied more extensively than other fishes (Table XVII). *Sardina pilchardus* from Barcelona and Palamos have higher DDT_t concentrations ($\sim 500 \,\mu g\, kg^{-1}$ FW) than sardines from Castellon ($125 \,\mu g\, kg^{-1}$ FW) and from the Spanish Atlantic coast (Vigo 84 and Santander 84 and $5 \,\mu g\, kg^{-1}$ FW, respectively; Fernandez and Franco, 1976). PCB concentrations diminish in the same way between Barcelona and Castellon as between the Mediterranean and the Atlantic coasts. Some data available from the French Mediterranean coast show medium to high values for Les Saintes Maries de la Mer in the Rhône delta, and a high PCB value for the Bay of Marseilles ($3000 \,\mu g\, PCB_t\, kg^{-1}$ FW) (De Lappe et al., 1972; Alzieu, 1976; Alzieu et al., 1976). DDT_t concentrations in sardines, anchovy and sprat between the Po delta and Cattolica had relatively high DDT and PCB concentrations (around $300 \,\mu g\, DDT_t\, kg^{-1}$ FW; $650 \,\mu g\, PCB_t\, kg^{-1}$ FW) (Viviani et al., 1969, 1973a, 1974). Anchovies caught in November 1972 in the Po delta, on the other hand, had much lower concentrations. DDT_t and PCB mean concentrations in sardines and anchovies from Rovinj were lower than all other values (Relevante and Gilmartin, 1975). Sharks have the highest DDT_t concentrations observed in Mediterranean marine organisms, with a mean concentration of $900 \,\mu g\, DDT_t\, kg^{-1}$ FW (range 760—1010). Near Barcelona, in the liver of one female shark, $18,100 \,\mu g\, kg^{-1}$ FW was determined (Baluja et al., 1973).

Only two tuna (*Thunnus thynnus*) have been analysed for chlorinated

TABLE XVII

Chlorinated hydrocarbons in some pelagic fishes from the Mediterranean Sea and the Atlantic Ocean

Species	n	PCB's	DDT_t	Dieldrin
Engraulis encrasicholus Mediterranean	4	14M (11—23)	235M (170—350)	—
	5H	525M (180—960)	275M (90—410)	—
Sardina pilchardus Mediterranean	2H, 8n	620 (570—700)	115 (80—150)	<1
	3H, 8n	4,700 (3,300—6,200)	490 (230—620)	48 (2—140)
	3H, 8n	4,200 (1,300—6,300)	<5	50 (20—90)
	?	1,240	500	<10
	4	7M (2—19)	90M (n.d.—200)	—
Atlantic	2H, 8n	320 (260—380)	90 (77—100)	7 (n.d.—14)
	3H, 8n	230 (n.d.—500)	1 (1—1)	n.d.
Squalus acanthias Mediterranean	2F	det.	930 (760—970)	<1
	2M	det.	865 (760—970)	<1
	1	720	—	—
Atlantic	4F	det.	15M (<5—95)	<1
	1M	det.	n.d.	<1
Sprattus sprattus Mediterranean	6H, <50n	760 (620—920)	350 (320—720)	—
	1H, 4n	25	425	—

H = composite sample; n = number of specimens, M = male; F = female; M (under data) = median, det. = determined, n.d. = non-detected.

hydrocarbons (Alzieu, 1976). In the muscle of the tuna (145 and 165 kg, 5—7 yr. old) 40—225 DDT_t $\mu g\ kg^{-1}$ FW was reported. DDT_t concentration in the liver was about 25 times higher. The PCB's were 407 and 95 $\mu g\ kg^{-1}$ FW in the muscle and 15—20 times higher in the liver. Detailed discussions of the

Aldrin	$\alpha + \gamma$ HCH	Length (cm) FW (g)	Sampling area and date	Reference
—	—	14.5 cm, 16.5 g	Rovinj, N. Adriatic June. 1973	Relevante and Gilmartin (1975)
—	—	(10—15 cm)	Po delta to Cattolica Jan.—Nov. 1970)	Viviani et al. (1973a)
n.d.	28 (27—29)	30.1 g (28.8—32.0 g)	Castellón Jun. 1970	Baluja et al. (1973)
<1	6 (4—7)	44.5 cm	Barcelona Feb. 1970	
3 (1—8)	20 (14—30)	36.1 cm (36.0—36.3)	Palamos Jan. 1970	
<1	—	13 cm	Po delta Nov. 1972	Viviani et al. (1974)
—	—	16.6 cm, 41 g	Rovinj, N. Adriatic Jun. 1973	Relevante and Gilmartin (1975)
n.d.	?	60 cm (56—63 cm)	Vigo Jan. 1970	
n.d.	3 (2—5)	84.7 cm (82.8—86.0 cm)	Santander Jul. 1970	
n.d.	(ND—9)	53.3 cm, 3 yr. (53—53 cm)	Barcelona Jan. 1968	Baluja et al. (1973)
(<1—1)	(ND—6)	58 cm, 3 yr. (56—60 cm)	Barcelona Jan. 1968	
—	—	50 cm	Po delta Nov. 1972	Vivani et al. (1974)
4M (<1—24)	7M (5—7)	58 cm M, 4—5 yr. (55—64 cm)	Vigo Jan. 1968	Baluja et al. (1973)
24	8	56 cm, 5 yr.	Vigo Jan. 1968	
—	—	(10—13 cm)	Po delta to Cattolica Jan.—Nov. 1970	Vivani et al. (1973b)
—	—	13.4 cm	Rovinj Jun. 1973	Relevante and Gilmartin (1975)

effect of chlorinated hydrocarbons in the Gulf of Mexico are presented by Sackett in Chapter 13 of this volume.

Coastal and benthic organisms

Mussels (*Mytilus galloprovincialis* or *M. edulis*) both from the Spanish Mediterranean (Cadiz, Castellon, Ebro delta, Barcelona and Blanes) and the

TABLE XVIII

PCB's and DDT's in *Mytilus galloprovincialis* (μg kg^{-1} FW)

Area	n	DDT$_t$	p,p-DDT	p,p-DDE	p,p-DDD	PCB's	Reference
Castellón Jan., Jul. 1976	3H	110 (±95)	—	—	—	—	Franco and Fernandez (1976)
Castellón Feb. 1970)	3H	25M (20—30)	9M (5—15)	1M (1—2)	15M (9—19)	620M (600—620)	Baluja et al. (1973)
Ebro delta Feb. 1970	3H	200M (130—200)	<0.05	140M (70—140)	60M (60—60)	1,400M (1,400—1,800)	
Barcelona Mar. 1968, Feb. 1970	8H	325M (50—700)	100M (7—200)	35M (20—75)	200M (15—460)	M n.d. (n.d.—1,500)	
Blanes Jan. 1970	3H	50M (50—75)	8M (8—20)	30M (22—55)	12M (3—12)	570M (400—780)	
Coast Sète—Genoa Jun. 1973, Dec. 1974 except:	57H	60$\bar{x}$* (7—400)	30$\bar{x}$* (3—180)	10$\bar{x}$* (n.d.—55)	20$\bar{x}$* (n.d.—130)	300$\bar{x}$* (35—1,000)	Marchand et al. (1976)
Marseilles	5H	260M* (80—2,100)	150M* (55—1,300)	40M* (5—200)	80M* (20—410)	1600M* (700—2,500)	
Toulon	5H	550M* (20—700)	250M* (25—250)	35M* (n.d.—170)	60M* (4—150)	1,200M* (520—2,700)	
San Remo	4H	600M* (90—800)	220M* (40—540)	50M* (20—285)	40M* (25—200)	200M* (40—490)	
Étang Thau (Sète) Mar., May 1976	5H	95M* (60—230)	20M* (10—170)	25M* (20—45)	20M* (15—55)	85M* (30—500)	Alzieu et al. (1976)
Étang Berre (Marseilles) Mar., May 1976	2H	(20—40)*	(3—4)*	(7—15)*	(9—20)*	(110—185)*	
Toulon	2—3H	35M* (25—45)	9M* (2—15)	12M* (10—25)	14M* (8—15)	350M* (300—650)	
Istrian coast to Losinj Oct. 1972, Dec. 1975	62H	50M (n.d.—625)	20M (n.d.—375)	13M (n.d.—88)	15M (n.d.—160)	35M (n.d.—390)	Picer et al. (1977)

M = median, $\bar{x}$ = mean, values in brackets are ranges; n.d. = not determined.
* FW values are calculated as 0.2 DW values.

Atlantic coast (Vigo and Santander) have been sampled extensively (Baluja et al., 1973; Franco-Soler, 1973; Franco and Fernandez, 1976). DDT_t is highest near the Ebro delta, followed by Barcelona, Blanes, Castellon and Cadiz (Table XVIII). The lowest values are from Vigo ($17 \mu g \, kg^{-1}$ FW) and Santander ($12 \mu g \, kg^{-1}$ FW) on the Atlantic coast. Alzieu et al. (1976) analysed DDT's and PCB's in *Mytilus*, *Crassostrea gigas* and *Patella vulgata* in the Atlantic and *Mytilus* in the Mediterranean. Local contamination resulted in high values in several locations on the Atlantic coast. In the Mediterranean, mussels from the Étangs de Leucate and de Thau showed very low to low concentrations of PCB's ($90—420 \mu g \, kg^{-1}$ DW). Relatively high values were found in the Sète Channel ($2465 \mu g \, kg^{-1}$ DW) and in the Toulon roads ($3160 \mu g \, kg^{-1}$ DW). Duursma et al. (1974), Marchand et al. (1975, 1976) collected samples from 17 stations between Sète and Genoa, and De Lappe et al. (1973) from three stations on the French Mediterranean coast (Grau du Roi, Marseilles and St. Tropez). DDT_t varied from about 6 to $1200 \mu g \, kg^{-1}$ FW. Marchand et al. (1976) observed a seasonal variation of residue levels at different stations. High values near large industrial centres and river outflows are not surprising, while the high DDT concentrations found near San Remo are probably due to the high-intensity flower cultivations because DDT may still be used since this crop is not intended for consumption. These authors suggest three reasons for this seasonal variation: (1) change in pollution level; (2) variation in the environment parameters; and (3) seasonal changes in the life cycle of the mussel. The latter is especially important since it is well known that the lipid content undergoes large fluctuations during the annual reproduction cycle. Except for the mussel data from Stirn et al. (1974) and Picer et al. (1977), who analysed mussels along the Yugoslavian coastline from the Losonj area to the Istrian area, no additional values on *Mytilus* have come to light.

Very few other benthic and coastal organisms have been analysed. In the Gulf of Cadiz, Franco and Fernandez (1976) found low concentrations in *Carcinus* sp. (DDT_t: $\sim 8 \mu g \, kg^{-1}$ FW; PCB_t: $45 \mu g \, kg^{-1}$ FW) and in *Mullus barbatus* ($\sim 22 \mu g \, DDT_t \, kg^{-1}$ FW; $67.5 \mu g$ PCB's kg^{-1} FW). In the Castellón and Barcelona areas *Carcinus* had considerably higher DDT_t ($65—130 \mu g \, kg^{-1}$ FW) and PCB ($1100—1300 \mu g \, kg^{-1}$ FW) body content; that of *Mullus barbatus* was only slightly higher: $115—230 \mu g \, DDT_t$, $290—865 \mu g$ PCB's kg^{-1} FW. The only other data on benthic fishes are from the Adriatic (Picer et al., 1977). These authors determined levels in a few specimens of fishes such as *Diplodus annularis*, *Oblada melanura*, *Boops boops*, *Pagellus erythrinus*, *Mugil auratus*, *Mullus barbatus*, etc. The DDT_t levels ranged from non-detected to about $900 \mu g \, kg^{-1}$ FW and the PCB levels showed a wide range of lower values. The highest contents were found in the organisms from the Losinj area. The few data on chlorinated hydrocarbons in benthic organisms clearly show the influence of local pollution sources, such as large industrial centres, towns and rivers. Picer et al. (1977) observed that even very small settlements could significantly contaminate local biota.

RELATIVE IMPORTANCE OF HEAVY METALS AND CHLORINATED
HYDROCARBONS

Several panels and working groups have classified heavy metals and chlorinated hydrocarbons according to their potential hazards to man and marine
organisms (e.g., N.A.S., 1973; G.E.S.A.M.P., 1976). The heavy metals
usually are ranked in the following order: $Hg > As$, Cu, Cd, Zn, Cr, $Mn > Fe$,
Ti, Pb. Chlorinated hydrocarbons and organo-mercury compounds rank
before other organics.

Estimating the effects of heavy metals and chlorinated hydrocarbons on
man is not easy. The data-base for an evaluation of potential hazards consists
of accidental contamination of humans and toxicity experiments on animals.
The exposure dose of accidental contamination is often difficult to estimate
and the extrapolation of toxicity data obtained from animal bioassays to
man carries a considerable uncertainty.

F.A.O. (1973) and W.H.O. (1973, 1976a, b, 1977) expert consultations
have critically reviewed the available data and suggested provisionally tolerable weekly intakes for several pollutants or have made data available for
their estimation (also Lu, 1973; Vettorazzi, 1975). From the average and
maximum amount of fish consumed and the weekly tolerable intakes, the
maximum tolerable concentration (MTC) of a pollutant in marine foods can
be calculated. The MTC of a pollutant in marine foods can then be compared
with the concentrations actually observed. If the MTC calculated is several
orders of magnitude greater than the actual concentration level observed, it
is very unlikely that the pollutant presents a hazard. A calculated MTC level
of the same order of magnitude as the concentration observed indicates that
this pollutant should be given priority in future research and monitoring
programmes in order to ascertain if the pollutant really presents a hazard or
indicates only a high natural level in the biota and the environment.

The effects on marine organisms can be assessed using the "safe concentration" or "minimal risk concentration" derived from laboratory experiments with marine organisms or field observations. Since at present only
very few toxicity data are available on Mediterranean species, the minimal
risk concentrations suggested by the Committee on Water Quality Criteria
of the U.S. National Academy of Sciences (N.A.S., 1973) are employed.
At a later date these data will have to be substituted with toxicity data from
Mediterranean species. Similarly to the assessment of potential hazards to
man, this preliminary assessment will show where the priorities for future
work may be set. It is not possible to discuss here all possible limitations in
the assessments used. However, it may be worth while recalling that environmental criteria are mostly derived from short-term (non-chronic) bioassays
carried out with unrealistically high pollutant concentrations which consider
only the effects of dissolved pollutants and not those ingested in food. The
most sensitive species are not used in these bioassays because their culture

conditions are unknown. Most bioassays are conducted on one particular phase of the life cycle, omitting the most sensitive reproduction phase. Synergistic and antagonistic effects of pollutants are not studied, or again only in unrealistically high concentrations. Finally, very little is known about effects in situ on ecosystems and communities.

INTAKE OF HEAVY METALS FROM MARINE FOODS

Marine foods do not constitute a large part of the diet of the general public in the Mediterranean area. Except for Spain ($\sim 17\%$) and Tunisia ($\sim 12\%$), less than 10% of the animal protein intake is of marine origin (Table XIX). Much fish is imported into the Mediterranean countries. From the few estimates available, the percentage of Mediteranean fish consumed by the general public in the area ranges from 10 to 50%.

Since good habits are variable, three categories of persons can be distinguished in a first approximation:

(1) The general public which consumes the national average of fish food of Mediterranean and non-Mediterranean origin as offered on a typical market.

(2) The general public which consumes the national average of fish, but exclusively of Mediterranean origin.

(3) The so-called critical groups which, because of special food habits and/or because of easy access to fish (fishermen, fish vendors, workers in fish-food processing industries and their families) consume considerably larger quantities.

In order to make the estimation conservative, only the two last population groups will be considered.

For the average consumption of F.A.O. estimates (Table XIX) are used. For the critical groups we have to rely on the few estimations available for Western Europe which range from 300 to 800 g of fish per day (Jokelainen, 1967; Preston, 1967; Bernhard et al., 1972; Paccagnella et al., 1973). The higher estimate corresponds to about 5.5 kg of fish per week. Taking Spain, the highest consumer of the Mediterranean countries, as an example the following estimations can be made:

(1) General public of Spain: 270 g of marine foods per week.

(2) Critical groups: 5.5 kg of marine foods per week.

Mercury

The higher than average concentration of Hg in large fishes and in fishes from the Mediterranean obviously raises the question of possible health risks. The maximum permissible legal Hg concentration (0.5—1.0 ppm FW) in fish and marine foods was established after the Minamata incident. This gives us the opportunity to discuss whether or not these limits are justified.

TABLE XIX

Consumption of fishery products in Mediterranean countries (data for 1972–1974 from F.A.O. (1977) and A. Crispoldi (pers. commun., 1977)

Country	Population 1974 (millions)	kg edible FW[*2] per year	kg edible FW[*2] per week	protein per day (9) total animal[*3]	protein per day (9) fish (%)	Sea food of Mediterranean origin (%)
Algeria	15.78	0.92	0.018	10.4	5.7	?
Cyprus	0.65	2.3	0.044	40.6	4.2	?
Egypt FshW[*1]	35.82	0.9	0.016	10.3	4.85	?
SW[*1]		0.65	0.0125		2.9	?
French FshW[*1]	52.115	0.15	0.003	61.2	<0.2	
SW[*1]		7.33	0.141		6.5	10
SWI[*1]		2.75	0.053		1.3	
Greece FshW[*1]	8.81	0.62	0.012	44.7	0.9	
SW[*1]		6.15	0.118		8.7	50
Israel FshW[*1]	2.0	2.3	0.044	52.8	7.2	
SW[*1]		5.2	0.1			
Italy FshW[*1]	54.9	0.21	0.004	42.8	0.2	
SW[*1]		4.38	0.083		7.2	?
Lebanon	2.7	1.1	0.023	18.0	5.0	?
Libya	2.15	2.13	0.04	20.1	9.45	?
Malta	0.32	4.3	0.083	40.6	8.6	?
Morocco	16.5	2.18	0.042	20.1	8.0	?
Spain FshW[*1]	34.6	2.31	0.045	44.9	0.2	
SW[*1]		13.91	0.268		16.7	10
SWI[*1]		4.91	0.095		4.5	
Syria	6.8	0.51	0.01	13.7	2.9	?
Tunisia	5.49	3.25	0.062	13.7	12.4	?
Turkey FshW[*1]	37.95	0.17	0.004	17.9	0.6	
SW[*1]		2.18	0.042		7.8	3
Yugosl. FshW[*1]	20.96	0.472	0.009	28.7	1.7	
SW[*1]		1.06	0.021		2.4	?

Kg FW equiv. = kg of fresh weight equivalent: processed fish products have been converted into the original fresh weight.
[*1] FW = freshwater fish; SW = marine fish; SWI = marine invertebrates. Countries without subidivision into SW, etc., marine fish only.
[*2] Edible portion of fish has been estimated as being 0.5 of live weight equivalent.
[*3] Total animal = total animal protein available.

The Hg contribution from terrestrial foods has to be subtracted from the tolerable intake of 200 µg Hg-M (methyl mercury) per week. Cigna-Rossi et al. (1976) estimated for the average Italian an intake of 50–85 µg Hg per week. This estimate can be compared with Schelenz and Diehl's (1973) value of 70 µg Hg per person per week for the F.R.G. and 35–70 µg Hg for the U.K. (Cohen, 1974). Bread and cereals contribute more than 50% of the Hg intake from terrestrial foods. Although on the average only 80% of the Hg in fishes is methyl mercury (p. 162), we will assume that all Hg is methyl mercury (Hg-M). So our allowance for marine foods is about 150 μg Hg-M and on these bases the following estimations can be made:

(1) General public (Spain): in 270 g of marine foods the Hg concentration shall not exceed 150 μg Hg-M, hence the MTC in marine food must not exceed 550 μg Hg kg^{-1} FW.

(2) Critical groups: in 5.5 kg of marine foods the Hg concentration shall not exceed 150 μg Hg-M, hence the MTC in marine food must be less than 40 μg Hg kg^{-1} FW.

Comparing the MTC in marine foods for the general public with the levels actually found in fishes, mollusks and crustaceans (Tables VI—XIII) shows that, in general, those levels are not exceeded. The MTC estimated in this way for the general public (Spain) corresponds at the same time to the lower legal limit (0.5 ppm FW) in force in many countries. However, an apparent health risk seems to exist for the critical groups, since most levels listed in the tables are well above the calculated MTC. In addition, Hg concentrations several orders of magnitude above the MTC for critical groups have been observed, particularly in large-size pelagic fish species, such as the commercially very valuable tunas and swordfish, but also in smaller fishes (*Mullus barbatus, Merluccius merluccius,* or *Pagellus erythrinus*).

The locally confined influence of the anthropogenic Hg sources so far investigated, the Solvay plant in Tuscany and the Athens outfalls, shows that large pelagic fishes and mammals (tuna, swordfish, porpoises and whales) which migrate over large parts of the Mediterranean and the Atlantic could not have been contaminated by anthropogenic sources. Also, these sources can have little or no influence on the body content of smaller pelagic fishes (anchovy, sardine, mackerel, etc.) and benthic fishes (mullet, hake, etc.) which are caught 15—30 km from the coast and in areas far from anthropogenic land-based sources. All results obtained on these and similar fishes in the Mediterranean and elsewhere (G. E. Miller et al., 1972; C. L. Peterson et al., 1973; Mackay et al., 1975; Schultz et al., 1976) strongly indicate that the relatively high Hg body contents are of natural origin and must have existed in these fishes throughout geological times. Consequently, fishermen and their families must have been exposed to these relatively high Hg concentrations for centuries. It is, therefore, of great interest that the possible health risk has been studied in some critical groups.

Paccagnella et al. (1973) selected the population of Carloforte (Sardinia) for such an epidemiological study, because its average consumption of marine foods was 3.5 times (300 g per week) the Italian national average (Table XIX) and because fresh tuna meat was available during the summer months. From 6200 residents 195 persons chosen at random agreed to give information about their food habits, to take a medical examination and to allow blood and hair analyses. About 65% of these eat fishery products more than three times a week: 11.7% consumed seven and more fish meals per week and 1.5% as many as 13—14 meals per week. Assuming 100 g edible fish as a portion, fourteen meals equal 1.4 kg edible fishery products. The Hg analyses of edible parts of tuna, fish and shellfish selected at random for the typical diet resulted in the following concentrations:

	Hg_t ($\mu g\, kg^{-1}$ FW)
Tuna	1,230 (50—2,800)
Fishes and shellfishes	330 (10—490)

During summer, a typical diet (including tuna) contained about 500 μg Hg kg^{-1} FW. During the rest of the year the Hg concentration should have been 330 μg Hg kg^{-1} FW, since no fresh tuna was available. In summer the average weekly intake of Hg is therefore equal to 0.3 kg × 500 μg Hg = 150 μg Hg per person per week and during the other seasons equal to 100 μg per week. The high consumers with fourteen meals per week have an estimated weekly intake of 700 μg Hg in summer and 460 μg Hg during the rest of the year. The higher than average Hg intakes of marine foods is reflected in the higher Hg concentration in blood and hair as compared to a control group from the city of Ferrara:

	n	Hg (μg per 100 ml blood)	Hg (μg per kg hair)
Carloforte	195	7 (0.001—25)	11,000 (n.d.—60,000)
Ferrara	15	2 (0.001—10)	1,850 (n.d.—10,000)

Three persons from Carloforte had blood levels between 20 and 25 μg per 100 ml blood and two persons levels in hair between $5 \cdot 10^4$ and $6 \cdot 10^4$ $\mu g\, kg^{-1}$ (= 50—60 ppm). Medical examinations did not reveal any symptoms of possible mercury poisoning. A similar comparison between persons consuming more than the average amounts of fish from the cities of Marina di Pescara and Pescara, and persons eating less than the average from Santa Lucia, located in the mountains of the Abruzzi, resulted in the following hair concentrations (Caracciolo et al., 1972):

	Hg (ppm)
Marina di Pescara	11.2 (3.7—23.85)
Pescara	4.3 (1.5—13.65)
Santa Lucia	0.8 (0.4—1.3)

Another interesting case concerns Japanese tuna fishermen and retailers. Unlike the population of Carloforte, they consume high amounts of tuna during most of the year and not only in the summer months (Doi and Ui,

1975). Of 34 retailers, 22 eat 100—200 g of tuna meat daily, as well as 70—300 g of shellfish and fish meat. One individual consumed daily 200 g of tuna plus 1000 g of other marine foods. The daily tuna consumption of tuna fishermen aboard ship ranged from 50 to 400 g for periods of between 130 and 180 days. Considering 500 μg Hg kg^{-1} FW an average concentration and 150 g tuna an average daily consumption rate, the weekly intake from tuna is about 500 μg Hg for tuna fishermen. The retailers obtain an additional Hg intake of about 140 μg per week from the other marine foods which contain less Hg than Mediterranean specimens — about 100 μg Hg$_t$ kg^{-1}. The high Hg intake, especially from tuna, is reflected in high hair and blood concentrations. The Hg in the hair of the tuna fishermen ranged from 25 to 45.7 ppm. The mean concentration in the hair of the retailers was 25.7, with a range from 6.4 to 44.4 ppm, while blood concentrations averaged 10 μg Hg per 100 ml with a range from 4.5 to 17.6 μg per 100 ml. One individual had at one time 64.7 ppm Hg in the hair. Again, no symptoms could be detected. Earliest effects in the most sensitive group in an adult population are estimated to appear following a long-term intake of 1500—3500 μg Hg per 70-kg man per week, with a blood level of 20—50 μg Hg per 100 ml and a hair concentration between 50 and 125 ppm (W.H.O., 1976a). In view of these estimates it is not surprising that no symptoms were observed, since the blood and hair concentration are below — and only in very few cases near — the lower range of the earliest effect levels.

The recent findings that Se acts antagonistically to Hg and that both in man and in some marine organisms high Hg concentrations are associated with high Se concentrations, may point to a protection mechanism against Hg poisoning. Ever since Parizek and Ostadalova (1967) had observed that small amounts of selenite could clearly contract the toxicity of HgCl$_2$, many other workers have investigated the antagonistic effects of Se on inorganic and organic Hg compounds (Iwata et al., 1973; Stilling et al., 1974; Stoewsand et al., 1974). In view of the high body levels found in several larger marine organisms, investigations which showed that methyl mercury intoxication is considerably reduced in experimental animals when tuna is supplied in the diet are of special interest. Ganther and Sunde (1974) and Ohi et al. (1976) suggest that the naturally occurring Se in tuna has a protective action against methyl mercury toxicity, even in "high-Hg tuna".

Kosta et al. (1975) and Cigna-Rossi et al. (1976) have shown that mercury mine workers from Idrija and Monte Amiata who had high Hg concentrations in their organs and tissues also have high Se concentrations (Table XX). The molar ratio seems to approach 1. Controls from the general "unexposed" population have Hg/Se molar ratios of 0.002—0.02. The most extensive data available on marine organisms are for the largest teleost fish *Makaira indica* from the Pacific (Mackay et al., 1975) and for marine mammals from the North Sea (Koeman et al., 1973, 1975). Again the molar ratio is about 1 for specimens with high Hg concentrations, while for fishes with lower Hg

TABLE XX

Mercury and selenium concentrations and Hg/Se molar ratio in humans and marine organisms

	n	Hg (μg kg^{-1} FW)	Se (μg kg^{-1} FW)	Molar ratio	Reference
Humans (Mediterranean):					
Mine worker (Idrija)					Kosta et al. (1975)
thyroid	5	26,500M (5,000—100,000)	12,300M (1,500—41,100)	0.95M (0.87—1.3)	
kidney	4	6,750M (2,300—11,400)	2,300M (1,600—5,100)	1.0M (0.6—1.2)	
liver	4	200M (40—370)	380M (360—400)	0.22M (0.04—0.4)	
Mine worker (Monte Amiata)					Cigna-Rossi et al. (1976)
kidney	3	2,770M (240—3,000)	525M (265—680)	1.76M (0.36—2.1)	
liver	3	185M (50—990)	125M (125—140)	0.6M (0.15—3.2)	
brain	3	80M (25—195)	60M (45—75)	0.5M (0.25—1.0)	
Idrija population					
82-yr.-old resident					Kosta et al. (1975)
thyroid	1	14,400	5,720	0.9	
kidney		760	700	0.44	
liver		65	210	0.12	
4-month-old infant					
thyroid	1	25	160	0.06	
kidney		610	470	0.5	
liver		85	120	0.3	
Control population					
thyroid	2	(2.5—40)	(460—790)	(0.002—0.02)	
kidney		(11—370)	(540—820)	(0.01—0.2)	
liver		(11—40)	(170—190)	(0.02—0.08)	

Marine organisms (Mediterranean):

Plankton	22	130 (50—680)	3,700 (1,900—6,400)	0.015	Kosta et al. (1978)
Nephrops norvegicus	5	1,650 (1,100—2,700)	1,430 (390—2,600)	0.47	
Murex sp., soft part	2H	(15—45)	(40—55)	(0.15—0.3)	Fowler and Benayoun (1977)
Mytilus galloprovincialis, soft part	1H	330	890	0.15	Fowler et al. (1976a)
Ostrea edulis, soft part	1H	40	610	0.02	Kosta et al. (1975)
Octopus vulgaris	1	70	370	0.08	
Mustelus vulgaris	3	1,850 (890—3,550)	460 (410—550)	1.6	
Raja clavata	1	670	450	0.6	
Torpedo marmorata, adult	1				
liver		1,150	1,980	0.23	
kidney		400	670	0.24	
tail muscle		650	260	1.0	
young	2				
liver		(150—180)	(220—230)	(0.27—0.3)	
tail muscle		(180—220)	(280—420)	(0.17—0.32)	

Marine organisms (non-Mediterranean):

Makaira indica					
muscle	42	7,300 (500—16,500)	2,200 (400—4,300)	1.3	Mackay et al. (1975)
liver	40	10,400 (300—63,000)	5,400 (1,400—13,500)	0.75	
Marine mammals					
liver	>25	(600—100,000)	(1,400—13,500)	1.1	Koeman et al. (1973)
Clupea heringa	12	?	?	0.02	
Scomber sp.	7	?	?	0.01	
High-Hg tuna	?	2,870	2,900	0.4	Ganther and Sunde (1974)
Low-Hg tuna	?	320	1,900	0.07	

For explanation of symbols see footnote to Table I.

concentrations, *Solea*, *Clupea* and *Scomber*, the molar ratios are about 0.02. Hg/Se ratios from the few Mediterranean samples so far analysed range from 0.01 to 0.3 for low Hg body contents and from 0.5 to 1.6 for higher Hg body contents.

The Hg concentrations in several fishes clearly surpass the legal limits of 0.5—1.0 ppm FW. Since the high concentrations in pelagic fishes and mammals are undoubtedly of natural origin, and since these organisms have been consumed for a long time without any ill effects, legal limits should not apply to foods prepared from them. If it is confirmed that these high-Hg foods also have high Se concentrations, one can rely on the antagonistic effects of Se further to reduce potential mercury hazards. The fact that consumers of large amounts of marine foods have not shown any symptoms is also reassuring. For coastal fishes, which may have high Hg body contents due to natural and anthropogenic sources, the Hg problem is more complicated. Only further research can tell us how to distinguish between anthropogenic and natural levels. Probably the direct release of organic Hg compounds into coastal environments is more hazardous than the discharge of inorganic compounds which resemble natural physicochemical forms.

Other heavy metals

The other heavy metals are not so well investigated as Hg. Only for As and Cu have provisional tolerable intakes (24.5 mg As per 70-kg man per week; 245 mg Cu per 70-kg man per week) been suggested. A calculation, following the Hg example, results in a MTC in marine foods for critical groups of about 4.5 mg As kg^{-1} FW and even as much as 45 mg Cu kg^{-1} FW. Although the As concentrations are relatively high, the safety margin between maximum concentration and observed concentration is great. This margin is still further increased since, like Hg, most of the As is in an organic form in marine organisms (Braman and Foreback, 1973; Edmonds and Francesconi, 1977). However, unlike methyl mercury, the methylated forms of As are much less toxic than inorganic As compounds on which the intake limits are based.

The provisional weekly tolerable intakes for Cd and Pb are about 400 μg Cd and 700 μg Pb for a 70-kg man. Until recently accepted Pb levels in marine organisms were in the range 0.2—1 mg Pb kg^{-1} FW. Now Patterson and Settle (1976) have shown that the actual Pb levels are much lower: 0.5—50 μg Pb kg^{-1} FW. The previously accepted values were much higher owing to contamination from terrestrial sources during sampling, sample preparation and analysis. Although marine foods still may contain considerable amounts of Pb, this Pb is not of marine origin and its source has to be sought in the procedure of food handling rather than in the marine food itself. With the development of new techniques, the concentration of Cd has also considerably decreased in marine organisms. Consequently, the

TABLE XXI

Summary of hazardous minimal risk concentration and application factor (AF) for marine organisms and "normal concentration" of inorganic substances in sea water (N.A.S., 1973; Robertson and Carpenter, 1976)

	Hazardous concentration	Minimal risk concentration	"Normal concentration"	96-dLC-50% AF	Note
Al	1,500	200	~5	0.1	
NH₃	400	10		0.1	
Sb	200	20	0.2	0.02	
As	50	10	10	0.01	
Ba	1,000	500	20	0.05	
Be	1,500	100	0.003	0.01	
Bi	?	?	0.000 02	?	
B	5,000	5,000	4,500	0.1	
Br (molecular ionic)		100 / 100,000	6,800	—	
Cd	10	0.2	0.2	0.01	synergetic effects of Cu to be expected
Cl (free residual)	10	?		0.1	
Cr	100	50	0.04	0.01	
Cr, oyster areas	50	10			
Cu	10	10	1	0.01	acts synergistically with Zn
CN⁻	5	5	?	0.1	
F	1,500	500	1,400	0.01	
Fe	300	50	~3	0.1	
Pb	50	10	0.02	0.02	
Mn	100	20	2	0.02	
Hg	0.1	0.1	0.1	0.05	acts antagonistically with Se
Mo	?	?	10	0.05	
Ni	100	2	5	0.02	
P (elemental)	1	?	~70	0.05	
Se	10	5	0.45	0.01	acts antagonistically with Hg
Ag	5 (0.25)*	1 (0.01)*	0.1	0.05	synergetic with Cu, 10 times as toxic as Hg
H₂S	10	?		0.1	
Th	100	50	~0.0004	(0.05)	20-d LC-50% recom. with AF = 0.05
U	500	100	3.3	0.01	
V	?	?	1.5	0.05	
Zn	100	20 (<1)*	10(1)*	0.01	when synergetic with Cu and Cd, AF = 0.001

* Values in brackets refer to the ionic form.

intake of Cd for marine foods is also negligible. The difficulties encountered in the accurate determination of Pb and Cd also exist for other heavy metals. It is likely, therefore, that the levels reported in the literature will decrease in future.

Effects of heavy metals on marine organisms

In Table XXI the "hazardous concentrations" and "minimal risk concentrations" are compiled for marine organisms, together with the "typical concentrations" of unpolluted seawater. A first evaluation of the possible hazards to marine organisms can be obtained by comparing the minimal risk concentration with the typical concentration. If the difference between these concentrations is great, the likelihood that seawater will reach the minimal risk concentration is small. For some elements minimal risk concentration and typical concentration are the same (Hg, As), for others the difference is small (Cd, Mn), and for some (Zn, Ni) the minimal risk concentration is even smaller than the typical concentration. In these latter cases one physico-chemical form of the element is more effective than the total concentration listed as "typical concentration". In general, the ionic form of heavy elements is more toxic than the complexed or organic bound form. Since toxicity assays are generally carried out with the soluble salts of the element, the minimal risk concentration determined can be lower than the total concentration of natural seawater, which comprises all physico-chemical states present. The toxic form is, in general, only a small fraction of the total amount present in natural seawater. Confronting the concentrations of heavy metals observed (Table I) in the open Mediterranean with "minimal risk concentrations" (Table XXI) shows that only in very few cases are the minimal risk concentrations reached. In coastal waters (Table II), on the other hand, minimal risk concentrations are often exceeded. Here, however, we have to take into consideration that in most cases the water samples for the determinations were taken in direct vicinity of heavy pollution sources and are not typical for large areas of the coastal waters. Further data, especially collected systematically, will only enable us to estimate the extent to which the Mediterranean coastal waters are actually polluted with heavy metals.

INTAKE OF CHLORINATED HYDROCARBONS FROM MARINE FOODS

In a way similar to the mercury example, the potential effects of chlorinated hydrocarbons on human health could be estimated. However, the health criteria for chlorinated hydrocarbons are much less certain. In W.H.O. (1976b) no tolerable intakes for PCB's have been suggested, since on the one hand in man, the most sensitive species, effects have been observed at intakes of about 3.5 mg per 70-kg man per week, while on the other hand occu-pationally exposed workers were apparently exposed to much higher doses without showing any symptoms.

For DDT the situation is similar. Here 2.45 mg of DDT has been pro-posed as the tolerable weekly intake (Vettorazzi, 1975), although no effects have become known on humans despite several decades of professional expo-sure. Therefore, an estimation based on this figure will give only a very rough idea. Assuming that 50% of the DDT in human diet is of terrestrial origin, the 270 g consumed by the general public (Spain) correspond to a MTC of 4.5 mg DDT kg^{-1} FW marine foods. For critical groups the MTC is 225 μg DDT kg^{-1} FW. Comparing these MTC's with the concentrations actually observed in fishes, shows that the MTC for the general public are about 20 times higher than the highest body content. For critical groups the MTC is of the same order of magnitude as the levels observed. For dieldrin, including aldrin, the MTC's for the general public and critical groups are 90 and 15 μg kg^{-1} FW, respectively. The few data available on concentrations in marine organisms (Table XVII) show that again the MTC is of the same order of magnitude as the levels observed.

The PCB effect cannot be evaluated, even approximately. It can only be pointed out that the PCB body content in fishes from certain part of the Mediterranean is remarkably high. Only further investigations can tell us if the chlorinated hydrocarbons, at concentrations encountered in these marine food really present a health hazard.

Effects of chlorinated hydrocarbons on marine organisms

The scarcity of data on levels of chlorinated hydrocarbons in seawater does not allow any real evaluation of possible risks. The open-sea data (maxi-mum value: 9 ng PCB l^{-1}) of Elder and Villeneuve (1977) (Table XV) can be confronted with the "minimal risk concentration" of PCB's which, depending on the degree of chlorination, ranges from 1 to 10 mg PCB l^{-1} (N.A.S., 1973). The only DDT levels available (Table XVI) are not representative. They illustrate only the situation in a very contaminated riverlet near Marseilles. For another approach to evaluating the effects of chlorinated hydrocarbons, the maximum "safe" body content in the prey of fish-eating birds and mammals (N.A.S., 1973) could be used. These limits can be compared with the more abundant data on levels in the Mediterranean biota. N.A.S. (1973) suggests the following values:

	(μg kg^{-1} FW)
total DDT's	50
total PCB's	500
sum of aldrin + dieldrin + endrin + heptachlor epoxide	5
Other chlorinated hydrocarbons such as lindane, chlordane, etc.	50

Turning to the levels in marine organisms shows that most levels are of the same order of magnitude as the "safe" body content in the prey. Some body contents exceed the "safe concentration". The data are, however, too few and the analytic results, the health and environmental criteria too uncertain, for an estimation of the environmental impact on Mediterranean ecosystems and communities.

CHAPTER 4

HISTORY OF HEAVY-METAL INPUTS TO MISSISSIPPI DELTA SEDIMENTS

John H. Trefry and Robert F. Shokes

INTRODUCTION

Rivers are the major pathway by which the products of natural geological processes and the pollutant inputs of man are added to the oceans. Garrels and Mackenzie (1971) estimate that rivers account for 90% of the total seaward transport of dissolved and suspended solids. The major river of the U.S.A., the Mississippi, drains more than 40% of the conterminous U.S.A., including numerous municipalities and scores of industrial facilities. Our study was undertaken to determine the present-day flux of selected heavy metals from the Mississippi River to the Gulf of Mexico and the history of metal inputs to Mississippi delta sediments.

The Mississippi River drainage basic stretches from New York to Montana and from Canada to the Gulf of Mexico (Fig. 1). The river carries about 60% of the total dissolved solids (Leifeste, 1974) and 66% of the total suspended solids (Curtis et al., 1973) transported to the oceans from the continental U.S.A. Mississippi River sediment, which accounts for about two-thirds of the total sediment delivered to the Gulf of Mexico, is deposited in an area at least as large as that outlined in Fig. 1. The massive sediment input of the river $(28 \cdot 10^7 \, t \, yr.^{-1})$ has built a series of deltaic complexes over the past 6000 years (Fisk et al., 1954; Coleman and Gagliano, 1964). The modern birdfoot (or Balize) delta is an 1800-km^2 platform on which 113 km^3 of sediment has been deposited since the early 1500's (Fisk et al., 1954). Approximately two-thirds of the present river sediment load is carried out three major passes of the main stem of the Mississippi River while the remaining one-third is diverted to the Atchafalaya River about 500 km upstream. The major passes of the Mississippi River have advanced to the edge of the continental shelf and are presently depositing sediment in progressively deeper water. Atchafalaya River sediment, however, is now filling Atchafalaya Bay where rapid subaerial land formation is observed.

Estimates of industrial, municipal and agricultural waste inputs to the Mississippi River (Everett, 1971) suggest that the river pollutant load is large and diverse. Organic contaminants have most often been cited as troublesome since the 1950's fish kills from the insecticides, endrin and heptachlor (L.W.F.C., 1958—1959). Recent data show that river phenol concentrations frequently exceed public water-supply criteria (U.S.A.C.E.,

194

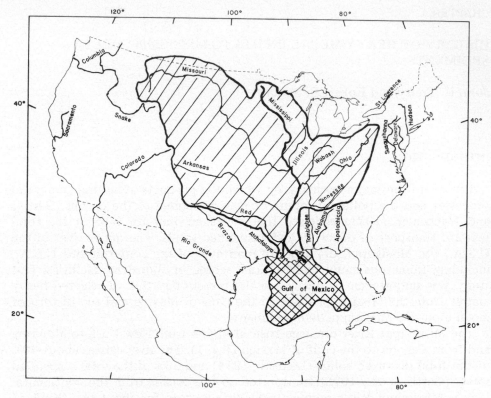

Fig. 1. Mississippi River drainage basin (from Judson and Ritter, 1964) and Mississippi distributive province in the Gulf of Mexico (from D. K. Davies, 1972).

1975) and that the river has a significant anthropogenic burden of light hydrocarbons (J. M. Brooks, 1976) and phthalic acid esters, DDT and PCB's (Giam et al., 1976). Added to this load is the domestic sewage from almost 2 million people in the lower Mississippi area alone. Because many of the organics use oxygen as they degrade in the river, a progressive decrease in dissolved oxygen has been measured downstream from Baton Rouge to the Gulf (Everett, 1971).

In addition to organic substances, an estimated $18 \cdot 10^6$ kg of inorganic wastes are discharged into the lower Mississippi River daily (Everett, 1971). This flux is equivalent to 7% of the total dissolved load at average river flow and 21% of the total at low flow. Much of this discharge is composed of industrial acids and such wastes are known to contain various heavy metals (N.A.S., 1975). Trefry and Presley (1976a, b) found evidence of Pb and Cd pollution in sediments from an extensive area of the Mississippi delta. Results from the present study confirm and expand the initial observations and provide a more reliable time scale for examining the history of the Delta metal pollution.

METHODS

Samples of suspended particulates, filtered water, and sediments were collected for this study from the Mississippi River, adjacent fresh- and saltwater marshes and the Mississippi's marine delta region (Fig. 2). Analysis of these phases for concentrations of selected heavy metals was augmented by measurements of water nutrients and dissolved solids, suspended particulate load, sediment and particulate grain-size distribution, sediment accumulation rate (using ^{210}Pb chronometry) and organic and inorganic carbon content. Nine stations were sampled for filtered water and suspended particulates within the Mississippi River and across the river water—seawater boundaries. Marine sediments were collected at eight locations within the delta "front" where most of the river-borne solids are deposited (Fig. 2).

River and marine waters were collected with 30-l Niskin bottles and their suspended particulates separated in gram quantities by 0.4-μm filtration of up to 120 l, using a closed vacuum-driven system. Waters for dissolved metal analysis were collected either by closed-system pumping well away from the ship or by hand-held plastic bottles from a small rubber boat. These samples were also 0.4-μm filtered, but this was done in smaller quantities and under more controlled conditions. Particulates were maintained

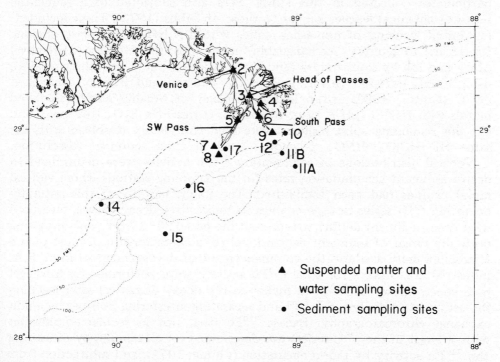

Fig. 2. Mississippi River and delta water, suspended matter, and sediment sampling sites.

by freezing whereas the water samples were acidified just after filtration. Sediments were collected with either a 7-cm diameter gravity corer or a 50-cm^2 NEL-type box corer.

River water dissolved metal concentrations were analyzed in most samples by direct injection into a flameless (graphite furnace) atomic absorption spectrophotometer. In some cases, pre-analysis concentration was performed by solvent extraction with ammonium pyrrollidine dithiocarbamate (APDC) and methyl isobutyl ketone (MIBK) and back extraction into nitric acid (R. R. Brooks et al., 1967). In the low-ionic-strength river waters, the metals Fe, Cr, Cu and Mn were determined quite well using direct injection, Pb and Cd less reliably and Ni not at all.

Total sediment and suspended particulate metal concentrations were determined by atomic absorption spectrophotometry after dissolution with HF, $HClO_4$ and HNO_3. Background matrix interferences were compensated for by the appropriate application of method-of-additions, instrumental background correction, matrix-matched standards and/or simultaneous monitoring of absorbance at nearby non-resonance wavelengths. Total sediment V and Al were determined by instrumental neutron activation analysis.

In addition to total metal determinations, sediments and suspended particulates collected in this study were also subjected to a sequential series of chemical leaches similar to those of Gibbs (1973). These included: (1) initial leaching of non-ashed solids with $1 N$ NH_4Cl (at pH 7, shaking for 18 hr.) to remove "exchangeable" metal ions; followed by (2) removal of Fe and Mn hydrous oxides (and the metals associated with these oxides), using citrate-buffered (pH 4.7) $Na_2S_2O_4$ solution and 2 hr. of shaking at 60° C (Coffin, 1963); (3) oxidative removal of organic-matter-associated metals with NaOCl (pH 7) prior to another citrate—$Na_2S_2O_4$ treatment; and (4) the remaining solid residues were then completely dissolved with the same HF—$HClO_4$—HNO_3 procedure used in total sediment dissolution.

Vertical distributions of sedimentary ^{210}Pb activity were determined to derive sediment accumulation rates for the sampling stations where vertical metal profiles had been established. The 22-yr. half-life of this naturally occurring ^{238}U series isotope enables its radioactive decay curves, measured with time (sediment depth), to represent the past 125—175 yr., encompassing both the range of sediment depositional rates characteristic of most of the Mississippi delta and also the commencement of development of the U.S.A. industrial era. Assays for total ^{210}Pb activity were performed by low-level beta spectrometry of ^{210}Bi (a higher-energy decay daughter) after leaching the sediments with hot, $6 N$ HCl and separating interfering isotopes by anion exchange chromatography. Excess ^{210}Pb (that not in secular equilibrium with the rest of the ^{238}U series) was determined for all samples by measuring the ^{226}Ra activity by radon emanation (Chung, 1971), and subtraction from the total ^{210}Pb activity.

More detailed analytical methodologies are outlined in Shokes (1976) and Trefry (1977).

MISSISSIPPI RIVER METAL FLUX

Mississippi River dissolved metal concentrations at locations shown in Fig. 2 are considerably below those established for water-quality criteria (Table I). Values from the U.S.G.S. (1975) and our study are also generally lower than those of previous workers (Durum and Haffty, 1961; Kopp and Kroner, 1967; U.S.G.S., 1972, 1973, 1974) and in line with Turekian's (1969) estimates for average river water. These low concentrations are most likely controlled by specific adsorption since the abundant fine-grained suspended material ($100-500 \, mg \, l^{-1}$) and high river pH (7.5—8.1) favor this process.

Temporal variations in Mississippi River dissolved metal concentrations

TABLE I

Mississippi River dissolved and particulate heavy metal concentrations

	Cd	Cr	Cu	Fe	Co	Hg	Mn	Ni	Pb	Zn
Dissolved metal concentrations ($\mu g \, l^{-1}$):										
Mississippi River (this study)	0.1	0.5	2	10		—	10	1	0.2	—
Mississippi River (U.S.G.S., 1975)	1	1	3	—		0.1	—	3	1	10
Average river water (Turekian, 1969)	—	1	7	—		0.07	7	0.3	3	20
Water quality criteria (U.S.E.P.A., 1973, 1976)	10	50	60	300		2	50	100	50	—
Particulate metal concentrations ($\mu g \, g^{-1}$):										
Mississippi River (this study)	1.3	80	45	46,100	21		1,300	55	46	193
Mississippi delta (this study)	1.5	84	56	46,400	21		1,290	56	49	244
Sediment disposal criteria (U.S.E.P.A., 1973)	2	—	50	—	—		—	—	50	75
Average crustal abundance (Taylor, 1964)	0.2	100	55	56,300	25		950	75	13	70

198

ranged from a factor of 2—> 20, depending on the metal. Cr concentrations, for example, varied from only 0.5 to 1.0 $\mu g l^{-1}$ for the periods sampled. Concentration ranges for Mn (~ 1—$30 \mu g l^{-1}$) and Cu (0.3—$4 \mu g l^{-1}$), were significantly greater. Fluctuations in Cd, Fe, Ni and Pb values were inter-mediate of the above extremes. In general, maximum concentrations for each metal were observed during the low river flow September and November periods.

Average metal concentrations in river particulates (Table I) were quite constant [< 10% variability; (standard deviation/mean) x 100%] for June, 1974, February, July and November 1975, sampling periods when the river was near or above mean flow. Differences in metal values among samples from several river locations (Fig. 2) and depths were also consistently below 10% (and frequently < 5%) for each period. During low river flow (September), however, an increase in particulate organic matter content from < 3% (for the other periods) to 25% brought about a corresponding decrease in Fe and Al concentrations due to dilution of aluminosilicates by organic matter. No significant change in particulate Co, Ni or Pb concentrations was observed during low flow, however, Cu, Mn and Zn concentrations were 30—40% higher and Cd values were increased by 80%. Enrichment of particulate Cd, Cu, Mn and Zn at low flow suggests an association of these metals with the increased organic matter and/or surface coatings on the very fine-grained particles being carried at this time.

Relative to average crustal abundances (Table I), Mississippi River par-ticulate Cd, Pb and Zn concentrations are quite high and indicative of an anthropogenic source. Yet, only Zn was found in significantly higher con-centrations than those acceptable for dredged sediment disposal (Table I).

Chemical leaching of the solid phase reveals, in a general way, the modes of particulate metal transport by a river and the relative availability of the metals. Fig. 3 compares data from this study of the Mississippi with that

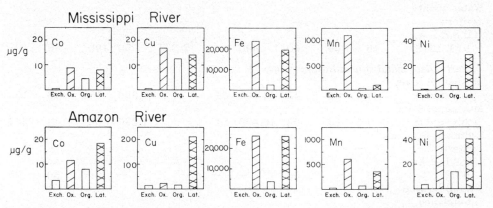

Fig. 3. Trace-metal partitioning in Mississippi and Amazon river suspended matter (Amazon data from Gibbs, 1977).

of Gibbs (1977) for the Amazon. In leaching the Mississippi River suspended matter, the initial 1 N NH_4Cl treatment (at pH 7) removed metal ions which were exchangeable (Fig. 3, Exch.) with the ammonium ion at an ionic strength slightly greater than that of seawater. In all cases $< 2\,\mu g$ Fe per gram of suspended matter were leached with the NH_4Cl, indicating that Fe hydrous oxides were not altered by this treatment. The second solution used, citrate-buffered sodium dithionite at pH 4.7, dissolved "free" Fe- and Mn-oxides (Fig. 3, Ox.). Sodium hypochlorite (pH 7), the third solution in sequence, oxidized organic matter and sulfides (Fig. 3, Org.) and metals associated with these phases. Finally, after particle surfaces have been thrice attacked, residual silicate and other crystalline minerals and their lattice-held (Fig. 3, Lat.) metals are dissolved in $HF-HNO_3-HClO_4$. Chemical leaching methodology used by Gibbs (1973) for Amazon samples is quite similar to that used for the Mississippi study.

Fig. 3 shows the Mississippi and Amazon river particulate metals were predominantly in the oxide and lattice phases. Data for Fe from both rivers were analogous, showing relatively equal oxide—lattice partitioning. Co and Ni distributions were also similar except that the absolute values were a factor ~ 2 higher for the Amazon particulates. Total Mn values are reasonably close for the Mississippi and Amazon, however, the Mississippi suspended matter is present predominantly in oxide coatings with a less significant lattice fraction than found for the Amazon material. Cu values show the greatest discrepancy, with Amazon totals six times greater than those for the Mississippi. This large excess Cu, which is mostly lattice held, may be indicative of significant Cu-bearing minerals in the Amazon basin. Partitioning of Mississippi River particulate Pb showed averages of < 0.5 ppm exchangeable, 20.3 ppm oxide, 14.4 ppm organic/sulfide and 14.8 ppm in the lattice phase.

Calculation of the total annual flux of dissolved and particulate metals from the Mississippi River to the Gulf of Mexico (Table II) shows the overwhelming importance of the particulates. Table II does not include bed-load transport, however, Mississippi bed loads have been estimated to be $< 10-20\%$ of suspended loads (Holle, 1952; Fisk et al., 1954). Since Mississippi River bed sediments have 2—3 times lower metal content than the suspended particulates, they probably contribute $< 10\%$ of the annual flux. As expected, only a small percentage of the metals are carried in solution and thus the primary modes of metal transport by the Mississippi are with the oxide and lattice phases of the river particulates.

Physico-chemical interactions involving heavy metals, reported to occur across the freshwater—seawater interface, may change both the ultimate area of metal deposition and their availability to near-shore marine organisms. Desorptive processes would make metals more available to organisms and delay their removal to the sediments whereas adsorptive processes would have an opposite effect. Comparison of our suspended matter data from the

TABLE II

Annual flux of metals from the Mississipi River to the Gulf of Mexico

Element	Particulate (10^9 g)	Dissolved (10^9 g)	Particulate (% of total)	Dissolved (% of total)
Fe	12,900	5.7	>99.9	0.02
Mn	364	5.7	98.5	1.5
Zn	54	5.7	90.4	9.6
Cr	22	0.3	98.7	1.3
Ni	16	0.6	96.4	3.6
Cu	13	1.1	92.2	7.8
Pb	13	0.3	97.7	2.3
As	4	0.6	87.0	13.0
Cd	0.4	0.06	87.0	13.0

Calculations are based on average water and suspended matter metal data from Table I with the sediment discharge data ($2.8 \cdot 10^{14}$ g yr.$^{-1}$) of U.S.A.C.E. (1950—1975) and estimated water flow at the river mouths ($5.7 \cdot 10^{11}$ m^3 yr.$^{-1}$) from Iseri and Langbein (1974).

Mississippi River with that in saline waters from immediately outside the river mouth (Table I) shows that, with the exception of Cu and Zn, concentrations are essentially the same. These observations argue against extensive desorption of any of these metals and suggest that Cu and Zn levels actually increase in particulates from salt water, perhaps due to the increased percentage of organic C.

HEAVY-METAL RECORD IN MISSISSIPPI DELTA SEDIMENTS

Fig. 4 shows the excess ^{210}Pb activity profiles measured at two locations (Stations 11A and 15) where sediments are accumulating at rates ideal for determination with this particular isotope (0.05—0.75 g cm^{-2} yr.$^{-1}$). These rates are found at the seaward extremities of the main depositional platform of the delta and rates increase in going toward the river mouth. In faster accumulating sediments, as the upper limit of resolution using ^{210}Pb is approached (determined by its half-life and sampling resolution), profiles of this isotope's activity become somewhat more difficult to translate into sedimentation rates. The near-ideal exponential decreases in activity seen in the two profiles in Fig. 4 indicate undisturbed, constant-rate sedimentation in one case (Station 15) and constant-rate sedimentation overlain by a surface zone of sediment mixing in the other (Station 11A).

Based on the sedimentation rates (S) derived in this study, the Mississippi delta can be segregated into three areas: (1) that immediately surrounding the river mouth ($S > 2$ g cm^{-2} yr.$^{-1}$), (2) the broad mid-delta area out to water depths of 100 m ($S = 0.5—2$ g cm^{-2} yr.$^{-1}$), and (3) the outer delta

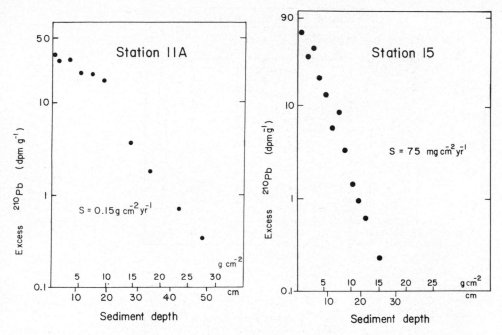

Fig. 4. Excess ^{210}Pb activity vs. depth in sediments from Mississippi delta, Stations *11*A and *15*.

TABLE III

^{210}Pb derived sediment accumulation rates, Mississippi River delta

Region	Station No.	Accumulation rate, S (g cm^{-2} yr.$^{-1}$)	Distance from river (km)	Water depth (m)
Outer delta	*14*	0.35	70	150
	15	0.08	50	550
	*11*A	0.15	26	400
	*11*B	1.1	14	130
	16	0.5	27	110
Mid-delta	*12*	1.3	8	60
Near river mouth	*10*	>2	7	50
	8	>2	8	30
	17	3—4	6	20

202

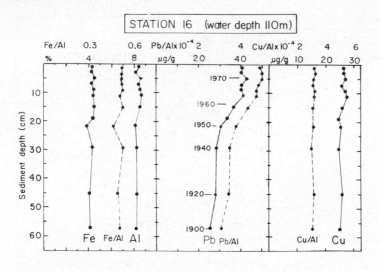

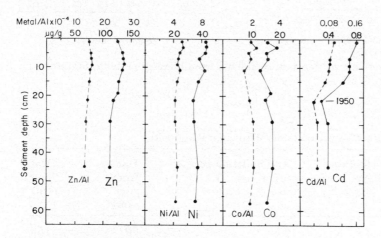

Fig. 5. Vertical metal profiles for mid-delta sediment, Station *16*.

where sedimentation rates drop to $0.1-0.5\,\mathrm{g\,cm^{-2}\,yr.^{-1}}$. Table III summarizes the ^{210}Pb-derived sedimentation rates for delta stations utilized in this study.

Having determined the sediment accumulation rate for each of the cores collected from the delta in this study, it is possible to assign approximate ages to the vertical sediment strata. This then makes it possible to evaluate the vertical distribution of heavy metals in these cores as historical records of elemental input.

In the mid-delta area, the characteristic range of sedimentation rates dictates that the 50—60-cm cores collected in this study will provide a record back to about the turn of the century. Fig. 5 for a typical mid-delta station indicates that Fe, Al, Cu, Zn, Ni and Co concentrations have been uniform over the past 75 yr. The same trend holds for Cr and V. Sedimentary Pb concentrations, however, have increased by 65% during this time interval. The major onset of pollutant Pb has occurred since the mid-1940's, a period of increased usage of leaded gasoline. Also, the surface sediment Pb concentrations (measured at this location) are comparable with those of present-day river particulates (46 ppm). The observed flattening of the upper portion of the Pb curve might be ascribed to vertical mixing of sediment; however, the comparability of Pb levels in river particulates and Station *16* surficial sediments, constant Pb concentrations over 50 cm (10—15 yr.) in near-river-mouth sediments, and no obvious mixing in the ^{210}Pb profile at this station do not support mixing in this instance. In addition to Pb, surface sediment Cd concentrations at this mid-delta station are also high at double their pre-1950 values. Moreover, the 0.4 ppm Cd baseline concentration is still in excess of values found in deeper (older) sediments on the delta.

Rapidly accumulating sediments ($S > 2\,\mathrm{g\,cm^{-2}\,yr.^{-1}}$) from the near-river-mouth area of the delta have relatively straight vertical metal profiles (Fig. 6). Pb and Cd concentrations, however, are respectively 70 and 200% above "base" values found in mid-delta sediments and are in good agreement with river particulate concentrations. Sediment Pb and Cd concentrations of > 40 and > 1 ppm, respectively, for 5—60-cm profiles in this area indicate that their pollutant levels have been relatively constant during the past 10—15 yr.

Outer-delta sediments, which are accumulating at rates on the order of 0.1—$0.5\,\mathrm{g\,cm^{-2}\,yr.^{-1}}$, also have relatively constant Fe, Al, Cu, Zn, Ni, V, Cr and Co concentrations (or at least uniform metal/Al ratios; Fig. 7). High surface Mn concentrations found in outer-delta cores are the natural geochemical result of post-depositional reduction (solubilization) of Mn, upward diffusion of dissolved Mn in interstitial water, and surface re-oxidation (precipitation). Sedimentary Pb concentrations at the outer-delta station shown (Fig. 7) exhibit a decrease from 43 ppm at the surface to 20—25 ppm in the lower core section. The time scale for this core, however, permits an extension of the historical record beyond that available in more rapidly accumulating sediments and shows that the initial onset of pollutant Pb occurred during the mid-1800's.

Although sedimentary Pb concentrations at the outer-delta station shown in Fig. 7 exhibit a similar decrease with depth (from 40—45 ppm at the surface) as do mid-delta sediments (Fig. 5), the decrease is to a concentration base of 20—25 ppm in the lower core section. Since the depositional rates are slower in the outer delta, the 50—60-cm core extends the historical record back to pre-1800, indicating that substantial amounts of pollutant Pb began being added to Mississippi delta sediments around the mid-1800's with an accelerated increase after 1940. This record corresponds remarkably with the advent of mining lead and other ores, extensive agricultural development,

204

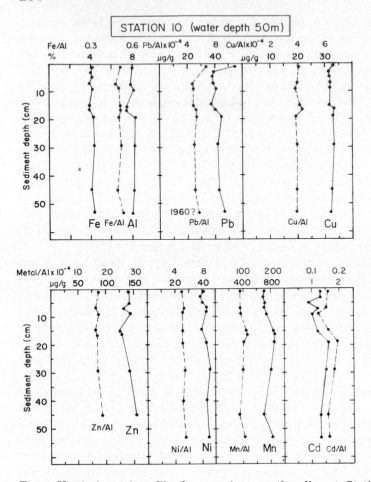

Fig. 6. Vertical metal profiles for near-river-mouth sediment, Station *10*.

fossil fuel burning and other industrial-age activities in the last half of the 19th century, followed by the accelerations in urbanization (and its waste disposal problems) and gasoline engine use in the post-1940 era. Given the magnitudes of the Mississippi River's drainage basin and its annual flow, it is difficult to relate the perturbations in the marine sedimentary Pb record to specific man-related activities, and an explanation for the observed Cd enrichments is even less obvious.

Similar historical records have been established for Pb in Lake Michigan sediments (Edgington and Robbins, 1976) and in southern California (Bruland et al., 1974). In the Lake Michigan study, where historic anthropogenic emissions are well documented, increased Pb fluxed have been attributed solely to a combination of coal and gasoline combustion. In southern California coastal sediments, which quite efficiently collect in a

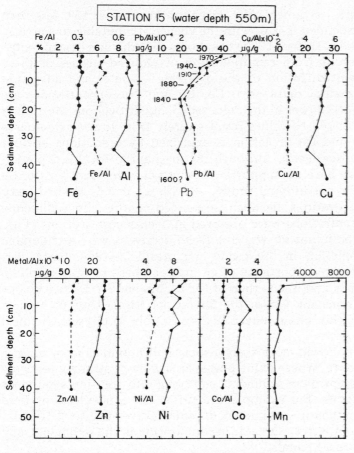

Fig. 7. Vertical metal profiles for outer-delta sediment, Station *15*.

series of near-shore basins, several metals including Pb, Cr, Cu, Ag and Zn have been found in excess of natural concentration levels since the 1930—1940 period and have been related to industrial/municipal discharges and runoff entrainments as well as to major contributions by atmospheric transport, especially in the case of Pb. With regard to the southern California data, it is interesting that only Pb and Cd are found to be anthropogenically enriched in Mississippi delta sediments. Evidently, the enormous annual detrital load of the Mississippi River grossly outweighs any man-induced amounts of the other metals (Cu, Cr, Zn and Ni in particular) to the point of their being insignificant.

Metals in Mississippi delta sediments were also chemically partitioned, using the previously described leaching scheme. Sediment total Fe concentrations were about 10% lower than those of the river particulates with the difference in absolute amounts accounted for by lower oxide-phase Fe in sediments. Similarly, sediment Mn concentrations were 45% less than those of the river suspended matter with the change attributable to a greatly lower level of Mn-oxide. Trefry (1977) has shown that the decrease in oxide-phase Fe and Mn results from remobilization of reduced species of these metals with subsequent diffusion to the overlying seawater and advection away from the Mississippi delta area. Co, Cu and Ni concentrations are correspondingly 20—30% lower in the delta sediments throughout the 1—2-m cores examined. Decreased Ni values result entirely from lower oxide-phase content whereas Co and Cu depletion is explained by reductions in both oxide and organic/sulfide levels. Although the mechanism for this depletion may be diagenetic, we have no supporting interstitial water data for Co, Cu and Ni. Sediment Pb partitioning shows, as with all metals studied, very similar lattice concentrations between suspended matter and sediments. Lower total Pb concentrations were observed in deeper core sections (Figs. 5 and 7) where less pollutant Pb was found. Pb decreases were attributable to significant Pb depletion in the oxide and organic/sulfide phases. Partitioning, then, shows the importance of an organic phase to Co, Cu and Pb transport to the oceans and a lack thereof for Fe, Mn and Ni. Lattice phases remain surprisingly constant across the Mississippi River—Gulf of Mexico transition with the oxide phase universally susceptible to physico-chemical alteration.

Pollutant metal inputs will most likely associate with oxide and/or organic phases as outlined above. Where oxide phases are involved, strongly reducing conditions in Mississippi delta sediments can bring into solution significant amounts of some metals. For example, interstitial Mn concentrations were as high as $10 \, mg \, l^{-1}$ in the top centimeters of sediment (relative to $< 10 \, \mu g \, l^{-1}$ in the overlying bottom water). Organic decomposition in the sediments may also release potential toxic metals.

Measurement of interstitial concentrations provides one means for estimating the availability of different metals to benthic fauna. A second approach is to measure the availability of sediment metals as a function of some analogue of the organisms' digestive fluid. One, perhaps oversimplified, procedure is to leach sediment samples at varying pH's, using a continuous buffer system. Fig. 8 shows the results of leaching individual samples of surficial sediment from Station 15 with phthalate buffers of varying pH. For Zn and Fe, a significant, exponential increase in metal released begins at pH $\sim$ 4. At pH 3, about 50% of the total sediment Zn is dissolved whereas $< 1\%$ of the total Fe is leached. Available data (E. A. Barnard, 1973) suggest that pH's of 6—8 are characteristic of invertebrate guts, with special cases probably no lower than 4. Vertebrate digestive systems may, however, have pH's on the order of 3. Our initial efforts in this area warrant further research as realistic metal bio-availability from polluted sediments remains a generally unsolved problem.

A detailed discussion of the role of heavy metals in the Mediterranean

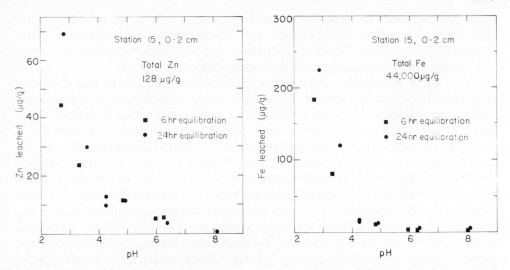

Fig. 8. Leachable Zn and Fe from outer-delta sediment (Station *15*) as a function of pH.

Sea is presented by Bernhard (Chapter 3), and again in the Gulf of Mexico by Sackett (Chapter 13); as well as in the New York Bight by Gunnerson (Chapter 11) and off the northwest coast of the U.S.A. by Gillespie and Vaccaro (Chapter 5).

CONCLUSIONS

Mississippi River delta sediments provide an ongoing historical record of pollutant discharge from the central U.S.A. This record shows no evidence of Co, Cr, Cu, Fe, Mn or Ni pollution, but strong indications of anthropogenic Cd and Pb additions. Sediment Pb concentrations have increased by 70% and Cd values by 200% over the past several decades. These man-derived inputs may be traced to the advent of mining, agricultural and industrial activities during the latter half of the 1800's with additional acceleration during the past 1940's era.

Pollutant Pb fluxes from the Mississippi River to the Gulf of Mexico are on the order of $6000 \, t \, yr.^{-1}$, or about 30 times higher than estimated for the southern California coastal zone (Bruland et al., 1974). Similarily, anthropogenic Cd inputs from the Mississippi ($\sim 300 \, t \, yr.^{-1}$) are about 35 times greater than those reported for southern California. The massive sediment load of the Mississippi ($\sim 3 \cdot 10^8 \, t \, yr.^{-1}$), however, dilutes these significantly higher fluxes to concentration levels comparable to those observed off southern California. Such dilution may completely obscure pollution inputs of some of the other metals. For example, no significant alteration in Mississippi delta sediment Cr flux ($2200 \, t \, yr.^{-1}$) would be observed by adding the

350 t of pollutant Cr found for the California coast, or even 3 or 4 times that amount. In this manner then, the Mississippi provides a viable outlet for carrying enormous loads of industrial and municipal metal wastes without significantly increasing natural (background) metal levels.

Since delta sediments give only a time-averaged picture of metal inputs, river dissolved and particulate fluxes have more value as real-time evaluators and prognosticators of metal pollution. Our data show that dissolved trace-metal concentrations are low and most likely non-problematic, a function of the high suspended load and pH of the river. River-particulate Al, Cd, Cr, Fe, Pb and V values are essentially equivalent to those for delta surface sediments; however, river suspended matter Co, Cu, Mn, Ni and Zn concentrations are 20 to 40% higher than those in the delta sediments. The 40% lower Mn in the sediments has been accounted for by a diagenetic loss of the reduced Mn^{2+} species from the delta sediments (Trefry, 1977). Variations in the other four metals may also be diagenetic; however, recent pollutant inputs and physical fractionation of river-borne particles may also prove important. Thus the usefulness of river particulates in up-to-date monitoring of small (on the order of 20—30%) anthropogenic increases in loadings of at least Co, Cu, Ni and Zn awaits further understanding of the near-shore geochemistry of these elements.

Using a more classical geochemical leaching scheme, we found oxide and lattice phases to be of primary importance for most metals, with Co, Cu and Pb having significant organic/sulfide partitioning. Preliminary work on a pH-varying series of leaches may yield a more helpful analog for evaluating biological metal availability.

In summary, sediment Pb and Cd pollution is observed throughout the Mississippi delta with origins extending back to the late 1800's. No clear evidence of Co, Cr, Cu, Fe, Mn, Ni, V or Zn pollution was found, possibly a function of the high river particulate flux which can dilute away large anomalous inputs. Most of the river particulates (which carry ∼ 90% of the river metal load) settle out very quickly upon entering the ocean, thus only a small percentage of the Gulf of Mexico surficial sediments (< 1%) are presently known to have measurable Pb and Cd pollution.

ACKNOWLEDGEMENTS

Our work was supported by the U.S. National Science Foundation, Office for the International Decade of Ocean Exploration (Grant No. GX-42576), and the Office of Naval Research (Grant No. N00014-75-C-0537).

CHAPTER 5

HETEROTROPHIC MICROBIAL ACTIVITY IN EXPERIMENTALLY PERTURBED MARINE ECOSYSTEMS

P. A. Gillespie and R. F. Vaccaro

INTRODUCTION

The classical role of the heterotrophic bacterioplankton in the oceans concerns the recycling of the inorganic nutrients required for primary production. However, the impact of sublethal concentrations of pollutants on heterotrophic microbial behavior remains largely unexplored. In addressing this question, the authors have reviewed data compiled by the Controlled Ecosystem Pollution Experiment* (CEPEX), a multidiciplinary effort which stresses the experimental manipulation of entire marine ecosystems.

An inability to duplicate natural conditions in the laboratory has adversely affected most past studies on the tolerance of microbial heterotrophs. Often the results obtained reflect only the unique behavior of single or limited numbers of isolated species while comprehensive response patterns dictated by entire marine communities have been ignored. To overcome this objection the CEPEX format has emphasized the capture of unusually large vertical columns of seawater within plastic enclosures called controlled experimental ecosystems or CEE's.

The site selected for CEPEX studies, Saanich Inlet near Sidney, British Columbia, Canada promised minimal exposure to excessive current velocities and afforded a high degree of shelter. Water depths, greatly in excess of the compensation level for photosynthesis are available within a few kilometers off shore yet the local rates of primary production are favorable being appreciably higher than those generally available in the deep ocean. The local food web tends to be coastal in character and has been intensively studied. Finally, the Saanich site is considered remote from obvious sources of pollution.

Variations of the CEE approach have been used previously to explore basic problems in aquatic biology (Strickland and Terhune, 1961; Goldman, 1962; Lund, 1972; J. M. Davies et al., 1975). Such studies support the feasibility of impounding natural biological assemblages for subsequent

---

* Supported by the National Science Foundation under the International Decade of Ocean Exploration. Grant Numbers GX-39147 and OCE73-05941; A01, A02.

experimental manipulations for periods in excess of 100 days. Procedural experience at Saanich Inlet has shown that biological events can be made to replicate under such conditions providing the enclosures are elevated simultaneously from a subsurface position and that they remain isolated from undue exchange with outside water during the experimental period (M. Takahashi et al., 1975).

MATERIALS AND METHODS

Important criteria for CEE construction was that the enclosures used be transparent to light and chemically non-toxic. The general construction decided upon was a plastic bag suspended from a flotation collar whose surface location was permanently fixed by anchoring. The enclosure itself consisted of two adjacent layers of plastic; an inner container made of 4-mil sheet polyethylene and an exterior enclosure made of 8-mil woven polyethylene. During the majority of experiments, however, the use of an inner bag was considered unnecessary and therefore omitted. Conical shaped bottoms were provided to collect and discharge sediment material via pumping through a hose which led to the surface along the exterior of the bag.

Flotation collars were constructed of either stainless steel or 0.4-m diameter clear lucite tubing fashioned into a hexagon. Besides providing suspension platforms these collars were also used as a bridge for sampling operations. Elastic shock cord was used to connect the flotation collars to the top of the plastic enclosures to help de-energize vertical motion from wave action. At the time of deployment the plastic bags were submerged directly beneath their anchored flotation collars. On signal, they were raised simultaneously and upon reaching the surface were attached to their collars thus entrapping entire and relatively undisturbed columns of water. Further details regarding the experimental design and deployment procedures are given in Menzel and Case (1977).

Two sizes of CEE's were launched and moored in Saanich Inlet during the summers of 1975 and 1976. The larger was designed for long-term experiments (up to 120 days) and measured 9.5 m in diameter, 23.5 m long and contained a seawater volume approximating $1300\,m^3$ (Fig. 1). A scaled-down version for shorter experiments had a diameter of 2.4 m, a depth of 16 m and a volume of $68\,m^3$. For a given experiment from 3 to 5 CEE's were lifted simultaneously.

The categories of perturbants used in these studies were: (1) plant nutrients (nitrate, phosphate and silicate); (2) hydrocarbons (water extract of No. 2 fuel oil and a naphthalene mixture); and (3) the heavy metals Hg and Cu. In all instances heterotrophic ^{14}C uptake patterns provided a common read-out parameter for evaluating sublethal microbial effects. For sampling purposes, integrated volumes of seawater were periodically obtained by

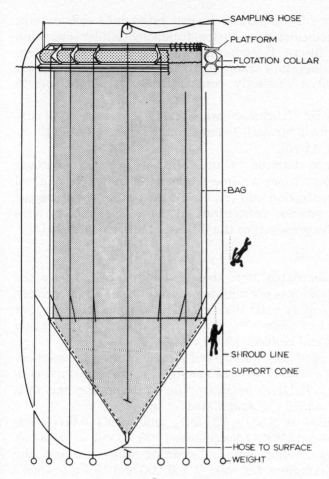

Fig. 1. Sketch of 1300-m³ controlled experimental ecosystem (CEE) from Grice et al. (1977).

pumping from preselected depth intervals. The sampling depth ranges selected for the large CEE's were 0—5, 5—10, 10—15 and 15—20 m. Preselected depths for the smaller, 68-m³ CEE's, were 0—5, 5—10 and 10—13 m.

Measurements of the kinetic parameters used to evaluate heterotrophic activity were obtained by observing ^{14}C-glucose uptake according to a simplified version of the technique described by R. T. Wright and Hobbie (1966). Additions of $0.1 \, \mu$Ci of ^{14}C-glucose (U) were made to a series of 25- or 50-ml subsamples augmented with ^{12}C-glucose to give a range of concentrations (in duplicate) not exceeding $10 \, \mu$g l^{-1} gluclose. The subsamples were routinely incubated in the dark for 1 or 2 hr. along with a formalin-killed control containing 1 ml of 10% buffered formalin. At the end of the incubation period the balance of the subsamples was also killed with formalin and the amounts of ^{14}C-glucose incorporated into cell material were determined by liquid scintillation counting, following cellular separation onto 0.45-μm membrane filters. Periodic assessments of ^{14}C respiration, made by measuring $^{14}CO_2$ liberation according to the method of Hobbie and Crawford (1969) indicated a relatively constant rate of carbon respiration.

Since the concentration of ^{14}C-glucose was initially constant in all subsamples, a single killed control for each heterotrophic series was used (P. A. Gillespie, unpublished data, 1975).

Kinetic parameters used to describe ^{14}C utilization were V_{max} (maximum uptake velocity), T_t (turnover time) or time required for complete removal of the in situ glucose concentration and $(K_t + S_n)$ (the sum of the transport constant and the in situ glucose concentration). These parameters were determined from graphical expressions of the following linear relationship:

$$C\mu t/c = [(K_t + S_n) + A]/V_{max}$$

where c = radioactivity assimilated plus that respired as $^{14}CO_2$ by the heterotrophic population in counts per minute (cpm); A = the added glucose concentration in $\mu g\,l^{-1}$; $C = 2.22 \cdot 10^6$ disintegrations per minute (dpm) per μCi of ^{14}C; μ = the amount of added ^{14}C per sample in μCi; t = incubation time in hours (R. T. Wright and Hobbie, 1966).

To facilitate the processing of large numbers of samples during the mercury experiment, the uptake of ^{14}C-glucose was measured at one substrate concentration only $(1.2\,\mu g\,l^{-1})$. Relative heterotrophic incorporation was then expressed as the percent of added ^{14}C assimilated $l^{-1}\,hr.^{-1}$.

Nutrient enrichment added as KNO_3, KH_2PO_4 and $NaSiO_3 \cdot 9H_2O$ was made to a series of four 68-m^3 CEE's (designated -J, -K, -L and -M) at 3-day intervals over a 30-day experimental period. Regarding nitrate-nitrogen, the amounts added on each enrichment day were 0, 1.6, 4.0 and 16.0 g to CEE's -J, -K, -L and -M, respectively. The remaining two elements, P and Si were added simultaneously in sufficient quantity so that the final atomic ratios were 10:1:10 for N:P:Si, respectively. On this basis, the nominal concentrations of added Si in the four CEE's were also 0.0, 1.6, 4.0 and 16.0 μgAt.l^{-1}, respectively, while the comparable phosphorus concentrations were 0.0, 0.16, 0.40 and 1.6 μgAt.l^{-1}.

Two types of petroleum derivatives were introduced into the 68-m^3 CEE's. These were a water extract of No. 2 fuel oil (CEE-G) and a mixture of naphthalene, methylnaphthylene and dimethylnaphthalene (CEE-I). Changes in the concentrations of these substances with time, as described by Lee and Takahashi (1975) and by Lee and Anderson (1977), indicate that the initial concentration of 40 $\mu g\,l^{-1}$ of No. 2 fuel was reduced within three days to 20 $\mu g\,l^{-1}$.

Heavy-metal additions of Hg or Cu were made by pumping concentrated solutions into selected depths with the enclosures in order to obtain a satisfactory distribution. Individual formulations will be described in greater detail in the following section.

RESULTS AND DISCUSSION

Eutrophication studies

The effects of N, P and Si on the fertility of marine systems have been well documented by subjecting similar habitats to variant nutrient regimes and by observations on the distribution of biological activity around point-sources of nutrient enrichment. In general, a causal relationship between excessive plant nutrient enrichment and adverse ecological consequences is to be anticipated.

A comparison between V_{max} for CEE-K, which received the least amount of nutrient addition, and the unenriched control, CEE-J, failed to reveal any significant differences in the observed levels of heterotrophic potential over the 34-day experimental period (Fig. 2a and b). However, at higher nutrient concentrations (Fig. 2c and d) CEE-L and CEE-M showed a positive increase in heterotrophic activity in response to the accumulated amounts of nutrients added. This tendency was most pronounced in the case of the 0—5-m observations for CEE-M where glucose uptake increased six-fold above that recorded for the comparable control situation.

Plant nutrient changes in CEE-J and CEE-M measured during August 1976 and summarized by Parsons et al. (1977a) showed a significant accumulation of excess N, P and Si as well as a three-fold increase in chlorophyll *a* and primary production (Table I). Thus it would appear that heterotrophic activity responded rapidly to variations in available organic substrates leading to an interchange of nutrient substrates between the phyto- and bacterioplankton.

Further comments by Parsons et al. (1977b) concerned the relatively high rate of primary production ($109 \, mg \, m^{-2} \, day^{-1}$ C) in CEE-J despite an absence of nutrient enrichment. The persistence of significant amounts of plant nutrients throughout this experiment also underlines the importance of the heterotrophic recycling process. This opinion is supported by the relative constancy of ($K_t + S_n$)-values (not shown) which are indicative of a lack of change in concentration of organic substrate. These results support previous findings of K. C. Morgan and Kalff (1972), Albright and Wentworth (1973), and P. A. Gillespie (1976) who have described aquatic systems wherein heterotrophic measurements reflect the level of eutrophication and provide a sensitive indication of nutrient enrichment.

Hydrocarbon studies

The extreme vulnerability of the marine environment to the uncontrolled

TABLE I

Changes in nutrients and primary productivity (July 28—August 28; 0—10 m) in CEE's
without (-J) and with (-M) prior nutrient enrichment (Parsons et al., 1977a)

	Time	CEE-J (control)	$r^{(*)}$	CEE-M (enriched)	$r^{(*)}$
Nitrate-N ($\mu g\,l^{-1}$)	initial	0.26	−0.39	0.33	0.75
	final	0.21		1.51	
Phosphate-P ($\mu g\,l^{-1}$)	initial	0.38	0.69	0.39	0.89
	final	0.79		2.44	
Silicate-Si ($\mu g\,l^{-1}$)	initial	1.90	0.08	2.60	0.76
	final	2.11		5.99	
Ammonium-N ($\mu g\,l^{-1}$)	initial	0.84	0.23	0.69	0.58
	final	1.22		1.39	
Chlorophyll a	initial	0.20	0.58	0.58	0.79
	final	0.46		1.22	
Average productivity (mg m^{-2} day^{-1} C) July 28—August 13:		109		300	

$^{*}r$ = correlation coefficient indicating the degree of correlation between the variable
shown and the time period of the experiment.

entrance of hydrocarbons into the sea has also been repeatedly demonstrated.
This subject, treated in greater detail in Chapter 12 of Volume 1, was also
investigated under the CEPEX format as described below.

Results from the No. 2 fuel oil experiment, shown in Fig. 3a, describe
time-related changes in heterotrophic analyses of V_{max}. The integrated depth
sampled were 0—5, 5—10 and 10—13 m and the level of heterotrophic
activity just prior to the time of fuel oil addition was utilized as the reference
control. During the week following fuel addition, a general decline in the
magnitude of V_{max}-values was observed. This decrease was most pronounced
at the 0—5- and 10—13-m levels where order of magnitude changes were
recorded between 4th and 7th day after spiking. During the next nine days
V_{max}-values showed a general increase and ultimately exceeded the minimum
values of day 7 by more than ten-fold.

The naphthalene experiment (CEE-I) with its control (CEE-H) was of
shorter duration than the No. 2 fuel oil study, being limited to alternate
day sampling over a 6-day period. The results shown in Fig. 3b and c do not
include data from the deepest sampling (10—13 m) thought to be anomalous

Fig. 2. Changes in heterotrophic potential for glucose uptake (V_{max}) in controlled experimental ecosystems after addition of nutrients. a. CEE-J, control. b. CEE-K, 1.6 g KNO$_3$.
c. CEE-L, 4.0 g KNO$_3$. d. CEE-M, 160 g KNO$_3$.

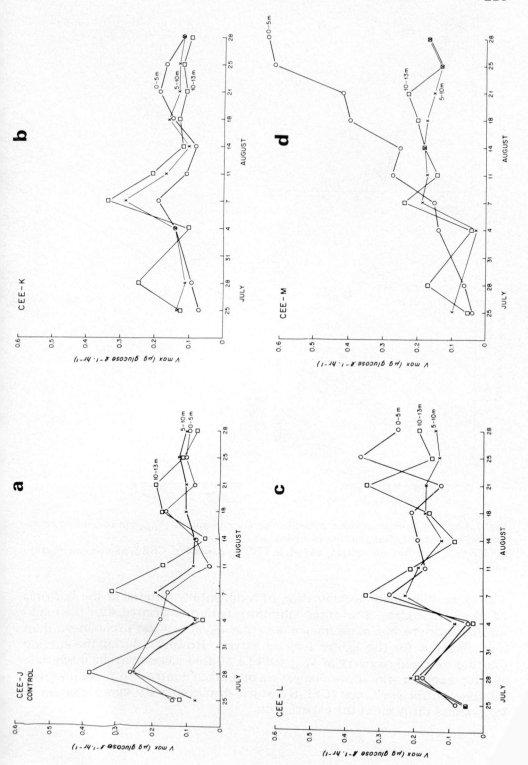

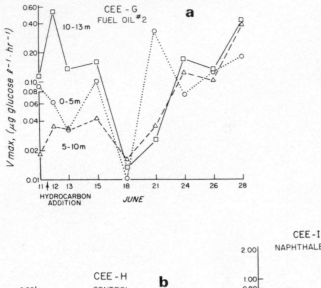

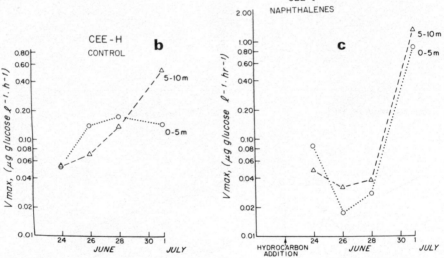

Fig. 3. Changes in heterotrophic potential for glucose uptake (V_{max}) in controlled experimental ecosystems after addition of hydrocarbons.
a. CEE-G, No. 2 fuel oil extract, added. b. CEE-H, control. c. CEE-I, napthalene mixture added.

due to an inordinate accumulation of sedimentary material at the bottoms of these enclosures. The initial inhibitory period associated with the naphthalene mixture was more immediate (four days) but less pronounced than that observed for the lighter fuel oil extract. However, during the ensuing four days a rapid recovery in V_{max} led to a 30-fold increase in the magnitude of this parameter. The above changes are in marked contrast to those observed for the control enclosure (CEE-H) which displayed only minor changes in V_{max}-values throughout the experiment.

Further information on the impact of naphthalenes on the heterotrophic process can be gained by examining variations in the additional kinetic parameters $(K_t + S_n)$ and T_t. As stated previously the parameter T_t, refers to the time required for complete utilization of an available heterotrophic substrate (S_n) present at time t, while $(K_t + S_n)$ corresponds to the sum of the cellular transport constant and the in situ concentration of the organic substrate being scrutenized. Between the first and third day following spiking, a period which corresponded to the early decrease in V_{max}, a notable increase in $(K_t + S_n)$ was observed (Fig. 4c) along with a simultaneous increase in T_t (Fig. 4d). At the 0—5-m depth interval T_t's for glucose increased from 40 to 900 hr. over the same 2-day period. In contrast to the above these two parameters remained more stable in CEE-H, the control enclosure

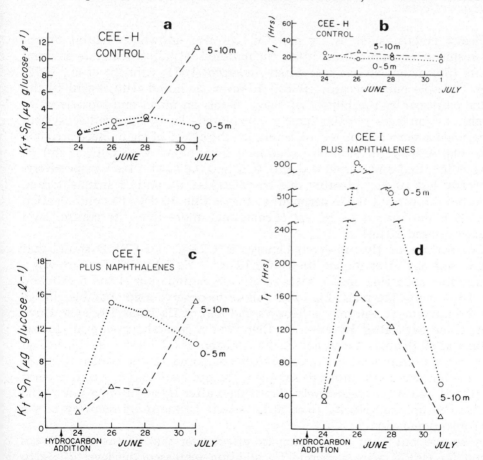

Fig. 4. Changes in heterotrophic parameters $(K_t + S_n)$ and T_t in controlled experimental ecosystems after addition of naphthalenes.
a. $(K_t + S_n)$-values, CEE-H, control. b. T_t-values for CEE-H, control. c. $(K_t + S_n)$-values for naphthalene-treated CEE-I. d. T_t-values for naphthalene-treated CEE-I.

(Fig. 4a and b). These results are consistent with an accumulation of organic substrate due to an early retardation in microbiological activity and a stress-induced increase in the rate of excretion of dissolved organic material from phytoplankton. However, the heterotrophic flora was able to adapt rapidly to both naphthalene and the increased release of substrate. Later, as the rate of substrate utilization increased, the T_t decreased to a level which was only slightly higher than that observed soon after naphthalenes were added.

HEAVY-METAL IMPACTS

Mercury effects

Heavy metals, such as Hg and Cu, are also known to inhibit micro-organisms at concentrations often encountered in polluted marine environments (Steemann Nielsen and Wium-Andersen, 1971; Albright et al., 1972; P. A. Gillespie and Vaccaro, 1978). However, as noted with regard to the other perturbants, the impact of heavy metals on intact and representative trophic assemblages remains largely unexplored. Two controlled ecosystem experiments were completed to determine the effect of Hg on heterotrophic microbial activity. In the first experiment three 68-m^3 CEE's, -C, -D and -E, were committed and spiked with 0.0, 0.12 and 0.67 $\mu g \, l^{-1}$ Hg^{2+}, respectively. A steady loss of Hg, possibly due to volatilization, and/or sedimentation, occurred throughout the experiment so that within 30 days its concentration in CEE-E diminished to 80 ng l^{-1} somewhat more than the control level which averaged 10 ng l^{-1}.

Comparison of the observed changes in CEE-C and CEE-D shows that within one day after the addition of 0.12 $\mu g \, l^{-1}$ Hg^{2+} to CEE-D there was a measurable reduction in the rates of glucose assimilation at the 5—10- and 10—13-m depth intervals (Fig. 5b). Failure to observe a comparable change for the uppermost 5 m may reflect a rapid loss of Hg from this layer. However, three days after Hg addition there was a general increase in glucose uptake at all depths. As expected, the addition of 0.67 $\mu g \, l^{-1}$ Hg^{2+} led to an even more pronounced inhibitory effect compared to the control (Fig. 5a and c). For the depth intervals sampled, glucose assimilation was depressed by 82, 81 and 43%, respectively. Three days after Hg addition, a substantial increase in uptake velocity once again attests to the rapid recovery of the microbial population.

A similar but even more dramatic effect over that of the control was noted for CEE-E after a second Hg addition increased the level of Hg to 4 $\mu g \, l^{-1}$ (Fig. 6a and b). In this instance the levels of heterotrophic activity at the three depth intervals decreased by 91, 80 and 77% within 12 hr. Once again, the subsequent recovery was rapid and spectacular, amounting

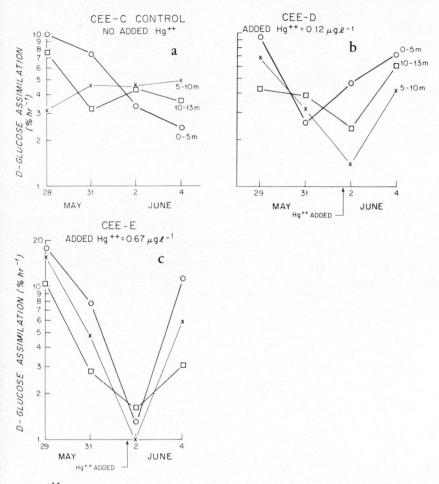

Fig. 5. ^{14}C-glucose assimilation in controlled experimental ecosystems after addition of mercury.
a. CEE-C, control. b. CEE-D, plus 0.12 μg l^{-1} Hg. c. CEE-E plus 0.67 μg l^{-1} Hg.

to a 10—100-fold increase over the minimum heterotrophic rates measured three days earlier.

Other CEPEX studies with Hg conducted in the larger, 1300-m^3, enclosures and described by Azam et al. (1977) have also shown the effects of a Hg concentration of 5 μg l^{-1} on microbial activity. In this instance dramatic decreases (2—3 orders of magnitude) in microbial heterotrophic activity and viable cell numbers were noted. However, bacterial biomass calculated from ATP (adenosine triphosphate) concentrations in the < 0.6-μm particulate fraction decreased by only a single order of magnitude. From this, the authors speculated that the effect of Hg on bacterial cells may be more bacteriostatic than bacteriocidal. During the latter stages of this experiment

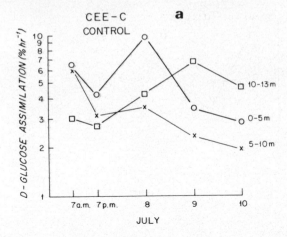

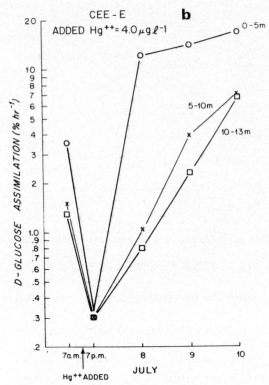

Fig. 6. ^{14}C-glucose assimilation in controlled experimental ecosystem after a second addition of mercury.
a. CEE-C, control. b. CEE-E plus $4\,\mu g\,l^{-1}$ Hg.

an increase in heterotrophic uptake rates once again indicated significant bacterial recovery and stimulation associated with the stress-related release of organic substrates following Hg addition. An inability to demonstrate concomitant increases in dissolved organic C concentrations during the microbial recovery period seems to suggest that in this case the heterotrophic response was sufficiently rapid to preclude a discernable accumulation of extracellular organic C. Recognition must also be given to the alternative possibility of a significant shift in dissolved organic quality during the time of stress. However, neither of the above alternatives contradicts a general leakage of organics from various components of the original ecosystem (Beers et al., 1977; W. H. Thomas et al., 1977b).

Copper effects

The effects of low levels of Cu on microbial activity were not as spectacular as those associated with Hg. Vaccaro et al. (1977) have described the effects of 10 and $50 \mu g l^{-1}$ Cu on bacterial populations in 68-m^3 CEPEX enclosures. They report a marked enhancement in ^{14}C-glycine uptake two days after the addition of $10 \mu g l^{-1}$ Cu and four days after the addition of $50 \mu g l^{-1}$ Cu. Here a lack of detailed sampling precluded information on the short-term microbial response. However, more timely data compiled by W. H. Thomas et al. (1977a) have described a rapid decline in chlorophyll a concentration and primary production rates during the time in question. In addition, the above authors also noted a simultaneous increase in the rates of ^{14}C organic excretion by the Cu exposed phytoplankton population.

A later CEPEX experiment designed to demonstrate inhibitory effects from Cu at the 10-$\mu g l^{-1}$ level gave results which were less conclusive. Only a slight and highly transitory impact on the heterotrophic flora occurred at this lower concentration of Cu. A tentative interpretation of this observation, based on the indirect evidence presented below, is that the $10 \mu g l^{-1}$ Cu very closely approximated the Cu complexation capacity for Saanich seawater. Stated differently, but equally relevant, the addition of $10 \mu g l^{-1}$ Cu to Saanich Inlet water may have been insufficient to establish a persistent pool of free Cu, hence Cu inhibition could not be subsequently demonstrated.

In support of the above explanation, Sunda and Guillard (1976) have shown that the potential toxicity of Cu in natural waters depends not only upon the total Cu concentration but also on the nature and abundance of available complexing ligands. In practice there appears to be a direct relationship between the inhibitory response as exhibited by glucose uptake in marine bacteria and the voltage response of an electrode selective for cupric ion (Sunda and Gillespie, 1979). As a result, the form of Cu which persists within the seawater is of overriding importance. Actually, P. A. Gillespie and Vaccaro (1978) recorded chelation capacities in Saanich Inlet water of $\sim 12 \mu g l^{-1} Cu^{2+}$. On this basis, the inhibitory effects from lesser concentrations of added Cu would be effectively diminished by the intervention of naturally occurring chelators. Furthermore, at higher levels of added Cu the concentrations of organic ligands may be augmented by a release of protective dissolved organic C from stressed organisms. Such considerations point out the complexity in assigning threshold perturbation levels for trace metals in natural systems and the need for a simultaneous assessment of other relevant water properties.

Additional detailed discussions of the effects of heavy metals are presented in this volume by Mearns (Chapter 1), involving the west coast of California, Bernhard (Chapter 3) for the Mediterranean Sea, Trefry and Shokes (Chapter 4) and Sackett (Chapter 13) for the Gulf of Mexico, and Gunnerson (Chapter 11) and Simpson et al. (Chapter 12) for the east coast of the U.S.A.

BACTERIAL TOLERANCE DEVELOPMENT

An important concern of the CEPEX microbiological effort was the ability of natural marine bacterial populations to undergo changes in heavy-metal tolerance. A mixed bacterial population isolated for such studies originated from the surface water of Saanich Inlet ~ 3 km offshore. The toxic heavy metals applied to the primary population were Cu and Hg, each being employed at three different concentrations in an effort to encourage the development of different extremes of metal tolerance.

Procedurally the aseptically collected surface water was first membrane-filtered (1-μm porosity) to remove the associated phytoplankton and the larger flora and fauna. The filterable bacterial population was then removed on 0.20-μm filters and resuspended in cell-free volumes of surface water so that a four-fold increase over the in situ bacterial concentration resulted. The bacterial suspension was then divided into six aliquots, three for Cu enrichment and three for Hg enrichment. The concentrations of metal added for the Cu series were 0.0, 25 and 50 μg l^{-1} Cu^{2+} as CuSO$_4$ · 5H$_2$O. The comparable concentrations prepared for Hg were 0.0, 0.25 and 2.5 μg l^{-1} prepared from desiccated HgCl$_2$. All of these subsamples were stored in darkness with refrigeration (5°C) for a period of six months before being used as sources of test bacterial populations possessing different prior histories of metal exposure.

As with other CEPEX microbiological studies, assessment of differences in metal tolerances by the above subcultures was based on changes in their ability to take-up ^{14}C-glucose (U) after a secondary exposure to known and different metal backgrounds. To develop adequate cell numbers for test purposes it was necessary to stimulate cellular densities via 48-hr. contact with added tryptone, dextrose, ammonium nitrate and yeast extract, 0.20, 0.50, 0.50 and 0.10 g l^{-1}, respectively, just before uptake measurements. The use of tryptone along with its amino acid complement for this purpose probably resulted in at least a partial, if temporary, complexation of free metal ions and a corresponding decrease in their inhibitory effects. Yet, these subcultures presumably consisted of a group of mixed species having the potential for unique and varied heavy-metal responses.

Ultimately each of the above cultures became a source of test cells when harvested via membrane filters (0.2-μm porosity) and transferred into glass vessels containing 6 l of prefiltered bacterial-free seawater. At approximately 2-day intervals the metal concentrations of these stirred vessels were gradually increased from 0 to 25—50 μg l^{-1} for the Cu series and from 0 to 0.25—2.5 μg l^{-1} for the Hg series. Superimposed on each of these basic concentrations were observations of the changes in ^{14}C uptake patterns resulting from ~ 15 above ambient concentrations of heavy metal. For both the Cu and Hg cultures the range examined was 1 to 200 μg l^{-1} and responses have been recorded in terms of the percentage change in ^{14}C uptake with and without the additional metal increments.

The results on development of Cu tolerance are shown in Fig. 7a—c and comparable information for mercury appears in Fig. 8a—c. Interpretation of these results is facilitated by comparing the number of instances wherein Cu or Hg additions had little or no effect on ^{14}C uptake, i.e. the uptake response approached 100% of a control shown on the abscissa which received no additional metal.

Regarding the Cu series, the greatest metal inhibition corresponded to the culture developed from cells stored for a 6-month period in the absence of added Cu (Fig. 7a). Subsequent exposure of this culture to as little as 15 μg l^{-1} Cu caused a 50% decrease in ^{14}C uptake. Later as the ambient Cu concentration was increased to 5 and still later 25 μg l^{-1} Cu, the concentration of Cu required for a 50% reduction in uptake showed a corresponding increase to 22 and 200 μg l^{-1}. Finally, a 2-day exposure at 50 μg l^{-1} produced a population which was essentially unaffected by a Cu concentration as high as 250 μg l^{-1}.

A contrast to the above behavior was that noted for the culture which had been stored for six months in the presence of 50 μg l^{-1} Cu. Descendents of these cells required well over 50 μg l^{-1} Cu to decrease ^{14}C uptake by 50% (Fig. 7c). Later this culture developed even greater resistance to Cu as the ambient Cu concentration was increased. Ultimately, a Cu concentration as high as 100 μg l^{-1} caused only a 10% reduction in ^{14}C assimilation. Similar analyses of ^{14}C utilization for the intermediate Cu concentration of 25 μg l^{-1} resulted in uptake patterns intermediate between the above extremes (Fig. 7b).

A similar sequence of changes in ^{14}C uptake patterns was observed for Hg (Fig. 8a—c) except that an intermediate effect from storage at 0.25 μg l^{-1} Hg was not so clear-cut. Nevertheless, in the extreme case, the bacterial survivors of a Hg exposure of 2.5 μg l^{-1} generated a population

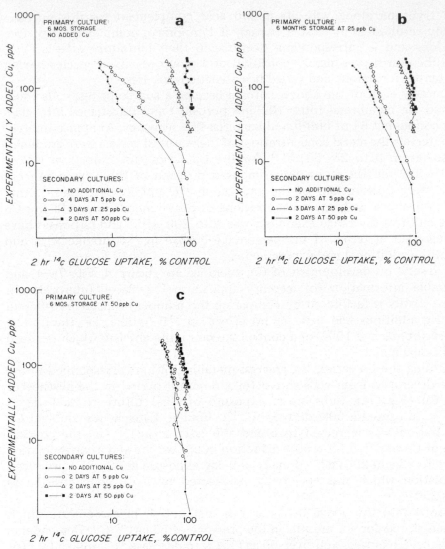

Fig. 7. The effect of previous copper contact on copper tolerance of a marine heterotrophic bacterial community after storage in excess of six months.
a. No added Cu. b. $25 \mu g \, l^{-1}$ Cu. c. $50 \mu g \, l^{-1}$ Cu.

whose Hg tolerance increased to a point where no appreciable inhibition
was observed from a concentration of $100 \, \mu g \, l^{-1}$ Hg (Fig. 8c).

CONCLUSIONS

Observations of heterotrophic microbial activity can be extremely useful
for evaluating changes in ecosystem stability following experimental

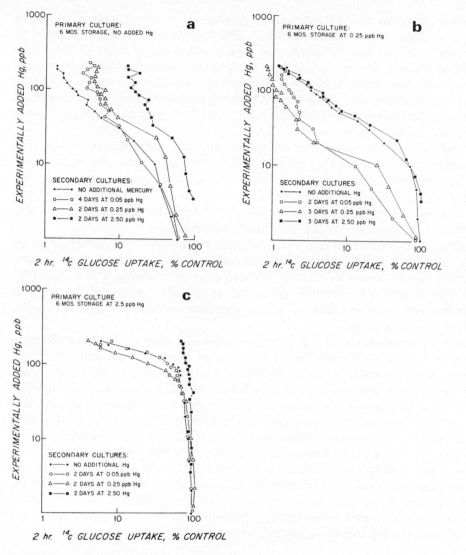

Fig. 8. The effect of previous mercury contact on mercury tolerance of a marine heterotrophic bacterial community after storage in excess of six months.
a. No added Hg. b. $0.25\,\mu g\,l^{-1}$ Hg. c. $2.5\,\mu g\,l^{-1}$ Hg.

perturbation of marine systems. The ecosystem approach clearly demonstrates how interactions between various trophic components help determine
the nature of subsequent biological response patterns. Indications are that
the microbial population is particularly responsive to perturbations affecting
all trophic levels.

Enrichment with N, P and Si, causes a stimulation in primary production by phytoplankton and a concomitant increase in released organic substrates which in turn affects heterotrophic activity. The microbial response to hydrocarbons and heavy metals also bears a striking relation to the above sequence except for an initial but temporary reduction in heterotrophic activity. In the latter instance a secondary buildup of readily utilizable organic substrates also occurs and the heterotrophic population shows an adaptation to the experimental pollutant which becomes increasingly obvious as organic availability increases.

A mechanism has also been described whereby microbial responses to a perturbant can be altered by properties inherent in a particular water mass, i.e. its chelation capacity for heavy metals. Such criteria underline the danger of generalizing regarding stress responses in different marine situations and suggest the need for more stringent methods for establishing realistic water-quality standards by the regulatory agencies.

Our data also demonstrate how the previous history of a particular water mass can affect the tolerance levels of resident species. Stressing of natural communities by heavy metals can initiate a series of events which markedly alter the nature of bacterial behavior. Inevitably, microbial recovery is accomplished by a population which exhibits increased resistance to the original source of stress.

Early interpretations of tolerance development have emphasized selective processes which ultimately led to dominance by fewer but more resistant species. More recently, cellular physiologists (Summers, 1972; Silver et al., 1976) have been able to describe the tolerance mechanism in more specific terms by invoking the exchange of genetic material between tolerant and non-tolerant species. For example, Hg tolerance has been associated with direct cellular conjugations which result in an exchange of extrachromosomal hereditary entities called plasmids. Tolerance develops following acceptance of an appropriately encoded plasmid which confers the ability to chemically reduce both inorganic and organic Hg to metallic Hg which is mobilized via volatilization. According to Azam et al. (1977), a shift to more resistant populations may exert a controlling influence on the biologically-mediated fate of many heavy metals in natural waters.

ACKNOWLEDGEMENTS

We would like to thank the CEPEX staff for their help in taking samples. We thank also Dr. R. F. Lee for collaboration during the hydrocarbon experiments; Dr. T. E. Parsons for collaboration during the eutrophication experiment; Dr. G. T. Wallace for the addition and monitoring of heavy metals; and Ms. Pamela Hindley and Mr. Mark Dennett for technical assistance.

CHAPTER 6

DREDGING EQUIPMENT AND THE EFFECTS OF DREDGING ON THE ENVIRONMENT

John B. Herbich

INTRODUCTION

The very nature of dredging operations results in disturbing bottom sediments, and depositing a mixture of water and solids in a disposal area. In case of polluted sediments, gas may be released from solution. It may escape from the point of contact of a draghead (in case of a hopper dredge), or a cutterhead (in case of a cutterhead dredge), or the gas may be sucked into the suction pipe of a hydraulic dredge. This causes a mixture of water—gas—solids to travel through the pipe into hoppers of a trailing-suction dredge or to the disposal area. The basic types of dredges, as well as methods of disposal and environmental impacts are discussed in this chapter.

The fate of dredged material will depend on the dredging method, location of disposal area, whether in open water, or on land. Disposal of the material in open water can occur in different environments: (1) tidal estuary, (2) non-tidal estuary, (3) open bay, (4) nearshore areas up to the 30 m depth of water, (5) on the continental shelf up to 200 m depth of water, or (6) in deep ocean in water depth greater than 200 m.

The fate of dredged material will greatly depend on the location of disposal area, and its environmental characteristics. Disposal of material on land can occur in three ways: (1) unconfined disposal; (2) partly-confined disposal; and (3) confined disposal.

In an unconfined or partly-confined method of disposal the effluent will find its way into a natural stream or a natural depression. There is no control of the effluent's quality and sufficient length of path will be required of the dredged material to permit it to settle. The natural slope formed by the settled dredged material will be very flat, possibly a slope of 1%, or less.

Confined area for disposal of dredged material is prepared prior to dredging operation. The main dikes and inner dikes as well as control weirs are constructed on land. The dredged material is then pumped into different compartments according to a detailed management plane. This plan includes a schedule of pumping the dredge material into a compartmentalized diked area. The material is pumped into the first compartment, then into the second, the third, etc. The effluent is released by gravity, or is pumped out. Drainage and trenching in partially-filled compartments may also be required.

DREDGING EQUIPMENT

All dredges are of two main types: mechanically operating and hydraulically operating (Herbich, 1975).

Mechanical dredges

Mechanical dredges were the first to be developed, because of their simplicity and analogy with land-based excavating machines. Mechanical dredges can be further classified into the grapple dredge, the dipper dredge and the bucket-ladder dredge.

The *grapple dredge* consists of a derrick mounted on a barge and equipped with a "clamshell" bucket. Its optimum use is in very soft underwater deposits.

The *dipper dredge* is the floating counterpart of the more familiar land-based mechanically operating excavating shovel. It works best in hard, compact material or rock, because of its great leverage and "crowding" action.

The *bucket-ladder dredge* consists of an endless chain of buckets. The top of the chain is thrust into the underwater deposit to be dredged so that each bucket digs its own load and carries it to the surface. Since the work cycle is continuous, bucket-ladder dredges are more efficient than either the grapple or dipper dredge. Bucket-ladder dredges are particularly useful to sand and gravel suppliers, because the end of the bucket-ladder can be terminated high above the supporting barge and the buckets made to discharge their contents onto vibrating screens. Thus, the different material sizes may be separated and stored on the barge by using gravity (Fig. 1).

Mechanical dredges are all characterized by their inability to transport the dredged materials for long distances, their lack of self-propulsion, and relatively low production. Their chief advantage lies in their ability to operate in restricted locations such as docks and jetties.

Effect of mechanical dredge operations. In all mechanical-type dredges there are three locations where fine sediment and pollutants (if any) are released to the surrounding waters: (a) at the digging location; (b) in moving the excavated material through the water column; and (c) at the disposal site (if unconfined).

In digging fine sediments, the finer fractions are disturbed and generally form a turbidity cloud, which may or may not move from the subaqueous digging site, depending on currents and wave action. In partially-closed areas, such as bays and estuaries, currents created by the astronomical tides may be quite strong and may carry the fine sediments out to sea through natural or artificial inlets. The wind-generated wave action in shallow water may also keep the turbidity cloud in suspension, or agitated. In near-shore

Fig. 1. Bucket dredge "Azië" (courtesy I.H.C. Holland, Papendrecht, The Netherlands): dredging capacity $1400 \, m^3 \, hr.^{-1}$, dredging depth: max. 22 m; dredging rate 28 buckets per minute; volume of bucket 835 l.

areas the wave action may be more pronounced than the littoral currents in moving the suspended-sediment cloud.

The presence of the suspended-sediment cloud in itself may not be detrimental and in some the digging operation may release nutrients to the water column.

As the excavated material is moved through the water column, either by a clamshell, dipper bucket or an endless chain of buckets, fine, as well as coarse sand and even chunks of clay are released to the water column. The coarse sands and chunks of clay will return to the bottom in the immediate vicinity of the site.

Hydraulic dredges

Hydraulic dredges which are more suitable for dredging in open water are self-contained units and handle both phases of the dredging system. The hydraulic dredges not only dig the material, but also dispose of it, this is done either by pumping the material through a floating pipeline to a deposit area, or by storing it in hoppers which can subsequently be emptied over the deposit area. It can be seen that hydraulic dredges are more efficient, more versatile and more economical to operate due to this continuous, self-contained digging and disposal operation.

With a hydraulic dredge, the material to be removed is first loosened and mixed with water by cutterheads, or by agitation with water jets, and then pumped as a fluid. The three basic units in a hydraulic dredge are dredge pumps, agitating machinery, and hoisting and hauling equipment. The latter is used primarily to raise and lower the cutter and suction drag-heads. Hydraulically operating dredges can be classified into three basic types: (1) dustpan dredge; (2) hydraulic pipeline cutterhead dredge and (3) self-propelled hopper dredge.

Dustpan dredge

The dustpan dredge is a hydraulic, plain suction, self-propelled type. It derives its name, because its suction head resembles a large vacuum cleaner or dustpan. It consists essentially of a dredge pump which draws in a mixture of water and dredged materials through the suction head. The head is lowered by winches to the face of the deposit to be removed. It is about as wide as the hull of the dredge, and is outfitted with high velocity water jets for agitating and mixing the material. After sucking the mixture of the surface, the dredge pumps it to a disposal area, either at sea or on shore, through a floating pipeline. The dustpan dredge (Fig. 2) is best suited for high-volume

Fig. 2. Dustpan dredge (courtesy of U.S. Corps of Engineers).

soft-material dredging, because it lacks a cutterhead, which loosens up hard, compact materials.

This type is particularly well suited for use in conjunction with a hopper dredge. The latter makes its cycle, returning to empty its hoppers next to a dustpan dredge. Next, the dustpan dredge sucks up the deposited material and pumps it ashore to a disposal area.

Environmental effect of dustpan dredge operations. The effect of a dustpan dredge operation is limited to creation of turbulence and suspension of sediment around the dustpan suction head. Since dustpan dredges operate mainly in large rivers to maintain the navigation channels and since the rivers, such as Missouri and Mississippi transport non-cohesive sediments, there is little fine material thrown into suspension by the dustpan head.

Hydraulic pipeline cutterhead dredge

This is probably the best-known dredging vessel as well as the most efficient and versatile. It differs from the dustpan dredge in that it is equipped with a rotating cutter apparatus surrounding the intake end of the suction pipe. This dredge is generally equipped with two stern spuds which are used to advance the dredge into the cut or excavating area. It can efficiently dig (Fig. 3) and pump all types of alluvial materials, and also compacted deposits such as clay and hardpan. The larger and more powerful machines are used to dredge rock-like formations such as coral and the softer types of basalt

Fig. 3. Cutterhead dredge (courtesy Williams—McWilliams Company, New Orleans, Lousiana).

Fig. 4. Floating discharge line (22-in. or ~56 cm) showing ball and bell connection at the stem of a dredge (courtesy Williams—McWilliams Company, New Orleans, Louisiana).

and limestone without blasting. Some of these dredges have been known to excavate and transport boulders in sizes up to 75 cm in diameter.

In a typical application of cutterhead dredging, the discharge line from a dredge is connected with a flexible pipe or ball-joints to a floating pipeline. The pipeline is supported on small pontoons and is made quite flexible to permit the dredge to advance and pump continuously. The additional sections may be added at the discharge end of the pipelines to increase the conveyance distance. Booster pumps may be added in the discharge line to permit the pumping of dredged material to greater distances. A 55-cm floating discharge line is shown in Fig. 4. A well-designed 75-cm dredge (size is stated in the diameter of the discharge pipe) with 5000—8000 hp. (3729—6000 kW) on the pump and 2000 hp. (1490 kW) on the cutter will pump 1530—3440 m³ hr.⁻¹ in soft material, and 150—1530 m³ hr.⁻¹ in soft to medium hard rock through pipeline lengths up to 4.5 km.

The cutterhead dredge is considered an American specialty. Nowhere else in the world has this type of machine been so highly developed and so widely used in submarine excavation. However, in recent years many foreign countries, notably The Netherlands and Japan have manufactured an increasing number of cutterhead dredges. This type of dredge is best suited for maintenance and improvement of waterways. The cutterhead dredge is currently the most widely used of all dredges.

Environmental effect of cutterhead dredge operations. The environmental effect of a cutterhead-dredge action is generally limited to the operation of the rotating cutter, and the resulting creation of turbulence and disturbance of bottom material, as the cutterhead swings from one side to the other (Fig. 5). The environmental effects depend almost entirely on the type

Fig. 5. Turbidity created by dredge plant equipment. (From Huston and Huston, 1976.)

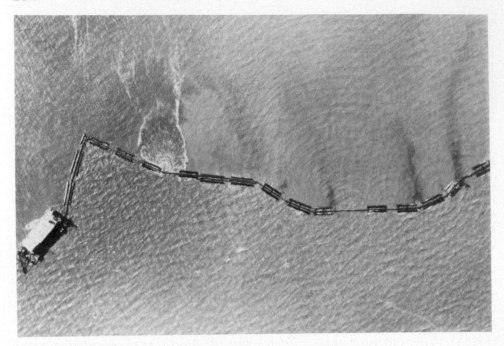

Fig. 6. Turbidity created by leaking floating line connections. (From Huston and Huston, 1976.)

of material being dredged, on the presence of pollutants in the material being dredged and on the presence of pollutants in the material. The effects of unpolluted medium-to-coarse sands is minimal, because the sediment will settle within a short time and within a short distance from the bottom cut being made.

The fine materials such as silts, and to some extent clays, will be thrown into suspension by the cutter. These will remain in suspension for some time, depending on the currents and wave action at the time of dredging. If these fine materials are polluted, the effect may be appreciable, depending on the type of pollutant, or where the fine material settles and on the marine or riverine biological life. The effect of moving the dredge by spuds or wires and anchors is minimal, although some fine sediments may be released to the water column as the spuds are pulled out of the bottom, as part of the "walking" action of the cutterhead dredge. Some turbidity may be created from leaks along the joints of the floating (Figs. 5 and 6), or submerged lines. However, leaks of this type can be corrected quickly by an alert operator (Huston and Huston, 1976) (Figs. 7 and 8).

Self-propelled trailing suction hopper dredge
A sea-going type hopper dredge has the molded hull and lines of an ocean vessel and functions in a manner similar to the cutterhead suction

Fig. 7. Broken floating line and resulting turbidity. (From Huston and Huston, 1976.)

type. The bottom material is raised by dredge pumps through dragarms which are connected to the ship by trunions. The dredge pumps lift the mixture of solids and water through the dragheads to the surface where it is discharged into hoppers. After the hoppers are filled, the dragarms are raised and the dredge proceeds at full speed to the disposal area where it empties the loaded hoppers through bottom doors. Fig. 9 shows a typical hopper dredge.

Hopper dredges require a greater draft than cutterhead dredges and cannot operate in bays or shallow-water areas. However, hopper dredges can be used in deeper, tributary ship channels.

Environmental effect of hopper dredge operations. The environmental effects of a hopper dredge operation are caused by the following: (1) draghead; (2) overflow from hoppers; (3) and dumping of material from hoppers.

As the draghead moves along the bottom the material is disturbed and the environmental effects will depend on the presence of pollutants in the material. A turbidity cloud will be generated in silty and in some clayey materials, although to a lesser extent than the cutterhead operation. If a water jet is used in dredging of consolidated fine material, the extent of suspended material will greatly increase.

The total volume of overflow from hoppers depends on the type of

Fig. 8. A. Shore line leakage from holes in pipe. (From Huston and Huston, 1976.)

Fig. 8. B. Shore line leakage two days following Fig. 8A. (From Huston and Huston, 1976.)

Fig. 9. Hopper dredge "Langfitt" (courtesy of U.S. Army Corps of Engineers).

material pumped, since the efficiency of the hopper dredging operations depends to a large extent on the amount of solids accumulated in the hoppers during the time that the draghead is in contact with the sea bottom. Therefore, an operator tries to maximize the volume of solids accumulated in the hoppers. This calls for a certain amount of overflow to permit settlement of sediments in the hoppers. The settlement rates depend on the type of material pumped, i.e. gravel and coarse sand will settle rapidly and the overflow will contain little or no solids whereas the silt—clay mixture will settle very slowly and the overflow may contain a large proportion of solids. Thus the environmental effect while pumping gravels and sands is minimal. However, the pumping of silts and clays will produce extensive turbidity which may remain in suspension for some time, depending on the currents and wave action at the time of dredging. If a highly polluted material is being dredged and the environmental impact at this particular location is considered undesirable, the pumping cycle will be restricted to a single filling of the hoppers, i.e. no overflow. Thus, silts and clays produce a very inefficient operation and increase the unit costs of dredging on this particular project.

Dumping of materials from hoppers at an open-water disposal site occurs very rapidly. R. B. Gordon (1973) estimates that the initial velocity of silty material from a scow in shallow water may be as high as $5\,m\,s^{-1}$, but it decreases to $1\,m\,s^{-1}$ near the bottom. Holliday et al. (1978) indicate that the dredged material disposed in open water on dredging projects may take from minutes to hours to be deposited at the bottom. Discharge of coarse material will initially be accelerated due to gravity and will reach terminal velocity in constant-density ocean water. The velocity rate for fine materials will also reach the terminal velocity, but it may decrease due to the entrainment of ambient water. While discharging in very deep water, the particles will segregate with the larger particles settling faster and the smaller particles settling at a slower rate. The terminal velocity will be affected by the density changes in the water column, by the currents and to some extent by the storm waves in the upper part of the water column. These factors are

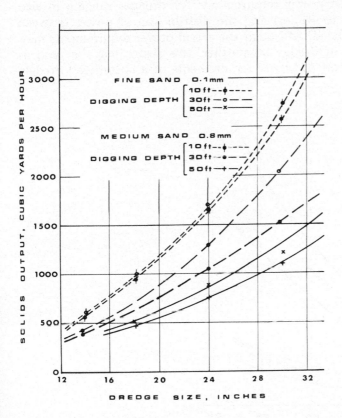

Fig. 10. Typical cutterhead dredge production rate ($1\,yd.^3\,hr.^{-1} = 0.7645\,m^3\,hr.^{-1}$; 1 in. = 2.54 cm; 1 ft. = 30.48 cm).

discussed in more detail in the next chapter. Additional detailed basic discussions on the role of particulates in the ocean are presented by Eisma (Chapter 9) and specifically with respect to ocean mining by Morgan (Chapter 14), Ozturgut et al. (Chapter 15) and Ichiye and Carnes (Chapter 16).

POWER REQUIREMENTS AND DREDGE PRODUCTION

To gain some perspective as to the volumes of material handled by hydraulic-cutterhead dredges, typical production rates are shown for dredges ranging in size from 12—30 in. (30.5—76.2 cm). The size refers to the discharge pipe from the dredge pump (Fig. 10). For example, a 20-in. (51-cm) dredge should deliver between 550 and 1200 yd.^{3} hr.$^{-1}$ (420 and 917.5 m^3 hr.$^{-1}$) of solids whereas a 30-in. (76-cm) dredge will have a solids output between 1200 and 2700 yd.^{3} hr.$^{-1}$ (917.5 and 2064 m^3 hr.$^{-1}$). The actual rates will depend on the material pumped and the dredging depth.

Fig. 11 indicates horse power requirements for dredges ranging in size from 12 to 32 in. (30.5 to 81 cm) and the distribution of power between the pump and the cutterhead. Although the actual power requirements may vary from one category of dredge to another, this figure may be used as a rough guide for selection of power requirements of a cutterhead dredge.

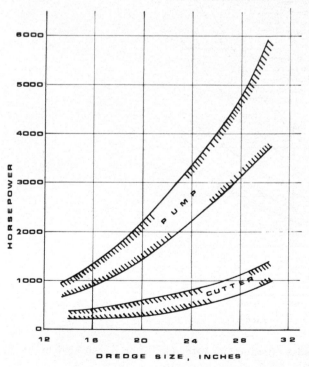

Fig. 11. Typical cutterhead dredge power requirements (1 hp. = 0.7457 kW; 1 in. = 2.54 cm).

CHAPTER 7

ENVIRONMENTAL EFFECTS OF UNCONFINED AND CONFINED DISPOSAL OF DREDGED MATERIALS IN OPEN WATER

John B. Herbich

EFFECT OF UNCONFINED DISPOSAL OF DREDGED MATERIALS IN OPEN WATER

Introduction

Approximately $2.3 \cdot 10^8$ m^3 of sediment is removed annually from the harbors and navigable waters in the U.S.A. as part of the maintenance program to maintain a minimum water depth for safe navigation. Some $6 \cdot 10^7$ m^3 are removed as part of the new work dredging operations. This important task is administered by the U.S. Army Corps of Engineers. In recent years the disposal of dredged material has proved to be one of the major problems facing the Corps of Engineers in meeting its responsibilities of maintaining and improving navigable waters.

In recent years many plans have been advanced to deepen U.S.A. ports in order to utilize more effectively the advantage of modern shipping technology, to remain competitive in the world trade market and to reduce the cost of goods and crude materials shipped to this country. In Texas, there are proposals to deepen the Galveston ship channel and Port to a 17 m depth; to deepen the outer portion of the Corpus Christi ship channel to a 24 m depth and dredge a turning basin and harbor at Harbor Island (near Port Aransas); and construct a 12-m ship channel and turning basin for a LNG (liquid natural gas) terminal in Matagorda Bay. These proposed projects will require very large dredging operations.

Case histories

Effect of disposal in open water

The effect of open-water disposal prior to the 1960's was generally considered minimal with the exception of disposal of toxic and radioactive materials. Few studies were conducted on the environmental impact and open-water disposal was common practice prior to the early 1970's. Several studies were conducted since then, particularly as part of the U.S. Army Engineers Dredged Material Research Program (D.M.R.P.) (DiSalvo et al., 1977; Oliver et al., 1977; W. D. Barnard, 1978; Brannon, 1978; Burks and Engler, 1978; Hirsch et al., 1978; Holliday, 1978; Neff et al., 1978; T. D. Wright, 1978), and by others (Bassi and Basco, 1974; T. D. Wright et al.,

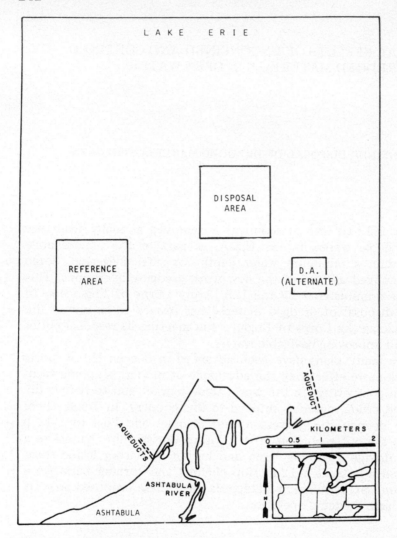

Fig. 1. Locations of disposal and reference areas, Ashtabula River disposal site, Ohio. (From T. D. Wright, 1978.)

1975; Blazevich et al., 1977; Pequegnat and Smith, 1977; Sly, 1977; Sullivan and Hancock, 1977). Five selected sites were evaluated as part of the D.M.R.P.: (1) Ashtabula, Ohio; (2) Columbia River, Oregon; (3) Duwamish Waterway, Washington; (4) Eatons Neck, New York; and (5) Galveston, Texas.

(1) Disposal site was in Lake Erie (fresh water) within 5 km from the shore. Dredged material consisted of both contaminated and uncontaminated sediments from the harbor and was disposed from a hopper dredge (Fig. 1).

(2) Disposal site was at the mouth of Columbia River draining into the

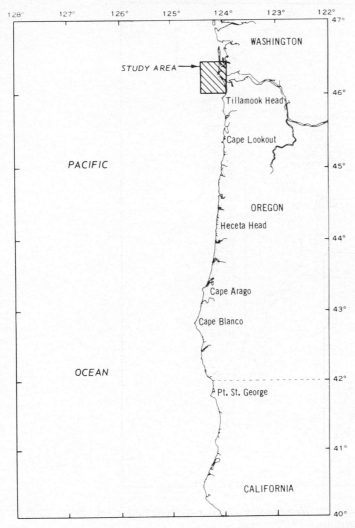

Fig. 2. Location of study area, Columbia River disposal site, Oregon. (From T. D. Wright, 1978.)

Pacific Ocean. Dredged material consisted of fine- to medium-grained clean sand and disposal was from a hopper dredge in shallow water (Fig. 2).

(3) Disposal site was in Puget Sound in 60 m depth of water. Highly-polluted dredged materials (containing polychlorinated biphenyls — PCB's, petroleum hydrocarbons and metals) were disposed from barges (Fig. 3).

(4) Disposal site was Long Island Sound in 12—55 m depth of water. Polluted, fine-grained, mechanically-dredged material was placed at this site for over 75 years (Fig. 4).

(5) Disposal site was in the Gulf of Mexico. Dredged material consisted

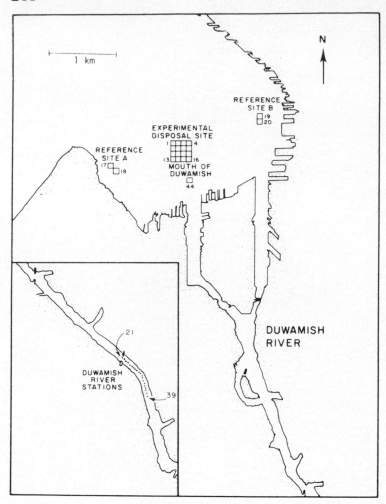

Fig. 3. Locations of Duwamish Waterway disposal and reference sites, Puget Sound, Washington. (From T. D. Wright.)

of clean fine- and coarse-grained sediments and disposal was from a hopper dredge in 20 m of water. Some contaminated sediments were also placed at this site (Fig. 5).

Results from the Dredged Material Research Program. The results of the five above-mentioned studies were somewhat unexpected. Because of the environmental concerns many felt that major short-term effects and substantial long-term environmental impacts would result from disposal of dredged material in open water. Instead, it appears that open-water disposal at all five sites had a negligible effect upon physical, biological and chemical

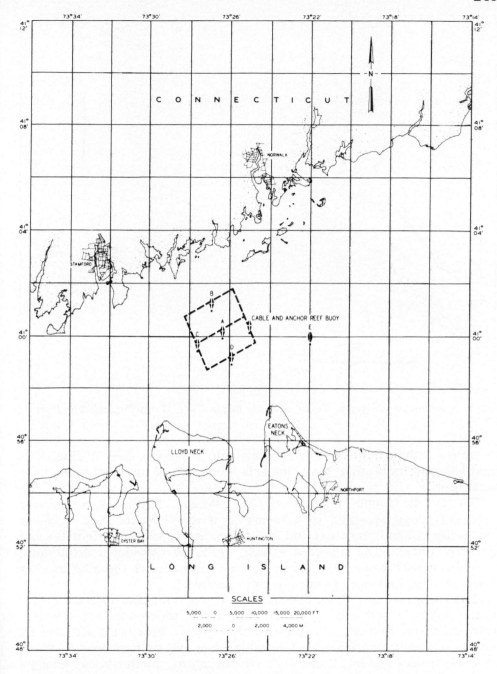

Fig. 4. Locations of the Eatons Neck Disposal site, the northern extended area, and the marker buoys. (From T. D. Wright, 1978.)

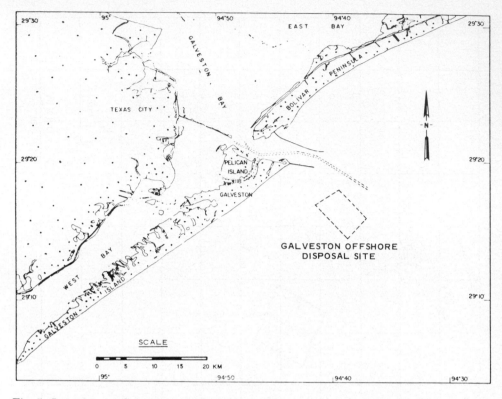

Fig. 5. Location of Galveston, Texas, offshore disposal site. (From Pequegnat and Smith, 1977).

variables. Although the results are based on studies at specific sites, it is significant to note that most impacts observed, if any, were relatively short term.

The specific results, reported by T. D. Wright (1978), are as follows:

"(1a) Disposal appeared to have minor, if any, impacts upon sediment and water column chemistry but did affect the benthic invertebrates and demersal finfish. The observed effects would be considered deleterious, with the impact being more severe for the invertebrates (because of slow recovery) than for finfish (relatively rapid recovery).

(1b) There may have been chemical changes and impacts upon the nature of the sediment and upon planktonic organisms which were not detected. This unknown, in part, is a result of the disposal area being affected by natural sedimentation from the Columbia River and of there being a variety of water masses present. Under such circumstances, the day-to-day changes are so great that unless an almost continual sampling effort is established, natural change and variation tend to obscure disposal effects. Likewise, because variations from a "normal" or average condition in such an area are large, the organisms present are adapted to cope with such variations, and large changes must often occur before there is a measurable organism response.

(2a) Disposal of material contaminated with PCB's in Elliott Bay during the Duwamish Waterway ADFI* appeared to have minimal impact. A disposal mound was created which gradually spread during the post-disposal period. There were minor changes in the chemistry of the water column. These appeared to be associated with a transitory increase in suspended particulate material, and, as soon as this material had settled, values for chemical parameters returned to predisposal conditions.

(2b) There was no significant uptake of PCB's or metals by organisms inhibiting the disposal area or by caged animals which were held in close proximity to the disposed material for up to three weeks. Some changes were noted in the abundance, diversity, and species composition of benthic invertebrates in the disposal area; however, similar changes in the reference area populations make it unlikely that disposal was wholly responsible for the changes.

(3) The Eatons Neck ADFI indicated that, 4 yr. after the cessation of 75 yr. of disposal, little in the way of chemical and biological effects could be observed although the disposal mounds were still evident. However, caution is advised concerning this conclusion because the times and/or locations of sampling were such that natural fluctuations and human perturbations (such as sewage, industrial waste, and river discharge) could not be adequately taken into account.

(4) The Galveston ADFI showed few impacts as a result of disposal operations. With the exception of Mn and NH_4-N, there was virtually no change in water column chemistry, and even these changes were relatively minor. Disposal resulted in the formation of distinct mounds; these were eroded by waves and currents, with the rate and degree of erosion being a function of water depth and the cohesiveness of the material. There was a suggestion of an impact upon benthic invertebrates. However, organism behavior in the reference areas was sufficiently similar to that in the disposal area that no definitive impacts could be firmly established. It was found that the disposal site was a highly variable and dynamic system with natural perturbations in the chemistry, physical aspects, and biota being very large. These large variations may have served to obscure impacts associated with dredged material disposal.

(5) There were but few important impacts as a result of dredged material disposal at the Ashtabula disposal site. Some chemical changes were observed, but these were of small magnitude and transient in nature. There were changes in the benthic community which persisted throughout the study; these primarily consisted of species replacement and an

---

* Aquatic Disposal Field Investigations (ADFI).

increase in the abundance of some organisms. Because these benthic organisms are of importance as food for fish, these changes would be of concern were it not that the feeding activities of fish in the area did not seem to be altered."

Results from earlier studies

Earlier studies included those by Gunter et al. (1964) and Briggs (1968) and the Mobile District of the U.S. Army Engineers (Anonymous, 1978). Briggs (1968) studying the fate of dredged material deposited in open water next to the Mississippi River Gulf of Mexico outlet reported that it spread as much as 1.2 km from the dump area under the influence of waves and currents. The median diameter of sediment was about 0.09 mm and the greatest movement rates were observed in 3 m of water. Gunter et al. (1964) evaluated the effect of dredged material deposited alongside a ship channel in Chesapeake Bay. They concluded that the disposal mounds did not change the total water transport or the sediment rates and the sedimentation was not detectable at a distance of about 305 m from the discharge pipe outlet. The total amount of fine sediment re-suspended in bay water from the dredging operations was only 3—5% of the sediment carried by the rivers into the bay. The Mobile District of the U.S. Army Engineers (Anonymous, 1978) evaluated the movement of suspended sediments from dredging operations near oyster producing beds. It was found that the sediment deposition on oyster beds was negligible and the lateral movement of sediment from the discharge pipe was about 365 m.

Other studies were conducted by Harrison (1967), Cronin (1970), O'Neal and Sceva (1971), Saila et al. (1972), Sustar and Ecker (1972) and Bassi and Basco (1974). Harrison (1967) studied the effect of deposited dredged material ~ 1.3—3.2 km from the oyster reefs on live oysters. Saila et al. (1972) concluded that adult lobsters were able to survive turbidities encountered near the dumping grounds for relatively short periods when the sediment remained in suspension. Survival rates for lobsters, moved later to clear water, were found to be high. Bassi and Basco (1974) described a field study of an open area disposal site in the Galveston Bay. They found tidal-generated currents to be the predominant force directing gross movements of dredged material. Dredged material was deposited up to 1 km from the point of discharge within one week following dredging operations. As much as 47% of the material placed inside the confines of the disposal area was carried out of the area within one week of deposition. Sixty-three percent of the material was carried out of the area in five months.

Pequegnat and Smith (1977) studied the potential effects of deep-ocean disposal of dredged material. Fig. 6 shows the effect of thermoclines, pycnoclines, and current on the discharge plume of dredged material from a hopper dredge. The authors also discussed the biological impacts of dredged material on phytoplankton, zooplankton, nekton, and on benthos. Their

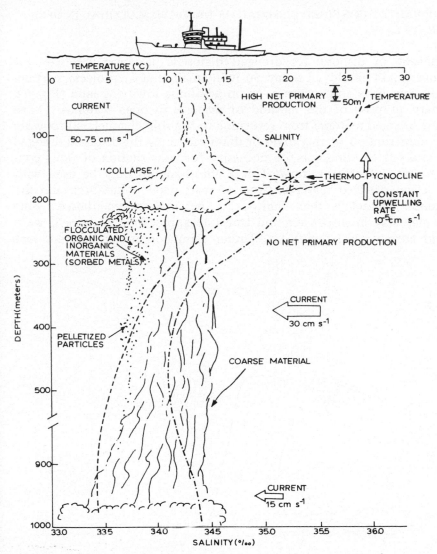

Fig. 6. A schematic view of hopper dredge discharging about 1500 m³ of dredged material in the deep water. Note the thermo- and pycnoclines and the current structure (from Pequegnat and Smith, 1977).

conclusions are that the impacts of dredged materials on the above are either negligible or that the disposal will not cause significant or long-lasting damage to marine organisms. Their final conclusion is that deep-ocean disposal of dredged material is environmentally sound and such action will reduce stresses on the more valuable environments in shallow water presently used as disposal sites.

EFFECT OF PARTLY CONFINED DISPOSAL OF DREDGED MATERIAL IN OPEN
WATER OR ON LAND

This method of disposal for dredged materials has been practiced for
decades and is still employed today on many maintenance projects, particularly in shallow bays and estuaries, or on adjoining low-lying land (Fig. 7).
The primary purpose of this method of disposal in open water is: (1) to
prevent the dredged material from returning to the ship channel; (2) to avoid
expensive construction of ring dikes or dikes confining the disposed dredged
material; and (3) to eliminate the necessity for construction of weirs controlling the rate of effluent flow. This method should only be used with
relatively clean dredged material (free of man-made pollutants), and the
environmental impact is then limited to increased sedimentation. Fine
sediment may stay in suspension and travel an appreciable distance if wave
action and currents are significant. Under calm conditions the sediment

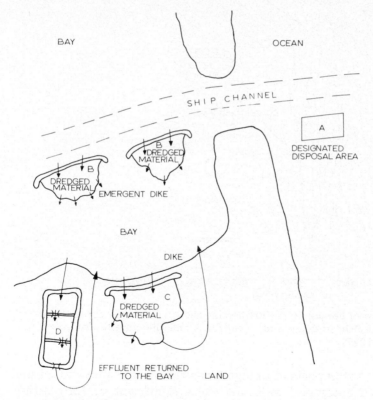

Fig. 7. Disposal of dredged material in unconfined and partially-confined areas in open
water or on land.
A = unconfined disposal in open water; B = partially-confined disposal in open water;
C = partially-confined disposal on land; and D = confined disposal on land.

settles in a fairly short time. Spreading of the dredged material under action of waves and currents causes the longer-term environmental impact. The deposited material forms flat slopes and consequently covers a large area of the original bay bottom. Some marine life burrows up through the freshly-deposited material during the dredging operations and/or during the significant wave action. In most cases marine life quickly repopulates the newly-formed slopes, as discussed subsequently on p. 259. This method of partly-confined disposal is generally limited to maintenance dredging operations, since the disposal areas have been designated as such, and the emerging islands have been used for decades. The disposal of dredged material from the construction of new channels in environmentally-sensitive areas is normally onland or in confined areas in open water.

Disposal in partly-confined areas on land is feasible since the effluent usually becomes of acceptable quality during its travel over land areas, through depressions and finally returning to the bay or open ocean through streams. The fine and coarse sediments are deposited along the way and relatively clear water is returned to the system.

Confined disposal of dredged material in open water

Dredged material may be disposed in open-water contained areas to: (a) reduce the area of bay bottom covered by dredged material, (b) prevent spreading of dredged material deposited in open water in high-energy environment, or (c) build an artificial island for future wildlife habitat or bird sanctuary.

Containment levees or dikes must be first constructed prior to the commencement of dredging operations, unless the dikes are to be built from dredged materials. For a successful operation, the dikes must be emergent and sufficiently high to prevent overtopping by high tides, storms, surges or waves. In addition, the exterior surface of the levees should be protected to prevent erosion by waves and currents. To prevent the dikes from assuming flat slopes, and thus covering large areas of bay bottom, they should be constructed from either gravel, coarse to medium sand, clay balls, etc. The dikes may be constructed by hydraulic dredge, if suitable materials are available, or by mechanical means. The exterior surface should be protected by rip-rap or artificial blocks placed on a filter cloth. The extent of protection required will depend on the wave climate and expected currents. It is important that dikes be designed and constructed well since, in case of failure, large volumes of dredged material will be released to the open body of water.

The method of disposal in open-water confined areas is expensive but the environmental effects are minimized to the areal coverage of bay bottom. The rate and quality of effluent is controlled by a weir system incorporated in the confining dike. The quality of effluent depends on many factors; however, it is important to note that it can be regulated.

Confined disposal of dredged material on land

Containment dikes are constructed prior to the commencement of dredging operations, as in the open-water case, but the procedure is somewhat simplified and the cost is reduced. Confining dikes reduce the land area required for disposal and allow retention of effluent until it attains suitable quality before release to the system. The environmental effect is limited to the coverage of land area by dredged material and the disposal area becomes temporarily unsuitable as a habitat for land animals and birds. However, some time after dredging is completed, when the dredged material is consolidated and revegetated, land will again regain productivity.

EFFECT OF CONFINED DISPOSAL OF DREDGED MATERIAL IN OPEN WATER

Dredged material may be disposed in open water inside a confined area created by a previously constructed ring (or rectangular-shaped levees). In this type of disposal only the area inside the levees is affected, but the dredged material cannot spread under the action of waves and/or currents as in the unconfined, open-water disposal discussed previously. For this method of disposal to be effective, the islands should be emergent, and the exterior surface of the levees should be protected to prevent erosion by waves and/or currents. To prevent levees from assuming flat slopes, they should be constructed by hydraulic dredging methods from either clay balls, coarse to medium sands; or by a dragline or clam-shell mechanical dredges. Levees must be properly constructed on suitable foundations to prevent failures which would allow the fluid dredged material to flow through a breach back to the surrounding waters.

Retention levees' or dikes' slopes can be as steep as 20% (one vertical on five horizontal) while maintaining an adequate factor of safety against slope and foundation failures. The height of dredged materials placed inside the diked area should be controlled to avoid overstressing dikes and/or marine foundations. Typically, the dredged material is pumped into the enclosed area for a given length of time, then the material is allowed to settle and the effluent is discharged over the weir structure installed in one section of the dike. The weir height is controlled to permit only the discharge of relatively clear water which conforms to the local environmental standards for quality of effluent. When most of the effluent is discharged from the confined area, dredging can resume and another layer of dredged material forms on the previously deposited layer. If the disposal area is large and the sediment coarse enough to permit a relatively fast rate of particle settlement, the waiting period may be short; or sometimes the dredged material may be pumped on one side of the disposal area, while the effluent is discharged onto open water at the other end of the disposal area. Proper management and inspection normally results in an efficient operation with a minimum impact on the environment.

From an economic point of view the maximum depth of water for confined disposal in open water subject to waves and/or currents would be $\sim 8\,\text{m}$. The actual optimum depth is a function of wave climate, magnitude of current and sediment size. Erosion protection of the outer portions of the dike is required to prevent sliding and dike failure. Such protection may be in the form of rip-rap of natural or artificial stone and the extent of protection depends on the location of the disposal island.

There are two main advantages of dredged material disposal in a confined area in open water: (1) the quality of effluent is controlled throughout the operation and after dredging operations have ceased; and (2) when stabilized, the disposal area in open water becomes a permanent island suitable for bird habitats.

Additional detailed discussions of the fundamentals of the role of particulates in the ocean are presented in this volume by Eisma (Chapter 9) and specifically with respect to deep-ocean mining by Morgan (Chapter 14) and Ozturgut et al. (Chapter 15), and on shallow-water disposal off the east coast of the U.S.A. by Gunnerson (Chapter 11) and Simpson et al. (Chapter 12).

EVALUATION OF DISPOSAL SITES

Unconfined disposal in open water

Prior to disposal, it is important to survey potential areas for disposal of dredged material in open water. A judicious selection of the disposal area prevents material from returning to the channel recently dredged and avoids rapid spreading of deposited material (particularly in shallow water) by waves and currents. A pre-dredging disposal area selection survey should obtain data on currents generated by tides and waves; on natural circulation currents; and on wave characteristics (height, period of length) and wave climate.

An example may be related regarding disposal operations in the Delaware estuary (Cable, 1969). It was estimated that maintenance dredging was reduced by 70% since changing from open-water disposal basins close to shore to pumping of the dredged material into "sump rehandler" barges. Formerly, the dredged material was pumped from the open-water deposition areas into confined disposal areas on land. The substantial reduction in maintenance dredging clearly indicates that currents were removing material from the open-water disposal areas.

Partly-confined or confined disposal in open water

A careful survey must also be made regarding the prevailing current, wind, and wave action, so that the retaining dike is placed in such a location as to minimize the loss of disposed material and shoaling of the ship channel. An evaluation should be made as to whether the outside portions of the retaining dike must be protected by vegetation, rip-rap, or other means.

The magnitude and extent of dikes must be limited on poor foundations such as those composed of silts and soft or compressible clays. If the soft marine soils are overstressed, large deformations may occur and may also cause dike failure. In addition, silt and clay foundation will consolidate as the dredged materials are placed in confined or partially-confined areas. Dikes themselves will consolidate under their own weight and the consolidation rates must be estimated prior to construction in order to allow for sufficient freeboard.

Partly-confined and confined disposal on land

Confining dikes are usually constructed with soils obtained from shallow excavations within the disposal site area, or from suitable dredged materials. Prior to construction, surveys should be made to estimate consolidation rates of foundation soils. The location of disposal areas on land is usually made on the basis of topography, type of land to be covered, and foundation strength. Marshy areas are usually excluded on the basis of environmental constraints. Discharge weir locations are selected on the basis of topography and distance from the body of water to which the effluents are discharged. The longest path for the effluent is preferred.

FATE OF DREDGED MATERIALS DISPOSED IN OPEN WATER AND ON LAND

As discussed previously, the fate of dredged materials disposed in open water depends to a great extent on waves and currents. In high-energy areas the dispersion, deposition and even re-suspension of sediment may occur. The environmental impact of clean coarse sediments is minimal; however, as the sediments become finer, some impact can be expected. Concern has been expressed and is considered significant by Brannon (1980) if the fine sediments contain potentially toxic chemical contaminants. Evaluation of the fate of fine-dredged materials is difficult and predictions may not be reliable.

Fluid mud

Fine-grained sediments accumulating at the bottom of streams, dredged channels, and in estuaries, tend to form layers of fluid mud over the more

TABLE I

Classification of water—solids mixtures

Solids concentration $(g\,l^{-1})$	Bulk density $(g\,cm^{-3})$	Type of fluid
0—10	1.0—1.01	turbid
10—200	1.01—1.13	low-density fluid mud
200—400	1.13—1.25	high-density fluid mud
> 400	> 1.25	bottom sediment

dense and sometimes coarser bottom sediments (Nichols et al., 1980). A fluid mud layer may also be formed by cutterhead or draghead operations. The classification in low- and high-density fluid muds by Nichols et al. (1980) is based on the solids concentration of the fluid mud. Following this classification the boundary between turbidity and low-density fluid mud is between 5 and $10\,g\,l^{-1}$, and between low- and high-density fluid muds the boundary is between 175 and $225\,g\,l^{-1}$. Table I summarizes the classification of fluid muds.

Another classification can be made on the basis of the physical process, i.e. independent settling is characteristic of turbidity and hindered settling occurs in low-density fluid mud.

Fluid mud flow generally propagates in all directions from the point of disposal of a hopper dredge or a pipeline. It was estimated by Schubel et al. (1978) that 97—99% of fine-grained dredged material slurry settles rapidly through the water column and impacts on the bottom. Coarser fractions of slurry settle directly under the point of discharge forming mounds, while the finer fractions form fluid mud. As sand content increases over 30%, mounding increases, and mud flow decreases. A mound also forms, or its original size increases for sediment concentrations $> 200\,g\,l^{-1}$. The side slopes of the mound are very flat for this type of material ranging from 1 on 50 to 1 on 2000. The mounds consolidate because of the sediment weight when the fluid—mud concentration exceeds $200\,g\,l^{-1}$.

Dispersion

Material dispersion in the dredging operations is a function of many variables, i.e.:

(material dispersion) $= f_1$ [(characteristics of the dredged material), (nature of dredging), (nature of disposal operations), (physical properties of fluid and sediment), (environmental climate)]

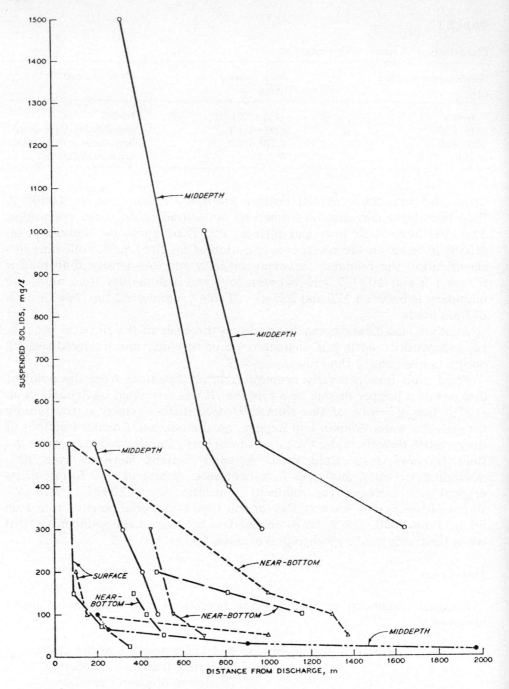

(characteristics of dredged material) $= f_2$ [(specific gravity), (shape), (size distribution), (concentration)]

(nature of dredging) $= f_3$ [(dredge type), (type of suction head), (rate of flow), (type of discharge)]

(nature of disposal operations) $= f_4$ [(type of discharge), (whether open water or on land), (whether confined or unconfined)]

(environmental climate) $= f_5$ [(magnitude of currents), (wave height), (wave period), (frequency of occurrence), (salinity), (temperature)]

EFFECT OF TURBIDITY

Turbidity may be described as a water—solids mixture where solids are in suspension, or settle at a very low rate. Table I indicates that when the solids concentration is between 0 and $10 \, \mathrm{g \, l^{-1}}$ the fluid may be considered turbid.

Field studies indicate that concentrations of turbid clouds decrease rapidly with increasing distance downstream from the discharge point (Fig. 8). (Schubel et al., 1978; Nichols et al., 1980). The concentration also decreases laterally from the centerline due to settling and horizontal dispersion. The turbid cloud (plume) is affected by currents and wave action. Fig. 9 shows a turbid plume generated by a 71-cm pipeline disposal operation in the Atchafalaya Bay (W. D. Barnard, 1978).

The characteristics of a turbid plume are a function of many variables:

Fig. 8. Relationship between suspended solids concentration along the plume center line and distance downcurrent from several open-water pipeline disposal operations measured at the indicated water depths (from W. D. Barnard, 1978)

Symbol	Location	Dredge size		Background suspended solids $(\mathrm{mg \, l^{-1}})$	Total water depth (m)
		(cm)	(in.)		
o———o	Atchafalaya Bay	71	28	30—100	1.8
□———□	Apalachicola Bay	41	16	25—50	3.3
△-------△	Corpus Christi Bay	76	30	15—25	3.3
▽—·—▽	Mobile Bay	60	24	15—25	3.5
●----●	James River	46	18	20—25	2.5

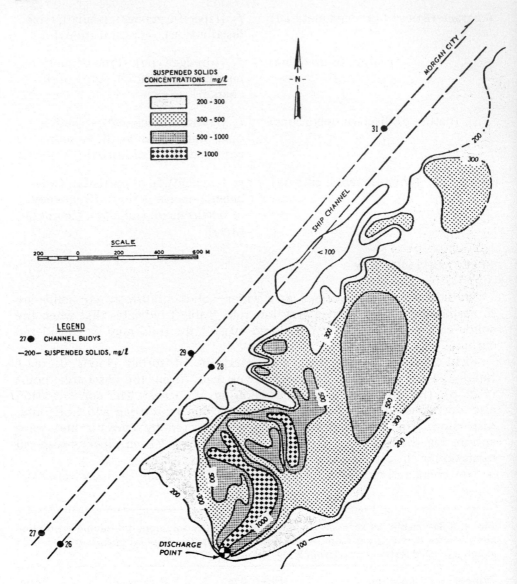

Fig. 9. Mid-depth (0.9 m) turbidity plume generated by a 71-cm pipeline disposal operation in the Atchafalaya Bay. Current flow is generally toward the northeast (from Nichols et al., 1980).

(turbid plume) $= f$ [(rate of flow), (characteristics of slurry), (water depth), (current regime), (wave climate), (type of discharge), (temperature), (salinity), . . .]

In fresh calm water turbid plumes may persist for a long time; however, in a saline environment, whether fresh water sediment is discharged in saline water, or whether the saline sediment is discharged in fresh water, flocculation occurs (Wechsler and Cogley, 1977; DiGeorge and Herbich, 1978). Increasing the salinity tends to intensify flocculation up to a limiting concentration, above which increase concentration has little effect. For example, as water salinity approaches $14^0/_{00}$, the clay particles flocculate and settle out of suspension much faster than in fresh water. One of the advantages of flocculation is that water above the flocculated material is very clear, thus in confined disposal areas, the clear water may be drawn off rapidly thereby exposing the flocculated layers to drying by wind action and evaporation.

CONCLUSIONS

Recent research results from a major research effort by the U.S. Army Engineer Waterways Experiment Station indicate that the original fears of water-quality degradation resulting from the re-suspension of the dredged material during dredging operations are for the most part unfounded.

The magnitude of release of the various contaminants to the water column by Jones and Lee (1978) found little or no correlation between the physical and chemical characteristics. The studies were based on the elutriate laboratory tests and field studies. Manganese was the only metal that was released from the sediments to the water column. Ammonia was also found to be consistently released from the sediments, but the release of phosphorus from the dredged sediments was highly site-specific. Most of the short-term chemical and biological impacts of dredging and disposal have generally been minimal. No significant long-term increase in water-column contaminant concentrations have been observed at any aquatic disposal field sites (Brannon, 1978).

The greatest impact of dredged material disposal is the potential effect on benthic organisms. Most dredged material studied, however, has not been particularly toxic. The toxicity of dredged material can be determined by benthic bioassay procedures (Hirsch et al., 1978).

Effects of dredging and disposal operations on aquatic organisms have been described (Hirsch et al., 1978). Data indicate that recovery of disturbed sites may take from weeks to years and recolonization may be by other, more opportunistic species than the original occupants of the disposal site. Most organisms studied were relatively insensitive to the effects of sediment suspensions, or turbidity. Release of heavy metals and their uptake into organism tissues have been rare. Similarly, the accumulation of oil and grease residues by organisms have been minimal.

For more detailed descriptions of environmental effects, the reader is referred to a series of D.M.R.P. Synthesis Reports listed in the Appendix.

APPENDIX

Some Technical Reports referring to the environmental effects of dredged materials in the series of Dredged Material Research Program (D.M.R.P.) Synthesis Reports of U.S. Army Engineering Waterways Experiment Station

Technical Report No.	Title
DS-78-1	*Aquatic dredged material disposal impacts*
DS-78-2	*Processes affecting the fate of dredged material*
DS-78-3	*Predicting and monitoring dredged material movement*
DS-78-4	*Water quality impacts of aquatic dredged material disposal (laboratory investigations)*
DS-78-5	*Effects of dredging and disposal on aquatic organisms*
DS-78-6	*Evaluation of dredged material pollution potential*
DS-78-7	*Confined disposal area effluent and leachate control (laboratory and field investigations)*
DS-78-8	*Disposal alternatives for contaminated dredged material as a management tool to minimize adverse environmental effects*
DS-78-9	*Assessment of low-ground-pressure equipment in dredged material containment area operation and maintenance*
DS-78-10	*Guidelines for designing, operating, and managing dredged material containment areas*
DS-78-11	*Guidelines for dewatering/densifying confined dredged material*
DS-78-12	*Guidelines for dredged material disposal area reuse management*
DS-78-13	*Prediction and control of dredged material dispersion around dredging and open-water pipeline disposal operations*
DS-78-14	*Treatment of contaminated dredged material*
DS-78-15	*Upland and wetland habitat development with dredged material: ecological considerations*
DS-78-16	*Wetland habitat development with dredged material: engineering and plant propagation*
DS-78-17	*Upland habitat development with dredged material: engineering and plant propagation*
DS-78-18	*Development and management of avian habitat on dredged material islands*
DS-78-19	*An introduction to habitat development on dredged material*
DS-78-20	*Productive land use of dredged material containment areas planning and implementation considerations*
DS-78-21	*Guidance for land improvement using dredged material*

CHAPTER 8

CHANGES IN A HARBOR ECOSYSTEM FOLLOWING "IMPROVED"
WASTE TREATMENT

Dorothy F. Soule

INTRODUCTION

If a group of marine scientists from several universities were asked to
study the impacts of a new industry, before and after the installation of a
waste outfall, and if the group found that the ecosystem had, in their considered judgment, been degraded, one might expect that there would be an
outcry from agencies and citizens alike. Consider then the following examples of impacts found (Soule and Oguri, 1979a, b) over a period of about
five years:

Birds

— All marine-associated birds decreased 2.5 times.
— California gulls decreased 23 times.
— All gulls decreased 4 times.
— Sanderlings were down 11 times.
— Surf scoters decreased 2.5 times.
— Pelicans increased slightly, probably due to cessation of DDT from a
neighboring sewer outfall.
— Least terns increased slightly, perhaps because the local harbor depart-
ment stopped running bulldozers and parking vehicles on their nesting sites.
— All other terns decreased slightly.

Fish

— Total fish, taken by otter trawl, decreased 4 times.
— White croaker were down 10—20 times, depending on area.
— Anchovy in the harbor were down about 100 times, as compared with a
decrease offshore of 4 times.

Benthic fauna

— Organisms per square meter were down about 4 times. (Since bottom-
feeding fish decreased, predation was reduced, so that actual decline would
have been greater.)

These impacts have not been decried by the public agencies but instead have been denied by them as valid. Why? Because the U.S. Environmental Protection Agency (U.S.E.P.A.) had required the fish cannery plants at Terminal Island in Los Angeles Harbor first to install pretreatment [dissolved air flotation (DAF)] units, and then to divert the pretreated wastes to the City of Los Angeles Terminal Island Treatment Plant (T.I.T.P.), which had been converted from primary to secondary waste treatment.

HISTORICAL BACKGROUND

The setting for these studies is San Pedro Bay, where development began in the 1800's on the south coast estuaries which drained Los Angeles basin and the San Gabriel Mountains to the north (Bancroft, 1884; Beecher, 1915; L.A.R.W.Q.C.B., 1969; Kennedy, 1975; A.H.F., 1976). Fig. 1 compares the 1872 map of the estuary with the 1972 configuration, in which Ports of the cities of Los Angeles and Long Beach occupy portions of a single hydrological harbor area.

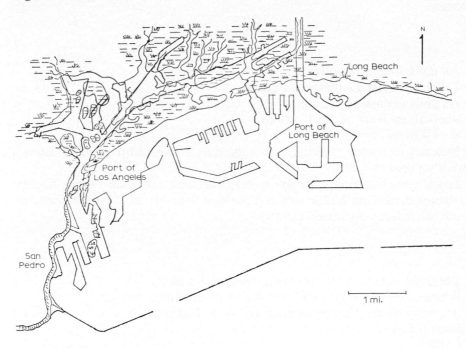

Fig. 1. West San Pedro Bay 1872—1972.

The climate of southern California is Mediterranean, so that rainfall is usually limited to the period between November and April, with an average of $33-35.5 \, \text{cm yr.}^{-1}$, but with a range of from 15 to more than

75 cm yr.$^{-1}$. The winter rains, combined with maximum semi-diurnal tides, create quite different harbor environments in winter than exist in the summer. Also, cool ocean waters from the north are sometimes replaced in winter by warm, subtropical waters from the south. In the years prior to urban development rivers flowed year around, but now that the basin is extensively paved and rapidly drained, the estuarine character of the harbor is limited because salinity gradients are largely transitory.

As development of the harbor and the surrounding urban area increased. pollution of the harbor became extreme. Flows came from adjacent oil fields and refineries, from other industries and from the raw sewage of dockside industry and boats. The rapid expansion of industry and of U.S. Navy activity during World War II culminated afterward in demands for cleanup, since inner-harbor sediments were largely anoxic, without macroscopic life (Reish, 1954) and sulfide fumes damaged boat hulls and sickened workers.

Pollution control

Passage of the California Environmental Quality Act and the National Environmental Policy Act in 1969—1970 provided local authorities with enforcement options to exert control on pollutant dumping in the harbor. Cessation of drainage of oil in industry wastes into inner Los Angeles Harbor at Consolidated Slip (Fig. 2, *C11*) resulted in almost immediate improvement (Reish, 1971), with biota moving into the area. Not an unmixed blessing, because the untreated wood pilings began to collapse from boring organisms, and boat owners, who previously did not have to use anti-fouling paints or scrape hulls, began to complain. However, animals that had previously been found only outside the harbor began to move into the outer harbor, and outer-harbor species started to colonize up-channel and even move into the previously barren blind-end slips.

HARBOR'S RESEARCH

With the coming of environmental legislation in 1970, public agencies and industries were required to produce environmental impact reports (E.I.R.'s) for the State of California and environmental impact statements (E.I.S.'s) for the Federal government. These had to include baseline studies and predict impacts of development of proposed projects. In spite of the presence of a large U.S. Navy base in Long Beach, there were no published oceanographic data, nor were there any harbor-wide biological or water-quality studies on which to base the necessary documents.

Harbors Environmental Projects (H.E.P.)

Since scientists cannot blame decision-makers for mistakes in

264

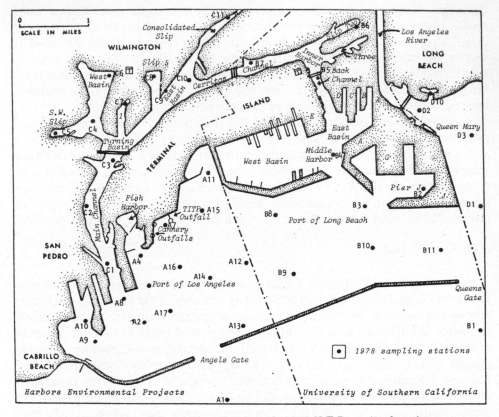

Fig. 2. Ports of Los Angeles and Long Beach with 1978 H.E.P. station locations.

environmental decisions if scientists refuse to come out of the ivory towers and help, Harbors Environmental Projects (H.E.P.) was formed at the University of Southern California (U.S.C.) in 1970 to investigate local marine problems. Rather than to wait for a single agency to arrive with large amounts of grant money, H.E.P. was able to convince a number of agencies and private entities to invest in cooperative programs such that the information gathered would not be proprietary. Also, tasks could be designed so as to integrate data into a baseline wherever possible. It was not possible to have continuous harbor-wide baseline data collected, but such inventories were made in 1973, 1974 and 1978. More limited areas were monitored from 1971 through 1978 — as it happened, in the area of the waste outfalls. This was entirely fortuitous, but it thus became possible to document the changes in the ecology under the changing waste treatment regimes that ensued. Other research projects on thermal tolerance, oil spill and dredging impacts, toxicity and mariculture have also been carried out.

Long-term funding for the harbor studies came from the Southern

California Gas Company, which once planned an LNG plant in the outer harbor near the waste treatment and fish cannery outfalls. Other participants have included the City of Los Angeles and the Port of Los Angeles, the City of Long Beach and the Port of Long Beach, the U.S.E.P.A. the U.S.C. Sea Grant Program (N.O.A.A., U.S. Department of Commerce), the Tuna Research Foundation, StarKist, Van Camp and Pan Pacific seafood corporations, and Union Oil, SOHIO, Shell Oil and ARCO corporations, as well as others. The baseline helped determine the design of the SOHIO terminal (Port of Long Beach, 1976) and to evaluate the impacts of the tanker "Sansinena" explosion and Bunker C spill (Soule and Oguri, 1978; Chapter 19 of Volume 1).

FISH CANNERIES AND THE ECOSYSTEM, 1971—1973

Fish canneries have been located in Los Angeles Harbor since the turn of the century. Prior to 1970, fish offal and scales, and boat hold, flume, thaw and process waters were all dumped into Fish Harbor (Fig. 2). That small slip had ~ 5 m of blackened fish scales for a benthos, with a few capitellid worms occasionally in residence. Zero oxygen readings were frequent. The growing reduction meal industry at the canneries, which processed anchovy and fish scrap for commercial animal feed, made feasible the salvage of all but the suspended particulate matter, instead of dumping the refuse in the harbor. The effluents from the canneries were diverted in 1970 to the east of Fish Harbor, near the T.I.T.P. primary waste outfall. Only ~ 5 m deep, the area was not ideal for waste disposal, but fortunately a large circulation gyre (Fig. 3) dispersed the wastes throughout the outer harbor. This gyre was discovered and mapped (Soule and Oguri, 1972; K.S. Robinson and Porath, 1974; McAnally, 1975). Subsequent research showed that the combined control of pollutants and natural circulation patterns made this the richest inshore, soft-bottom habitat in southern California in 1973—1974 (A.H.F., 1976; Soule and Oguri, 1976).

There were problem periods, however; in the autumn, when the days are very hot and nights are cool, the deeper harbor waters which were warmed up during the summer can "turn over", just as freshwater lakes do. This brings normally anaerobic sediments, bacteria and organic material into the water column and exerts a large chemical or biological oxygen demand (COD, BOD). Sometimes changes in normal wind direction bring strong desert winds (tee so-called Santa Ana winds in this area), which can drive the shallow outer-harbor waters and also result in stirring or turnover. Unknown factors also stimulated periodic red-tides (the non-toxic *Gonyaulax polyedra*). Phytoplankton blooms of various species resulted in dissolved oxygen readings of up to 20 ppm, well above saturation of ~ 9 ppm, followed by near-zero or zero readings when the bloom died. Yet with all these perturbations, blamed on the waste loads, the numbers of

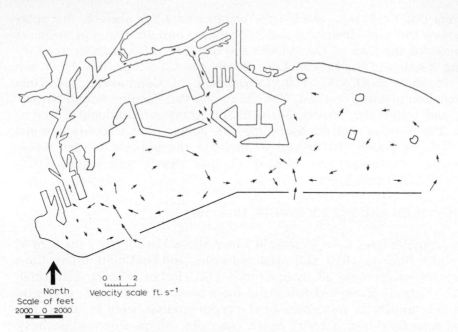

North
Scale of feet
2000 0 2000

0 1 2
Velocity scale ft. s⁻¹

Fig. 3. Surface current patterns — base test, spring tide 6ʰ00ᵐ (after McAnally, 1975).

species and numbers of organisms increased dramatically between 1970 and 1973.

Just as a newly-exposed substrate or other ecological niche is colonized, where numbers increase rapidly to a peak and then tend to level off, the harbor species and numbers increased and then began to level off slightly. However, the juggernaut of regulation had been set in motion by then, and, although the results are disputed by the responsible agencies, the trends in the harbor for the most part have been downhill ever since.

CHANGES IN WASTE TREATMENT, 1973—1978

The thrust in enforcement on the national level was to replace environmental receiving water criteria with technology-based standards, primarily, because the criteria and the technology used in 1970 and thereafter failed, for the most part, to achieve the water quality desired in freshwater lakes and rivers. The old, traditional primary and secondary treatment of domestic wastes, and the monitoring of effluents by drinking water and public health standards did not remove or measure the modern generation of synthetic chemicals, trace metals and other toxic components in urban wastes. Neither did laboratory tests of single chemical entities on single species of organisms predict results that would occur in the field.

It was particularly inappropriate to impose drinking water effluent standards on the marine environment. Nutrients in the sea are largely terrigenous in origin, even as upwelling. Removal of nutrients by feeding them to bacteria grown in secondary waste treatment tanks results in reducing the detested BOD, but denies the complex organic nutrients (proteins, carbohydrates and fats) to marine organisms. Instead, sludge, composed of the dead bacteria that ate the nutrients, is created and must be disposed of on land, according to the U.S.E.P.A. regulations. This concentrates some toxic materials in small areas on land, possibly to re-enter the water table. On the other hand, dilution and dispersion of non-toxic nutrients in marine waters can and do serve to enrich the ecosystem.

Assimilation capacity

Just as a farmer does not fertilize a field by dumping a sack of chemicals on a single plant and expecting both the plant and the field to flourish, in a similar manner too much undispersed waste would be damaging in a limited area of marine waters. Thus, efforts were begun in 1972 to measure the dissolved oxygen in the harbor waters, examine weather conditions, look at the BOD of the cannery wastes and suggest control of the amounts released by the canners. This work on assimilation capacity of the receiving waters has led to modelling efforts (Kremer and Chiang, 1976; Kremer, 1978) and could eventually have led to managed loading. However, management of effluent loads and calculations of assimilation capacity are not options under the present U.S.E.P.A. policy. As a consequence resources are wasted.

Under U.S.E.P.A. technology-based requirements, the fish canneries at Terminal Island began installation of DAF units in 1974. These units produce a wet, brown, cottage-cheese-like sludge, but reduce the BOD and oil significantly in the effluent. Since proteins "salt out" in seawater, suspended coagulum in the effluent was fed on by juvenile fish and gave the water a "café au lait" color (and an odor!) at the outfalls. However, BOD was at ambient levels within 200—300 m of the outfalls. Low-income shoreline anglers fished from outfall pipes with unbaited gang hooks, as larger fish fed on the juveniles in the area.

Meanwhile, public agencies had urged construction of secondary waste treatment facilities at the City of Los Angeles' T.I.T.P. under the Clean Water Act construction grants (really loans) program. It was the E.I.R. for that outfall that led to the 1978 H.E.P. monitoring program. Thus monitoring in the area of the waste discharges had been underway under the following conditions:

1971—1974. Screening of fish cannery wastes; primary treatment of urban wastes at T.I.T.P.
1975—1977. DAF-treated cannery wastes; primary T.I.T.P. wastes.
April—October 1977. DAF-treated cannery wastes, secondary T.I.T.P. effluent.

268

October 1977. One cannery outfall connected to T.I.T.P., one DAF-treated; T.I.T.P. attempting secondary treatment.
January 1978. Both canneries in T.I.T.P., all wastes secondary treatment.
January—May 1978. Variable quality secondary T.I.T.P.
March 9—August 30, 1978. T.I.T.P. chlorinated.
June—August 1978. T.I.T.P. malfunctions; primary treatment, excessive suspended solids, released bacterial floc lost.
September—December 1978. Variable quality secondary T.I.T.P.

THE ECOSYSTEM — 1978

Although it was predicted by H.E.P. that the traditional secondary treatment plant with biological (bacterial) digestion could not tolerate the variations in salinities that characterize the fish cannery effluents, engineers insisted that it could. Operating at only $38,000\,\text{m}^3\,\text{day}^{-1}$ instead of its $114,000\,\text{m}^3\,\text{day}^{-1}$ design capacity, T.I.T.P. has been plagued by loss of the bacterial floc, by higher than estimated suspended solids, fats, BOD, ammonia and sulfide in the effluent. Canners are forced to hold the wastes longer, in an effort to create uniform flow, which has increased the ammonia and sulfide levels. T.I.T.P. has been threatened with "cease-and-desist" penalties.

And what of the ecosystem? When the data for the six-year period 1972—1978 were plotted (Soule and Oguri, 1979a, b), scientists involved in the baseline and ancillary experimental studies were shocked to see the results. However, personnel in regulatory agencies, who have no comparable data base or investigations, have blamed the changes in drought or excess rainfall, cold years, or warm years, and on poor data collection and/or poor analysis techniques. Yet the number of coincidences in trends is persuasive that the trends are substantive.

Methods

Since a number of separate studies were involved in the 1971—1978 period, there were some changes in methodology and some gaps in the harbor-wide coverage. The reader is referred to table I of Chapter 19 in Volume 1 (Vol. 1, p. 500) for the methods of earlier studies; a series of publications by H.E.P., Marine Studies of San Pedro Bay, California, details experimental work as well. The only other long-term investigations in the harbors were carried out from 1973 to March 1978, a pre- and post-discharge monitoring of the Southern California Edison plant thermal discharge in inner Long Beach Harbor. That plant went on line in 1977 and is capable of circulating $2.8\cdot10^6\,\text{m}^3\,\text{day}^{-1}$ of heated effluent. Plant operation is intermittent and variable so that fish attracted to warm waters containing killed plankton one day, may find conditions changed the next. The data are therefore difficult to compare with the T.I.T.P. effluent receiving waters in outer Los Angeles Harbor, especially when the studies at the Edison plant were stopped in March 1978.

RESULTS

The summary of impacts in the Introduction of this chapter presents the more startling results of our research. Probably more important, however, is that the research has revealed a dynamic detrital food web, which has been severely eroded.

The detrital food web

To demonstrate the basis of the detrital food web, radioactive uptake studies were conducted at the microheterotrophic level in 1978 as an extension of earlier uptake studies on fish and invertebrates (Chamberlain, 1976; Bever and Dunn, 1976; Brewer, 1976; Emerson, 1976a, b; McConaugha, 1976a, b; Morey-Gaines, 1978).

There can be little question that bacteria and other micro-organisms recycle carbon rapidly in the marine environment. In addition, Soule and Oguri, (1979a) reported:

"A considerable research effort has been devoted to determine the role of microautotrophs (diatoms, dinoflagellates, monads, etc.) as a nutritional resource for higher trophic levels (zooplankton, suspension and deposit feeders) in aquatic ecosystems. It is now well known that these organisms play a significant role and are considered to be the primary food base for many ecosystems. However, Fenchel and Jørgensen (1977) recently estimated that 40% to almost 100% of the carbon fixed in primary production is utilized by the secondary producers or microheterotrophs (bacteria, yeasts, fungi and Protozoa), depending on the ecosystem in question. Indeed, it does seem reasonable that organisms which rapidly cycle dissolved carbon and produce particulate biomass, such as bacteria, will not go unexploited as a food resource for higher trophic organisms. The role of the detrital food web may be more significant in ecosystem which are either organically enriched or deficient in a necessary component for photosynthesis, i.e. light, nitrogen or phosphate. Pomeroy (1974) has recently pointed out that in the open ocean microheterotrophs play a highly significant role, both in the nutrition of higher trophic organisms and in their long known role in nutrient regeneration."

and

"It is now becoming well established that bacteria serve as a nutritional resource for many aquatic organisms, including planktonic and benthic feeders. The early workers Doflein and Reichenow (1928) stated that some Protozoa, and ciliates in particular, feed upon bacteria. It seemed likely to them that free and attached bacteria were consumed and metabolized in planktonic ecosystems. More recently several investigators (ZoBell and Feltham, 1937; Jørgensen, 1966; Fenchel, 1969, 1972, 1975; Barsdate et al., 1974) have obtained good evidence that bacteria do play a substantial role in the nutrition of deposit and filter feeders. Wavre and Brinkhurst (1971) have demonstrated experimentally that bacteria are digested from the bolus as it passes through the gut of tubificid oligochaetes. Duncan et al. (1974) by means of a simple radioassay, showed that bacteria are ingested and assimilated by the aquatic nematode *Plectus palustris*. Sorokin (1973, 1978) has reported that filter feeders in coral communities, such as sponges, ascidians, sabellid polychaetes, and oysters, are capable of filtering bacterioplankton from the water. He also found that some specias of coral, gastropods, and holothurians were capable of ingesting and assimilating bacterial biomass."

Results demonstrated that:

" ... bacteria do serve as a food source for higher trophic organisms found in Los Angeles Harbor, included Protozoa and some invertebrates. These findings support the proposed importance of the detrital food web in marine ecosystems currently found in the literature. It has been suggested that in many ecosystems the microheterotrophs (bacteria, etc.) play a substantial role in the nutritional support of higher trophic organisms."

The study suggested that:

" ... a shift in dominant species towards species that can utilize bactivory for nutrition will be seen in waters which are organically enriched. There is also evidence which suggests that bactivory is in steady state with bacterial production; i.e. bactivore standing stock and feeding activity will be balanced with bacterial production which, in effect, is dependent on organic input. For example, areas of Los Angeles Harbor which receive organic wastes have a comparatively high rate of bacterioplankton production; and one would expect high rates of feeding and production among the bactivorous plankters and benthic invertebrates in these waters. From the evidence presented here and in the literature, one would expect the microheterotrophs to play a more substantial role as a food base in these organically enriched systems than in phytoplankton-based systems. In conclusion, ecosystems which are bacterially enriched and yet perhaps poor in phytoplankton production may be as productive overall as other phytoplankton-based ecosystems."

It seems clear that a 30-fold decrease that was found in bacterial counts in the harbor following secondary treatment would seriously affect the formerly rich, soft-bottom benthic community that had sometimes numbered more than 60 species and $6 \cdot 10^4$ individuals per square meter.

Fig. 4 shows the long-term trends in annual mean numbers of benthic species and numbers per square meter. The rapid increase between 1971 and 1973 is clear, and a normal leveling off in 1974 could have been expected. However, populations decreased rapidly almost to 1971 levels; conversely, species increased under DAF treatment in 1976 but decreased rapidly in 1977 and 1978. The end result was a 4-fold decrease in mean populations and a 1.4-fold decrease in species.

The decrease in fish species ranged from 4-fold to more than 20-fold for omnivores such as white croakers (the poor man's "sewer trout", sometimes sold illegally as "butter-fish"). The trends in H.E.P. fish trawl data are shown in Figs. 5—7, and compared with the Edison generating plant studies where they overlap.

The question of what happened to the northern anchovy *Engraulis mordax* is well debated; populations offshore peaked in 1973 and declined

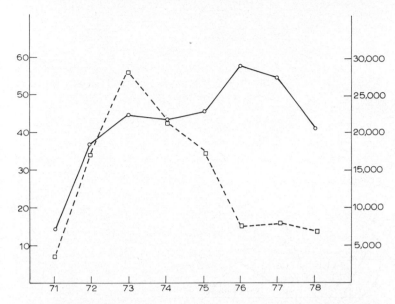

Fig. 4. Annual means of benthic species and number of benthic individuals per square meter for stations sampled, 1971—1978 (multiple grabs enhanced diversity in 1975—1977) (from Soule and Oguri, 1979a).
——o—— = mean number of species per square meter; — —□ — — = mean number of individual square meter.

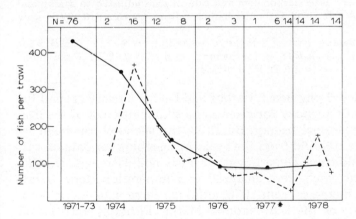

Fig. 5. Mean number of fish per trawl in outer Los Angeles Harbor, 1971—1978.
——●—— = annual mean; — —+ — — = semi-annual or quarterly mean; N = number of trawls. *Asterisk*: trawl records from J. S. Stephens further lowered means from those reported in Soule and Oguri (1979a).

4-fold in 1978 (N.O.A.A., 1978; Humpert, 1979; Stauffer, 1979). This was blamed on warmer-than-normal waters offshore or on "down-welling" phenomena.

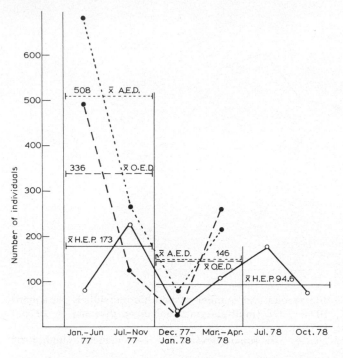

Fig. 6. Comparison of H.E.P. outer-harbor data and Edison data adjusted to season and daytime for all stations (A.E.D.) and for outer-harbor Edison trawls (*T13*, *T15*) only (O.E.D.).
——○—— = H.E.P. $\bar{x}$ trawl data; |————| = H.E.P. $\bar{x}$ for period; · · · ● · · · = A.E.D. $\bar{x}$ for all Edison stations; | · · · · · · | = A.E.D. $\bar{x}$ for period; – – ● – – = O.E.D. $\bar{x}$ for outer-harbor Edison stations; | – – – – | = O.E.D. $\bar{x}$ for period.

The outer Los Angeles—Long Beach Harbor had been a nursery ground for the 0- to 1-yr. age class of anchovy for years and a source of most of the live-bait catch for the recreational fishery. By 1978, the anchovy caught in research trawls was down 100-fold from the peak. In searching for data on offshore fish catches, California Department of Fish and Game (C.D.F.G.) statistics for party boat catch were analyzed to compare long-term trends (Fig. 8). These showed increases from 1976, although these data are not considered to be as reliable as the U.S. National Marine Fisheries statistics on anchovy.

Lasker (1975) has shown that anchovy larvae require dense aggregates of phytoplankton in order to survive the crucial period of limited ability to forage. However, no studies have been made of the role of detritus as larval food. Laboratory experiments by H.E.P. have shown that anchovies can be fed on dried cannery sludge, but gut-content studies of anchovies do not identify the nature of amorphous material in their guts. Anchovies feeding on benthic animals in shallow water and on cannery scum at the surface

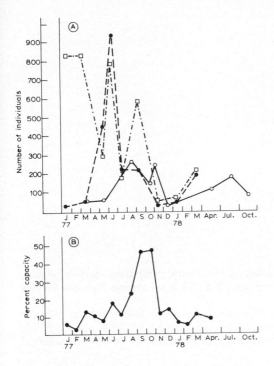

Fig. 7. Comparison of thermal output with means of all Edison trawls, and outer-harbor Edison *T13* and H.E.P. trawls (daytime).
A. Number of individuals; ——○—— = H.E.P. $\bar{x}$ Mar. 1977—Oct. 1978; – – ● – – = Edison trawl *T13* $\bar{x}$ Jan. 1977—Mar. 1978 – · –□ – · – = all Edison trawls $\bar{x}$ Jan. 1977—Mar. 1978.
B. Percent capacity (data from E.Q.A.—M.B.C., 1978).

have been observed casually in the field; unfortunately, anchovies had disappeared from the harbor when uptake studies were begun.

The role of turbidity, due to phytoplankton, sediments, wastes or detritus, has not been considered by agencies so far. Larval and juvenile fish are better protected from predators during critical growth stages in cloudy waters. Clear waters may be esthetically pleasing to the public, but may represent lack of food and cover for fish.

Temperatures were higher than normal in 1977 and 1978, which is proposed as the cause of lack of anchovies and other fish in the harbor. However, plots of mean temperatures (Fig. 9) at the sea buoy outside the harbor entry show that temperature curves have not necessarily coincided with low or high fish catch. Since the anchovy fishery was not well developed until the sardine fishery disappeared, older records from offshore commercial catch are not useful for comparison.

Would anchovies return to the harbor if cannery effluents were resumed? They did return to the sewer outfall, as did white croaker, in July 1978

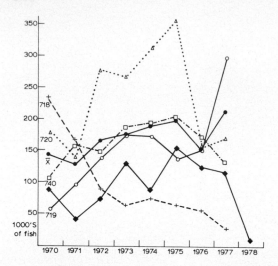

Fig. 8. California Department of Fish and Game data on total party boat catches by year, by block and mean of four blocks, compared with U.S. National Marine Fisheries Service commercial anchovy catch data.

———○——— = block *719* outside Los Angeles Harbor; — — + — — = block *718* off Orange County; — • —□— • — = block *740* Southern Californian Bight; • • △ • • = block *720* off Palos verdes; ———●——— = x, mean of four blocks, compared to: ———◆——— = commercial anchovy catch.

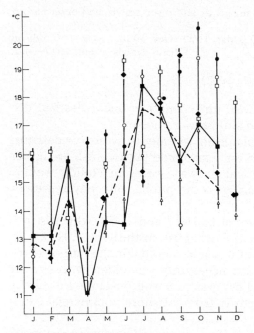

Fig. 9. Surface temperatures at the sea buoy, Station *A1*, outside Los Angeles Harbor.
◆ = 1972; ■ = 1973; ▲ = 1974; △ = 1975; ○ = 1976; □ = 1977; ● = 1978.

when the treatment plant put out quantities of suspended solids. This is of course only a case of attracting fish; recruitment and survival would depend on a steady source of nutrients, preferably organic rather than mineralized nutrients available only to phytoplankton.

Birds

Just as white croaker is called trash fish by some, sea gulls are also called trash gulls, scavengers and nuisances.

Ironically, the C.D.F.G. and environmentalists are currently charging the City of Los Angeles with damaging the nesting site of gulls by pumping water from the Mono Lake basin, some 500 km inland. The halt to pumping which they demand could reduce city water supplies by ~ 20%. In the harbor, a 23-fold drop in California gulls, and a 4-fold drop in all gull species, due to harbor ecology changes, has been written off by that same agency as not important because the gulls are only scavengers. Positions expressed relate perhaps to the fact that environmentalists are in full cry to save Mono Lake, but it seems nobody is concerned about saving harbor gulls.

Intermediate consumers

The detrital food web is apparently more important to higher consumer levels in the harbor than phytoplankton and zooplankton are, if the obvious, large-order decreases are used as criteria. Results of phytoplankton and zooplankton studies tend to confirm this but are more difficult to interpret. There have been no harbor-wide red-tide blooms, nor any along the southern California coast, since 1974. The drop in blooms has reduced values in the harbor for mean chlorophyll *a*, productivity, and assimilation ratios, all phytoplankton measurements (Fig. 10). It is difficult to determine the extent to which a decrease in the consumers of phytoplankton might result in an increase in the standing phytoplankton crop. Although decreases occurred in complex organic nutrients, increases in mineralized nutrients (nitrate, nitrite, ammonia and phosphate) might have been expected with secondary treatment, but little change could be seen other than the usual cyclic winter increases and summer decreases.

Methods for zooplankton sampling were changed at the conclusion of LNG studies in 1977 from horizontal tows to vertical tows. This was done after informal discussions with colleagues and agency personnel about comparability with vertical sampling data taken by plankton pump in the Edison power plant studies. Oblique tows, such as are used in open waters, cannot be used in the confines of the harbor, where wreckage and debris litter the outfalls area and adjacent bottoms. Slight increases in numbers of species were obtained by vertical tows as compared with prior horizontal tows.

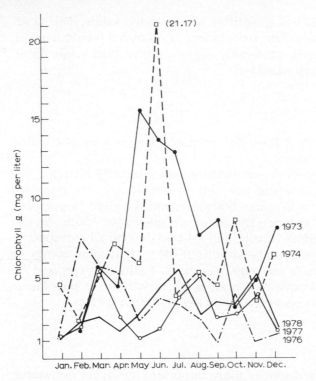

Fig. 10. Mean chlorophyll *a* for Stations *A1—A4*, *A7* and *A8* compared for 1973, 1974, 1976, 1977 and 1978.

However, it is unfortunate that the historic data on total concentrations could not be directly compared with the 1978 data with confidence, although comparison tests were run. (Agencies might have been happier if none of the studies had gone back farther than the fall of 1977, as required for the E.I.R., for results would not have been so dramatic.) It appeared that zooplankton (copepod and cladoceran) concentrations were reduced greatly in 1977, as compared with peaks of 84,000 and 20,000 individuals per cubic meter in the summers of 1976 and 1977, respectively.

The decrease in predator fish could have resulted in an increase in total zooplankton, but all indications are that this was not the case.

CONCLUSIONS

This case history of fish processing and urban wastes has provided some interesting insights. First of all, a complete baseline, representing physical, chemical and biological aspects of the water column and the benthos, is essential to understanding an ecosystem. Without the long-term sampling of multiple parameters no real trends could have been seen under the unstable conditions that followed in one year after conversion to secondary effluent.

Secondly, it seems fairly clear that the blanket imposition by regulatory agencies of traditional wastewater treatment for freshwater and marine environments alike does not offer any flexibility to evaluate the impacts of that imposition. At present, the law allows for alternative approaches, but the regulations selected to implement the law have not been responsive to that position.

Thirdly, it seems that a valuable non-toxic resource in the form of cannery wastes must be destroyed rather than used to enhance the natural biological environment. The C.D.F.G. has responded that they approve of single-species aquaculture (they raise trout to restock freshwater streams, but both U.S.E.P.A. and C.D.F.G. objected to a proposed application for an open-water mariculture (polyculture) permit; so far only power plants have received these apparently. Meanwhile, most of the reduction plant meal that is made from anchovies and creates part of the wastes, is used for feed at C.D.F.G. and other freshwater trout hatcheries! They also have indicated that if H.E.P. could raise fish experimentally on liquid waste or sludge, it should be marketed and not returned to the sea.

The California Bays and Estuaries Policy allows for effluents only if "enhancement" can be demonstrated. To date, enhancement has been permitted only as it involved release of treated secondary effluent to a river that is normally dry in summer. Biological enhancement (bioenhancement) has been discussed at length but U.S.E.P.A. has so far refused to define or accept definitions or to accept criteria developed for the concept. Meanwhile, expensive new treatment for U.S.A. fish processors is proposed. The cost could halt efforts by the U.S. Department of Commerce and National Marine Fisheries Service to develop the U.S.A. fishing fleet under the 200-mile limits. Thus, both economic and environmental degradation may be the result of inflexible regulations.

In physical and biological systems the principle of feedback is essential to midcourse correction. In environmental regulation there seems to be no provision for feedback, perhaps because of the potential for loss of control or territoriality. In an article in the *Bulletin of the California Water Pollution Control Association* (Soule et al., 1978) on the situation in Los Angeles Harbor, the editor changed the authors' title from "Urban and fish-processing wastes in the marine environment: bioenhancement studies at terminal Island, California" to read "You can tailor effluent BOD to fit the receiving-water ecosystem . . . and enhance the environment (BUT YOU'D BETTER NOT)".

Section B — Ocean Disposal and Mining

CHAPTER 9

SUSPENDED MATTER AS A CARRIER FOR POLLUTANTS IN ESTUARIES AND THE SEA

D. Eisma

INTRODUCTION

Material in suspension in estuaries and the seas comes from a variety of sources: rivers, the atmosphere, direct runoff from land, ice flows and ice rafting, coastal erosion, sea-floor erosion, organisms and waste discharges in particulate form. Pollutants interact with the particulate matter in suspension: the interaction depends largely upon the conditions in the direct vicinity of the particles as well as upon the nature of the particles and the pollutants involved. Therefore the pollutants contained in suspended matter can be expected to have a very mixed history. The possible harmful effects of pollutants are very much related to their degree of accumulation, either through adsorption to particles and other chemical as well as sedimentary accumulation processes, or through biological accumulation. Particulate suspended matter plays a large role in most of the inorganic processes but may also be an important link in biological accumulation.

Until recently, the study of particulate matter in suspension in the sea and in estuaries was mainly concerned with the total mass of suspended material and its bulk composition, with its distribution and deposition in river mouths, harbours and in the sea, and with its importance as a source of food for organisms. Particle size was only imperfectly known but particle size studies improved greatly with the development of the Coulter counter (Sheldon and Parsons, 1967). Also it was recognised that suspended particles are important regulators of seawater composition as well as a means of transport of substances from coastal area to the open sea or from the surface of the ocean to the deep-sea floor (Bewers and Yeats, 1977; Bishop et al., 1977; Turekian, 1977). After 1960, when pollution problems began to receive widespread attention and the development of analytical techniques made it possible to carry out reliable pollution studies in estuaries and the sea, it was realized that particulate matter in suspension is also an important carrier for pollutants, inorganic as well as organic (W.E. Odum et al., 1969; Turekian, 1977; Duinker, 1980).

Interaction between pollutants, suspended particles, bottom sediments, organisms and the surrounding water is especially strong where sharp gradients in chemical and physico-chemical conditions occur as in estuaries. These are typically transitional areas where freshwater from land with a very low salt content [estimated average $0.12^0/_{00}$, Livingstone (1963)] mixes with seawater of $\sim 35^0/_{00}$ salinity. Estuaries are complex systems due to the strong (and variable) gradients, and also to the often shallow depth which implies a strong influence of the bottom, and the exceptional biological activity. Particulate matter, supplied from the land or transported inward from the sea, is very much influenced by these conditions, resulting in changes in composition and physical characteristics.

The adjacent sea, normally a shelf sea, is a far more uniform and stable environment. In the shallower parts the bottom deposits can easily be disturbed by storm waves and (tidal) currents, which results in reworking of bottom material and prevents deposition, but quiet conditions often prevail already on the inner continental shelf. Most sources of suspended matter and pollution are situated along the coast or in shallow water and transport along the bottom is often in landward direction. Therefore, most suspended material and associated pollutants are deposited near shore instead of being transported farther out toward the shelf edge and the deep ocean. This may be different where rivers have built deltas out towards or over the outer shelf, where canyons are present near the river mouth, or where the continental shelf is very small.

At greater distances offshore, where the influence of coastal and near-shore sources of suspended particles diminishes, the supply of dust from the atmosphere and the admixture of biogenic particles become of increasing importance. The admixture of atmospheric dust is especially strong where winds blow out of dry-land areas into the ocean. In total 10—30% of the lithogenous material deposited in the deep sea is estimated to be supplied by wind (Windom, 1975). Biogenic particles consist of organic matter, (calcium) carbonate or opal (amorphous SiO_2), produced by planktonic or benthic organisms. In the open ocean more than 90% of the suspended particulate material can be of planktonic origin, 30—70% being organic matter and 25—50% being skeleton parts (Lal, 1977). Near shore there is a large admixture of biogenic particles from benthic organisms (Rhoads, 1973).

The association of pollutants with suspended particles is only crudely known and leaves many questions to be asked. In this chapter some basic ideas relating to particulate matter in suspension and its interaction with pollutants will be reviewed. In addition, some general problems of transport and deposition of suspended particles will be discussed in relation to the dispersion of pollutants.

PARTICULATE MATTER IN SUSPENSION

Particulate suspended matter can generally be divided into a non-living fraction and a living (planktonic) fraction with both varying to a large extent independently. The non-living fraction is, strictly speaking, not entirely non-living because of bacteria and other micro-organisms that adhere to the particle surfaces but their amount is small in relation to the bulk of non-living material. Common use defines particulate matter in suspension as the material filtered off with a 0.45-μm filter. Electron microscope investigations by Harris (1977) have shown that in oceanic water particles down to 0.02 μm can be distinguished and similar small particles were found by the author in coastal water from the southern North Sea. There is not a sharp but a gradual transition from material in true solution, by way of complexes and large organic molecules, to colloids and true particles, as is shown for metals in Table I.

Non-living suspended particles are single grains or aggregates. In the southern North Sea more than half of the particles in suspension larger than a few micrometers are aggregates, and particles larger than 70 μm are almost exclusively aggregates. A predominance of aggregates over single particles has also been observed in oceanic surface water (D. C. Gordon, 1970; G. A. Riley, 1970). However, in deep-ocean water and in the Arctic Ocean the percentage of aggregates is lower (Jacobs et al., 1973; R. E. Peterson, 1977). Single grains vary in composition from nearly amorphous hydroxides, organic matter and opal, to carbonate and crystalline silicates and oxides, the most common being quartz, feldspars and clay minerals. Composition changes with size. Clay minerals, organic matter and coccoliths are pre-dominantly fine-grained ($< 5 \mu$m). Quartz, feldspars and hard exoskeletons of organisms are usually larger; and the latter may be strongly reduced in size because of fragmentation. The single particles are almost exclusively litho-genous (detrital) or biogenic, but can also be formed by precipitation of dis-solved material. Metal-hydroxide particles were formed in experiments simu-lating estuarine mixing (Sholkovitz, 1978) but have not yet been observed in nature, possibly because they are small and easily adsorbed onto other particles. There is also evidence for the formation in estuaries of particles of $\sim 2 \mu$m containing Ca^{2+}, SO_4^{2-}, Cl^- and probably OH^- and CO_3^{2-} (Eisma et al., 1980). These particles have only been found at salinities below 20‰ and probably dissolve at higher salinities.

Aggregates can be formed in a variety of ways, through: (1) charge reduction of charged particles (either through increasing ionic strength of the surrounding medium or through adsorption); (2) enmeshment in hydroxides or in fecal matter, or mucus, secreted by organisms; (3) bridging by mole-cules adhering to the particle surfaces; (4) temporary deposition on the sea floor followed by resuspension; and (5) the action of bubbles in the water (G.A. Riley, 1970; Burton, 1976; Zabawa, 1978). Aggregation and the influence of particle composition, particle concentration and environmental conditions have been studied by field observations as well as by laboratory experiments and theoretical approach. Examination of aggregates in their natural state is hampered by the technical means that are available. Collect-ing aggregates on filters gives some distortion because of settling and drying out. Even more distortion will take place when the particles are subsequently

TABLE I

Specific forms of metals (after J. M. Martin et al., 1976)

Free ions	Inorganic complexes	Organic complexes (e.g., chelates)	Metal complexes associated with organic molecules of high molecular weight	Dispersed metal colloids	Metals associated with colloids	Particulate
		10Å		100Å	1000Å	
Examples:						
Cu^{2+}	$Cu_2(OH)_2^{2+}$	Me-OOCR 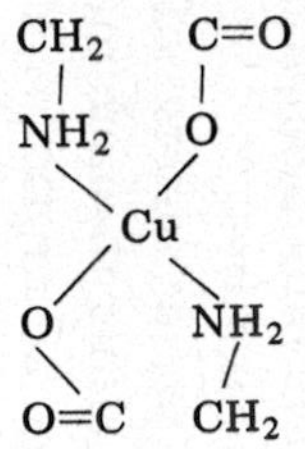	Me-lipides	FeOOH	*on clays:*	Me-precipitate organic and inorganic detritus
Fe^{2+}	$PbCO_3$		Me-humic acis	$Fe(OH)3$	$Me_x(OH)_y$	
Pb^{2+}	$CuCO_3$		Me-polysaccharides	$Na_4Mn_{14}O_{27}$	$MeCO_3$	
	AgSH			Ag_2S	MeS	
	$CdCl^+$				*on oxides:*	
	$Zu(OH)_3^-$				FeOOH	
	$Ag_2S_3H_2^{2-}$				Mn^{4+}	

Organic complexes structure:

$$CH_2 \;-\; \underset{NH_2}{} \quad \overset{C=O}{\underset{}{O}} \quad NH_2 \;\; CH_2$$

(Cu chelate: CH₂–NH₂ and C=O–O coordinated to Cu, with O=C and NH₂–CH₂)

in real solution

separated off by dialysis

separated off with membrane filters

separated off with standard filters

Me = metal; R = organic radical.

removed from the filter for electron microscope examination (Harris et al., 1972). Particles larger than ~ 5 μm can be examined with an inverted microscope as is commonly used for plankton studies. This should be done directly on board before the particles have time to settle and collide.

For electron microscope examination the sample can be brought directly on a sample grid and when small amounts are used collision can be reduced to a minimum. Actually not much has been done in this field. Jacobs et al. (1973) found that aggregates in deep water from the Atlantic Ocean contain almost exclusively mineral grains and few exoskeleton parts of organisms. Feeley (1976) describes similar aggregates from the oceanic nepheloid layer where aggregation may occur because of the higher particle concentration. Floccular structures, their size diminishing in a seaward direction, were observed in the James river estuary, whereas in Chesapeake Bay only organically bound aggregates were observed (Meade, 1972). According to Zabawa (1978) inorganic aggregation in Chesapeake Bay produces aggregates with an angular network of mineral grains, whereas fecal matter produced by organisms results in pellet-shaped aggregates that may be very loosely compacted. Mineral grains are attached to such aggregates. Aggregates from a variety of coastal waters along the east coast of Canada were examined by Kranck (1973) with an inverted-phase-contrast microscope: they appeared to be irregular masses or clumps of numerous individual grains, occasionally of organic origin but usually transparent subrounded mineral grains. Only the largest mineral grains were never part of an aggregate.

In the southern North Sea and the Eastern Channel the aggregates larger than ~ 5 μm (examined with an inverted microscope) were very similar to those described by Kranck (1973). The individual grains were held together by a semitransparent, usually brownish-greenish substance. In the fraction smaller than 5 μm (examined with a transmission electron microscope) three types of aggregates could be distinguished: a floc-type aggregate (usually containing some mineral grains), irregular more or less round aggregates of mineral grains glued together by a structureless substance, and small chain-like aggregates. The irregular and chain-like aggregates dominated, the floc-type aggregates and single mineral particles were far less common (Eisma et al., 1980).

Very little is known on the precise nature of aggregates and how they may react with pollutants or to what extent pollutants influence aggregate formation. The experimental and theoretical work has centered chiefly on the formation of aggregates by inorganic (physicochemical) processes in estuaries. There is some confusion with regard to the terms aggregation, agglomeration, coagulation and flocculation, which are often indiscriminately used. Here aggregation is used for any process that glues particles together, aggregates being the result.

River water contains dispersed particles but also some aggregates. The latter were observed by E. F. Belevitch (in Meade, 1972), as well as by the present author in the Rhine and Zaire rivers. The dispersed particles are usually charged. Most crystalline particles like quartz and feldspars have a very low surface charge and probably are covered with an amorphous outer layer, as can be observed in sand-sized grains with a polarization mircoscope. Cohesive particles like clay minerals have a high surface charge and exhibit ion-exchange properties (Grim, 1968). Usually they are negatively charged due to cation substitution in the crystal lattice and broken bonds at the particle edges. In water the negative charge is balanced by a double layer of positive ions: the thickness of this layer depends primarily on the ionic strength of the water (as well as on temperature and pH). When the charged particles come into a solution of higher ionic strength (or when the surface charge is partly or wholly shielded by adsorption), the electrical double layer is compressed and the surface charge more neutralized. A decreasing negative particle charge with increasing salinity and a charge reversal from negative to positive at salinities of $2-6^o/_{oo}$ was found by Pravdić (1970). J. M. Martin et al. (1971) found a reversal at a salinity of $\sim 20^o/_{oo}$, and Neihof and Loeb (1972) and Aston and Chester (1973) observed only a gradual decrease of negative charge. Reduction of the repulsive barrier between the particles results in a closer contact between particles, which have a greater chance of being glued together by Van der Waals forces (Van Olphen, 1963). Depending on the number of successful collisions (which is a function of particle concentration), aggregrates are formed. Collision can be due to the Brownian movement, to differential settling of the particles (due to differences in size, shape and density), and to internal shear of the flowing water. It is still debated whether aggregates are formed in estuaries in this way and whether aggregation contributes to the formation of the turbidity maximum that is present in many estuaries (Zabawa, 1978). Aggregates of increasing size and decreasing density and shear strength were experimentally produced by Krone (1976) from cohesive sediment material at decreasing sharing.

An upper size limit of aggregates is not known. In seawater aggregates up to a millimeter or more have been observed and sheet-like aggregates of ~ 2 m length have been seen at the water surface (G.A. Riley, 1963). In coastal waters stable log-normal size distributions of aggregates seem to be formed, the mode being related to the mode of the size distributions of the individual grains (Kranck, 1973). Stirring experiments with dispersed particles indicated that the formation of aggregates diminishes when the particle concentration decreases, and stops when the settling velocity of the largest aggregates equals the settling velocity of the largest grains. In this way artificial aggregate distributions with a mode up to 4 μm could be produced from dispersed suspensions, involving increasingly larger grains. In nature aggregate modes up to 64 μm have been observed. Since the experiments were made in the absence of organic matter (which had been removed by low-temperature ashing) this may indicate the influence of organic matter on aggregate formation.

Organic aggregates in ocean water have been intensively studied because they are a source of food for planktonic organisms and a means of transport of material to the ocean floor through settling. That granular organic aggregates — as well as apparently homogeneous flakes — can be formed in the sea has been ascertained from observations as well as from experiments (G.A. Riley, 1970). The flakes can form very quickly on bubbles in the water and at large depths well below the surface water flakes are the most abundant particles in the size range of 15—35 µm. They occur fresh as well as in various stages of decay: experiments have shown that they are readily degraded by bacteria (G.A. Riley, 1970). Bacteria and other, larger, organisms can form the granular type of aggregate, incorporating particles as well as adsorbing dissolved material. Large (50—700 µm) aggregates of fecal matter and fecal pellets, which have a much higher rate of settling than the more common finer particles, are mainly responsible for downward transport of material in the ocean (Bishop et al., 1977). They comprise only 5% of the total mass of suspended material but account for 99% of the downward flux in the upper 400 m. Such large particles are not collected with the conventional water-sampling techniques but only with large sediment traps or by filtering large volumes of seawater.

Fecal pellets of bottom fauna may range from several tenths of micrometers to more than a millimeter. They are dropped on the bottom but reworking by currents results in a turbid layer at some meters above the bottom, containing entire pellets as well as irregularly-shaped loose aggregates of mineral grains and organic matter (Rhoads, 1973). Fecal pellets are shaped by a surface membrane. The membrane of marine copepod pellets (of 200—600 µm) is degraded within a few hours at 20°C but only after 20 days at 5°C (Honjo and Roman, 1978). The pellet then breaks up into small amorphous aggregates.

INTERACTION BETWEEN POLLUTANTS AND SUSPENDED PARTICLES

Studies of pollution in relation to particulate matter have the drawback that the nature of the particles is usually poorly known, so they tend to emphasize the behaviour of the pollutants. These pollutants are introduced into estuaries and the sea in dissolved, colloidal or particulate form. Leaching techniques give some indication in what form the pollutant is present in the particles (adsorbed to the surface or in a lattice position, in the form of carbonate, organic compounds or hydroxide) but the interpretation of leaching data can be rather problematical (Burton, 1978). Pollutants cover a wide range of substances: inorganic and organic chemicals, oil, city sewage, sludge from factories and purification plants and radioactive elements. Commonly, they are introduced into estuaries and the sea through rivers, direct discharges, offshore dumping and through the atmosphere. Pollutants introduced in dissolved of colloidal form can become particulate, chiefly through adsorption onto particulate matter. Conversion of pollutants from a particulate form to a dissolved or colloidal state (chiefly through desorption) also occurs, but is of minor importance for trace metals in estuaries (Duinker, 1980) and most probably for the other forms of pollution as well. The exceptions are direct discharges of city sewage and sludges. Since the highest concentrations of pollutants enter the sea from the land or from near-shore dumping sites, the greatest effects are to be expected in estuaries and the near-shore sea. Notable exceptions are pollutants predominantly supplied through the atmosphere (trace metals, e.g. Pb, some halogenated hydrocarbons, radioactivity, and CO_2 which interacts with the carbonate system in the oceans).

In studies of the behaviour of dissolved components in estuaries it has become customary to relate their concentration to chlorinity (or salinity) as a conservative index of mixing. Theoretical mixing of freshwater and seawater gives a straight-line relation, with the concentrations in the river and in the sea as end-points. Significant deviation from this line indicates removal or release [see Liss (1976) for a full discussion]. The processes involved as well as the end-products have been experimentally studied by mixing (filtered) river water with seawater.

Measurements in nature and the interpretation of results can be complicated by variations in the end-member concentrations and in the speciation of the pollutants involved, as well as by secondary sources within the area of study and because of problems with the representativity of sampling. In many estuarine studies it was found that Fe is usually removed from solution and removal was also found for N, Mg, Al, Si, Mn, Co, Ni, Cu, Zn, Cd, Sb, Hg and U. Al, Cu and Zn can also be released into solution (Zn at high salinities) whereas Mn in a number of estuaries is cycled with release at low salinity and removal at high salinity. Conservative behaviour has been found for B, N, F, Na, Mg, Si, SO_4, Cl, K, Ca, Mn, Fe, Zn, Se, Mo, Pb and U, release for Ba and Ra (besides Al, Cu and Zn) whereas P (as PO_4) is extensively buffered [summaries in Liss (1976) and Duinker (1980), additional data from Edmond et al. (1978), Fanning and Maynard (1978), Figuères et al. (1978), J. M. Martin et al. (1978), Sholkovitz (1978), Van Bennekom and Jager (1978), Borole et al. (1979) and Li and Chan (1979)].

A complex series of processes can be responsible for the removal of dissolved trace elements into a particulate form. Mixing of filtered river water and aliquots of seawater results in rapid flocculation of Fe and to a lesser extent of Mn, Al, Cd and Cu, together with humic acid colloids, while other trace elements may be removed simultaneously by coprecipitation (Sholkovitz, 1976, 1978). Iron removal is greatly enhanced by the presence of suspended particles (Aston and Chester, 1973) and iron hydroxide is probably precipitated as a coating on the sediment particles. Precipitation is also influenced by the type of Fe in solution, uptake of Fe from fresh solutions being much stronger than uptake from aged solutions. Besides adsorption and the formation of coatings, ion exchange may play a role but this is probably not very important because the major elements (chiefly Ca in freshwater, and Mg and Na in seawater) occupy most of the exchange sites. Experiments of Kharkar et al. (1968) and Van der Weijden et al. (1977) indicate that some desorption of trace elements (Cr, Mn, Co, Ni, Cu, Zn, Se and Ag besides Ca) may take place at low salinity. Desorption processes, however, are very sensitive to changes in pH and at a pH normal for estuaries the amount of desorbed Co and Cu is very small compared to the total amount that is adsorbed to the particles (T.P. O'Connor and Kester, 1975). Organisms can play a large role in the removal of substances in solution, especially of N, P and Si, but in the open sea adsorption (scavenging) of dissolved or colloidal material onto non-living suspended particles is more important for removal than uptake into planktonic organisms. In the ocean ^{210}Pb is transported to the bottom only for less than 5% by plankton settling (Turekian, 1977). The large particles of fecal matter with a comparatively high settling velocity are the most effective means for removal (Bishop et al., 1977).

Organic pollutants include thousands of different compounds, many of which are readily adsorbed to suspended particles and incorporated in bottom sediments: organochlorine compounds (DDT and derivatives, dieldrin, aldrin, endrin, endosulfan), PCB's, carbamates, aromatic hydrocarbons and a number of herbicides are all relatively insoluble in water and will evaporate or become adsorbed to particles. This can result in a concentration in particulate suspended matter or in bottom sediments of $10^4 - 10^5$ times the concentration in water (W.E. Odum et al., 1969). Humic particles are especially effective in adsorbing organic pollutants as was found for DDT (R.H. Pierce et al., 1974). Adsorption may lead to a greater stability of the compounds involved but also to a more rapid break-down, depending on the nature of the particle. Thus endolsulfan, adsorbed to sediment particles, can be hydrolysed only for 70—75%, but adsorbed on fresh iron hydroxides it is broken down rapidly and completely because the hydroxide acts as a catalyst (Greve, 1971). Another example is the herbicide paraquat, which is for almost 100% adsorbed on sediment particles or on organic matter. In that state it is not further broken down but it also has lost its herbicide properties.

Organic coatings on particles may concentrate other organic waste products as was found for DDT on particles exposed to oil pollution (Hartung and Klinger, 1970). Degrading oil gradually aggregates into tar balls which on further degradation become smaller and denser, eventually sinking to the bottom and incorporating suspended particles. The concentrations of pollutants in bottom sediments to a large extent reflect the history of pollution in the area concerned. A decreasing content of pollutants (metals, PCB's, DDE) with increasing depth in the sediment has been found in a number of sea areas near to industrial centers (Baltic, North Sea, coastal basins off

California; Hom et al., 1974; Suess, 1978). These data make it possible to determine the period and flux of contamination as well as the degree of increase in concentration but the interpretation depends on a reliable determination of the sedimentation rate during the past 200 years, which is complicated by reworking of the sediment by organisms (Benninger et al., 1979).

In estuaries and the sea, removal of pollutants from solution into particulate form seems to dominate for a number of dissolved substances but the degree of removal may vary considerably. Repeated determinations in the Rhine estuary show strong variations in removal rate even for the same elements (Duinker and Nolting, 1976) (Table II). The importance of removal in particulate form for the dispersal of pollutants is nevertheless clear in many cases. This makes it important to know the flux of suspended matter through estuaries and the near-shore sea to the deep ocean. Since the processes of transport and deposition of suspended matter in estuaries and the near-shore sea are strongly interrelated, suspended matter dispersal within the estuaries and the near-shore sea is here taken as a whole and treated separately from: (a) the flux of suspended material from the continental seas to the ocean; and (b) the flux of material from the ocean surface waters to the ocean floor.

TABLE II

Percentage removal from solution into particulate form

Metal	Rhine estuary[1] (%)	St. Lawrence estuary[2] (%)	Mixing experiments[3] (without suspended particles) (%)
Fe	50, 50, 0	65	95
Cu	50, 40, 20	0.7	41
Mn	20, 0 , 0	11	25—45
Zn	60, 50, 50	0.2	—
Co	—	4	11
Ni	—	1.3	43
Cd	—	0.02	5

[1] Duinker and Nolting (1976); [2] Bewers and Yeates (1977); and [3] Sholkovitz (1978).

FLUXES OF SUSPENDED MATTER

Two approaches can be followed to estimate the flux of suspended matter: calculation of the mass balance from suspended matter and bottom sediment data, and determination of the flux of natural or artificial tracers. Although qualitatively the circulation of suspended matter and the formation of a turbidity maximum in estuaries is rather well known (Postma, 1967), flux

calculations are uncertain because of uncertainties concerning the representativity of sampling and the response of sediment dynamics to flow dynamics
(Dyer, 1978). For the net transport of suspended material from continental
sources to the deep ocean a simple calculation shows that this is relatively
small. The annual deposition of lithogenous material in the deep ocean
(> 3000 m) is estimated to be $1.2 \cdot 10^9$ t yr.$^{-1}$ (Judson, 1968) which is
about 6% of the total continental supply of $18.3 \cdot 10^9$ t yr.$^{-1}$ (Garrels and
Mackenzie, 1971). About 10—30% of the lithogenous material in the deep
ocean has been supplied from the continents by wind (Windom, 1975) so
5—8% of the total supply from surface runoff reaches the deep ocean. This
picture is complicated by the erosion of older deposits on the continental
shelf. In the North Sea the supply from bottom erosion is of the same magnitude as the supply from rivers (Eisma, 1980) and on the shelf along the
U.S.A. east coast between Cape Hatteras and Cape Cod relict sediments are
the principal source for suspended matter (Meade et al., 1975). This reduces
the contribution from present runoff to a still lower percentage. Estuaries
and the nearby sea floor are therefore considered to be the main areas of
sedimentation, accommodating up to 90% of the supply of suspended
sediment and its associated pollutants (Postma, 1980).

The actual flux of suspended matter from rivers to the deep ocean is very
localized. Nearly 40% of the supply by continental runoff is carried seawards
by nine rivers, of which only two (the Amazon and the Mississippi) are not
located in Southeast Asia (Table III). In total ~ 80% of the continental

TABLE III

Suspended load of major rivers in the world

	Suspended load (10^6 t yr.$^{-1}$)	Percentage of total supply
Hoang Ho[1]	2,080	11.4
Ganges[1]	1,600	8.7
Amazon[2]	850	4.6
Brahmaputra[1]	800	4.4
Yang tse Kiang[1]	550	3.0
Indus[1]	480	2.6
Mississippi[1]	340	1.9
Irrawaddy[1]	330	1.8
Mekong[1]	187	1.0
		39.4
Total supply to the sea	18,300	100

[1] Holeman (1968); and [2] Meade et al. (1979).

supply comes from East Asian rivers (Holeman, 1968). The distribution of recent deposits and of suspended matter concentrations on the East Asian continental shelf indicates that from the material supplied by the large Chinese rivers, by the Mekong and by the rivers that discharge into the Gulf of Thailand, only a very small amount escapes from the continental shelf (Niino and Emery, 1961; Emery and Niino, 1963; Honjo et al., 1974). High suspended matter concentrations and recent deposits are confined to the inner shelf, separated from the continental slope by a belt of relict sediments and low suspended matter concentrations near to oceanic values on the outer shelf. The same probably also applies to the Irrawaddy (Rodolfo, 1969).

From the Amazon and the Zaire (the second largest river entering the Atlantic Ocean) only less than 5% of their total suspended matter supply reaches the deep ocean in suspension (Milliman et al., 1975; Eisma et al., 1978). Only where the river mouth is located near to or at the shelf edge, as is the case with, e.g., the Mississippi, the Niger and with the small rivers along the Indian west coast, or where a canyon goes from the river mouth to the deep ocean, as at the Ganges, the Zaire and the Irrawaddy, suspended sediment as well as bed-load material, supplied by the rivers, may reach the deep ocean in appreciable quantities directly or through the canyon.

Also suspended material eroded from shelf sediments may not reach the deep sea except in minor amounts. Bottom material is primarily stirred up in the shallower parts of the shelf and deposited in estuaries and nearby deeper parts, where the effects of waves and (tidal) currents are smaller. In the North Sea in- and outflow of suspended matter from and to the North Atlantic Ocean are about equal: less than 10% of the suspended material supplied to the North Sea by bottom erosion, rivers, primary production, influx from the Baltic Sea and influx from the atmosphere escapes towards the North Atlantic Ocean (Eisma, 1980). Processes along the shelf edge — downslope flow, slumping and turbidity currents through gullies and canyons — are probably the most important transport mechanisms of shelf-eroded material to the deep ocean (Southard and Stanley, 1976; Feeley et al., 1979; MacIlvaine and Ross, 1979).

Since pollution from rivers and other continental runoff sources, including direct discharges and near-shore dumping, to a large extent follows the dispersal pattern of the suspended material, a large percentage of the pollutants will be deposited in the estuaries and the near-shore sea. In some coastal seas trapping of trace elements is virtually complete (like, e.g., in Long Island Sound; Turekian, 1977) but for many elements removal in the estuary is not complete or is absent. Offshore, further lowering of their concentration can be due to dilution, to adsorption on suspended particles (scavenging) and to uptake by organisms. However, higher concentrations are possible because of pollutants supplied through the air, which in coastal seas can be an appreciable amount. In the North Sea the yearly atmospheric input of trace metals is for most metals of the same magnitude as the input from the Rhine, which is the principal river source of pollutants in that area (Cambray et al., 1979; Table IV).

For Pb and Cu the atmospheric input is much higher than the river input, for Hg and As it is much lower. The atmospheric input affects the North Sea as a whole and remains in solution or can be adsorbed on large numbers of particles that do not settle in the North Sea area but are removed to the North Atlantic Ocean. Coastal seas near to industrial centers may therefore be a more important source of pollutants for the nearby ocean waters than estimated on the basis of coastal and near-shore supply alone.

A second question involves the particle size of the material in suspension. In estuaries the estuarine circulation is assumed to lead to a concentration of suspended particles with a narrow size range. The larger particles are deposited and the finer particles are removed to the open sea (d'Anglejean and Smith, 1973). Labeyrie et al. (1976) assumed the presence in coastal waters of very small particles containing Fe released from bottom sediments. These small particles would only slowly become associated with larger particles and would therefore disperse Fe to the open sea. The presence of fine particles with a high trace-element content that do not settle in the estuary was also postulated by Duinker et al. (1974) to explain the lower trace-element content in bottom sediments as compared to suspended matter. Such particles are retained by filters of $0.5 \mu m$, but their actual size may be less because of clogging of the filters and adsorption.

Coulter counter measurements in the southern North Sea and the Eastern Channel indicate that there is a variable admixture of particles of $\sim 1 \mu m$ (which is near to the limit of detection), but this admixture is by no means ubiquitous (Eisma et al., 1980). The bulk of the small particles therefore may be smaller but the very small particles usually escape attention because they are filtered off and are not measured by a Coulter counter. Up to now they have only been (imperfectly) investigated by electronmicroscopy (Harris, 1977; Eisma et al., 1980) and occasionally filtered off with very fine filters [e.g., by Figuères et al. (1978) who used filters of 0.22, 0.05 and $0.025 \mu m$].

In the ocean the downward flux of suspended particles almost entirely concerns biogenic particles. During transport through the water column they change in composition through adsorption of trace elements like Fe, Cu, Pb, Th and Pu, and through solution of Ca, Si and C. Unsolved problems concern these processes in relation to aggregates of fecal matter and to fecal pellets. Also unexplained is the constancy of the particle size distributions in the oceans with depth, apart from an admixture of particles of organic matter in surface water (Brun-Cottan, 1976; Lal, 1977). For ^{210}Pb the downward flux in the ocean, based on large particle transport, matches the atmospheric input (Buat-Menard and Chesselet, 1979). The behaviour of particulate Pb appears to be consistent with the behaviour of ^{210}Pb, so more than

TABLE IV

Atmospheric input of trace metals in the North Sea, compared with input from the River Rhine, the total dissolved content in the North Sea and the total content in the North Sea (after Cambray et al., 1979)

Metal	Input from the Rhine (t yr.$^{-1}$)	Atmospheric input 1974—1976 (t yr.$^{-1}$)	(annual atmospheric input)/ (total content in the North Sea) × 100 (%)	(annual atmospheric input)/ (dissolved content in the North Sea) × 100 (%)
Cr	1,000	740	—	34*
Mn	6,000	4,100	10	< 20
Fe	80,000	110,000	30	55
Ni	2,000	1,900	< 6	18
Cu	2,000	5,600	40	35
Zn	20,000	16,000	21	20
Cd	200	390	—	< 25
Hg	100	7.1	—	0.53*
Pb	2,000	5,800	23	46
As	1,000	400	—	0.35*

*Based on average seawater contents.

90% of the particulate Pb in the deep water of the North Atlantic is considered to be due to pollution, moved downward in large aggregates.

SUMMARY

Particles suspended in seawater range in size from 0.02 μm up to several millimeters. Bulk studies and particle size determinations of suspended matter are biased towards a central range of 0.5 to ~ 100 μm and under-estimate the very fine and the very large particles. The latter are important for downward transport in the ocean because of their relatively high settling velocity, the very fine particles are important for horizontal dispersal because of their low settling velocity and their large specific surface. Little is known about the nature of suspended particles besides their bulk properties. Aggregates form a major and often dominant component of suspended matter but their formation and distribution, their composition, their surface character-istics and their stability are imperfectly known. This makes it difficult at present to evaluate their role in the dispersal of pollutants.

The rate of removal from solution into a particulate form in estuarine and coastal waters varies for the various kinds of pollutants and may also, for un-known reasons, vary with time for one single pollutant. Further lowering of their concentration in solution in the sea is due to dilution, adsorption onto suspended particles and uptake by organisms. Pollutants associated with particulate matter will be deposited primarily in estuaries and in the near-shore sea. Only less than 10% (and probably less than 5%) of the continental supply of suspended particles will reach the deep ocean, diluted by supply of material from sea-floor erosion. The fraction that reaches the deep ocean, however, may be relatively important for transport of pollutants because of its small size and a relatively high content of adsorbed material. In coastal seas near to industrial centers additional supply of pollutants through the air can be comparatively high. Supply of pollutants from these areas to the ocean may therefore be more important than estimated on the basis of continental and near-shore supply (by runoff and direct discharges) alone.

Further details on the specific applications of the basic principles of the role of particulates can be found in this volume — with respect to dredge spoils — in Chapters 6 and 7 by Herbich, in Chapter 11 by Gunnerson and in Chapter 12 by Simpson et al., and — with respect to deep-ocean mining — in Chapter 14 by Morgan, in Chapter 15 by Ozturgut et al. and in Chapter 16 by Ichiye and Carnes. Similarly, their role in heavy-metal dispersal is dis-cussed by Bernhard (Chapter 3), Trefry and Shokes (Chapter 4), Gillespie and Vaccaro (Chapter 5) and Sackett (Chapter 13), and for dispersal of radioactive elements by Angel et al. (Chapter 10).

CHAPTER 10

MARINE BIOLOGY NEEDED TO ASSESS THE SAFETY OF A PROGRAM OF DISPOSAL OF HIGH-LEVEL RADIOACTIVE WASTE IN THE OCEAN

M. V. Angel, M. J. R. Fasham and A. L. Rice

INTRODUCTION

Several mathematical models have been developed to explore the dispersal of radioactive isotopes back into man's immediate environment following their disposal in or on the deep-oceanic sea bed (e.g., Grimwood and Webb, 1976). These models have been almost entirely based on physical processes and have tended to ignore biological mechanisms by which the flux of dangerous isotopes may be accelerated back towards the surface because the fluxes can not be adequately quantified. Furthermore, since the underlying aims of the models have been to estimate the potential maximum direct dosage to man, other environmental effects such as the possible damage to fragile deep-sea communities have been ignored. The importance (if any) of the deep-sea communities to global ecological health is unknown and should be assessed before there is any hope of persuading the conservational lobby of the acceptability of deep-sea dumping.

All the data that are available at present on the fate of high-level radio-active waste isotopes have been derived from fallout from weapon tests or outfalls from power or reprocessing plants. These contaminants are injected into the marine environment via the surface layers and either diffuse through the water column in solution or are absorbed onto particulate matter and follow the natural gradient of sedimentation down through the water column to the sea bed. If radioactive waste is disposed by emplacement of the radionuclides on, or preferably, in the sea bed, the question arises whether the nuclides will move counter to the general flux of material up through the water column in sufficient quantities to present an immediate danger to man as a direct dosage or cause indirect danger to man through environmental damage. This last factor has been almost entirely ignored in the literature on disposal of radioactive waste.

THE OCEANIC ECOSYSTEM

The oceanic ecosystem can be divided into four main zones; the epipelagic, mesopelagic, bathypelagic and benthic zones. Although the boundaries between them are not always clearly defined, there are important differences in ecological processes between the various zones. Organisms living in each zone often show significant and characteristic morphological adaptations so the environmental differences between zones must be significant.

Epipelagic zone

The epipelagic zone extends from the sea surface to ~ 200—250 m depth. This coincides roughly with the bottom of the seasonal thermocline and also the depth at which the light field becomes totally symmetrical; the brightest intensity light is observed coming vertically down, and there is no indication of the bearing or elevation of the Sun. The epipelagic contains the photic zone within which the total primary production of the ocean is produced, and which is the prime food source for all animals in the oceans from the surface to the greatest depths. Nearly all oceanic plants are microscopic; there is generally a small standing crop, but a high turn over rate. In areas of high productivity where the surface layers are enriched by upwelling, large diatoms form an important constituent of the phytoplankton. Similarly, at higher latitudes the spring bloom which develops at the onset of stratification is also dominated by large phytoplankton species. However, in oceanic areas at lower latitudes where the surface layers are permanently stratified, (and also in the conditions that prevail after the spring bloom in temperate regions), the majority of the phytoplankton standing crop is composed of coccolithophorids, flagellates and other nannoplankton forms in which the cells are $< 10\,\mu$m in size.

The herbivorous animals that feed on these minute cells either have to be extremely small themselves or have specialized filtering systems to extract the minute cells from relatively large volumes of water. The herbivores in this oceanic habitat are typically planktonic, ranging in size from ~ 100 μm to 10—20 mm. Other planktonic animals are either carnivores or detritivores. The larger animals (nekton) are predominantly carnivores feeding on plankton or other nektonic animals. At the top end of the size range are the fishes, squid and marine mammals that are commercially exploited by man. It is the movement of radioactive waste products into these organisms that is the main route for direct dosage to man. However, the anchovetta fishery off Peru and the potential krill fishery in the Antarctic Ocean are examples of herbivorous stocks which are, or may be, exploited, and which may in future shorten some of the routes to man so far considered by the modellers.

From all these organisms dwelling in the epipelagic zone falls a rain of detrital material ranging in size from faecal pellets which sink at a rate of 36—376 m day^{-1} (Smayda, 1969) to the bodies of larger animals that may sink to abyssal depths within hours rather than days. The slower the sinking rate the greater the proportion of labile organic material which is degraded by micro-organisms before it reaches the sea bed, and the greater the chance of the pellet being located and eaten by a detritivore. The predictability of

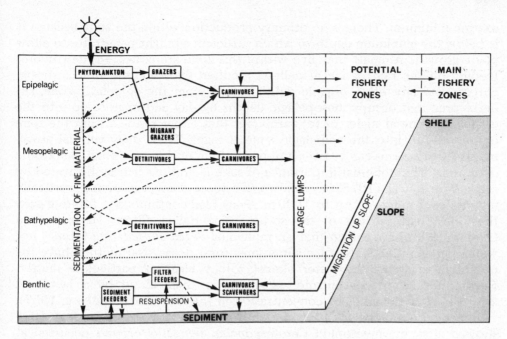

Fig. 1. Schematic diagram of the main pathways of the movement of biomass through the bentho-pelagic ecosystem.

encountering a particle will decrease rapidly with size because not only are larger particles less abundant, but also their residence time in the water column is considerably shorter. This is clearly an important route whereby production at the surface is transferred to deeper-living communities.

In Fig. 1 the main pathways of the movement of biomass are schematically represented. Assimilation rates are often 50%, the remainder either being lost while manipulating the food prior to ingestion or being passed out in the faeces. Metabolic efficiencies are seldom > 20%, i.e. only a fifth of what is consumed is converted onto new biomass; the remainder is either excreted or is burnt up by the consumer's metabolism. Isaacs (1976) has also pointed out that whereas terrestrial organisms retain their reproductive products within their trophic level, marine organisms which produce large numbers of minute eggs and have planktonic larval stages leak biomass back down into lower trophic levels.

Mesopelagic zone

The mesopelagic zone extends from the base of the epipelagic zone to ~ 1000 m. Its lower boundary tends to coincide with the greatest depth to which detectable daylight penetrates, and, in many parts of the oceans, the oxygen minimum. There is no primary production within the zone because it is below the maximum depth to which sufficient sunlight penetrates to allow plant growth. Animals that live within this zone are either: (a) detritivores feeding on the rain of faecal pellets, moulted exoskeletons, mucus sheets, corpses and any other organic debris falling from the epipelagic zone, (b) herbivores that occupy mesopelagic depths by day and migrate up into the epipelagic zone at night, or (c) carnivores that either reside within the zone or migrate up into the epipelagic zone at night. These daily vertical movements of organisms can be over hundreds of meters by even small planktonic organisms. The potential importance of such migrations is best illustrated by salps. Wiebe et al. (1978) have shown that these gelatinous animals vertically migrate over ranges of up to 1000 m. They feed continuously, filtering even the finest particles out of the water, using mucus filters. Harbison and Gilmer (1976) have shown that an individual *Pegea confederata* filters 1 l of water per minute. Since salps sometimes form dense swarms they have the potential to sweep the water almost totally clear of particulate matter. Furthermore they concentrate vanadium in their blood by a factor of $2.5 \cdot 10^5$ over the ambient concentration in the seawater (Goldberg, 1957).

Repeated sampling at 1000 m depth around 42°N,17°W in May 1978 showed that the myctophid *Ceratoscopelus elongatus kroyeri* occurred at 1000 m by day and could be caught at the surface at night ("Discovery", unpublished data, 1978). If this fish or extensively-migrating species of salp accumulate radioactive isotopes, they could provide a very rapid transport route from deep water up into the surface layers.

The mesopelagic zone represents the greatest depth to which commerical fisheries are likely to operate economically. Pelagic fisheries for the most part are concentrated in the epipelagic zone, and it is only in a few places that deep fishing techniques are economic such as the espada fishery (for *Aphanopus carbo*) off Madeira.

Bathypelagic zone

The bathypelagic zone is considered here to extend from the base of the mesopelagic zone at 1000 m to the sea bed. There is little if any diurnal vertical migration from the bathypelagic zone up into the mesopelagic zone. All animals inhabiting this zone are either carnivores, scavengers or detritivores. The decrease in available food with increasing depth means that very large organisms are unable to find enough food in mid-water at bathypelagic depths to survive. The energetic cost of active movement and even the maintenance of well-developed muscle and sensory systems becomes relatively large. Bathypelagic organisms tend to be rather fragile and flaccid because they have poorly developed body musculature and skeletal elements. They adopt low-energy food-gathering methods of luring and waiting. Neutral buoyancy is maintained by mechanisms such as storing massive quantities of lipid, or using ionic pumps. Musculature is often very watery and gigantism in individual species is often achieved without increasing dry weight of the organisms much above the norm for the group, this factor possibly explains the conclusion of some authors that body size increases with depth (e.g., Mauchline, 1972).

Benthic zone

The benthic zone is associated with the sea bed and benthic animals live within or in the immediate vicinity of the bottom. The sea bed acts as a collector for the whole range of particles and lumps falling from the layers above. As a result there is a sharp increase in the standing crop of biomass at the bottom, particularly in large scavengers. Benthic standing crop decreases with increasing depth and also with the decrease in the productivity of the overlying water, which indicates a link between the two. Benthos, for practical reasons, is categorized into approximate size classes: (a) the meiofauna which passes through sieves with 300—500 μm mesh; (b) the macrofauna which is collected by grabs and corers and is retained on sieves with 300—500 μm mesh; and (c) megafauna which consists of larger, often more mobile organisms, that may be sampled using large trawls and traps or seen in photographic surveys, but more often may not be effectively sampled by any technique at present available to oceanographers.

The feeding of these organisms can be categorized into:

(a) Suspension feeding by sessile immobile forms like sponges, corals, sedentary worms and some mollusks, which is common in shallow areas where the water carries a high suspended load of organic material. It is also thought to be important in the oligotrophic regions in the centers of the great current gyres (Sokolova, 1965). Here a low-cost existence can still be achieved by spreading a sieve and using the energy from the current to accumulate what little food is available.

(b) Deposit feeding by slow-moving organisms which may burrow and eat their way through the deposit or skim off the surface layer of the sediment—water interface. These are the organisms that are important in bioturbation of sediments, particularly the burrowing forms.

(c) Scavenging and predation is usually carried out by larger, more active organisms like fish, cephalopods and various crustacean groups particularly the decapods and amphipods. In terrestrial environments such carnivores form the top of the ecological pyramid and have a relatively low productivity. However, in the deep-sea benthic environment it seems possible that a major input of organic material into the ecosystem is in the form of large lumps, i.e. carcases of whales, large fish and squid. The scavengers would then be the base of an ecological pyramid redistributing this resource to the other benthic feeders via their messy feeding habits or inefficient assimilation.

Another important group associated with the sea bed are the benthopelagic predators, for a variety of fishes normally caught on the bottom have been shown to feed extensively in mid-water, particularly over the continental slopes.

THE INFLUENCE OF RADIOACTIVE WASTE DISPOSAL ON THE OCEANIC ECOSYSTEM

There is an extensive literature on the biological effects of organic and heavy-metal pollutants on shallow-water marine organisms (e.g., Bernhard, 1976; and Chapter 3 of this volume). There are also data on the accumulation of isotopes in marine organisms (e.g., Pentreath, 1978a, b). But there is little information on what effects these isotopes have on the viability of the organisms or on the structure and functioning of the ecosystem. Preston (1975), in reviewing the knowledge of the potential damage to marine resources, concludes that so long as man does not receive a damaging dose:

"The consequential dose to aquatic organisms will carry with it a negligible risk of biological damage."

However, this conclusion is based on results such as those of Woodhead (1970) who showed that chronic radiation effects could reduce the viability of fish eggs and embryos. The levels of contamination required were far higher than observed levels from the reprocessing plant at Windscale (which are similar to the natural levels of radioactivity in seawater). It has been argued that because of the enormous egg production by commercial species such as the cod and the plaice, and the subsequent high levels of natural mortality, even a 50% reduction in egg viability would not be significant. However, if as seems likely at present, the reproductive strategy of most deep-sea species involves the production of relatively few large eggs, such a reduction in the numbers surviving could be catastrophic for the population.

Williams and Murdock (1973) in their study on the effects of continuous low-level gamma radiation on sessile shallow-water benthic invertebrates, found marked variations in the sensitivity of different organisms to the radiation. New growth in sponges was reduced at 0.82 rad hr.$^{-1}$ but the filter feeding gastropod *Crepidula fornicata* grew well at levels of 8.5 rad hr.$^{-1}$ which killed coral and oysters. There is no information available on the effects of pollutants on the deep-oceanic communities.

Assuming that dumping would be restricted to a small number of sites and that diffusion will rapidly dilute toxic concentrations, dramatic effects such as the total annihilation of an animal community will be very localized. In-bed disposal, where a barrier of several meters of sediment is interposed between the cannisters of waste and the ecosystem, is obviously safer than on-bed disposal where no such barrier exists. In highly variable environments, typical of shallow-water and terrestrial communities, localized damage might be very serious because populations of rare species whose distributions happened to coincide with the affected region might be considerably reduced or made extinct. Abyssal species tend to be widely distributed and so any devastation would need to be very extensive to cause the extinction of a species.

Whereas biological communities in shallow water and particularly in estuarine environments are already adapted to highly-variable stressful environments, deep-sea communities are not. The abyssal environment has a far longer geological history of constancy than any terrestrial or shallow-water habitat. The benthic communities are unusually rich in species. There are two main hypotheses to explain this high species diversity. Sanders (1968) suggested that the long-term stability of the physical environment has led to high diversity through extreme niche specialization. As there is little environmental heterogeneity, the niche specialization must be in relation to feeding. Dayton and Hessler (1972) pointed out that because food is in such short supply most benthic organisms are generalist feeders. Pointing to the large populations of predators, they suggest that the high predation pressures keep the populations of the smaller organisms so low that they do not compete with one another for either food or living space. The corollary of this theory, that the predators will show a lower diversity than the other species, does seem to be true.

May (1973) has shown theoretically how more diverse communities are more fragile, only small perturbations being needed to produce irreversible changes. Connell (1975) has discussed the results of field experiments and models that suggest that predation at an intense level can reduce competition and hence maintain a high diversity. The removal of a single key-predator can produce a surprisingly large decrease in the numbers of other species. It can be predicted that even if there is no annihilation of life in the area of a dump site, there will be large irreversible changes in the sea-bed community. Neither the type of change nor the geographical extent of the change is predictable at present. Neither is it possible to say if it matters to our global ecological health whether or not the deep-oceanic communities exist. *If* the deep-sea communities were not linked with the mid-water communities and abyssal depths were a sink for organic matter, then the destruction of these communities would be relatively unimportant except from a moralistic standpoint. Then the purely physical models of dispersal of isotopes could be acceptable; however, *such links exist.*

LINKS BETWEEN THE DEEP-OCEANIC COMMUNITIES AND THE SURFACE

Bioturbation

The initial link would be via the benthic deposit and suspension feeders. The activities of the deposit feeders are most important because of bioturbation, the churning of the surface layers of sediment by their activities of burrowing and feeding. This activity rarely extends deeper than 20—30 cm beneath the sediment—water interface, and never more than ∼ 1 m. In shallow water the reworking is a relatively rapid process, burrows and craters are formed and disappear all within a few days or weeks (Chapman and Rice, 1971; Rice and Chapman, 1971). It might be expected that in the deep sea, where the biological activity is slower and the physical environment more stable, bioturbation would proceed more slowly. But a recent photographic survey suggested that features produced by bioturbation are no less ephemeral than those in continental shelf regions (A. Paul et al., 1978).

Bioturbation may affect the mobilization of isotopes in three ways: (1) by bringing shallowly buried material to the sediment interface; (2) by active ejection of suspended material up into the water; and (3) by roughening the bottom texture so that water currents erode the surface and carry off sediment in suspension. Furthermore, because the sediment feeders come into intimate contact with the sediment both externally and internally, any tendency they may show to absorb and accumulate isotopes will be important. Carey et al. (1966), for example, showed significant concentration of ^{65}Zn from sediments by ophiuroids.

The direct effects of bioturbation on radioactive wastes should not be overstated. The vertical transport by biological processes of isotopes buried in sediments will occur only in the superficial few tens of centimeters. If the wastes are buried in the sea bed, *at worst* bioturbation will effectively reduce the depth of burial by the few tens of centimeters to which the animals burrow. For "on sea-bed" disposal the biological activity may help to accelerate the movement of isotopes downwards. However, as many of the sediment feeders feed at the sediment—water interface, they will feed where the sediment has its highest load. Bowen et al. (1976) have observed the concentration of Pu by various members of the infauna.

It is important to contrast the conditions prevailing in the vicinity of the Windscale outfall with those of the deep ocean. Off Windscale there is a high deposition rate of sediments which removes 94% of the Pu absorbed onto the sediment. Most of this removal occurs in the immediate vicinity of the outfall where bioturbation, mostly by the tiny crustacean *Corophium*, is restricted to the top centimeter or so of the mud which becomes anoxic just below the surface (Hetherington et al., 1975). In the deep ocean, sedimentation rates may be as low as 1—2 mm per century, so that although the rate at which the sediment is reworked may be much slower, the time over which it is reworked may be as many as three orders of magnitude longer than the sediments near Windscale. Thus, in deep-oceanic conditions sediments may not be as effective in removing Pu from the biotope as off Windscale.

Movement by benthic organisms

The resuspension of sediment can occur by the activities of deposit feeding organisms or by the erosion of the sediment surface by water currents.

Resuspended sediments may be sieved out of suspension by the filter feeders; for example, Hetherington et al. (1975) found that the common mussel *Mytilus edulis* concentrated Pu 200 times over the concentrations in the surrounding water. Both deposit and suspension feeders will be consumed by at least some part of the large benthic carnivore/scavenger population. These carnivorous scavengers must have relatively high mobility to be able to exploit the unpredictable large lump resource. They will transport biomass and hence radioactive isotopes both vertically and horizontally. Vertical movement can occur by: (a) the upward movement of the sediment feeders; holothurians normally considered to be benthic, and with sediment in their guts, have been caught 1000—1500 m above the sea bed (Barnes et al., 1976; "Discovery", unpublished data, 1978), while McGowan (1974) records the capture of an amphipod *Eurythenes gryllus* with sand in its stomach at a depth of 800 m over a sounding of > 5000 m; (b) the feeding migration of bentho-pelagic carnivores; the rattail fish *Coryphaenoides rupestris* has been caught 270—1440 m above the sea bed in the Denmark Strait where they were feeding on mid-water fishes (Haedrich, 1974; numerous other examples are summarized by Marshall and Merrett, 1977); (c) the release of reproductive products which float up to shallow depths — rattail fishes and some mollusks are amongst the benthic groups already known to produce pelagic eggs and larvae (Haedrich, 1974; Bouchet, 1976a, b; Pearcy, 1976).

Evidence for upward movement of sediment being large enough to measure is two-fold:

(1) Honjo (1978) found that a very large proportion of the particles caught in a sediment trap set 114 m above the bottom at a sounding of 5367 m in the Sargasso Sea consisted of red clay resuspended from bottom sediments. Spencer et al. (1978), in discussing the chemical fluxes associated with this sedimentation, concluded that the fluxes of ^{232}Th, ^{238}U and ^{226}Ra were consistent with 60% of the total mass flux being derived from bottom material resuspended by a bentho-pelagic population which produces distinct red faecal pellets consisting of about 80% clay. They further suggest from GEOSECS data that the influence of the nepheloid layer can be detected in the trace-element profiles at least 1000 m above the sea bed.

(2) Noshkin and Bowen (1975) found that the sedimentation of Pu resulting from fallout can be explained by a simple model suggesting that the Pu is associated with particles that sink at 70—390 m yr.$^{-1}$. Yet the high incidence of the use of mucus for feeding, which is highly efficient at extracting even the finest particles from the water (Gilmer, 1972), and the results of other sediment trap experiments (e.g., Wiebe et al., 1976), imply that the main particle flux occurs in the form of faecal pellets. Smayda (1969), pointed out above, observed sinking rates of 36—37 m day^{-1}, i.e. flux rates 2—3 orders of magnitude faster than the settlement rate of fallout plutonium. This suggests that although the net flux is down, there is considerable

recirculation of particulate material and this is borne out by the observations of Bishop et al. (1978) that 90—99% of the particulate matter in the surface 400 m of the Benguela Current region is recycled within the layer. They found that particles contributing most to vertical flux reach 4000 m in less than 30 days. Thus the slowness of the movement of Pu down through the water column could in itself be an indication of a considerable reflux of particulate matter up towards the surface.

Horizontal movement of radionuclides may also be brought about by the migration of the bentho-pelagic scavengers. Many of these species have very extensive zoogeographical and wide bathymetric ranges. So their horizontal movements could take them high up the continental slopes. Extrapolating to deep-sea fishes on the basis of Pentreath's (1978a) results on the accumulation of Pu in plaice, teleost fishes can be expected to loose relatively quickly the small quantities of the element that they may accumulate from their food in liver, kidney and spleen. In the thornback ray, Pentreath (1978b) found that the biological half-life of Pu is 60—104 day. Deep-sea elasmobranchs are abundant, and do move up into shallow waters where they are caught during commercial trawling operations, e.g., Greenland sharks (*Somniosus* sp.) caught by trawlers working in Icelandic waters have also been photographed at great depths off California. Thus, the sharks and rays could prove to be important links in a critical path to shallow water and commercial species.

The movement of these benthic fishes up the slope also seems to be associated with an increase in the tendency for them to swim up into midwater to feed (Okamura, 1970; Haedrich and Henderson, 1974; Pearcy and Ambler, 1974). These feeding migrations will link up with the migrations of mid-water organisms moving down from the epipelagic zone. The influence of the proximity of the continental slope on the movement of both the benthic species and the distribution and migrations of the mid-water species is poorly known.

Migrations by mid-water organisms

The general gradient of biomass from the surface layers to the abyssal depths means that by swimming up the gradient either along the sea bed or by swimming towards the surface in mid-water that the availability of food tends to increase. This generalisation is reversed in the vicinity of the sea bed where there may be an increase of biomass as the bottom is approached. Very little is known of this near-bottom enrichment in deep water although preliminary results of "Discovery" sampling suggest that it is limited to within 100 m of the bottom. Associated with these gradients of increased availability of food is a gradient of increasing predation pressure. Thus, at times of vulnerability, e.g. at moulting in crustaceans or during reproduction, animals may find refuge deeper down, or in the case of benthic organisms they may swim up into mid-water to find refuge.

Specialist feeders may make unexpectedly extensive migrations in search of food. Sperm whales, for example, specialize in feeding on deep-sea squid and sometimes on fish; indeed, virtually the only source of material and data on the distribution of many of the larger squid species has been the stomach contents of commercially captured whales (e.g., Clarke and MacLeod, 1976). Lockyer (1977) observed that only 5% of dives by sperm whales off South Africa were to depths in excess of 800 m. However, Clarke (1972) recorded the capture of a whale caught after resurfacing from a 1-hr. 52-min. dive over a sounding of 3200 m which had a fresh bottom living shark in its stomach. Weddell seals are also known to dive to depths of 600 m (Kooyman and Andersen, 1969). So some of these deep diving marine mammals could form important links between the deep-living fauna and the surface.

Harden Jones (1968) has described a conceptual model of the movement of commerical fishes from feeding grounds to spawning grounds, the dispersion of the eggs and juveniles to nursery areas, and a gradual movement on maturation to the adult feeding grounds. A similar model can be applied in a modified form to mid-water organisms. The migrations are normally vertical and geographical displacement may occur because of the three-dimensional structure of the current field. For example, the copepod *Calanus finmarchicus*, an important food organism of commercial fish species in the North East Atlantic particularly north of 50°N, overwinters in deep water as a stage-V copepodite. It probably uses the deep water as a refuge during the period of low surface productivity in the winter. In the spring it migrates back up to the surface to mature and breed at the optimum time for the larvae to exploit the spring bloom. During the winter the copepod occurs at depths where the residual current drift is southwards, but in summer when it occurs in the surface layers where the drift is generally northwards (Zaripov and Rzheplinskiy, 1977) these vertical movements help to keep the species within its normal range at fairly high latitudes in the North Atlantic (Jaschnov, 1970).

Such seasonal vertical migrations occur in many organisms at high latitudes where there is a marked seasonality in the pulsing of the production cycle. They are much less common within the bounds of the subtropical convergences where the production cycle is of lesser amplitude. However, ontogenetic migrations are still important in tropical and subtropical plankton (e.g., Angel, 1979). Juvenile stages which may be herbivores occur at shallow depths, and as maturation proceeds they move deeper and often change diet.

Superimposed on this pattern are the diurnal migrations of animals up at dusk and down again at dawn. The timing of migrations tends to be associated with the light cycle, animals either following isolumes or being stimulated to migrate by the rate of change in light intensity. The younger juvenile stages are often non-migrants, but as they age they start to migrate, extending the amplitude progressively as they mature. The full migration is seldom undertaken by the whole population in plankton organisms, whereas the whole of migrating nekton populations usually move synchronously. The function of these diel migrations has not been firmly established (e.g., Longhurst, 1976; Enright, 1977); theories include escape from predations, search for food, more efficient use of metabolic energy and improved reproductive performance. The timing of feeding in association with these migrations is particularly important to the potential fluxes of biomass and hence radionuclide pollutants, since if no feeding occurs deep down there will be no upward flux. There is some evidence of feeding occurring at depth in euphausiids (Roger, 1975), decapod crustaceans (Foxton and Roe, 1974) and mid-water fishes (Merrett and Roe, 1974). Present data are mostly from the surface 500 m of the water column. Published evidence of diurnal migrations occurring from below 1000 m is sketchy (Waterman et al., 1939), but unpublished "Discovery" data (1978) from the North East Atlantic around 42°N,17°W imply that there are only a few species that migrate at such depths.

QUANTIFICATION

So far this chapter has been purely descriptive and no attempt has been made to quantify the amounts of biomass involved in these movements. Generally the data available are inadequate. This is partly because the inaccessibility of the deep-sea habitat has resulted in: (a) our present lack of knowledge of the taxonomy of deep-oceanic communities which often inhibits meaningful cross-programme comparisons of results: (b) a lack of quantitative biomass samples, particularly for deep benthic communities; (c) a lack of knowledge of the basic ecology of these communities such as the structure of the food webs and the reproductive strategies of the constituent species; and (d) an almost total ignorance about the rates of ecological processes.

Some of the inherent difficulties in measuring rate processes can be used to illustrate the reasons for our present inadequate knowledge. Organisms living at depth are very sensitive to the increase in temperature and the decrease in hydrostatic pressure which they are subjected to when they are brought to the surface in trawls, so that attempts to conduct physiological experiments on them are often misleading (Childress, 1977). Cod-end devices designed to keep animals at in situ temperature have been used successfully, but those designed to maintain the animals at in situ pressures have proved too cumbersome for widespread routine use. Technological problems have kept in situ observations of respiration rates to a minimum, but they do imply that metabolic rates of organisms from deep water may be about two orders of magnitude lower than those occurring on the shelf

(Jannasch and Wirsen, 1973; K. L. Smith and Hessler, 1974). However, data on the activity of micro-organisms is not always consistent; for example Schwarz et al. (1976) found the metabolic activity of the intestinal micro-flora from trap-caught amphipods from the Aleutian Trench was not significantly lower at in situ pressures than at atmospheric pressure. Attempts to age a small deep benthic bivalve mollusk suggested that it reached maturity at the age of 50—60 yr. and had a total longevity of about a century (Turekian et al., 1975). A recolonisation experiment of defaunated sediment at almost 2000 m indicated that recovery of the community even in relatively small areas was much slower than on the continental shelf (Grassle, 1977).

In deep-living midwater animals the metabolic activity is slower, the longevity greater and the reproductive strategy quite different from that in shallow-water species. Childress and Price (1978) compared the life history characteristics of the bathypelagic mysid *Gnathophausia ingens* with a shallow-water species *Metamysidopsis elongata*. The shallow-water species produces about 340 larvae in about 14 broods, lives $\sim$ 0.5—1 yr. and has an oxygen consumption of $4 \mu l \, O_2 \, mg^{-1}$ DW hr.$^{-1}$ (DW = dry weights). The bathypelagic species produces about 350 larvae in a single brood, lives $\sim$ 8 yr. and has an oxygen consumption of $0.2 \mu l \, O_2 \, mg^{-1}$ DW hr.$^{-1}$. Childress and Price conclude that *Gnathophausia* has all the characteristics of a species that is conserving energy and maximizing fecundity. They attributed the selection of this reproductive strategy to the effect of the great stability of the deep-sea environment, i.e. the very low level of biologically significant variability effecting mortality.

In Fig. 2 the variability in the displacement volumes (in milliliters per $1000 \, m^3$ water filtered) of micronekton from 24 repeated hauls at 1000 m depth collected at 42°N,17°W, mostly within a 36-hr. period, illustrates the low degree of sampling variability. This result also shows: (a) the low abundance of biomass at this relatively shallow depth; (b) the stability of the distribution of biomass; and (c) it indicates that single sample estimates of biomass abundance are within a factor of 2 of the mean abundance. This last factor should be borne in mind when examining the profiles of plankton and micronekton in Fig. 3 from the same position in the North East Atlantic

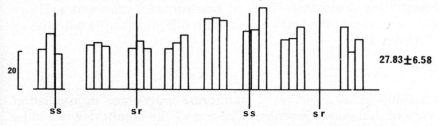

Fig. 2. Variability in micronektonic biomass standing crop (milliliters displacement volume per $1000 \, m^3$ water filtered) in 24 repeated tows at a depth of 1000 ± 10 m in the vicinity of 42°N,17°W during May 1978. *ss* = sunset; *sr* = sunrise.

310

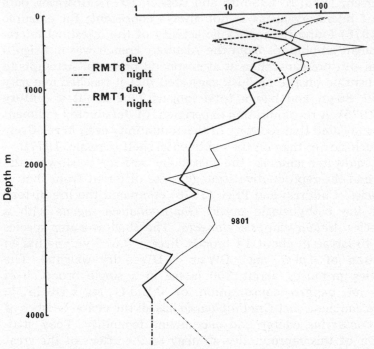

Fig. 3. Profile of standing crop (milliliters displacement volume per 1000 m³ water filtered) from a series of RMT1 (plankton) and RMT8 (micronekton) hauls taken in the vicinity of 42°N,17°W during May 1978 (for further details see Angel, in press).

(Angel, in press). These profiles illustrate the very low standing crop of both plankton and micronekton in the bathypelagic zone, and the redistribution of biomass occurring because of diurnal migrations in the surface 900 m.

A low standing crop does not necessarily indicate low production; for example the high turnover rates of populations of many micro-organisms and phytoplankton results in high production despite their low standing crops. However, as discussed above, what little evidence is available suggests that turnover rates in deep-living communities are likely to prove to be about an order of magnitude lower than those of near-surface communities. Hence the low standing crop of the bathypelagic zone will probably be reflected by a very low productivity.

MONITORING

The establishment of a biological monitoring programme in association with the start of dumping is essential. The aim of the monitoring would be to: (a) assess the impact of the dumping on the deep-sea communities (b) check that movements of isotopes into man's biosphere are as predicted by the models, or (c) are close enough for the deviations of the predictions from reality to be ignored. The limits of acceptability should be set prior to the onset of dumping, and contingency plans should be drawn up to demonstrate what action should be taken if the limits are exceeded. The final aim of the monitoring programme would be to explore the possible consequences of interference with the deep-oceanic communities on the ecological health of the oceanic ecosystem.

It is also essential to start the monitoring of both benthic and mid-water communities as soon as any dump site is selected. Despite our belief that the deep-oceanic environment is very stable we have little knowledge as to its heterogeneity in time and space at any scale of variability. Adequate monitoring (i.e. the ability to recognize significant changes soon enough to take steps to reverse the trends or minimize their severity) is only possible if there has been a thorough baseline study to establish the amplitude and scale of variation. Patchiness in communities in space across the whole spectrum of scales has been a notorious problem in making quantitative measurements of both mid-water and benthic communities. The techniques of sampling and analysis of these patchy communities presents considerable difficulties that are only just beginning to be resolved in mid-water communities (e.g., Fasham, 1978; J. H. Steele, 1978) but are still to be adequately tackled for benthic communities, particularly at depths beyond the continental shelves. Until recently the high level of long-term variability at time scales at orders of magnitude in excess of 5—10 yr. was unrecognized (e.g., Colebrook, 1978) in populations occurring in shallow water. It is unknown how such changes link with the deep-sea community and what degree of lag or smoothing in the amplitude of variations the effect of increasing depth has on the linking.

Another fundamental requirement for monitoring to be effective is an adequate set of control observations at another similar site where no dumping is proceeding. Such control observations would allow any long-term fluctuations at scales longer than the original baseline study in the dump-site area to be filtered out. The cost of such a monitoring programme will be considerable, although small compared with other costs associated with dumping. Offset against this cost, should be the fallout benefit in pure science terms of our knowledge of the deep-oceanic environment. In addition, the information that will accrue on the effects of perturbation on potentially fragile ecosystems will be useful in the management of future exploitation of the deep-sea and other environments.

Set against this need for an adequate monitoring programme is the doubt as to whether the technology to carry out such a programme exists, and as to whether there are enough trained experts on the taxonomy of the great variety of deep-sea organisms (training such a team may in itself take several years).

CONCLUSIONS

(1) Important elements in routes whereby biological processes may accelerate the transport of radioactive waste have been identified as: (a) bioturbation of benthic sediments, leading to the resuspension or the accumulation of isotopes making them more available to other organisms; (b) the horizontal and vertical movements of benthic organisms that may link the benthic ecosystems with the mid-water ecosystem, particularly if movement occurs up the continental slope onto the shelf; (c) the vertical movement of biomass associated with the reproductive strategies of benthic and deep-living organisms; (d) the vertical movement of mid-water organisms, ontogenetically, diurnally, seasonally and aperiodically for feeding.

(2) The importance of a sound taxonomic and zoogeographic foundation to interprogramme comparisons and the assessment of lateral spread of isotopes is stressed.

(3) Our ignorance of rate processes in the deep ocean makes it impossible to model the deep-oceanic ecosystem. However, what little is known tends to suggest that biological processes are not intensive enough to alter significantly the transport rates of radionuclides by physical processes. This applies particularly if the absence of accumulation of high-level radioactive isotopes up the trophic chain found in freshwater biota (Wahlgren et al., 1976) is confirmed for marine organisms.

(4) The importance of the deep-sea ecosystems to our global ecosystems is unknown. Since there is every indication that these ecosystems are adapted to a highly stable state they can be expected to be extremely fragile. Any minor perturbation is likely to produce a large and irreversible change in the structure of the ecosystem.

(5) Public acceptability of a radioactive waste disposal programme in the sea will only be achieved if the consequences of the programme is unequivocably demonstrated to be insignificant.

(6) The need for an adequate monitoring programme is stressed.

CHAPTER 11

THE NEW YORK BIGHT ECOSYSTEM

Charles G. Gunnerson

INTRODUCTION

The New York Bight (NYB) stands among the classic examples of solving waste disposal problems by transportation. Wastes dumped into the Hudson and Raritan river estuaries flush downstream into the NYB. Dredge spoil contaminated by previously dumped wastes, sewage sludge, chemical wastes and other materials are barged directly to offshore dump sites. Emphasis has been given to local and immediate financial costs to the neglect of the long-term economic ones.

However, times and attitudes change. Even though the ocean can and does accept large quantities of wastes, its capacity is not unlimited. Consequences of overfishing, loss of shellfisheries because of pollution, the stranding of solid wastes on recreational beaches, and the recent development of ecological concern at local, U.S.A. State and Federal levels have resulted in major research efforts. These include describing and understanding the natural and man-made environments, the characteristics, fates and effects of wastes discharged to the NYB and the remedial programs needed to maintain or restore living ocean resources. More recently, better definition of problems and of the research, development and monitoring protocols needed to solve them is resulting in intensive studies of the NYB which began with the National Oceanic and Atmospheric Administration's Marine Ecosystem Analysis (M.E.S.A.) New York Bight Project in 1973.

THE NATURAL BACKGROUND

The location of the New York Bight is shown in Fig. 1. It lies seaward of most of the people in the Boston—Washington corridor whose predicted future by the year 2000 includes a tripling of the urban area (Koebel and Krueckeberg, 1975). This future is the continuation of five centuries of immigration, growth and development of natural resources such as climate, harbors and soils, and is clear evidence of the increasing need for effective land- and water-resources management. Seventy-three percent of the fresh-water entering the NYB is from the 35,000 km^2 of the Hudson River drainage basin. It strongly affects water quality and circulation in the western portion of the NYB, especially in the Bight Apex. Table I presents average annual inputs. Estimated average outflows vary from 210 m^3 s^{-1} (7500 ft.^{3}

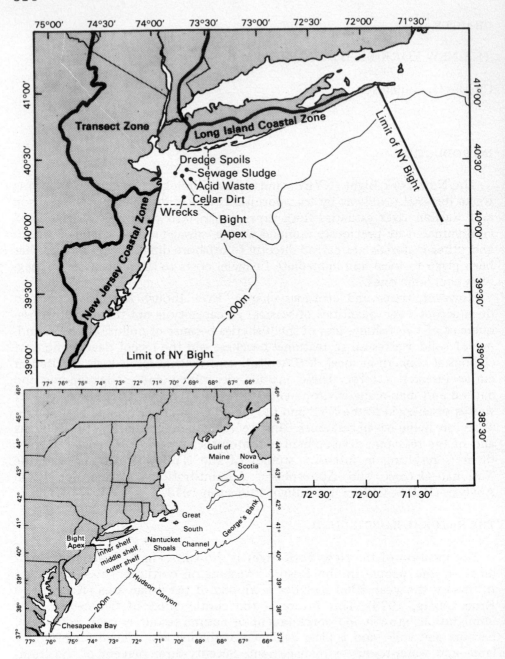

Fig. 1. New York Bight, Bight Apex, and ocean dumping locations (after Bowman and Wunderlich, 1977; Mueller and Anderson, 1978).

TABLE I

Freshwater inflows to the New York Bight (after D.J. O'Connor et al., 1977)

	Average annual runoff		
	$(m^3\ s^{-1})$	$(ft.^3\ s^{-1})$	$(km^3\ yr.^{-1})$
Hudson River basin	580	20,300	183
New York City and Long Island	80	2,900	26
Northern New Jersey			
(into New York Harbor)	110	4,000	35
New Jersey coastal zone	20	700	6
Total	790	27,900	250

s^{-1}) in summer to $1300\,m^3\ s^{-1}$ ($47,400\,ft.^3\ s^{-1}$) in April (D.J. O'Connor et al., 1977).

Geological oceanography

The major sea-floor features of the NYB have formed in response to glacial eustatic changes in sea level that have taken place over the past $4 \cdot 10^6$ yr. The last significant event, which occurred at the end of the Pleistocene glacial epoch, began $\sim 75,000$ yr. ago when the Laurentide Ice Sheet advanced from its Canadian center to a line passing lengthwise through what is now Long Island, and continuing westward across northern New Jersey. During the time of maximum glacial advance, sea level worldwide was at least 125 m (410 ft.) below its present level. The NYB then was subjected to subaerial erosional processes. Stream erosion, amplified by runoff of glacial melt water, dissected the uppermost, semi-consolidated sedimentary strata, forming ancestral northern New Jersey, Long Island, and the Hudson shelf valley (Fig. 2).

The conclusion of the last ice recession $\sim 15,000$ yr. ago was accompanied by glacial melting which produced a rapid rise in sea level and inundation of the gradually sloping surface of the shelf. Clean sand from the retreating shore face and the shelf surface blanketed the NYB. This sand blanket is undergoing continuing transport in response to storm-generated currents. Seismic reflection profiles show drowned river valleys (the Hudson in particular) only partially buried and filled with Holocene (Recent) sediments. River mouths were drowned, thereby creating present estuarine environments which act as sediment traps for sediments from both river and the ocean. Details of the evolution of the New York—New Jersey shelf surface, which is 180 km wide off New York, have been developed by Swift et al. (1976).

316

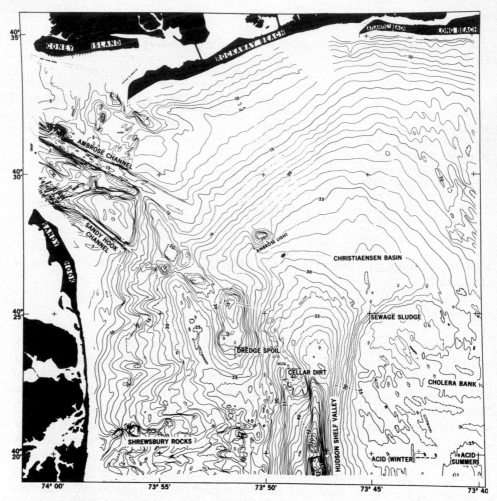

Fig. 2. Bathymetry of the New York Bight Apex, contours are in 1-m intervals (from Freeland and Swift, 1978).

Comparison of bathymetric surveys in 1936 and 1973 reveal that the most significant change has occurred at the dredged material dump site where $93 \cdot 10^6 \, \text{m}^3$ of sediment have accumulated resulting in up to 10 m of shoaling (Fig. 2). Elsewhere over the 718-km² Bight Apex, there has been general erosion so that the net change is $10^6 \, \text{m}^3$ of deposition (Freeland and Swift, 1978). The present distribution of surficial sediments in the Bight Apex is shown in Fig. 3. Muds apparently originating in the dredge spoils accumulate in the Christiaensen basin and move downslope into Hudson canyon. This movement corresponds with sediment transport calculations based on energy requirements (Swift et al., 1976) and with distribution of heavy metals in sediments discussed on pp. 345—349.

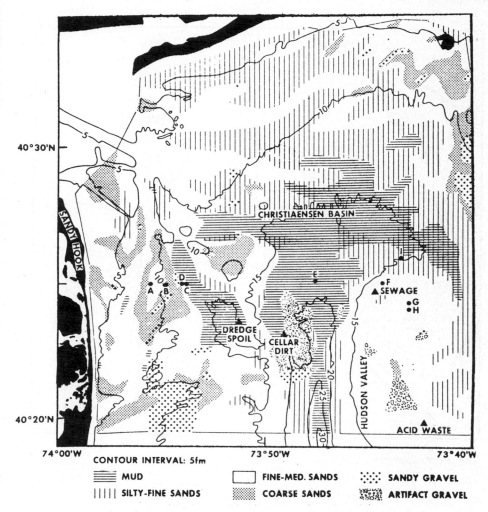

Fig. 3. Surficial sediments in the New York Bight Apex (from Freeland and Swift, 1978).

Physical oceanography

Research and observations of water levels in the NYB have been summarized by Swanson (1977). Mean semi-diurnal tidal ranges are from 0.8 to 1.4 m (2.8 to 4.6 ft.) along the coast and up to 2.2 m (7.2 ft.) in Long Island Sound. These are superimposed upon the annual cyclical variations of ~ 25 cm (8 in.) and a secular increase in mean sea level of ~ 35 cm (1.1 ft.) per century. Coastal features, structures and shipping are also vulnerable to storm surges with heights up to 4 m and periods of 3 hr. to 6 days (Pore and Barrientos, 1976).

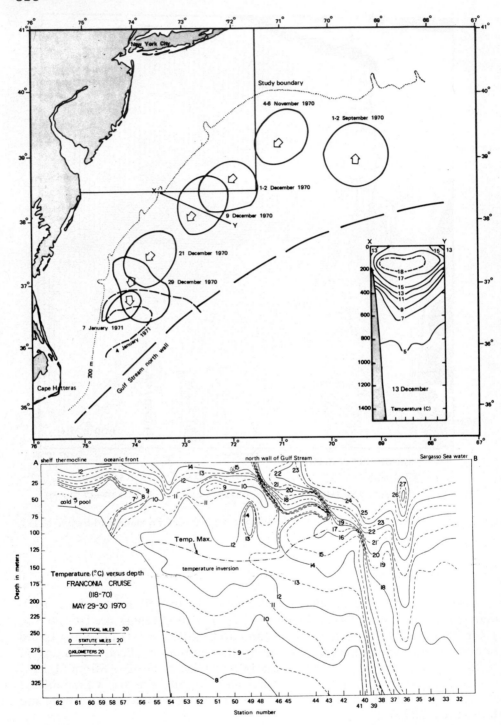

Large-scale current and circulation patterns include the northeasterly flow of the Gulf Stream some 400 km (220 n.m.) off New York. Shoreward of the shelf—slope front, generally southwesterly flows contain slow-moving eddies ($5-10$ cm s^{-1} or up to ~ 9 km day^{-1}) which developed from meanders of the Gulf Stream. Fig. 4, taken from Bowman and Wunderlich (1977), shows the distributions of temperatures and the movement of one such eddy which occurred along the boundary of the two flows. Within the NYB, flows reflect the influence of drainage from the Connecticut and Hudson river basins (Fig. 1), totalling 64,000 km^2. Net surface flows are generally southwesterly but highly variable. Along the bottom near the river mouths estuarine circulation results in a net shoreward and upstream movement. In the Hudson canyon and Hudson shelf valley, aperiodic up- or down-slope movements result from regional meteorological conditions; monthly average shoreward velocities in the valley are as much as 5 cm s^{-1} (0.1 kt.) up-valley (Hansen, 1977). Mean velocities over the 180 km (100 n.m.) wide continental shelf, calculated by Beardsley et al. (1976), are shown in Fig. 5; transport estimated at $2 \cdot 10^5$ m^3 s^{-1} within the 100-m isobath implies residence times of $\frac{3}{4}$ yr. Residence times in the Apex vary from ~ 4 to 12 days (Swanson, 1979) with slower flushing during the September—October minimum Hudson River outflow period.

Currents over the shelf are strongly affected by local and regional meteorological factors which have been summarized by Lettau et al. (1976). Surface wind speed and direction over the NYB have pronounced seasonal variations but generally little spatial variation, suggesting that observed air movements are governed by large-scale meteorological systems rather than by small, local systems. However, storm fronts which are approximately the scale of the NYB result in significant spatial variability. Wind persistence, which is the ratio of resultant or vector displacement of an air parcel over time to the average wind speed, reaches a maximum in winter with westerly to northwesterly winds generally averaging $16-25$ km hr.$^{-1}$ ($9-13$ kt.). A secondary persistence maximum occurs in July with southwesterly to southerly winds $10-16$ km hr.$^{-1}$ ($7-10$ kt.). Strong winds of over 50 km hr.$^{-1}$ (28 kt.) occur mostly in winter (December—February) when they exceed 50 km hr.$^{-1}$ (28 kt.) more than 10% of the time. The maximum recorded wind occurred at the Battery in 1954 during hurricane Hazel when it reached 180 km hr.$^{-1}$ (98 kt.).

Mean air and water temperatures are presented in Table II. Coastal water temperatures are generally $1-4°C$ cooler than air temperatures in spring and early summer, and up to $6°C$ warmer than air in winter (Lettau et al., 1976). The springtime air—sea temperature differentials, local wind mixing, and outflows from the Hudson River influence the rates of development and the

---

Fig. 4. Major offshore current systems, eddy movements, temperatures and fronts, May 1970—January 1971 (from Bowman and Wunderlich, 1977).

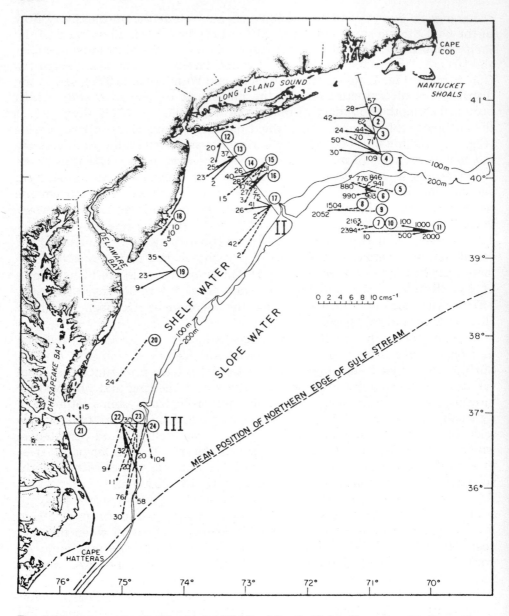

Fig. 5. Mean current velocities in the Middle Atlantic Bight, November 29—December 4, 1975. Winter measurements are indicated by *solid arrows*, summer by *dashed arrows*. Measurement depths (in meters) are shown near the *head of the arrows* (from Beardsley et al., 1976).

TABLE II

Mean air and surface water temperatures (°C) in the New York Bight (after Lettau et al., 1976)

Month	Coastal zone		Offshore zone (continental margin)	
	air	water	air	water
Jan.	0—2	4—6	4—6	10—12
Feb.	0—2	~ 2	4—6	8—10
Mar.	~ 4	~ 4	~ 6	8—10
Apr.	8—10	~ 6	8—10	8—10
May	14—16	~ 12	~ 12	~ 12
Jun.	18—20	~ 18	~ 19	17—18
Jul.	23—24	22—23	23—24	22—23
Aug.	22—23	22—23	~ 24	~ 24
Sep.	19—20	19—20	~ 21	~ 22
Oct.	~ 14	~ 18	~ 18	~ 20
Nov.	~ 8	~ 12	12—13	~ 16
Dec.	2—4	~ 8	~ 8	~ 12

depth and strength of the pycnocline over the inner shelf. Anomalies in these rates are discussed on pp. 367—369.

Measurements reported by Swift et al. (1976) and by Hansen (1977) of the effects of tides and surface winds on currents in ~ 15 m of water off Long Island are shown in Fig. 6. Currents responded rapidly to the strong N-NW storm winds of 10—20 m s^{-1} (18—36 kt.) and maintained their direction and speed for about a half-day after the winds subsided. Tidal components may be filtered out to reveal longer-term current variability and displacements shown in Fig. 7.

Tidal components reported by Hansen (1977) increase near the entrance to the harbor as shown in Fig. 8. Net surface outflows average > 16 cm s^{-1} near the center of the Sandy Hook—Rockaway Point transect while average inflows reach 12 cm s^{-1} near Rockaway Point and 4—6 cm s^{-1} within the deeper channels off Sandy Hook.

Chemical oceanography

Seasonal distributions of surface and bottom temperature, salinity and dissolved oxygen (D.O.) are shown in Figs. 9—11. Average values for other properties are listed in Table III. Distributions of trace elements are discussed below in connection with waste discharges.

Living marine resources

Impacts of pollution upon living resources and their support systems in

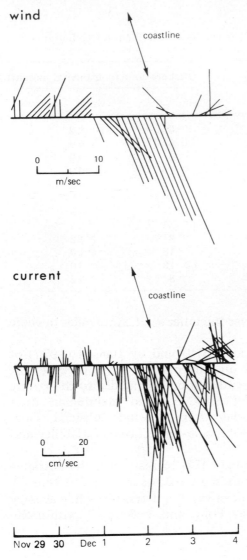

wind

coastline

0   10
m/sec

current

coastline

0   20
cm/sec

Nov 29 30   Dec 1   2   3   4

Fig. 6. Surface winds and currents off Long Island during a 1974 winter storm (from Hansen, 1977).

Fig. 7. Spring and summer currents (A and B; *lines* projecting from time baseline denote direction and speed of observed currents) and displacement (C and D; *dots* are one-day time intervals), April—September 1974 (from Hansen, 1977).

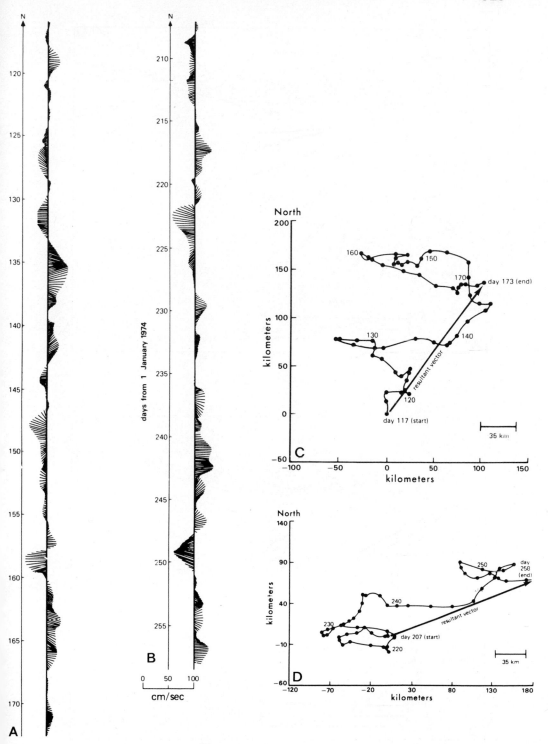

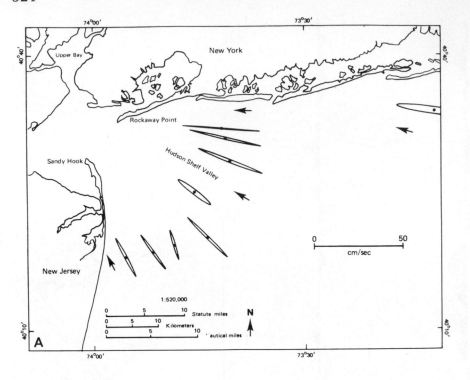

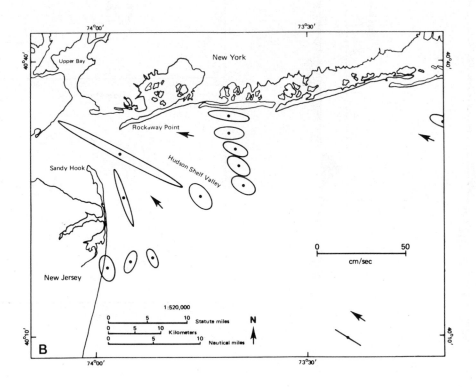

the NYB have been a major concern throughout the current investigations. Current knowledge of the plankton in the NYB has been summarized by Malone (1977) and Yentsch (1977). These and other sources provide benchmark data for evaluating fish kills and other episodes discussed on pp. 359—371.

Fisheries are sensitive to both pollution and to fishing pressures. Edwards (1976) has shown recent reductions due in large part to overfishing of standing crops of important commercial species off the northeastern U.S.A.; these are listed in Table IV. Depletions in stocks of haddock and herring are particularly noticeable. Mackerel, some of which spawn in the NYB, increased. One result of fishing pressures has been the recent imposition of a 200-mile limit within which fishing by foreign vessels is controlled, although the timing of the decrease in landings makes it clear that foreign fishing was not the cause (McHugh, 1977). Some of the decline in commercial finfishing has been offset by sportfishing (McHugh and Ginter, 1978). In this regard, A. C. Jensen (1979) notes that increasing competition for particular species such as striped bass is leading to habitat destruction by fishing gear and disputes over areas, access, traditional rights and legal prerogatives.

Other recreational activities in the area include public and private beaches, parks, marinas, recreation areas and wildlife refuges (Carls, 1978). Beaches and their bathing waters are susceptible to pollution and to regulation on aesthetic and public health grounds discussed on pp. 359—371 in connection with a 1976 beach pollution episode.

Shellfisheries also change with time. Commercial landings of the American oyster in the NYB have decreased from $\sim 7000\,t\,yr.^{-1}$ in 1950 to a low of $46\,t\,yr.^{-1}$ in 1967, followed by an increase to $960\,t\,yr.^{-1}$ in 1975. Hard clam landings changed from 6000 t in 1950 to 2500 t in 1960 to 4700 t in 1975 with a value of US $16 million. The surf clam fishery has grown from $\sim 3000\,t$ in 1950 to $\sim 16,000\,t$ from 1968 through 1975 (McHugh, 1977).

ECONOMIC DEVELOPMENT

Rapid population, commercial, and industrial growth of the New York metropolitan area began with 24,000 people in 1786 in lower Manhattan. By 1970, it included Long Island, coastal and northeastern New Jersey, Dutchess County in New York, Fairfield and New Haven Counties in

---

Fig. 8. Tidal currents: (A) 8 m (26 ft.) above bottom; and (B) 3 m (10 ft.) or less above bottom. Rotation is predominantly clockwise at 8 m above bottom and counter-clockwise at 3 m or less above bottom. The *centers of ellipses* are the station locations. *Arrows* indicate direction of progress of the maximum M_2 flood current velocity. Current velocity vectors rotate $360°$ in 12 to 42 hr. (from Hansen, 1977).

Connecticut, and 19.3 million people. This figure is estimated to reach 25.8 million by the year 2000 (Koebel and Krueckeberg, 1975). 1977 population densities range approximately from $1\,km^{-2}$ ($3\,mi.^{-2}$) in the lower Hudson River valley to $21,000\,km^{-2}$ ($67,000\,mi.^{-2}$) in New York City. In 1972, 1.8 million manufacturing employees in the metropolitan area contributed over US $28 billion worth of value added, approximately half from the chemical, publishing and textile industries (Mueller and Anderson, 1978).

WASTE LOADINGS

Gross (1976a, b), Mueller et al. (1976a, b), and Mueller and Anderson (1978) have compiled information on waste loadings to the New York Bight. In 1972, 127 major municipal discharges contributed $114\,m^3\,s^{-1}$ ($2.6 \cdot 10^9$ gal. day^{-1}) of sewage of which $21\,m^3\,s^{-1}$ ($0.48 \cdot 10^9$ gal. day^{-1})

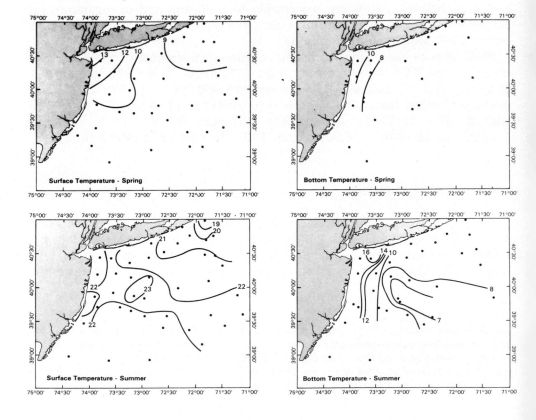

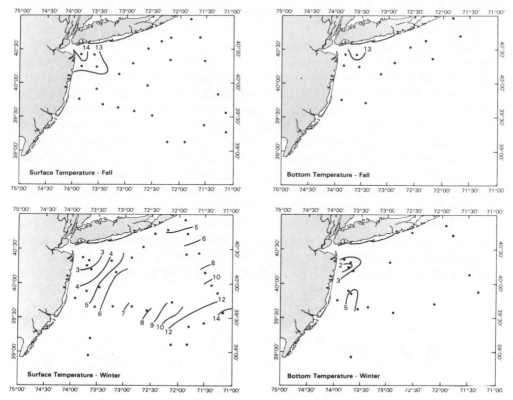

Fig. 9. Seasonal distribution of surface and bottom temperatures (°C) (after D. J. O'Connor et al., 1977).

from Manhattan, Staten Island and Brooklyn were untreated, reduced by 1978 to $9 \, m^3 \, s^{-1}$ $(0.30 \cdot 10^9 \, gal. \, day^{-1})$. Industrial discharges amounted to $27 \, m^3 \, s^{-1}$ $(0.61 \cdot 10^9 \, gal. \, day^{-1})$ of which 47% went through municipal systems.

Ocean dumping at locations shown in Fig. 1 is the most visible discharge to the NYB; it even can be seen from satellites (Fig. 12). Mueller et al. (1976a, b) reported an annual total of $27.6 \cdot 10^6 \, m^3$ $(21.7 \cdot 10^6 \, yd.^3)$ dumped of which 53% is dredge spoils, 26% is sewage sludge with ~ 5% solids, 15% is acid waste, 3% is chemical wastes, and 3% is building and construction debris. Dredge spoils account for most of the sediments entering the NYB; of these, about half derive from sands moved by littoral drift and estuarine bottom flow into the harbor (Gross, 1976b). Because of sediment contamination by waste discharges to upstream dredged channels shown in Fig. 13 (Hammon, 1977), at least half of the Cd, Cr, Cu and Fe entering the NYB is associated with dredge spoils (Mueller and Anderson, 1978).

Tables V and VI list sources and loadings of materials presently entering the NYB. Their effects range from unmeasurable to significant. Estimates

328

of future quantities of wastes depend upon assumptions of growth, funding, effectiveness and timing of remedial projects and programs and upon technological and institutional developments in resource conservation and recovery. Sewage treatment is being upgraded to provide for 85% removals of suspended solids (SS) and biological oxygen demand (BOD). This will result in a three-fold increase in the volume of sludge to be handled. Since current regulations require stopping the present practice of sludge dumping by the end of 1981, alternative land-based methods are being considered. Meanwhile, additional areas are being sewered and sewage treatment plants and effluent outfalls are being constructed within the drainage basins tributary to the NYB. A similar analysis of mass loading is described for offshore southern California by Mearns (Chapter 1 of this volume) and for the Gulf of Naples by Oppenheimer et al. (Chapter 2 of this volume).

A. R. Anderson and Mueller (1978) have estimated future loadings of some waste constituents, assuming secondary (biological) treatment and effluent chlorination of municipal sewage; they further assume the mass loadings of industrial, atmospheric, ocean dumping and urban runoff loadings will

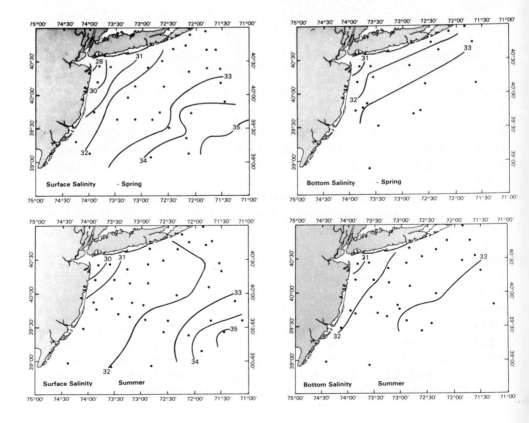

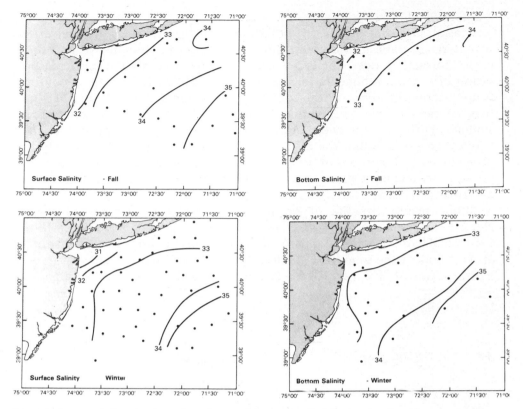

Fig. 10. Seasonal distribution of surface and bottom salinities in parts per thousand (after D. J. O'Connor et al., 1977).

remain the same as at present. While it can be argued that implementation of regulations which limit discharge of toxic or hazardous substances will reduce the amounts of materials entering the NYB, the large reservoir of these materials in estuarine and river sediments and on-land surfaces is expected to be a source of pollution of the NYB. The Anderson and Mueller results are summarized in Tables VII and VIII which, by comparison with Tables V and VI, predict little overall change in loadings to the NYB. These results are consistent with those of the N.Y.C.D.E.P. (1978) as well as others made for major U.S.A. inland streams (e.g., Mackenthun, 1961). Clearly, if stresses of waste discharges on the NYB are to be reduced, quite different waste management strategies will be required.

Chronic effects of waste discharges

There are shoreline, nearshore and offshore impacts of wastes discharged to the NYB. These are discussed in turn.

Shoreline effects

Coastal surf zone and beach pollution is characterized by concentrations of enteric bacteria in the water and by floating and stranding of anthropogenic flotsam. It does not come about by the resuspension and transportation of bottom sediments (M.E.S.A., 1975). 1973—1977 data on coliform bacteria concentrations in the surf zone are summarized in Table IX. Numbers of samples per year varied from 6 to 48, occurring mostly in the summer. Differences between stations reflect proximity to waste discharges (Coney Island is within the estuary) and differences in sampling, analytical procedures, and methods of data analysis and reporting. Differences between years at individual stations are real and reflect changes in waste discharges and treatment and in runoff.

Public health benefits of ocean bathing water protection are elusive. Nevertheless bacteriological standards have been applied, beginning with the California bathing water standard of ≤ 1000 total coliforms per 100 ml for at least 80% of the time (G. C. Gillespie, 1943). This was equivalent to a geometric mean of 250 per 100 ml (Gunnerson, 1955) and is roughly compa-

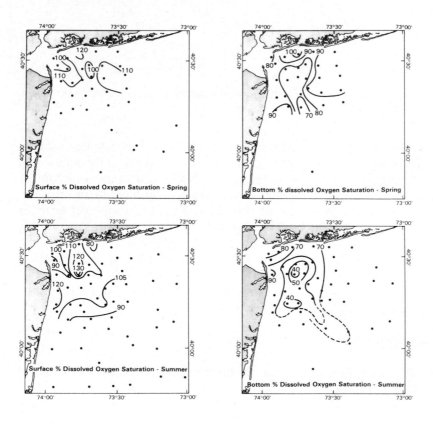

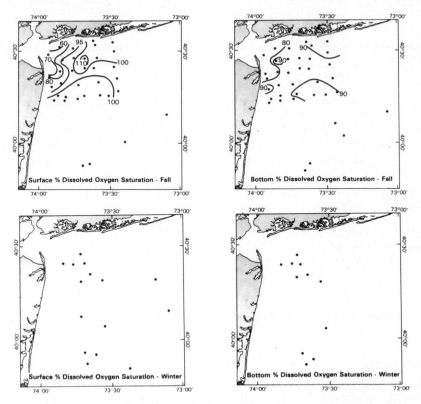

Fig. 11. Seasonal distribution of surface and bottom oxygen saturation (after D. J. O'Connor et al., 1977).

rable to the current U.S.E.P.A. guideline of a mean fecal coliform density of 200 per 100 ml. The California standard has been copied or adapted throughout the world although it is strictly applicable to populations with low incidence of disease. Recent work by Cabelli et al. (1974, 1976) in the NYB and elsewhere provides a measure of the risk involved in swimming in polluted water. They report that subclinical respiratory and gastro-intestinal symptoms were directly correlated with beach pollution at Coney Island and at the Rockaway beaches. Their results are summarized in Table X. This table indicates that enterococci (fecal streptococci) are generally, though not always, more reliable than total or fecal coliforms in assessing subclinical health effects of pollution. Levin et al. (1975) have described the enumeration technique for enterococci, which is simpler and correspondingly cheaper than that for coliforms. While enterococci survival in laboratory experiments exceeds that for coliforms, comparative in situ studies of their survival in the sea are needed before enterococci data can be substituted for

TABLE III

Average concentrations of materials in New York Bight (after D. J. O'Connor et al., 1977)

	Spring		Summer	
	surface	bottom	surface	bottom
NO₃-N Apex (mg l⁻¹) outer NYB	0.01—0.05 0.000—0.004	0.01—0.04 0.002—0.01	0.001—0.08 0.000—0.007	0.01—0.04 0.02—0.2
Inorg-P Apex (mg l⁻¹) outer NYB	0.005—0.03 0.002—0.01	0.01—0.02 —	0.01—0.06 0.001—0.004	0.03—0.06 —
Chlorophyll a Apex (mg l⁻¹) outer NYB	1—10 0.8—1.6	— —	0.4—8 0.1—0.3	— —
Fe (μg l⁻¹) Apex	20—150	30—100	20—70	20—100
Cu (μg l⁻¹) Apex	2—9	2—20	2—18	1—5

coliform data now used for system design and monitoring (Gunnerson, 1974).

Near-shore effects

The most significant effect of waste discharges into the NYB has been the bacteriological contamination of shellfisheries, resulting in a quarantine being placed upon the area shown in Fig. 14. The configuration of the area shows the effects of ocean dumping of sewage sludge and of Hudson River discharges. Within New York Harbor, shellfisheries and some finfisheries have long been stressed, abandoned, or quarantined because of pollution problems (Squires, 1979).

Other near-shore effects of waste discharges are most evident in the sediments. Detailed bathymetric change due to 35-yr. dumping of dredge spoils is shown in Fig. 15. Finer materials move down-slope into and along the Hudson shelf valley (HSV). Figs. 16 and 17 show concentrations in sediments of total organic carbon (TOC) and heavy metals which clearly follow this path. The question is raised as to whether portions of the NYB are beginning eutrophication. If so, this would be the first such case in open coastal waters (Swanson, 1979). In 1974 it was observed that in the bottom waters at the head of the HSV oxygen concentrations were as low as 30% saturation (Segar and Berberian, 1976). In 1976 oxygen-deficient ($< 2.0\,ml\,l^{-1}$) bottom waters sometimes with zero dissolved-oxygen levels, occurred over an extensive area off the coast of New Jersey, resulting in mass fish and shellfish mortalities. Lesser mortalities also were noted in 1968, 1971 and 1974. In 1977 oxygen depletion again occurred but was limited to a band 6—9 km off the New Jersey coast.

Fall		Winter	
surface	bottom	surface	bottom
0.01—0.05	0.04—0.08	0.01—0.1	0.01—0.01
0.01—0.04	0.05—0.07	0.01—0.1	0.01—0.2
0.02—0.06	0.02—0.04	0.02—0.03	0.01—0.02
0.007—0.01	—	0.007—0.02	—
3—5	—	1—10	—
0.5—2.9	—	0.5—1.2	—
10—100	10—200	10—100	10—80
1—7	2—6	3—9	3—8

TOC, shown in Fig. 16A, has been used as a crude indicator of human impact. The highest values occurred in the topographic depression at the head of the HSV and in the Hudson valley itself. Concentrations of TOC of 3—5% dry weight have been found in this same area (Gross, 1972). Small pockets of naturally-occurring black muds with similarly high values of TOC have also been identified within 2.8 km of Long Island beaches. Some muds with values of $\sim 2\%$ TOC occur within 0.9 km of these same beaches (M.E.S.A., 1975).

Several investigators have examined distributions of trace-metal concentrations in NYB sediments. Higher trace-metal concentrations in sediments have been observed at and near topographic depressions, decreasing with increasing grain size and distance from the depressions. The distribution of Pb in sediments is typical and has been plotted in Fig. 16B. Fig. 17 shows corresponding concentrations of Hg, Cd, Cu and Zn. The mobility of Hg is qualitatively different from that of the other metals, possibly because it is bound to finer sediments which are influenced by currents nearer the bottom (Fig. 9).

Polychlorinated biphenyls (PCB's) are also found in association with the head of the HSV. Valley sediments between the sewage sludge and dredge spoil dumping sites have been found to have concentrations of 1500 ppb. The primary source of PCB's here is apparently ocean dumping of sewage sludge with dredged material also contributing substantially. Up to 6200 ppb total PCB's has been found in analyses of sewage sludge samples directly from the treatment plant (M.E.S.A., 1978). PCB's and heavy metals are also discussed for the Mediterranean by Bernhard (Chapter 3 of this volume) and for the Gulf of Mexico by Sackett (Chapter 13 of this volume).

TABLE IV

Estimated biomass (in 10^6 lb.[*1]) of selected fish species for 1963—1965 and 1972—1974 (after Edwards, 1976)

	1963—1965		1972—1974	
	estimated standing crop	average landings	estimated standing crop	average landings
Silver hake	2,084	638	1,215	312
Atlantic herring	4,000	529	1,500	774
Spiny dogfish	1,373	+	978	+
Haddock	980	337	161	42
Red hake	694	176	443	131
Pollock	596	54	576	72
Thorny skate	579	—	550	+
Cod	489	156	399	123
Redfish	399	54	406	51
Little skate	386	—	275	+
Butterfish	309	10	400	+
Big skate	295	+	84	+
Argentine	187	9	23	+
Blackback	185	32	113	+
Yellowtail	185	93	174	70
Barndoor skate	178	+	48	+
Ocean pout	171	12	6	+
Goosefish	142	+	140	+
American dab	125	9	66	(56)
Mackerel	100	+	1,800	794
Alewife	87	42	8	+
White hake	72	7	149	+
Longhorned sculpin	71	+	34	+
Sand flounder	68	+	100	+
Scup	66	10	183	+
Witch flounder	32	4	34	(56)
Fourspot flounder	16	+	12	+
Sea raven	10	+	5	+
All others	308	460[*2]	455	386[*2]
Totals	14,187	2,630	8,837	2,811

[*1] 10^6 lb. = 453.59 t = 453,590 kg).
[*2] Includes landings of all species in table as well as species in "all others" category.

Biological impacts of waste disposal are more diffuse (Swanson, 1979). Benthic invertebrates, because of their sedentary nature and contact with sediments, are indicators of coastal contamination. Most species found in the Apex in the late 1960's were again found in 1973 and 1974. However, both species diversity and density of individuals (Fig. 18) are clearly de-

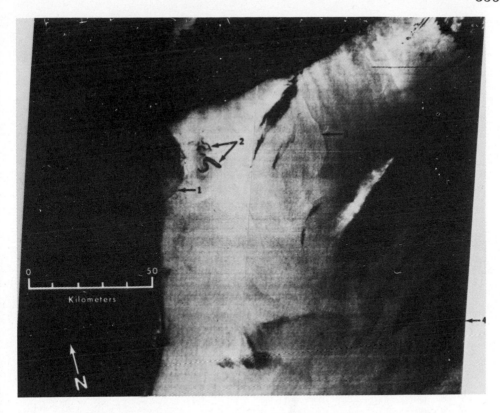

Fig. 12. New York Bight as seen by ERTS-*1* satellite, August 16, 1972. Turbid discharge plume (*1*) of Hudson River can be seen near New Jersey shoreline. *Distinct wavy line (2)* is discolored water from waste acid disposal; *less distinct lines* to north may be discolored water from earlier disposal operations, perhaps of sewage sludge. Some of relatively *sharp lines (3)* are naturally occurring water mass boundaries unrelated to waste disposal. Surface slicks probably due to internal waves are seen at lower right (*4*). (Photo from N.O.A.A.'s Atlantic Oceanographic and Meteorological Laboratory, courtesy of R. L. Charnell.)

pressed in contaminated and poorly flushed areas. The confusing patterns of diversity index and density of individuals are not surprising. Density of individuals is expected to vary considerably over the multisource contaminant impacted area. On the fringes, the generally enriched sediment probably supports a greater than normal population. Nearer the center of contamination, the populations fall below the normally expected levels, perhaps explaining the close geographic association of both the maximum and minimum density of individuals within the topographic depression at the head of the HSV. A similar relationship between diversity and the contaminated sediments is expected. Opportunistic species would pre-

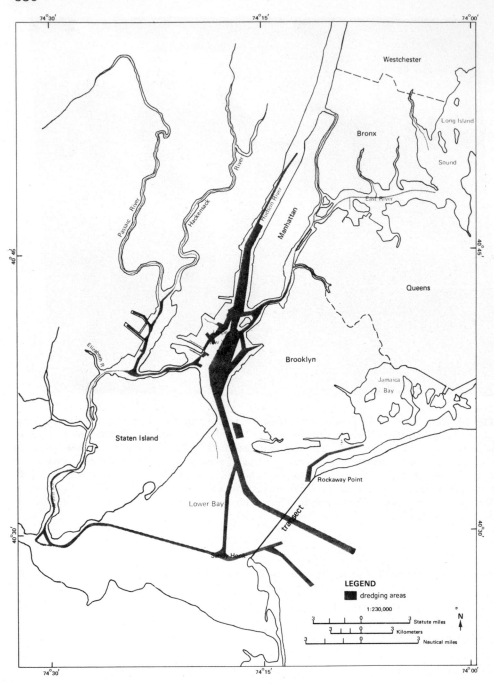

Fig. 13. Dredging areas in New York Harbor (from Mueller and Anderson, 1978).

TABLE V

Total mass loads into the New York Bight (after Mueller et al., 1976b)

Parameter	Mass load (t day^{-1})	Percentage contribution by location			
		direct Bight	Hudson and Passaic River basins	New Jersey coast	Long Island coast
Flow, ft.^{3} s^{-1} (m^3 s^{-1})	82,700 (2,340)	59	36	4	1
SS	24,000	68	31	0.6	0.1
Alk	5,100	1	96	2	1
BOD$_5$	2,100	30	67	3	0.5
COD	10,000	42	48	9	0.5
TOC	2,600	37	58	4	0.6
MBAS	59		95	4	0.5
O & G	870	38	53	9	0.6
NH$_3$-N	210	28	67	3	2
Org-N	190	27	68	3	2
TKN	400	27	68	3	2
NO$_2$+NO$_3$-N	120	33	55	10	2
Total N	520	29	65	4	2
Ortho P	51	1	91	3	5
Total P	138	51	45	2	2
Cd	2.4	84	15	0.5	0.07
Cr	5.0	51	44	0.6	4
Cu	13.8	54	45	0.9	0.2
Fe	230	82	16	2	0.2
Hg	0.30	9	85	6	0.6
Pb	12.7	53	46	0.5	0.2
Zn	33	47	52	1	0.3
Fecal coli {winter	$5.6 \cdot 10^7$	< 0.01	100	0.2	< 0.001
{summer	$4.9 \cdot 10^7$	< 0.01	100	0.2	< 0.001
Total coli { winter	$21 \cdot 10^7$	< 0.01	100	0.1	< 0.001
{summer	$11 \cdot 10^7$	< 0.01	100	0.2	< 0.001

SS = suspended solids; Alk = alkalies; BOD$_5$ = 5-day biological oxygen demand; COD = chemical oxygen demand; TOC = total organic carbon; MBAS = methylene blue-active substance; O & G = oil and grease; and TKN = total Kjeldahl nitrogen (sum of organic and ammonic nitrogen).

dominate in the most highly contaminated regions. Of considerable exonomic interest is that commercial-size surf clams (*Spisula solidissima*) are unusually rare in the area surrounding the Apex dump sites. Adult surf clams are found in the shallow margins of the Apex (Swanson, 1979).

Offshore effects
Sources of offshore contaminants include ocean dumping of chemical

TABLE VI

Total New York Bight loads by source (after Mueller et al., 1976b)

Parameter	Percentage contribution						
	direct Bight		coastal zone wastewater		runoff		groundwater
	barge	atmospheric	municipal	industrial	gaged	urban	
Flow	0.02	59	5	0.4	33	2	0.4
SS	63	5	4	0.2	16	12	nil.
Alk	1	nil.	35	0.3	59	5	0.03
BOD$_5$	21	9	48	2	11	9	0.01
COD	32	10	35	1	13	9	0.01
TOC	25	12	29	1	18	15	0.02
Mba's			86		5	9	0.05
O & G	38		22	0.7	16	23	
NH$_3$-N	24	4	55	3	10	4	0.04
Org-N	19	9	45	2	21	5	0.02
TKN	21	6	51	2	15	5	0.02
NO$_2$+NO$_3$-N	0.07	33	6	0.3	60	0.6	0.7
Total N	16	13	40	2	25	4	0.2
Ortho-P		1	72		18	9	nil.
Total P	50	0.7	35	1	9	4	nil.
Cd	82	2	5	0.6	5	5	0.001
Cr	50	1	22	0.8	10	16	nil.
Cu	51	3	11	9	10	16	0.006
Fe	79	3	5	0.5	6	6	0.01
Hg	9		71	2	13	5	
Pb	44	9	19	3	6	19	0.004
Zn	29	18	8	2	21	22	0.009
Fecal coli {winter	<0.01	nil.	87	0.2	0.01	13	nil.
Fecal coli {summer	<0.01	nil.	85	0.2	0.01	15	nil.
Total coli {winter	<0.01	nil.	91	0.1	0.05	9	nil.
Total coli {summer	<0.01	nil.	84	0.2	0.1	16	nil.

TABLE VII

Present and future municipal influent, effluent and sludge loads in New York Bight (after A. R. Anderson and Mueller, 1978)

Parameter	Present (1972—1973) loads (t day^{-1})			Future (2000) loads (t day^{-1})					
	influent	effluent[1]	sludge[1]	influent		effluent		sludge	
				load	% change	load	% change	load	% change
Flow (10^6 gal. day^{-1})[2]	2,600	2,600	3	3,200	+ 23	3,200	+ 23	8	+ 120
Suspended solids, SS	1,700	854	450	2,200	+ 29	220	− 74	1,220	+ 171
Total organic carbon, TOC	1,070	735	110	1,340	+ 25	400	− 46	290[4]	+ 164
Total N	230	208	17	290	+ 26	220	+ 6	72	+ 324
Total P	52	48	4.7	70	+ 35	49	+ 2	21	+ 347
Pb	3.4	2.7	0.72	4.07	+ 20	3.26	+ 21	0.81	+ 12
Fecal coliform without municipal chlorination[3]	—	4.9·10^7	3,600	—	—	1.3·10^7	− 73	9,800	+ 172
Fecal coliform with municipal chlorination[3]	—	4.2·10^7	3,600	—	—	1,900	− 100	9,800	+ 172

[1] Mueller et al. (1975).
[2] 10^6 gal. day^{-1} = 3780 m^3 day^{-1}.
[3] Coliform load (=) 10^{10} organisms/day.
[4] Based on TOC/SS ratio (0.24) in present sludge (Mueller et al., 1975).

TABLE VIII

Total future waste loadings by source (after A.R. Anderson and Mueller, 1978)

Parameter	Mass load ($t\ day^{-1}$)	Percentage contribution						
				wastewater		runoff		
		barge	atmosphere	municipal	industrial	gaged	urban	groundwater
Flow ($ft.^3\ s^{-1}$)*1	83,800	0.03	58	5.9	0.5	33	2.5	0.3
Suspended solids, SS	24,000	67	4.9	0.9	0.4	14	13	nil.
Total organic carbon, TOC	2,400	34	13	16	2.1	18	17	nil.
Total N	590	24	11	37	2.0	22	3.6	nil.
Total P	155	55	0.6	32	1.5	7.7	4.1	nil.
Pb	13.6	42	8.8	24	2.2	5.7	18	nil.
fecal coliform without chlorination*2	$2.0 \cdot 10^7$	0.05	nil.	65	0.5	0.04	36	nil.
Fecal coliform with chlorination*2	$7.4 \cdot 10^6$	0.1	nil.	0.03	1.3	0.09	99	nil.

*1 $1\ ft.^3\ s^{-1} = 0.028\ m^3\ s^{-1}$.
*2 Coliform load (=) 10^{10} organisms/day.

TABLE IX

Annual geometric mean coliform concentrations in waters along Long Island south shore (a)

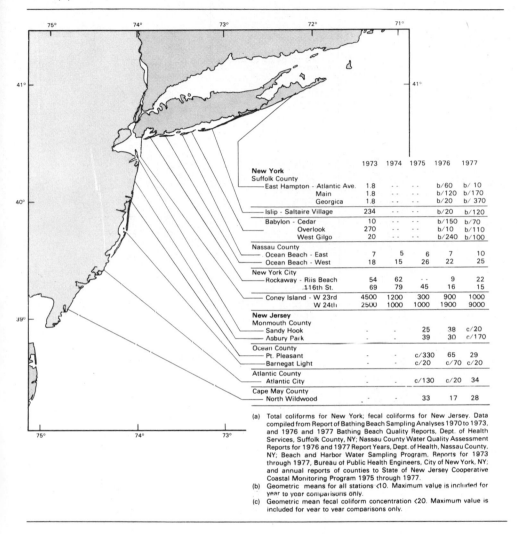

	1973	1974	1975	1976	1977
New York					
Suffolk County					
East Hampton - Atlantic Ave.	1.8	--	--	b/60	b/10
Main	1.8	--	--	b/120	b/170
Georgica	1.8	--	--	b/20	b/370
Islip - Saltaire Village	234	--	--	b/20	b/120
Babylon - Cedar	10	--	--	b/150	b/70
Overlook	270	--	--	b/10	b/110
West Gilgo	20	--	--	b/240	b/100
Nassau County					
Ocean Beach - East	7	5	6	7	10
Ocean Beach - West	18	15	26	22	25
New York City					
Rockaway - Riis Beach	54	62	--	9	22
116th St.	69	79	45	16	15
Coney Island - W 23rd	4500	1200	300	900	1000
W 24th	2500	1000	1000	1900	9000
New Jersey					
Monmouth County					
Sandy Hook	-	-	25	38	c/20
Asbury Park	-	-	39	30	c/170
Ocean County					
Pt. Pleasant	-	-	c/330	65	29
Barnegat Light	-	-	c/20	c/70	c/20
Atlantic County					
Atlantic City	-	-	c/130	c/20	34
Cape May County					
North Wildwood	-	-	33	17	28

(a) Total coliforms for New York; fecal coliforms for New Jersey. Data compiled from Report of Bathing Beach Sampling Analyses 1970 to 1973, and 1976 and 1977 Bathing Beach Quality Reports, Dept. of Health Services, Suffolk County, NY; Nassau County Water Quality Assessment Reports for 1976 and 1977 Report Years, Dept. of Health, Nassau County, NY; Beach and Harbor Water Sampling Program, Reports for 1973 through 1977, Bureau of Public Health Engineers, City of New York, NY; and annual reports of counties to State of New Jersey Cooperative Coastal Monitoring Program 1975 through 1977.
(b) Geometric means for all stations <10. Maximum value is included for year to year comparisons only.
(c) Geometric mean fecal coliform concentration <20. Maximum value is included for year to year comparisons only.

wastes at "Deepwater Dump Site *106*" located 196 km (106 n.m.) offshore, bounded by $38°40'N$, $39°00'N$, $72°00'W$ and $72°30'W$, in depths of more than ~ 2000 m (Fig. 1). The bulk of this material is $2 \cdot 10^6$ m^3 yr.$^{-1}$ of waste acids, pigments and Fe-hydroxides (Gross, 1976b). Field studies begun in 1974 included photographic and visual reconnaisance of the slope, using a small research submarine; assessment of benthic and mid-water fisheries resources, plankton and micronutrients in the water column, physical and

TABLE X

New York Beach pollution and health effects (from Cabelli et al., 1976)

	Coney Island				Rockaways			
	1973		1974		1973		1974	
Pollution indicator concentration [1]:								
Total coliforms/100 ml	983		1,213		39.8		43.2	
Fecal coliforms/100 ml	165		565		21.5		28.4	
Enterobacter coli/100 ml	174		15.3		24.8		2.4	
Klebsiella/100 ml	122		59.2		13.7		3.5	
Enterobacter + citrobacter/100 ml	530		434		11.1		6.6	
Entercocci/100 ml	91.2		16.4		21.8		3.5	
Epidemiological effects [2]:	S	NS	S	NS	S	NS	S	NS
Respiratory symptoms (%)	12.9*3	10.2	18.0*3,*4	11.7	7.2*3	6.4	8.3	7.8
Gastrointestinal symptoms (%)	7.2*3	2.4	8.1	4.6	4.2*3	2.6	3.9	3.5
(Sample size, n)	(474)	(167)	(1,961)	(1,185)	(484)	(197)	(2,767)	(2,156)

*1 Geometric mean.
*2 Incidence in percent as reported by swimmers (S) and non-swimmers (NS) in follow-up interviews.
*3 Significantly ($P = 0.5$) higher than nonswimmers.
*4 Significantly ($P = 0.5$) higher than at other beach.

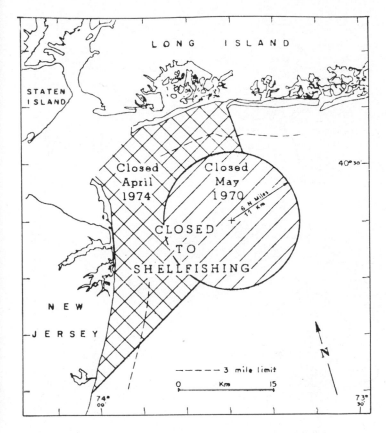

Fig. 14. Shellfish quarantine area (from Verber, 1976).

geological oceanography, and an investigation of the effects of previous radioactive waste disposal operations in a 10-n.m. (~ 1.85-km) square centered on 38°30'N, 72°06'W. Initial findings were reported as follows (N.O.A.A., 1975):

Physical oceanography. Temperature inversions were common in the upper 200 m throughout the area, due to interaction of the relatively cool, less saline shelf water and the warmer, more saline shelf water and the warmer, more saline Gulf Stream water and by the occasional southwestward movement of eddies through the region on the landward side of the Gulf Stream. There is some accumulation and spreading of particles along the pycnocline. Freshwater runoff and predominantly northwesterly to westerly winds enhance the outflow of shelf water over slope water.

The shelf—slope front in May 1974 was located over the shelf break. A warm water eddy was observed between 50 and 350 m in the northeast part of the area. Below 350 m North Atlantic Central Water was present, and at

344

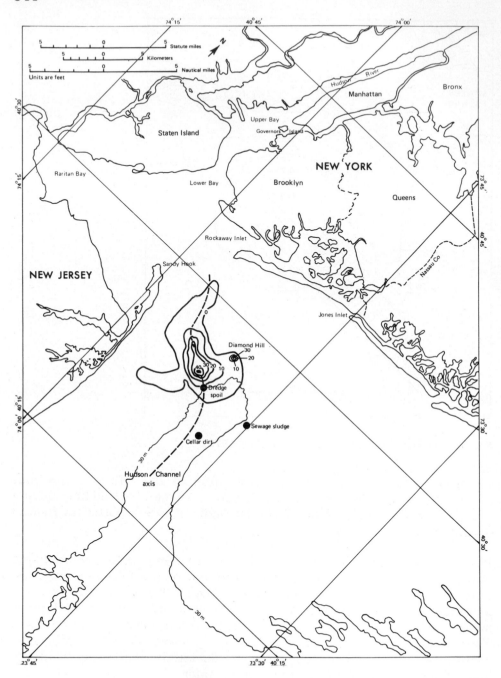

Fig. 15. Accumulation of dredge spoil (from Gross, 1976b).

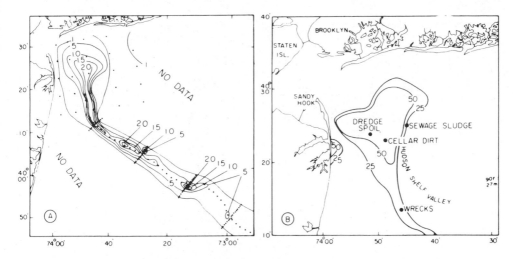

Fig. 16. Total organic carbon and lead in New York Bight sediments.

1500—2700 m (very near the bottom) North Atlantic Deep Water was present.

Geology. Erosion is occurring in the transition zone between the base of the continental slope and the upper continental rise, on portions of the upper and lower continental rise, and in submarine canyons that are incised in the slope near the dump site. The upper continental rise and areas between canyons have a uniform sediment cover and negligible current activity.

Metal concentrations. Metal concentrations in sediment samples from the continental shelf were significantly less than in samples from deeper areas. They were higher in and near the Hudson canyon. This can indicate a possible inshore source of metals, and their transport seaward through the canyon.

Most heavy metals in the finfish and invertebrates sampled showed little variation, with Pb showing greater variation than other metals. One species, *Alepacephalus agassize*, had the highest concentration of heavy metals. Concentrations in the other species sampled were consistent with values noted in previous studies in the area.

Fishes. Diversity of benthic species was greater and less variable on the slope than on the shelf. The biomass of demersal species increased at the shelf break and remained constant to 2000 m where it decreased. Numerical abundance of individuals caught showed an expected decrease with depth. Crustaceans appear to be the dominant food item for dermersal species in

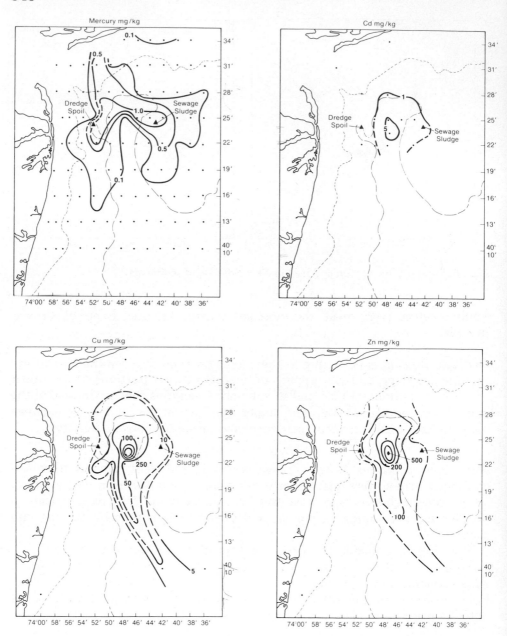

Fig. 17. Mercury, cadmium, copper and zinc in New York Bight sediments.

Fig. 18. Species diversity (A), and density (B) in Bight Apex (from Swanson, 1977).

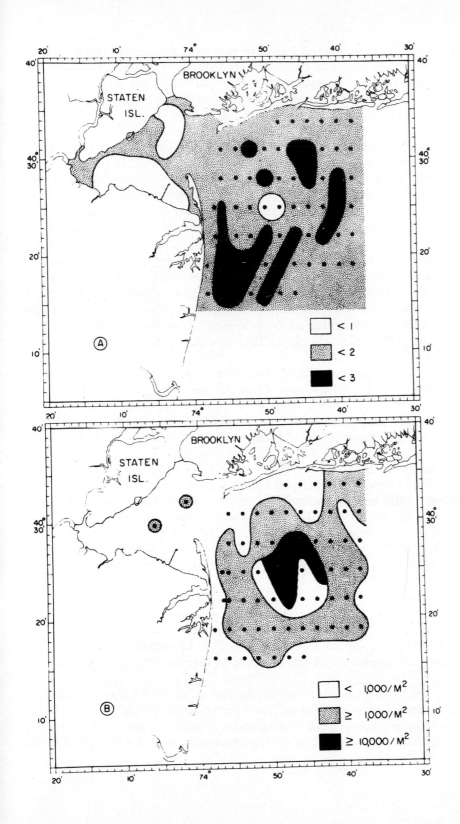

the area, based upon fragments found in fish stomachs. Amphipods and iso-
pods were the other principal foods. Thirty-three species of mid-water fishes
were identified as possible spawners in the vicinity of the dump site. Four
species were identified as eel leptocephali and 29 as mesopelagic species,
gonostomatids and myctophids.

Plankton. Diversity of identifiable shelf species of plankton was found to be
greatest at stations near the Hudson canyon. Juvenile fish and fish larvae of
tropical origin were found in surface waters, but the highest standing crop
was dominated by shelf spawning species. Copepods were the dominant
species of the invertebrate zooplankton collected. Both boreal and subtropical species were found; boreal species were predominate in shelf and
slope waters.

These initial findings were reassuring: except for previously reported
waste diffusion, discoloration of the water, and attraction of bluefish to the
area of the DWD-*106* site (Ketchum and Ford, 1948; Redfield and Walford,
1951), no impacts of the dumping were revealed. The findings were also
consistent with the reporting of no evidence of impacts on the plankton
(Vaccaro et al., 1972; Grice et al., 1973). Nevertheless, the research effort
is continuing.

In an even more rigorous effort, data have been summarized on trace elements in fish and shellfish from Cape Charles, Virginia, to Cape Cod,
Massachusetts. The data bases were those of the National Marine Fisheries
Service College Park, Maryland, and Milford, Connecticut, Laboratories,
including concentrations of 15 elements in 2309 samples from 50 species
of fishes and shellfishes (M.E.S.A., 1978). Essentially all the species analyzed
are of commercial importance for food. Mixed geographical trends were
found; in some cases, highest metal concentrations were found in fish from
Long Island Sound and the inner NYB, with levels decreasing with increasing
distance from the inner NYB. The species with highest metals concentrations
include crabs, lobsters, eels, clams, butterfish, herrings and sharks. The
highest values in fish were found in the livers of seven sandbar sharks where
the mean and high values were 6.6 and 30 ppm, respectively (M.E.S.A.,
1978).

Regional trends are well revealed by data listed in Table XI for lobster.
Concentrations were low along the Maine coast, higher in harbors, and
highest in Hudson canyon. The latter is consistent with movement of contaminated sediments shown in Figs. 16 and 17. In contrast, regional trends
for the low concentrations of trace metals in fish are mixed, as indicated by
Fig. 19.

Fish and shellfish stressed by contamination of the NYB also are diseased.
Winter flounder and other fishes, rock crabs, lobsters and other shellfish
suffer in large numbers from contaminant-related diseases. The problem
of fin-rot in fishes shown in Fig. 20 has previously gained widespread public

TABLE XI

Concentrations of mercury along the east coast of the U.S.A. in lobster (*Homarus americanus*)

	Number of observations	Mercury concentration (ppm)		
		mean whole body	range	
			whole body	tail
Nearshore:				
Jonesport, Maine	50	0.13	0.05—0.22	0.05—0.27
Portland Harbor, Maine	50	0.27	0.13—0.60	0.13—1.24
Boston Harbor, Mass.	25	0.52	0.28—0.87	0.41—1.38
Green Harbor, Mass.	37	0.44	0.27—0.79	0.34—1.23
Offshore:				
Veatch canyon[1]	82	0.30	0.14—0.49	0.21—1.55
Block canyon[2]	93	0.44	0.16—1.21	0.25—1.90
Hudson canyon[3]	89	0.56	0.16—1.51	0.22—2.31
Norfolk canyon[4]	78	0.53	0.16—1.07	0.20—1.49

[1] ~ 200 km (110 n.m.) south of Cape Cod, Mass.
[2] ~ 150 km (80 n.m.) south of Newport, R.I.
[3] ~ 125 km (79 n.m.) south of Long Island, N.Y.
[4] ~ 110 km (60 n.m.) east of Chesapeake Bay, Va. & Md.

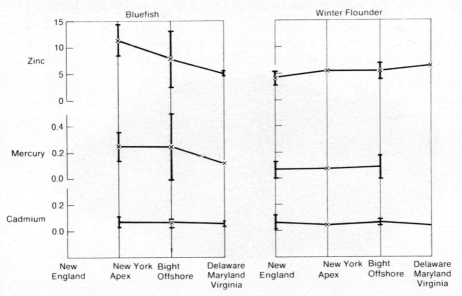

Fig. 19. Concentration gradients of zinc, mercury and cadmium in muscle from bluefish and Atlantic flounder.

attention. Fortunately, the occurrence of fin-rot in the last few years has declined (Murchelano and Ziskowski, 1977). A less widely known group of shellfish diseases include "black gill' and "exoskeletal erosion". These are diseases of lobsters, shrimp, rock crabs, horseshoe crabs, and other organisms with hard outer skeletons. For example, the incidence of "black gill" disease is linked to the organic content of underlying sediment (Sawyer et al., 1980). The incidence of diseases in these commerical and potentially commercial organisms may be a constraint upon fishing and marketing (M.E.S.A., 1978). This problem is also discussed in this volume for the California coast by Mearns (Chapter 1) and for the Mediterranean by Oppenheimer et al. (Chapter 2) and Bernhard (Chapter 3).

A number of synthetic organic contaminants in some natural waters are known to be hazards for phytoplankton. Current laboratory work on the effects of chlorinated hydrocarbon pollutants on marine plant cells reveals that growth rates, photosynthesis and chlorophyll a concentrations were all suppressed by $\mu g\,l^{-1}$ concentrations or PCB's. Dieldrin at 10 ppb inhibited growth, caused cell disintegration, and reduced the average cell size of a marine dinoflagellate. Both Chlordane and heptachlor were toxic to the same dinoflagellate, but trichloroethylene and hexachlorobenzene (widespread in the environment) were not toxic to a sensitive, mixed culture of estuarine diatoms and green algae (M.E.S.A., 1978).

Koditschek (1974, 1976) and Timoney et al. (1978) have reported abnormal tolerances of bacteria from sediments for unusually high concentrations of some metals and antibiotics. Efforts concentrated upon *Bacillus*, by far the most common genus found in all sediments. Large num-

Fig. 20. Fin-rot in summer flounder.

bers of *Bacillus* from near the dump sites of the inner NYB were resistant to Hg and the antibiotic ampicillin in concentrations which would kill normal *Bacillus* strains. From an average of two cruises, 33% of the *Bacillus* samples were resistant to Hg at $20 \, g \, ml^{-1}$ and 26% were resistant to ampicillin at $50 \, g \, ml^{-1}$. *Bacillus* from inshore stations over 6.5 km from the dump sites had significantly less resistance to Hg and ampicillin, and still less resistance was found in samples from near the outer edge of the continental shelf. Mercury-resistant strains were also commonly resistant to ampicillin. Strains less often had multiple resistance to additional heavy metals and antibiotics.

Evidence from this and other studies suggests that the genetic material determining resistance to a few metals and antibiotics is often transmitted as a single unit during reproduction. Because the concentrations of several heavy metals are high enough in the more contaminated sediments, resistant strains have a selective advantage over strains more susceptible to metal toxicity. Antibiotics are presumed to be essentially absent from the sediments. The observed antibiotic resistance is probably perpetuated because its genetic basis is often linked with that of heavy metals. This study documented a clear association between levels of sediment contamination and incidence of *Bacillus* resistance to metals and antibiotics. An additional finding was that one-third of the *Bacillus* strains from the dump-site area metabolized mercury and released it as elemental Hg. Because *Bacillus* is such a dominant bacterium of NYB sediments, this could be a significant mechanism for promoting a flux of Hg from sediments to the water column (Timoney et al., 1978).

Segar (1976) reviewed existing information on dredged material dumped in the NYB. He concluded that the oxygen demand of the dredged material has no significant involvement in summer oxygen depletion of the bottom waters. Mn and Cd (and, perhaps some other trace metals) appear to be released from the dredged material into the bottom waters at the dump site. Cr, Fe, Co, Ni, Cu, Ag and Pb are significantly enriched in the surface sediments at and around the designated dredged material dump site. Ammonium and sulfide ions are released to the water column upon dumping and may create local hazards, especially in summer when the oxygen concentration is low. The bulk of sandy sediments disposed in a 37-yr. period at the NYB designated dump site still remain within the mound. The amounts of fine materials transported away from the mound cannot be estimated with certainty; however, any increase in turbidity of the surface waters of the NYB due to dredged material dumping is negligible.

Cumulative effects of environmental stress upon marine life in the NYB are known to include genetic abnormalities of eggs and larvae of the Atlantic mackerel (*Scomber scombrus*) which develop in surface waters; contaminant-related diseases in fish and shellfish; environmental stress of synthetic organic contaminants; and the development of bacterial resistance to toxic metals and antibiotics in wastes. Longwell and Hughes (1980) have found

352

geographic patterns in the distribution of genetic abnormalities revealed in stages of cell division (I — cleavage; II — morula; III — blastula; IV — gastrula; V — early embryo; VI — tail-bud embryo; and VII — tail-free embryo) during spring spawning. These early life stages are spent in surface waters of the North West Atlantic each spring, and are subject to some natural mortality through genetic aberrations. However, the numbers and geographic pattern of genetic abnormalities are somewhat unusual if natural factors alone are involved. This is particularly true for the earliest developmental stages of mackerel eggs in the NYB. Stein estimates computed for stages IV and V mortality—moribundity are shown in Fig. 21. Generally lower effects are seen in new water entering the NYB along the south shore of Long Island and offshore waters. Greater effects are seen near the dump sites and in the Hudson River plume along the New Jersey coast (Fig. 12). Detailed discussions of the effect of various trace metals occurring in the Mediterranean are presented by Bernhard (Chapter 3 of this volume), and in the Gulf of

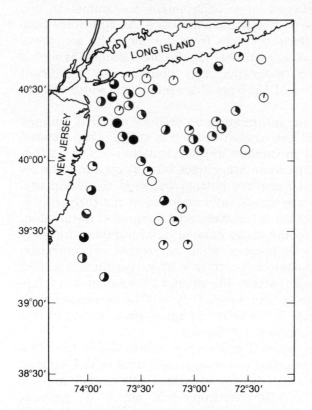

Fig. 21. Stein estimates of cell division mortality—moribundity for Atlantic mackerel eggs and larvae, May 1974 (after Longwell and Hughes, 1980).

Mexico by Trefry and Shokes (Chapter 4 of this volume) and Sackett
(Chapter 13 of this volume).

Acute effects of waste discharges

Although the chronic, often subtle effects of pollution of the NYB are
more important in both their impacts and the effectiveness of remedial
measures for them, the occasional severe episodes attract public attention
and demand scientific and engineering response. Two recent examples provide insight into the events themselves and to the problems of predicting and
controlling them. The first was the sudden stranding of grease balls and
plastics from sewage on Long Island beaches; the second was the slow development of anoxic bottom waters and an attending fish and shellfish kill
of the New Jersey shore, initially thought to be due to nutrients in waste
discharges.

June 1976 Long Island Beach pollution
The following summary is drawn largely from M.E.S.A. (1977) and
Swanson et al. (1978).
In June 1976 almost all of Long Island's major public ocean beaches were
closed to swimmers for varying periods because of floating trash and pollutants. Water-borne debris has been a constant irritant to beach users for
years, but the concentrations during June were the heaviest ever known. The
unprecedented closings began with the restriction of 32 km of Fire Island
beaches on June 15, and most of Long Island's south shore beaches were
closed during the third week of June. By July 1 these beaches were again
open, but during the interval normal summer beach use decreased causing
inconvenience and annoyance to prospective swimmers and economic loss to
local business. Additional strandings have recurred, but on a lesser scale, and
resulting closings have been comparatively brief. Locations of events are
shown in Fig. 22 and their chronology is as follows:

— In early May a medium-sized oil spill in upper NYB resulted in large quantities of
  black oil balls (tar balls) washing up on beaches from Jacob Riis Park to Fire Island.
  The U.S. Coast Guard (U.S.C.G.) immediately began observations and cleanup operations. U.S.C.G. analysis also confirmed the source of the pollution and used a New
  York Harbor spill forecast study to verify dispersal patterns.

— Throughout May, the Hudson River discharge was far above normal. On May 26, a
  storage tank ruptured at Jersey City, New Jersey, and large quantities of oil were
  spilled into the Hackensack River and into the wetlands of the Hackensack Meadows.

— On June 2, two sewage sludge tanks on Pearsalls Hassock exploded; $\sim 3800\,m^3$
  (10^6 gal.) of sewage sludge flowed into the water, and $\sim 4400\,m^3$ ($1.1 \cdot 10^6$ gal.)
  spilled onto the land. The tanks were used to store sludge prior to barging it to sea.
  The tanks were emptied from the bottom and an unknown amount of floatable

material had accumulated in the upper portion of the tanks during the twelve years of their operation. The next day, U.S.C.G. observers sighted material floating out to sea through East Rockaway Inlet.

— On June 3 and 11, pier fires broke out at Weehawken, New Jersey, and Manhattan, New York, dumping tons of wreckage and debris into the water. Large amounts of this debris could not be recovered in cleanup operations because of the small size. Finally, winds which had been variable became southerly in early June and continued to blow predominantly from this direction throughout most of the month. Meanwhile, the usual sources of wastes entering the NYB continued unabated. These included raw and treated waste water discharges, occasional spills from refuse disposal operations, commercial shipping, recreational and commercial boating and ocean dumping.

— At approximately noon, June 14, there was a report of unusual amounts of materials washing up on the Davis Park Beach. Tar balls, with diameters as large as 10 cm (4 in.), and grease balls, with diameters ranging from 3 to 5 cm (1 to 2 in.), mixed with a variety of other debris were found along the high waterline of beaches from Sunken Forest to Smith Point (Fig. 22).

— On the afternoon of June 15 the Suffolk County Department of Health recommended that all county beaches be closed to swimmers. The next day, the New York State Department of Environmental Conservation closed waters to shellfishing for seven days; state beaches were also closed.

— The problem intensified during the next week as most beaches on the south shore of Long Island were closed. On June 23, the Governor of New York declared Nassau and Suffolk Counties as disaster areas. Two days later, the President of the U.S.A. assigned the Job Corps to clean up the Long Island south shore beaches, under U.S.C.G. supervision. Cleanup efforts were completed by the end of the month, and by July 1, the last beach was reopened. Beach cleanup operations may have cost more than US $100,000 while the Long Island beach-related recreational industry lost $15 million to $25 million during and shortly following the event.

Initial reports of the materials accumulated on the beaches suggested that raw human feces were observed. Concern with public health led to official closing of the affected beaches to swimming. Later examination showed these materials to be tar balls and grease balls, with high concentrations of both total coliform and fecal coliform bacteria. The grease balls were generally of lighter outside colors and of smaller size than the tar balls. Most tar balls contained materials that looked like congealed kitchen grease and oil. About 75% of the waste balls were formed around foreign substances, generally pieces of plastic, including tampon applicators. The waste materials generally formed a swath ~ 0.3—3 m (1—9 ft.) wide, parallel to the beach. The swath was the apparent swash line of the preceding high tide and contained seaweed, driftwood, small pieces of burned wood [~ 3—20 cm (1—8 in.) long], considerable trash, garbage and waste balls.

The waste materials were almost all floatable. Plastics far exceeded all other materials in quantity, except the burned wood. The materials found included: plastic tampon applicators [about one per 3 m (10 ft.) of beach], decomposed condoms, thin pieces of plastic sheeting from sanitary napkins, and disposable diapers. Other materials were plastic straws, pieces of Styro-

355

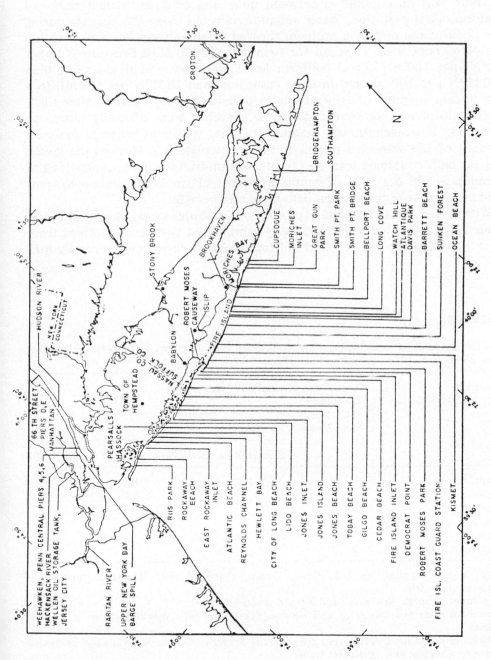

Fig. 22. Long Island beaches (M.E.S.A., 1978).

foam cups, plastic bottle caps, pieces of man-made sponges, corks and soft-balls, small plastic toys and household wares, and plastic cigar and cigarette tips. Volumes of waste materials are listed in Table XII.

There was no correlation between quantities of stranded materials and bacteriological pollution. Water samples collected along the beaches during the peak of the pollution problem showed total coliform levels well within the New York State standard for swimming [2400 Most Probable Number (MPN) per 100 ml]. Fecal coliform levels were low — in all cases less than 100 MPN per 100 ml, and in most cases less than 10 MPN per 100 ml. In a few instances where concentrations in individual water samples were high, specific local sources were responsible. Beaches were officially closed to swimming as a precautionary measure, pending results of bacterial analyses, because of the suspected origin of the waste materials. Most of the grease and tar balls contained total and fecal coliform bacteria in concentrations of as much as 106 per 100 g of sample. However, attempts to isolate pathogens such as *Salmonella* from grease balls were unsuccessful.

Within Hempstead Bay, the explosion of the sludge storage tanks had little impact upon bay water quality. Daily sampling at twenty-six stations from June 3 to 7, 1976, revealed that coliform bacteria levels were generally nor-

TABLE XII

Relative volumes of waste materials removed from three stretches of Long Island beaches, June 28—July 1, 1976 (Swanson et al., 1978)

Materials	Percentages of total observed (by volume)		
	Jones Beach	Fire Island	
		Robert Moses State Park	Kismet to Watch Hill
Wood[1]	95	80	60
Seaweed	} 4	17	10
Beach litter[2]			20
Sewage-related debris[3]	} 1	} 3	9
Tar balls			1
Fecal material	0	0	0

Observations supplied by U.S. Coast Guard, Third District, Marine Environmental Protection Branch.
[1] Includes dunnage, driftwood and burned wood bits.
[2] Includes soda and beer cans and bottles, Styrofoam cups, drinking straws and paper articles.
[3] Includes floatable materials that are generally acknowledged to reach sewage treatment plants on a regular and significant basis.

mal or below, except on June 3, when coliform levels were 3—5 times above normal. D.O. levels were even less affected. A smaller sludge accident occurred on June 18, when ~ 38 m³ (10⁴ gal.) of sludge was spilled from a storage trough at the Lawrence sewage treatment plant and eventually entered the western end of Reynolds Channel. This accident also produced a temporary increase in coliform bacteria levels in the bay, with a quick return to normal levels soon after the incident (N.C.H.D., 1977).

Estimates of major oil and grease loadings to the NYB and estuary include 191 t day⁻¹ from municipal discharges, a portion of 570 m³ (1.5 · 10⁵ gal.) of No. 6 oil spilled into the Hackensack River on May 26, 23 t day⁻¹ from sludge dumping, 300 t day⁻¹ from dredge spoil disposal, an undetermined amount from urban runoff and from combined sewer overflows from rainfalls in excess of 1 mm (0.04 in.) that occurred on 19 days during May and June, and an undetermined amount from the ruptured sludge tanks on Pearsall's Hassock.

Floatable transport mechanism include wind-driven surface currents and surface currents driven by outflows from the Hudson River. Both were unusual during May and June. Fig. 23 shows wind vectors immediately preceding and during the floatables episode. Fig. 24 shows the effects on surface salinity of the record high river outflows during May.

A U.S.C.G. model based on winds, river outflow and surface currents was used to hindcast floatables movement from East Rockaway inlet to Hempstead Bay near the sludge tanks and from the 66th Street pier fire. The results shown in Figs. 25 and 26 indicate possible movement of floatables from these sources onto the beaches.

A second model developed by the Brookhaven National Laboratory is basically a one-layer vertically integrated model that computes currents, using the topography of the area and the observed or predicted winds as the driving force. It assumes that the water surface (and floating material) moves as a vector sum of the vertically integrated current and 3% of the wind velocity. The effects of waves, tides and estuarine discharges are not included. Fig. 27 presents the results. Anything floating within the shaded area from June 1 to 24, regardless of original source, would have stranded on Long Island.

Summarizing the June 1976 floatables episode — there were multiple sources for the grease balls and for the plastic sanitary materials found on the beaches, and there were wind and surface current mechanisms to transport them there. The absolute contributions from individual sources is unknown. Nevertheless, the timing of the sludge tank explosions followed by reporting of unprecedented amounts of floatables, such as tampon applicators and condoms whose sewage origin was incontestable, brought to public and official attention the narrow limits within which present waste disposal systems operate satisfactorily. Minimizing the future probability and social and economic effects of such incidents will require a combination

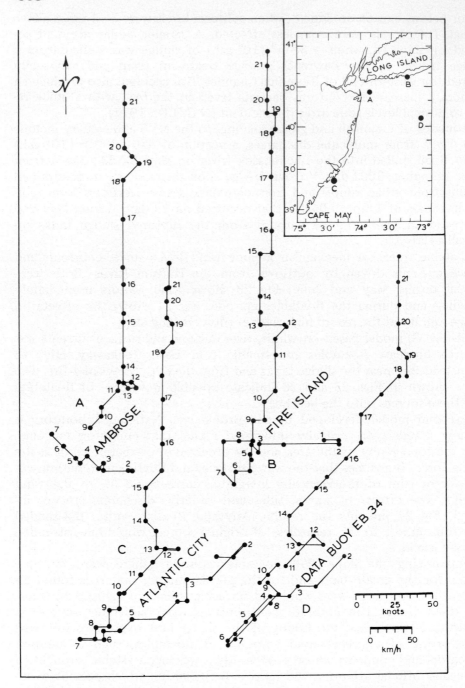

Fig. 23. Surface wind vectors from observations at 08^h00^m and 20^h00^m EDT, June 1—21, 1976 (from Swanson et al., 1978).

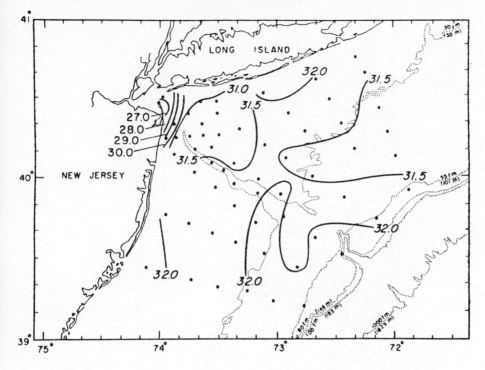

Fig. 24. Sea surface salinity (‰), May 17—24, 1976 (from Swanson et al., 1978).

of materials selection (such as readily decomposable tampon applicators), segregation, and removal at the sources of such materials as grease discharged through garbage grinders. Meanwhile, existing sources of information and models for describing or predicting movement of floating materials from either continuous, intermittent, or episodic sources are adequate.

These problems are by no means confined to the NYB area, but occur world wide. A detailed discussion is presented in Chapter 2 describing the Gulf of Naples, as well as in a number of chapters in Volume 1 on the effects of hydrocarbons on the pollution of the marine environment.

1976 anoxia and fish kill

Although not the first incident of its kind, the 1976 anoxia and fish kill event was the most severe on record. An area of $\sim 8600 \, \text{km}^2$ along the continental shelf off the New Jersey coast experienced mass mortalities of benthic organisms from July through October 1976 (Swanson and Sindermann, 1979). The observations suggest a north to south progression of

360

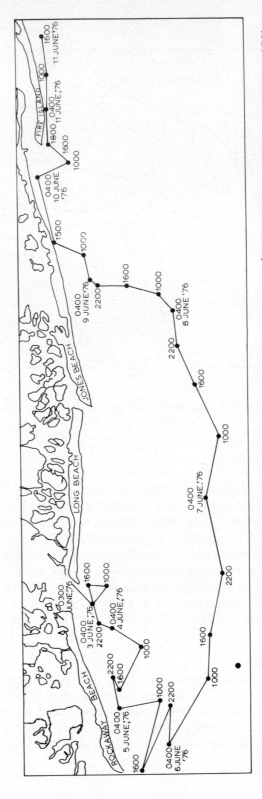

Fig. 25. Predicted movement of floatable material originating from East Rockaway Inlet at 03^h00^m EDT, June 3, 1976 (M.E.S.A., 1978).

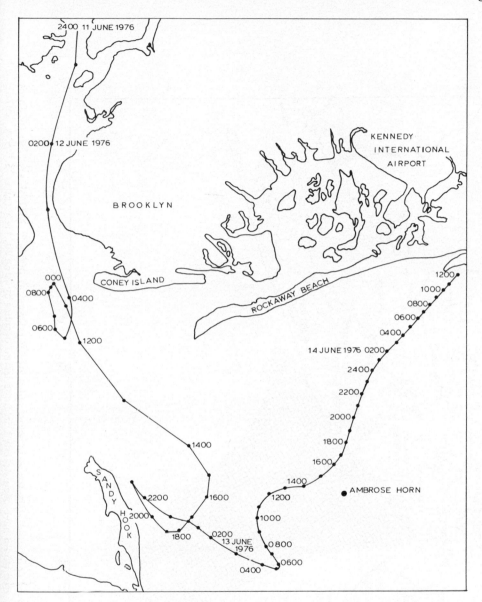

Fig. 26. Predicted movement of floatable material originating from 66th Street pier at 24^h00^m EDT, June 11, 1976 (M.E.S.A., 1978).

stressed or dead organisms from the vicinity of Monmouth Beach in early July to Atlantic City by late July.

D.O. in bottom waters in late June was low ($< 1.0 \, ml \, l^{-1}$) in the Apex region off Asbury Park south to Barnegat Inlet. During July and August low oxygen concentrations ($< 2.0 \, ml \, l^{-1}$) shifted to the south, occupying an

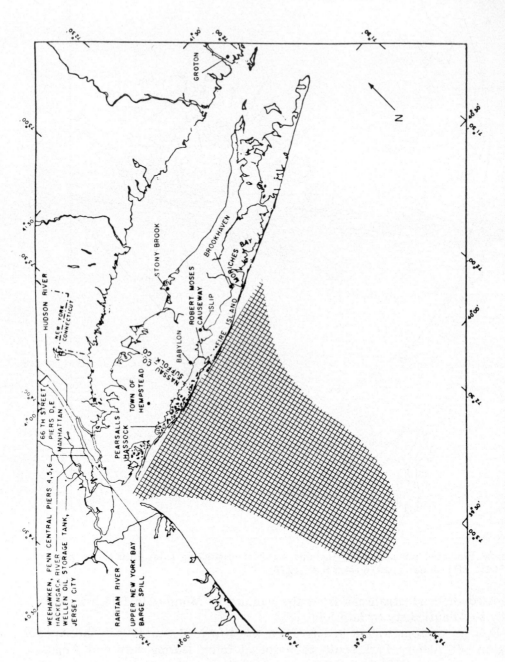

Fig. 27. Summary of computed trajectories in which all material released at 00ʰ00ᵐ EDT, June 1, 1976, from within the *shaded area* would have been stranded on Long Island beaches by June 24, 1976 (after M.E.S.A., 1978).

area of 6200 km^2 from 10 to 120 km offshore between Barnegat Inlet and the Delaware—Maryland boundary. By September (Fig. 28), D.O. concentrations in the northern portion of this area were again generally in excess of 2 ml l^{-1}. 1500 km^2 of bottom waters apparently had no oxygen at all during September, and hydrogen sulfide was observed at a number of locations. Surf clams, ocean quahogs, finfishes, lobsters and sea scallops (in decreasing order) were the commercial species most affected. There were minor mortali-

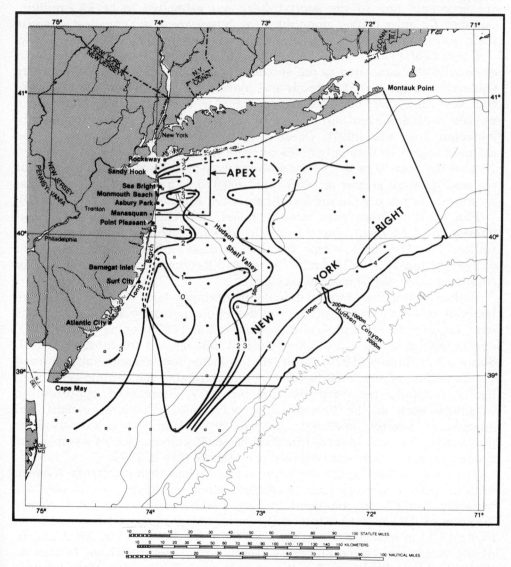

Fig. 28. Oxygen depletion, August—September 1976 (after Swanson and Sindermann, 1979).

ties of demersal fishes, but mostly they moved out of the area and were not therefore affected.

Figley et al. (1979) state that actual losses experienced by fishing, processing, and marketing industries were ~ US $7.9 million during the year and that the estimated losses of the resource were ~ $62.5 million. Until the stocks are replenished through recruitment, which may take up to seven years for surf clams, potential losses could exceed $550 million. The immediate economic loss to the sport fishery was estimated to be $3.7 million. Summer flounder fishing along the New Jersey coast was excellent, because their normal movement to offshore areas was prevented by the anoxic water.

Ropes et al. (1979) point out that surf clams off New Jersey occur in essentially the same area as the anoxic area (Fig. 29) and were consequently vulnerable to the low D.O. levels and hydrogen sulfide. Over 60% of the entire New Jersey surf clam biomass, ~ $1.8 \cdot 10^5$ t, was lost as a result of the oxygen depletion event; up to 85% of the resource within the most heavily impacted area was killed. Ocean quahogs were mostly seaward of the affected region so that only ~ 6% of them were lost. Similarly, an estimated 10% of the total New Jersey sea scallops were lost. Nevertheless, landings of sea scallops were greater in 1976 than in 1975 as a result of their serving as an alternative resource to the severly impacted surf clam (Ropes et al., 1979).

Few adult finfishes were actually killed, but Azarovitz et al. (1979) emphasize most of them moved from the affected area and normal migration routes. Their evidence is that few live or dead finfishes were taken from extremely low-D.O. ($< 0.4 \, \text{ml} \, l^{-1}$) waters. Repopulation occurred rapidly following the replenishment of oxygen.

Reduced American lobster catches in northern New Jersey in 1976 are believed to be the result of oxygen depletion. Lethargic and dead lobsters were found near wrecksites, and Azarovitz et al. (1979) point out that D.O. concentrations of at least $0.3-0.4 \, \text{ml} \, l^{-1}$ are needed for lobster survival. Steimle and Radosh (1979) report that finfishes, lobsters and crabs returned to the region of the 1976 low-D.O. event, although the benthic ecosystem had not totally recovered by summer 1977. Blooms of opportunistic organisms such as the tube-dwelling polychaetes, *Polydora socialis* and *Spiophanes bombx*, occurred. *Asabellides oculata*, also a tube-dwelling polychaete, was also observed in abundance. Recolonization of juvenile surf clams and sand dollars was observed off Atlantic City in 1977.

The cause of the oxygen depletion is less obvious than its effects. Average bottom D.O. concentrations in August along northern New Jersey are typically $3 \, \text{ml} \, l^{-1}$, as compared to only $1 \, \text{ml} \, l^{-1}$ observed in 1976. Nearly 30 years of data reveal bottom D.O. concentrations near critical conditions ($< 2 \, \text{ml} \, l^{-1}$) in localized areas; some of these are shown in Fig. 29. A matrix of the occurrence (when known) of seven potentially significant factors vs. time has been developed for the past decade (Table XIII). More unusual conditions occurred in 1976 by far. Although at least two anomalous factors

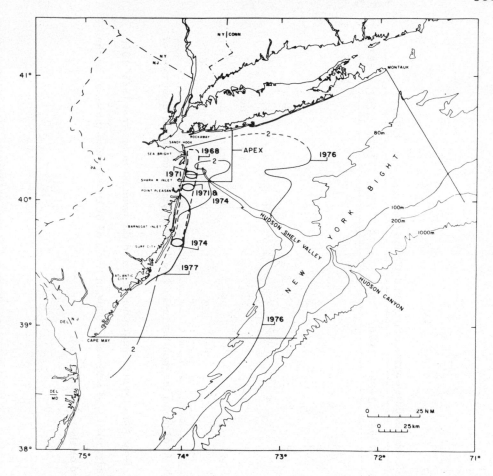

Fig. 29. Locations of oxygen depleted bottom waters, 1968—1976 (after Swanson and Sindermann, 1979).

were present during the other years when mortalities were observed, none approach the six anomalous factors in 1976. Some of the pertinent information needed to complete the table in the 1960's is not available. Future monitoring based on the 1976 sequence of events can be useful for assessing the likelihood, on a time scale of about two months, of future mass benthic mortalities. These events were according to Swanson (1979) and Swanson et al. (1979) as follows:

(1) In January, there was a large bloom of the dinoflagellate *Ceratium tripos* as early as January. The bloom occurred throughout NYB and over much of the northeastern continental shelf, peaking between April and June (Malone et al., 1979). The bloom was of such a geographic scale that anthro-

TABLE XIII

Occurrences of potentially significant factors contributing to development of oxygen depletion (Swanson et al, 1979)

	1966	1967	1968	1969	1970	1971	1972	1973	1974	1975	1976
High Hudson River flow *1	+			+				+		+	+
Early stratification *2	+								+		+
Late stratification			+			+					
Persistent southerly winds *3			+	+	+					+	+
Dumping inputs *4				+		+	+	+	+	+	+
Current reversals *5	?	?	?	no	?	?	?	no	no	no	+
Extensive phytoplankton bloom	?	?	?	?	?	no	?	no	no	no	+
Benthic mortalities	2	0	2	3	1	2	1	2	2	1	6
			×			×			×		×

*1 Years when the mean flow for February, March, April minus the monthly mean flow for that year exceeds the mean value of the computation over the 1965–1976 period.
*2 From Armstrong (1979).
*3 Years when the southerly component of winds from April through September exceeded the resultant mean from 1966–1975 for April through September.
*4 Years when ocean dumping (sewage sludge, dredge spoil, and acid waste) input in terms of BOD$_5$ exceeded the mean for the 1965–1976 period.
*5 Years when major current reversals were observed on the continental shelf.

pogenic nutrient inputs from the NYB are not considered influential. The bloom may have been a response to the relatively warm winter reported by Armstrong (1979) and Diaz (1979), or to relatively warmer temperatures and higher salinities in the offshore cold pool (J. J. Walsh et al., 1980).

(2) There was an early and warm northeast spring initiated by atmospheric circulation patterns that inhibited the transport of arctic air into the region (Diaz, 1979). During February—May, there was an extremely large and continuous flow of freshwater from the Hudson River and surrounding drainage basins, and an early warming of the sea-surface waters. The resulting lens of relatively warm freshwater contributed to density stratification in coastal waters some six weeks earlier than usual. This in turn contributed to isolation of bottom waters from re-aeration by turbulent mixing at the air—sea surface (Armstrong, 1979; Starr and Steimle, 1979). However, normal residence time of bottom waters in the affected region is short compared to the seasonal cycle of oxygen depletion (Han et al., 1979).

(3) The wind field generated by the anomalous atmospheric pressure patterns of early spring may have been the most significant indirect factor in the establishment and maintenance of oxygen depletion. Diaz (1979) points out that winds had a southerly component over much of the time from February and at least into June. Ordinarily, the wind does not shift to the south until April. In addition, the persistence of the wind was greater than normal during May and June. The character of the wind field was such as to establish a potential for upwelling of bottom waters along the New Jersey coast. Evidence of upwelling off the New Jersey coast was also observed in the bottom current meter data analyzed by Mayer et al. (1979).

(4) Mayer et al. also noted a relationship between the persistent southerly and southwesterly winds and the interruption of the expected southwestward flow of bottom water over the New Jersey shelf. Comparisons to summer 1975 show that offshore (100 km) the southward flow was considerably less than in the previous year, whereas the alongshore component on the inner shelf (50 km) actually reversed, flowing northward from mid-May through July. In early August the net bottom water displacement was small.

(5) The effect of the persistent southwesterly winds on the overall pattern of bottom flow was shown by Han et al. (1979). During most of June the net circulation near the bottom was northwestward along the New Jersey coast, into the Apex region, along with strong onshore flow and probable associated upwelling. Intensification of the flow occurred along the HSV. This flow transported oxygen, and in all likelihood *C. tripos*, into the impacted area. The net utilization rate of oxygen was calculated by Han et al. (1979) to be three to ten times larger in the Apex and the New Jersey coastal segment than in other segments of the NYB.

(6) *C. tripos* were observed to aggregate at the base of the pycnocline, which was below the photic zone (1—3% light level) — a location in the water column subject to shoreward transport. Because photosynthetic pro-

cesses were certainly reduced at the depths inhabited by *C. tripos*, the organisms were possibly living heterotrophically (Malone et al., 1979), thereby using oxygen through respiration. Stimulation of this aggregating bloom in May and early June could have been enhanced by the large concentration of particulate organic material in the near-shore waters from the large Hudson—Raritan estuarine runoff during spring. Subsequent stratification, possible reduction of particulate organic material, and reduction in Hudson River runoff may have limited the availability of nourishment for continued growth of *C. tripos* below the pycnocline during late June. In any event, the bloom declined rapidly during July, creating an organic floc at the bottom in or near the area of depressed bottom D.O.

(7) The respiration of the *C. tripos* can be calculated from the data of Malone et al. (1979) and then compared to net utilization rate in each segment from Han et al. (1979). The average *C. tripos* cell counts below the pycnocline were $0.64 \cdot 10^8$ cells per m^3 off Long Island and $1.1 \cdot 10^8$ cells per m^3 off New Jersey. These correspond respectively to segments *L1* and *J1* as shown in Fig. 30. The respiration rates for these cell concentrations are 0.37 and $0.22 \, ml \, l^{-1} \, day^{-1}$ for *J1* and *L1*, respectively (Fig. 30). If the mean benthic respiration rate of $11 \, ml \, m^{-1} \, hr.^{-1}$ (J. P. Thomas et al., 1976) is distributed over the measured thickness of the lower layer — 16 and 10 m for segments *L1* and *J1*, respectively — then the benthic respiration rates are 0.22 and $0.03 \, ml \, l^{-1} \, day^{-1}$ for segments *L1* and *J1*, respectively. The total respiration of the *C. tripos* and benthic communities is then 0.24 and $0.40 \, ml \, l^{-1} \, day^{-1}$ for segments *L1* and *J1*, respectively. The net utilization rates used to produce the observed oxygen decline, as calculated by Han et al. (1979), were $0.05 \, ml \, l^{-1} \, day^{-1}$ for segment *L1*, $0.17 \, ml \, l^{-1} \, day^{-1}$ for segment *J1*, and $0.15 \, ml \, l^{-1} \, day^{-1}$ for the Apex segment. These values and the observed oxygen concentrations are shown in Table XIV.

In mid-May the bottom D.O. concentrations were nearly the same in all segments. The time rates of change in concentration in the Apex and along the New Jersey coast were comparable, as were the rates of net utilization. The Long Island coast, however, was much less affected. The major differences were between the estimated rates of respiration off New Jersey and off Long Island during this period of time. If there were no advective inputs of oxygen into the system, respiration alone could have caused anoxic conditions along the New Jersey coast in ~ 10 days. Thus, simply the respiration of the large population of *C. tripos* was sufficient to account for the observed oxygen decline through June. It is less obvious why the anoxic conditions initially occurred in the southern Apex and about three weeks later east of Atlantic City. Diagnostic model computations and direct current observations indicated that southward advective processes did not transport low-D.O. water from the Apex.

Garside and Malone (1978) conclude that the average carbon loading to the Apex, including anthropogenic inputs, is insufficient to cause anoxic

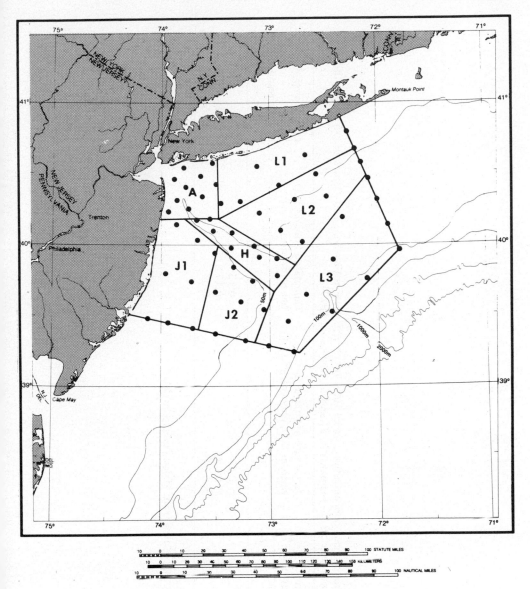

Fig. 30. Sectors for diagnostic modelling (after Han et al., 1979).

conditions. However, the normally large carbon load, in addition to the respiration of *C. tripos* (which to some degree may have been concentrated more in the southern Apex by flow up the HSV), likely created a greater total load in the southern Apex.

(8) J. P. Thomas et al. (1979) found that *C. tripos* were not present in the

TABLE XIV

Comparison of mean bottom oxygen concentrations, local rates of change, new utilization, and *Ceratium tripos* plus benthic respiration between May 18 and June 29 for segments *A*, *J1* and *L1* (Swanson, 1979)

Segment [1]	Mean bottom D.O. concentration [2]		Local rate of change in bottom D.O. concentration	Calculated bottom D.O. net utilization	*C. tripos* respiration plus benthic respiration
	May 18	June 29			
A	5.5	1.9	− 0.086	− 0.15	
J1	5.2	1.3	− 0.093	− 0.17	− 0.41
L1	5.4	4.5	− 0.020	− 0.05	− 0.25

Data from Han et al. (1979).
[1] Segments are shown in Fig. 30; [2] data from Malone et al. (1979).

water column in August and September, though Mahoney (1979) found a decaying floc of *C. tripos* near the bottom in July. Malone et al. (1979) estimated BOD from the oxidation of the *C. tripos*. Assuming that the average depth below pycnocline was 10 m off New Jersey and the oxidation took place over 60 days (July and August), we find that the oxidation rate is equivalent to $0.16 \, \text{ml} \, \text{l}^{-1} \, \text{day}^{-1}$. Thus, the oxidation of *C. tripos* alone was probably sufficient to maintain anoxic conditions throughout the remainder of summer.

(9) Anaerobic conditions and sulfide generation in the anoxic area were observed by J. P. Thomas et al. (1979), along with high rates of sea-bed oxygen consumption at the periphery of the area. The decaying floc increased stress on benthic organisms and the continuous yet typical anthropogenic loadings and oxidation of dissolved and particulate matter and carbon could only intensify and maintain the already depressed conditions.

(10) Although density stratification may not have been the primary cause in oxygen depletion, the anoxic conditions were not relieved until the breakdown of the pycnocline in autumn. Even the disturbing effects of hurricane Belle were only transitory. During autumn and winter, bottom oxygen values continued to increase, once again reaching saturation.

Existing evidence suggests that waste inputs probably did not start the anoxic event. There is slight evidence of a gradual decrease over the last ten years in bottom D.O. so that J. S. O'Connor (1979) has suggested that:

" ... the sensitivity of the system might therefore be changing such that a slight imbalance (either due to natural causes or increases on waste loadings) in the "normal" cycle of environmental conditions is sufficient to drive the system towards anoxia with increasing frequency. If this is true and if the slight decreasing trend in bottom D.O. over the last three decades is real and monotonic, then there is an increasing probability of low D.O. events in any year ... "

ECOSYSTEM MODELLING IN THE NEW YORK BIGHT

The geological, historical, and current development of the Hudson and Raritan river basins and the New York Bight provide a framework within which sub-systems models have been developed. The larger ecosystem approach considers interactions and equilibria between air, land, water and living things (including man) within certain boundaries of time and space. These boundaries are arbitrarily defined, technically and institutionally, by fluxes of living and non-living materials and energy and may be extended to include the global biosphere. Continental, oceanic and global events are clearly reflected in movement of acid rain into the drainage basin or in movement of continental slope waters up Hudson canyon. Some of the conditions and fluxes have been considered in the present summary. Most have been identified in other publications listed in the selected bibliography in the Appendix of this chapter.

The New York Bight M.E.S.A. project was, to an important degree, developed in response to ocean dumping (C.E.Q., 1970) The ecosystem approach (I.J.C., 1978):

> "forces us to speak of discharging wastes in the ecosystem and biosphere of which we are a part, rather than to an external environment around us."

Descriptive, diagnostic and predictive models applicable to the marine portions of the NYB ecosystem have been developed. Two of these, the U.S. Coast Guard model and the Brookhaven National Laboratories model (Tingle and Dieterle, 1977) have been discussed in connection with Long Island beach pollution. McLaughlin et al. (1975) developed a model using H. T. Odum's (1972) energy—mass flow notation shown in Fig. 31. Their necessarily simplified model is shown in Fig. 32. It is not possible to specify all of the required rates, constants, forcing functions, initial and boundary conditions over any appropriate modelling time scale. Thus the level of abstraction of the subsystem models (Fig. 32) must be further simplified and

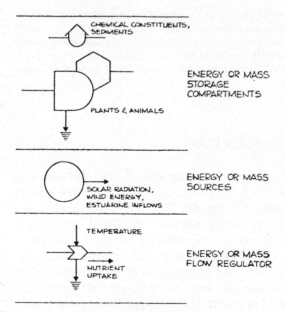

Fig. 31. Basic components of energy—mass flow models (after H. T. Odum, 1972; McLaughlin and Elder, 1976).

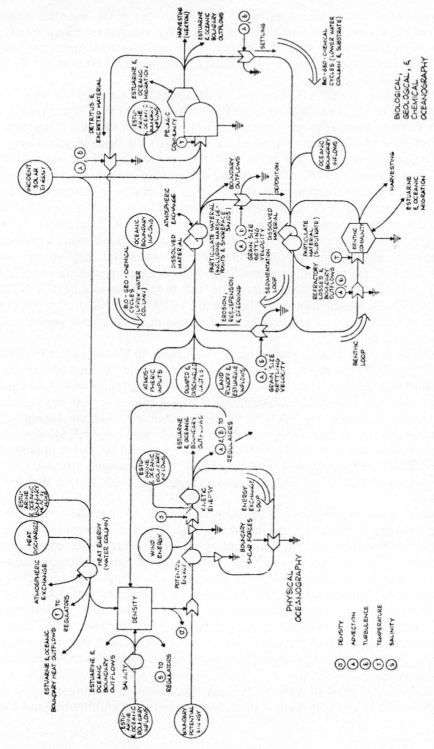

Fig. 32. Marine component of the New York Bight ecosystem (from McLaughlin and Elder, 1976).

consideration of interactions must be limited to the most essential ones. Regardless of the selected partitions of the marine ecosystem components, water motion (advection, A; and turbulence, δ) and physical state (temperature, T; salinity, S; and density, σ) regulate processes and control the flow of mass and energy.

The conceptual frame of reference of McLaughlin et al. (1975) and the transport model of Beardsley et al. (1976) shown in Fig. 5 was used to design a current measurement program. Statistical hindcasts and diagnostic analysis of over 4000 meter-months of data revealed the basic physical interactions which characterize the NYB. The analysis was based on adaptation by Han and Hansen (1978) of the steady-state diagnostic model developed by Galt (1975). The adaptation permits use of all available information on density, currents, sea level and meteorological factors. Currents and other computed properties are interpolated within a dynamic framework to a high degree of resolution. The process is analogous to inferring currents from dynamic height calculations except for a continental shelf where the classical level of no motion is unrealistic. The derived model was applied to analyzing the oxygen mass balance in the bottom waters of the NYB during the 1976 anoxic episode (Han et al., 1979). Fig. 33 illustrates the improved specification of velocity in bottom waters during four periods of the anoxic episode.

The model is now being used to diagnose current fields for earlier periods during 1973—1979 to drive a subsystem model of the cycling of carbon, oxygen and nitrogen (J. J. Walsh et al., 1980). The richness in temporal variability of currents due to local scale forcing events and to propagation of outside energy through the NYB, is being handled by the time-dependent formulation of Hsueh and Peng (1977) and Hsueh and Lee (1978). Results from this sub-model will be used in the C—O—N model in which chemical and biological components will be followed in the Lagrangian mode as they are allowed to change in an Eulerian grid (Tingle et al., 1979; J. J. Walsh et al., 1980).

In near-shore water less than 20 m deep, diagnostic models and our understanding of current dynamics are unsatisfactory. Here the influences of solid lateral boundaries, bottom friction, and tidal forces increase exponentially as the shore is approached. In the NYB, estuarine circulation further complicates the analysis. Statistical methods presently are being used to develop hindcast and predictive capabilities. Autoregressive relations with local wind forcing coupled with tide predictions account for much of the observed current variance. The hindcast match degrades with distance from shore where waters deepen and friction becomes less dominant due to greater exogenous influence and where, fortunately, classical ocean dynamics provides answers.

Eventually, the greatest utility in such diagnostic modelling will lie in being applied to broad continental shelf sediment quality and benthic animal

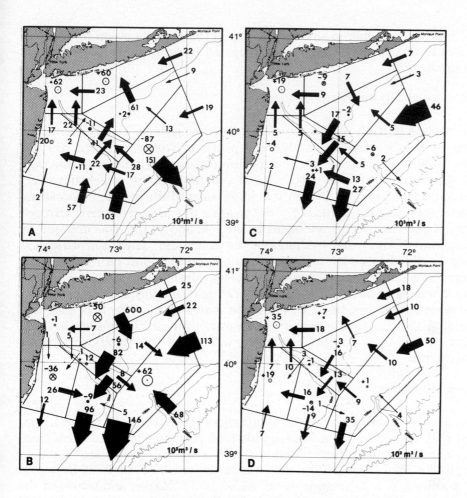

Fig. 33. Diagnostic modelling in the New York Bight, May—June, 1976: (A) May 18—23;
(B) May—June 3; (C) June 3—13; and (D) June 13—29. (After Han et al., 1979.)

response predictions such as those developed for the west U.S.A. coast by
Hendricks (1979). Off California there is a narrow shelf and continental
margin so that average currents necessarily follow isobaths and resuspension
of sediments results in their downslope movement. Biological models based
on an infaunal feeding index developed by Mearns (Chapter 1 of this
volume) for ~ 25 taxa in 20—200 m waters in the Southern California Bight
and 43 taxa in Puget Sound can then be used to predict response to changes
in waste discharges. Bascom (1979b) describes one such prediction. Major
physical elements in these models are advection and sedimentation processes
with averaging intervals of, say, one day for microbiological response to a
few weeks for development of anoxia.

The use of modelling in this area in regard to using the ocean for disposal of wastes in the same general area is discussed by Simpson et al. in the next chapter (Ch. 12).

CONCLUSIONS AND RECOMMENDATIONS

The marine portion of the New York Bight ecosystem clearly shows the impacts of man's activities on the land portion. Even Manhattan still discharges large quantities of raw sewage. Interactions include environmental, public health and fisheries responses to pollution, all of which have economic impacts. There are gaps in our knowledge of the marine physical and biological interactions, particularly in the areas of carbon—oxygen—nitrogen cycling, of the survival of pathogens in the sea, and of the risks imposed by man on himself and other elements of the ecosystem. These scientific gaps need to be filled in order to better monitor, predict, and respond to both chronic and acute effects of waste discharges. Meanwhile, there is a need to predict probable impacts on the New York Bight of alternative resource and waste management practices. Our best estimates of the effects of past and present conventional practices and of expensive proposed extensions to them do not show that continuing them will result in significant future improvements.

ACKNOWLEDGEMENTS

This largely descriptive chapter summarizes the findings of research conducted and supported by a large number of governmental, industrial, academic and individual efforts. Methods and analytical details are found in their publications. The author is particularly indebted to R. L. Swanson, H. M. Stanford, J. S. O'Connor, C. A. Parker, G. F. Mayer, and others of the M.E.S.A. New York Bight Project and of N.O.A.A. whose written and verbal contributions, not always properly acknowledged, form much of this chapter.

APPENDIX — SELECTED BIBLIOGRAPHY

1* *Hydrographic Properties.* Malcolm J. Bowman, with cartographic assistance by Lewis D. Wunderlich, Marine Research Center, S.U.N.Y., Stony Brook, N.Y., 1977.
2 *Chemical Properties.* James and Elizabeth Alexander, Institute of Oceanography, S.U.S., Miami, Fla., 1977.

---

* New York Bight Atlas Monograph Series number, published by New York Sea Grant Institute, State University of New York, Albany, New York.
★ Also listed in the References in the back of this volume.

378

31   *Marine and Coastal Birds.* Marshall A. Howe, Roger B. Clapp and John S. Weske, U.S. Fish and Wildlife Service, Washington, D.C., 1978.

32   *Environmental Health.* Joseph M. O'Connor, Victor J. Cabelli and Merril Eisenbud, Institute of Environmental Medicine, N.Y. University, Medical Center, New York, N.Y., 1981.

Bumpus, D. F. and Lauzier, L. M., 1965. *Surface circulation on the continental shelf off eastern North America between Newfoundland and Florida.* Serial Atlas of the Marine Environment, No. 7, Am. Geogr. Soc., New York, N.Y., 2 pp., 16 plates.

Colton, J. B. and St. Onge, J. M., 1974. *Distribution of fish eggs and larvae in continental shelf waters, Nova Scotia to Long Island*, Serial Atlas of the Marine Environment, Folio 23, Am. Geogr. Soc., New York, N.Y., 2 pp., 11 plates.

*Gross, M. G. (Editor), 1976. *Middle Atlantic Continental Shelf and the New York Bright.* Am. Soc. Limnol. Oceanogr., Spec. Symp. Vol. 2, Univ. Mich., Ann Arbor, Mich., 441 pp.

M.E.S. (Marine Experiment Station), 1973. *Coastal and Offshore Environmental Inventory, Cape Hatteras to Nantucket Shoals.* Graduate School Oceanogr., Univ. Rhode Island, Kingston, R.I.

N.M.F.S. (National Marine Fisheries Service), 1974. *Anglers Guide to the United States Atlantic Coast, Sect. I: Passamaquoddy Bay to Cape Cod; Sect. II: Nantucket Shoals to Long Island Sound; Sect. III: Block Island to Cape May.* U.S. Dep. Commer., Natl. Oceanic Atmos. Adm., Washington, D.C.

Walford, L. A. and Wicklund, R. I., 1968. *Monthly sea temperature structure from the Florida Keys to Cape Cod.* Serial Atlas of the Marine Environment, Folio 15, Am. Geogr. Soc., New York, N.Y., 2 pp., 16 plates, app. 28 pp.

CHAPTER 12

DEEP-OCEAN DUMPING OF INDUSTRIAL WASTES

Duane C. Simpson, Thomas P. O'Connor and P. Kilho Park

INTRODUCTION

Strategies to control man's toxic wastes are: (1) detoxifying by such means as incineration; (2) diffusion and dispersion for assimilative substances; and (3) isolation and containment or ejection from our planet for nondestructive wastes. Proper use of these strategies with contingency to cope with unforseeable accidents is needed for human species to continue life on Earth. Consider the role of the ocean for waste management, with the final receptacle of many wastes being the ocean. Industrial and municipal liquid wastes, after being treated, run into the ocean. Agricultural pollutants, such as pesticides and fertilizers, are often carried into the ocean by runoffs. Oil spills are common occurrences in recent years and air pollutants, such as Pb, rain over the ocean which covers 70% of the Earth's surface. Because the ocean occupies the lowest topographic domain in the hydrological cycle of the Earth's surface, it is not possible to transfer the polluted ocean into other domains (Park and O'Connor, 1980). Conversely, the ocean is an international resource and is an important food source for many countries. It stabilizes the climate, oxygen–carbon dioxide balance in the atmosphere, and provides water for the Earth's hydrologic cycle.

The resource concept of the ocean is incompatible with that of a receptacle, but due to the assimilative capacity of the ocean, in theory, these two may coexist. However, unassimilative substances, such as long-life radioactive wastes and synthetic toxic organic substances which nature is not capable of altering, will accumulate in the ocean. In recent years there has been increasing pressure on both industry and government to dispose safely of the ever-expanding quantities of waste materials. This pressure has been and is challenging the competence of human technology. Regulated ocean dumping of industrial wastes, sewage sludge, dredged materials and other material have been practiced in recent years. Concurrently, ocean dumping research and monitoring have begun (D. C. Simpson et al., 1978).

Under recent U.S. Statutes the ocean disposal of harmful waste is to cease by 1981, but this raises the question as to what is harmful. Sewage sludge has been designated as one of the harmful substances. A large problem arises with industrial waste. With an ever-expanding number of man-made chemical compounds, the concurrent waste management research becomes overwhelming. The number of chemical compounds registered by November

1977 with the American Chemical Society was ~ 4,000,000. Of this number some 70,000 are produced commerically. Some of the chemical wastes from the production of chemical compounds are now disposed of in the ocean.

EXISTING LEGAL FRAMEWORK

Until recently, most legislation has dealt with the impacts of dumping on navigation and/or sea-floor jurisdiction, but this legislation has little to do with the environmental impact of waste products. The first laws enacted did not deal directly with solid-waste disposal at sea. Instead, the major part of antipollution legislation, as well as all international treaties and agreements which dealt with pollution, were concerned with two separate problems: (1) oil pollution in ocean waters; and (2) pollution resulting from the exploration and exploitation of the natural resources of the sea bed and the subsoil of the continental shelf.

In the last two decades, there has been an increase in the concern for the health of the ocean. This is due to many considerations, among which is a recognition that the marine environment and living organisms which it supports are of vital importance to humanity. Also, all people have an interest in assuring that it is so managed that its quality and resources are not impaired. In addition, it is recognized that the capacity of the sea to assimilate wastes and render them harmless, and its ability to regenerate natural resources is limited. To this end in the international arena, a multi-nation Ocean Dumping Convention was convened and produced a document called *Convention on the Prevention of Marine Pollution by Dumping of Wastes and Other Matter*. It has been signed by 39 nations and became effective in 1975. This convention deals with marine pollution and listed substances which are illegal to dump. It also allows for a member nation to set up a system to issue permits to accomplish the regulation of dumping within its jurisdiction. This convention is called the London Ocean Dumping Convention, and its secretariat is the Intergovernmental Maritime Consultative Organization (I.M.C.O.).

On October 23, 1972, the U.S.A. enacted the "Marine Protection, Research, and Sanctuaries Act of 1972" (P.L. 92-532). It is commonly called Ocean Dumping Act. This law declares that:

"it is the policy of the United States to regulate the dumping of all types of materials into ocean waters to prevent or strictly limit the dumping into ocean waters of any material which would adversely affect human health, welfare, or amenities, or the marine environment, ecological systems, or economic potentials."

To accomplish this, the law also provided for "Comprehensive Research on Ocean Dumping" under its Title II. Recently, another law has been enacted by the U.S.A. This law is referred to as the "National Ocean Pollution

Research and Development and Monitoring Planning Act of 1978" (P.L. 95-273). This Act states that:

"there is a need to establish a comprehensive Federal plan for ocean pollution research and development and monitoring, with particular attention being given to the inputs, fates, and effects of pollutants in the marine environment."

This provides the framework for continued research in the field of the marine environment. The U.S. Department of Commerce, National Oceanic and Atmospheric Administration (N.O.A.A.), National Ocean Survey (N.O.S.), Ocean Dumping and Monitoring Division (O.D.M.D.), operates under the authority of Title II, Section 201, of the Marine Protection, Research, and Sanctuaries Act of 1972. The initial objective of ocean dumping research was to establish baselines. The purpose of baseline information was to establish the statistical definition of the natural variation, both spatial and temporal, against which monitoring could then determine deviations or trends from the norm.

The technical objectives which are presently being pursued are as follows:

(1) *Waste characterization*. To determine the most important constituents of the wastes, and how they interact and mix with seawater. This involves both laboratory and field studies.

(2) *Waste dynamics*. To determine the physical movement and dispersion of the wastes after they are dumped.

(3) *Biological effects*.

(a) To observe and measure, when possible, biological effects in the field. This was generally attempted in conjunction with a direct observation of a waste dump.

(b) To determine acute and chronic effects in the laboratory, and relate these findings to potential effects in the field.

When objectives (1)–(3) are achieved to the greatest extent possible, it will then be possible to:

(4) Determine the most stressed components of the ecosystem, and use these biological indicators to develop monitoring techniques and strategies.

(5) Model the stressed ecosystem sufficiently to predict the effect of continued or increased dumping and the conditions under which damage could occur.

The first three of these objectives have been studied extensively and will be summarized here. The deep-water sites investigated to date are referred to as "106-mile" Deep-Water dump site (DWD-*106*) and the Puerto Rico dump site.

SITE LOCATIONS AND GENERAL CHARACTERISTICS

In 1974 N.O.A.A. initiated studies at DWD-*106* (Fig. 1). Those studies were initiated to assess the impact of present dumping activities and to

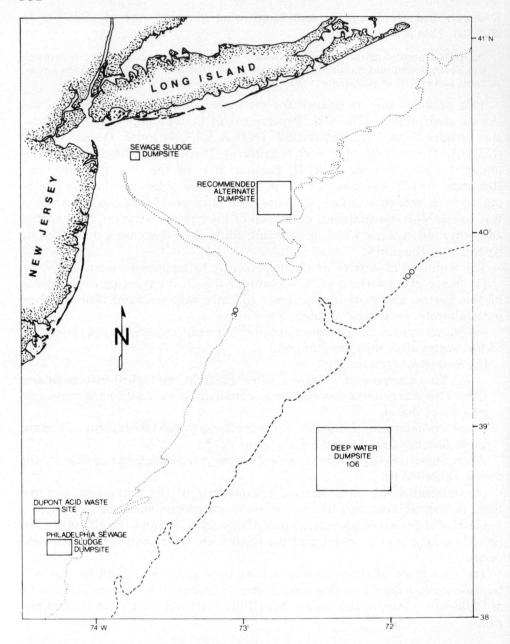

Fig. 1. Location of the "106-mile" Deep-Water dump site (DWD-*106*). *Contours* are in fathoms (1 fathom = 1.8288 m).

provide baseline data for future dumping impact. The dump site is located 106 n.m. or 196 km southeast of Ambrose Light Tower and 167 km east of Cape Henlopen, Delaware. The site has been used by more than 25 different dumpers but is presently used by only a few. During 1978 $\sim 8 \cdot 10^8$ l of waste were dumped there. The dump site is bounded by 38°40'N, 39°00'N, 72°00'W and 72°30'W, covering ~ 1700 km² and is over the U.S.A. continental slope and rise. Depths range from 1550 m in the northwest corner of the quadrangle where relief is rugged and the slope dissected by small canyons, to 2750 m in the southeast corner, where relief is slight with gentle-sloping plains of $\sim 1\%$. Any of several water masses can occupy different levels in the water column at a given time.

In 1978 N.O.A.A. initiated studies at the Puerto Rico dump site to complement the studies at DWD-*106* and determine similarities and differences of impact on the marine environment. The dump site is located 74 km north of Arecibo, Puerto Rico. The site covers a 500-km² area overlying depths of 6–8 km over the Puerto Rico trench. During 1978, it received $\sim 3.6 \cdot 10^8$ l of liquid waste primarily generated by pharmaceutical industries that include penicillin production. The dump site is bounded by 19°10'N, 19°20'N, 66°35'W and 66°50'W, and is over the Puerto Rico trench. The area is generally occupied by one water mass which is controlled by the Antilles Current, which flows generally to the west-northwest but its direction is variable. In this chapter the results of DWD-*106* site study are presented as a overview. When much is learned from the Puerto Rico pharmaceutical site study, the results will be presented elsewhere.

DWD-*106* SITE STUDY

Waste characteristics

Approximately $8 \cdot 10^8$ l of wastes were dumped at DWD-*106* in 1978 (Table I). Three major industrial dumpers are DuPont plant in Edge Moor, Delaware, DuPont–Grasselli plant in Linden, New Jersey, and American Cyanamid Warner's plant in Linden, New Jersey. An additional four dumpers use this area, but are only minor contributors. One was the city of Camden, New Jersey, which dumped sewage sludge. This chapter will confine itself to the first three major dumpers.

(1) The DuPont–Edge Moor waste comes from titanium dioxide production. About $4 \cdot 10^8$ l were dumped in 1978. The waste is highly acidic, 2–4M HCl, with 0.5–1.0M Fe. Other metals include Cr and V in the hundreds of parts per million (ppm) range. Cu, Zn, Ni and Pb are between 10 and 100 ppm, and Cd is ~ 1 ppm.

(2) The DuPont–Grasselli waste is derived from the manufacture of N,O-dimethyl hydroxyl amine (DMHA) and anisole (methyl phenyl ether). About

TABLE I

Waste inputs to DWD-*106*

| Source | 1976 (10³ wet tons) | 1977 (10³ wet tons) | 1978[1] | |
			(10³ wet tons)	(10⁶ l)
DuPont–Edge Moor	–	418	409	384
DuPont–Graselli	180	118	189	178
American Cyanamid	131	143	122	115
Modern Transportation[2]	69	91	79	74
Camden, N.J.	–	53	59	55
General Marine[2]	5	1	0	0
Total	385	824	858	806

[1] The 1978 input is given in terms of litres as well as wet tons to indicate the size of liquid volumes being dumped. The conversion is not exact since it assumes all wastes to have a specific gravity of $1.0 \, \mathrm{g \, cm^{-3}}$. Actual specific gravities vary from 1.02 to $1.16 \, \mathrm{g \, cm^{-3}}$.
[2] Modern Transportation and General Marine are concerns which dump waste at DWD-*106* for a variety of individual industrial waste producers. The inputs of each individual are small, none exceeding 40,000 wet tons. (From U.S.D.C., 1979).

$2 \cdot 10^8$ l were dumped in 1978. The waste is an alkaline solution, pH 12–13, of sodium sulfate with small amounts of dissolved metals and 0.5–1 wt.% organic C. Organic compounds in the waste include methyl sulfate, methanol, phenol and small amounts of product of unknown compounds.

(3) The American Cyanamid Warner waste is derived from a large and diverse manufacturing operations which include the production of organophosphorus pesticides and water treatment chemicals. Additional wastes result from the production of intermediates, surfactants and chemicals used in the rubber, mining and paper industries. In 1978 $\sim 10^8$ l were dumped in the ocean. The waste is slightly acidic, pH 5, and contains 1–4 wt.% organic C. No significant particulate phase is associated with the material. Of the three major wastes, it is the most difficult to characterize because it is derived from a large and diverse array of plant operations.

Waste–seawater interaction

Although the wastes are dumped at the sea surface, their initial dilution often is rapid due to barge generated turbulence, and reach a factor of 5000 (Csanady, 1980; T. P. O'Connor and Park, 1981).

(1) The DuPont–Edge Moor waste is quickly neutralized by the oceanic carbonate buffer system, and little effect on oceanic pH is realized. When mixed with seawater, Fe in the waste precipitates as a hydrous iron floc. These newly created particles have been shown to be efficient scavengers of

waste-derived lead. Cu was not as closely associated with Fe as expected and almost all the Cd, presumably because of chloride complex formation, remained independent of the particulate phase (Kester et al., 1980).

(2) The high alkaline content of the DuPont—Grasselli waste causes the precipitation of Mg from seawater as magnesium hydroxide (Kester et al., 1978). This precipitate is theoretically unstable at normal oceanic pH conditions which prevail almost immediately behind a dumping barge. Nevertheless, particles do persist and have been observed acoustically in the ocean after dumping events (Orr and Hess, 1978). Laboratory mixing studies Kester et al. (1978), have noted a non-conservative behavior of this precipitate consistent with dissolution, but have noted also that following some dissolution there is an increase in suspended material. Possibly, the magnesium hydroxide flocs serve as nucleation sites for calcium carbonate precipitation. It has been observed that if strong mixing occurs subsequent to introducing this waste to seawater, the precipitate dissolves completely (C. H. Whitlock, pers. commun., 1978).

(3) The American Cyanamid Warner waste generally mixes into the surface waters of the ocean and does not possess or generate any significant particle concentration. At present, studies have been initiated to determine, through gas chromatographic (GC) analysis, spectra or "fingerprints" which allow their detection in the ocean. Similarly, this technique is being used to reveal any alteration of waste which may occur in the ocean due to its volatization or its exposure to seawater, sunlight, or biodegradation.

Waste dispersion

These dispersion measurements have been based on Fe analyses, dye which has been added to waste, acoustic sensing of particulate phases that form upon dumping, color changes on the sea surface as seen from satellites or aircraft and, most recently, GC analysis for organic compounds which are unique to waste. Wastes are delivered to DWD-*106* site in $\sim 4 \cdot 10^6$ l lots, usually by barges. The wastes are discharged by gravity from moving barges, or by pumping from moving tankers. The rate is not to exceed a U.S. Environmental Protection Agency (U.S.E.P.A.) regulated maximum which varies from $7 \cdot 10^4$ to $12 \cdot 10^4$ l km^{-1}. The result is that during 1978, for instance, ~ 200 approximately U-shaped 45 km long ribbons of waste were laid down within DWD-*106*. The critical questions are (T. P. O'Connor and Park, 1981):

"What are the concentrations of waste in the ocean, where in the vertical structure of the ocean does the waste lie, how wide are the waste plumes, and is there any accumulation or additive effect due to repeated dumping events?"

Water-column stability, acoustically-sensed particle distribution, and Fe concentrations as measured after dumping events during summer and winter

386

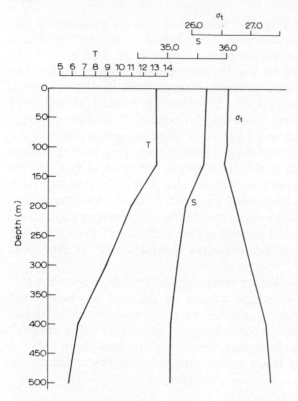

Fig. 2. Temperature, T ($^\circ$C), salinity, S ($^0/_{00}$), and water density, σ_t (g cm^{-3}) of "Albatross IV", Station 2 data for February 2, 1978 (Orr et al., 1980).

are given in Figs. 2–4 and 5–7, respectively. The point to be made here is that waste is distributed between the sea surface and the pycnocline. This layer, the mixed layer, is shallower in summer than in winter. A seasonal pycnocline forms between 10 and 30 m due to warming of the sea surface (Ingham et al., 1977). High-frequency (200 kHz) acoustic sensing (Figs. 3, and 6; Orr et al., 1980), indicates that particles derived from waste are constrained to the mixed-layer depth. Chemical data on Figs. 4 and 7 (Kester et al., 1979, 1980) do not confirm the presence of a vertical limit to waste penetration because in neither case was a sample from below the pycnocline available. However, more recent data as well as older information (EG & G, 1977a) do show by Fe analysis that waste is primarily distributed within the mixed layer.

The presence of a permanent pycnocline in deep water may preclude accumlation of waste on the sea floor. This contrasts with the situation over the U.S.A. continental shelf where, although a seasonal pycnocline is present during the warmer months, the mixed layer during colder months extends to the sea floor.

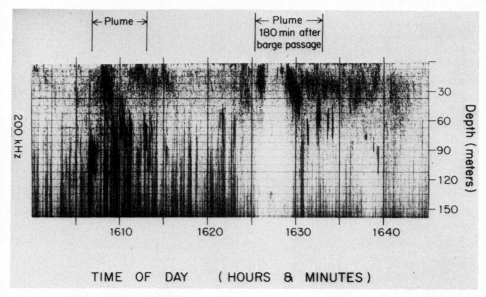

Fig. 3. Acoustic signal at "Albatross IV" of DuPont–Edge Moor waste on February 2, 1978. *Dark patches* extending to ~ 100 m are due to back-scatter from hydrous iron oxide floc. *Vertical lines* appearing throughout record are an artifact of transducer movement (Orr et al., 1980).

Dissolved waste constituents and particles could become incorporated into biological detritus or fecal material which could enhance its vertical migration. However, the longer the time required for such descent to occur, the more dilute the waste becomes and the less likely is its detection in sediment samples. Recently, bottom samples were taken at and near DWD-*106* for analysis to determine the presence of organic compounds indicative of waste. These compunds, if present, would confirm the descent of waste because they have no natural source. Inorganic chemical analysis (Greig and Wenzloff, 1977) has not been so diagnostic because in fine-grained sediments, heavy-metal concentrations are naturally quite high, and an anthropogenic input must be relatively large to be observed.

As stated in the Section "Waste–seawater interactions", on the basis of available data, Csanady (1980) has indicated that dilution created by barge generated turbulence is quite extensive by a factor of 5000 (T. P. O'Connor and Park, 1981). It is ~ 1.5—2 orders of magnitude more than is achieved by a well-designed sewage outfall (Csanady, 1980), and requires a relatively reasonable dumping time of only 5—6 hr., at a ship's speed of 5 kt. for $4 \cdot 10^6$ l of waste. Subsequent dilution is due to oceanic mixing processes which in approximation are independent of the waste characteristics. The rate of descent of particles through the pycnocline may be waste specific in that it depends on particle size and density. Other factors affecting descent, such as incorporation into biological material or surges through a pycnocline (Orr

388

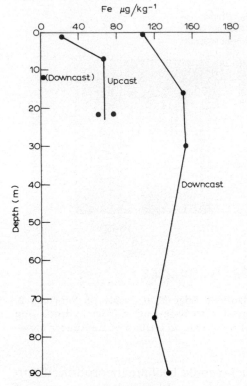

Fig. 4. Iron concentrations at 4 hr. post-dump on February 2, 1978. Concentrations before dump or away from plume, ranged from 0.2 to 0.9 μg kg^{-1}. (Kester et al., 1979.) Ship drifted $\sim$ 100 m between times of first and second downcasts.

and Hess, 1978), are not necessarily waste source dependent. The basic principles of the role of particulates in the ocean is discussed in detail in this volume by Eisma (Chapter 9) and are pertinent to the observations made in the previous paragraphs.

Dispersion measurements in terms of maximum measured concentrations, converted to minimum dilution factors, are summarized in Fig. 8. All of these data were obtained in the presence of a seasonal pycnocline so wastes were vertically distributed over the upper 10–30 m of the water column. These data, taken as a whole, indicate that: (1) dilution factors within 1 hr. after a dump are $\sim 10^4$; and (2) subsequent dilution can be a slow process. In two cases, where wastes were followed for about a day, dilutions measured at the end of the experiment were not 10 times more than those measured $\sim$ 4 hr. after the dump.

A corollary of waste dilution is the width of a plume. The image of a plume of DuPont–Edge Moor waste taken from the LANDSAT-1 satellite $\sim$ 12 hr. after the plume was created (Fig. 9) and indicates a width of 750 m. Csanady (1980) has considered waste dispersion during the summer, where

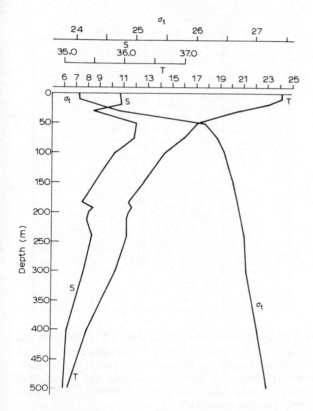

Fig. 5. Temperature, T ($^\circ$C), salinity, S ($^0/_{00}$), and water density σ_t (g cm^{-3}) data at "Albatross IV", Station 2 on 19^{h}40^m EDT July 27, 1977 (Orr et al., 1980).

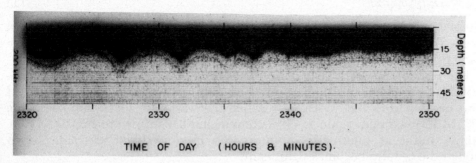

TIME OF DAY (HOURS & MINUTES).

Fig. 6. Acoustic record at "Albatross IV" showing DuPont–Edge Moor waste on July 26, 1977 riding an internal wave 480 min. after barge passage (Orr et al., 1980).

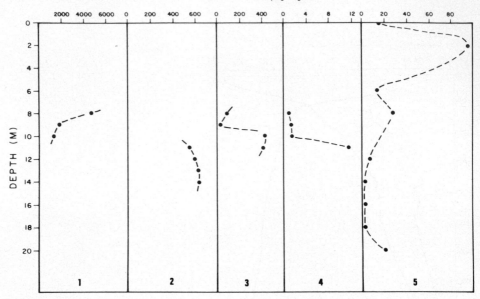

Fig. 7. Iron concentrations in DuPont–Edge Moor plume during July 1977. Background concentrations in the mixed layer ranged from 0.2 to 0.6 μg kg^{-1}. Samples taken within plume were not obtained exactly simultaneously so apparent vertical structure may include horizontal variations in concentration. 1 = Station 8, during the dump; 2 = Station 9, 4 hr. after dump; 3 = Station 10, 5 hr. after dump; 4 = Station 11, 8 hr. after dump; 5 = Station 13, 27 hr. after dump (Kester et al., 1980).

vertical waste distribution due to initial mixing coincides approximately to the depth limit of the mixed layer, to be basically a one-dimensional process whereby:

$$D_t = D_0 l_0^{-1} \sigma_t \sqrt{2\pi} \tag{1}$$

where D_t is the dilution factor at time, t, after a dump; D_0 is the dilution factor due to initial mixing; l_0 is the initial width of plume; and σ_t is the standard deviation of plume width at time, t.

The horizontal waste distribution is assumed to be normal with the maximum concentration at the center. For a normal distribution a width of 4.3σ incorporates that width over which the waste concentration rises from 10% of its maximum value to its maximum and back down to the 10% value. Plume widths at times after a dump can be considered to equal $l_0 \pm 4\sigma$ though the l_0 term becomes significant.

Rates of growth of plume width, σ, are due to oceanic mixing. We illustrate six very simple models below (T. P. O'Connor and Park, 1981):

I	$\sigma_t = \sqrt{2Kt}$	with $K = 0.011 l^{4/3}$	$(\mathrm{cm^2\,s^{-1}})$
II	same as I	with $K = 0.011 l^{1.15}$	$(\mathrm{cm^2\,s^{-1}})$
III	same as I	with $K = 10^3$	$(\mathrm{cm^2\,s^{-1}})$
IV	same as I	with $K = 10^4$	$(\mathrm{cm^2\,s^{-1}})$
V	$\sigma_t = \omega t$	with $\omega = 0.2$	$(\mathrm{cm\,s^{-1}})$
VI	same as V	with $\omega = 2.0$	$(\mathrm{cm\,s^{-1}})$

In models I and II, the eddy diffusion coefficient, K, is assumed to increase as the plume widens and is progressively mixed by larger and larger eddies. The exponent of $\frac{4}{3}$ in Model I is the classical "$\frac{4}{3}$ law" while the value of 1.15 in Model II was chosen by Okubo (1971) to best fit a collection of

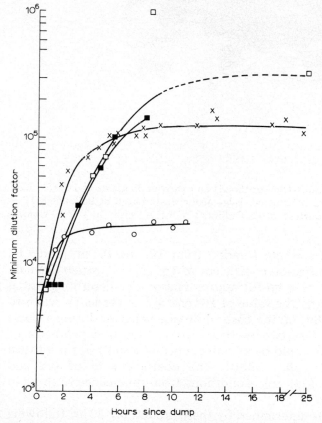

Fig. 8. Measured minimum dilution factors: $\circ$ = DuPont–Grasselli, September 1976 (EG & G, 1977b); $\blacksquare$ = DuPont–Edge Moor, September 1976 (EG & G, 1977a); $\square$ = DuPont–Edge Moor, July 1977 (Kester et al., 1980); and $\times$ = American Cyanamid, June 1978 (unpublished data, 1978). Wind and sea conditions were as follows: September 1976 ($\circ$ = DuPont–Grasselli) 4–5 m s^{-1} and calm; September 1976 ($\blacksquare$ = DuPont–Edge Moor) 10–12 m s^{-1} and 1–2 m; July 1977 ($\square$ = DuPont–Edge Moor) 10–15 m s^{-1} and 1–2 m (wind and sea diminished during study); and June 1978 ($\times$ = American Cyanamid) 5 m s^{-1} and 1 m or less except during the first 2 hr. (T. P. O'Connor and Park, 1981).

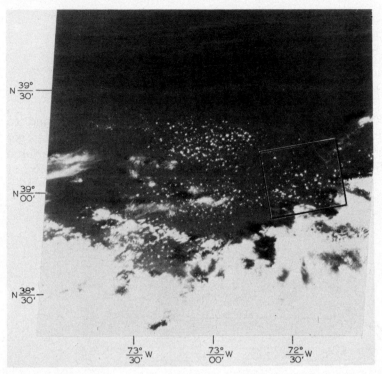

Fig. 9. LANDSAT-2 image of DWD-106 (outlined) and surrounding area at 10^h15^m EDT on September 9, 1977. A dump of DuPont–Edge Moor waste began at 20^h50^m EDT in the NE quadrant of the site (courtesy of C. Ohlhurst, N.A.S.A., cited in T. P. O'Connor and Park, 1981).

diffusion experiment results where l varied from 10^2 to 10^5 m. Model III employs a constant-eddy diffusion coefficient of 10^3 cm^2 s^{-1} which Csanady (1973) has recommended as a useful approximation for describing initial plume behavior in the ocean. The value of 10^4 cm^2 s^{-1} in Model IV was estimated by Ichiye et al. (1980) on the basis of drogue behavior during a waste tracking experiment. The last two models invoke a linear dependence of plume width on time which would be a consequence of a shear in the vertical distribution of waste (Csanady, 1980). The coefficient ω of 0.2 and 2.0 cm s^{-1} covers a range of oceanic conditions from quiescence to those in the presence of strong wind.

Minimum dilution factors determined by these models for 20 hr. following a dump, assuming a uniform vertical waste distribution, over 15 m are shown in Fig. 10. Comparison of these curves with data in Fig. 8 would indicate that Models III, IV, or V are reasonable approximations. They predict rather slow changes in concentration and concentrations during the 4—20-hr. post-dump period in the 5–50-ppm range. These same models, of course, yield values for plume width as a function of time. As seen in Fig. 11: models

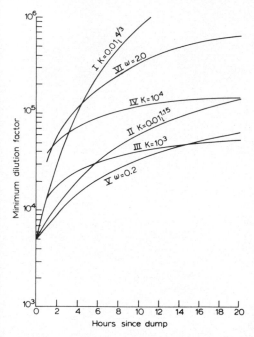

Fig. 10. Minimum dilution factors as calculated for Models I–VI (see text), assuming uniform waste distribution over upper 15 m (T. P. O'Connor and Park, 1981).

III–V indicate widths in the 600–1200 m range at 20 hr. after a dump. There seems to be sufficient data to conclude that crude models which describe the formation of narrow plumes with maximum waste concentrations in the 5–50 ppm range are relevant to waste dispersion over the first day after a dump. A most important question is, for how long do these models apply? The results of extending Model III, for example, for prolonged times is shown in Table II which gives, at 1 month, a maximum concentration of 2 ppm and a width of 3000 m. Csanady (1980) considers that slow dispersion and narrow widths are the fate of plumes until a storm is encountered. Then diffusion coefficients will increase and, because of the effect of horizontal shear on splitting up a plume, horizontal diffusion will be effective in two dimensions. Other model studies dealing with similar problems in the same general area are discussed in this volume by Sackett in the next chapter (Ch. 13).

Waste plume advection

A single waste plume may be compounded by subsequent dumpings to add to its contaminant load if no advection exists. We have undertaken, therefore, a study on advection at DWD-*106* site. Estimates on the residence

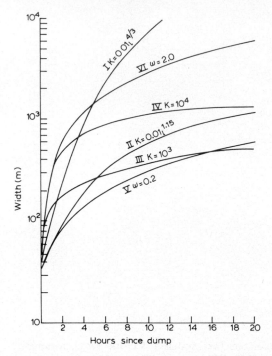

Fig. 11. Plume widths ("10%" widths) as calculated for Models I–VI (T. P. O'Connor and Park, 1981).

TABLE II

Long-term history of plume width, minimum dilution factor and maximum waste concentration if plume grows in one horizontal direction with a constant-eddy diffusion coefficient, K, of 10^3 cm^2 s^{-1} (T. P. O'Connor and Park, 1981)

Time since dump	Dilution factor ($\times 10^5$)	Concentration (ppm)	Width (m)
1 day	0.78	13	550
2 days	1.1	9	770
1 week	2.1	5	1,400
2 weeks	2.9	3	2,000
1 month	4.3	2	2,900

time of water and waste plumes within the dump site and their trajectory upon leaving the site are influenced very much by the presence or absence of a warm-core Gulf Stream eddy at the dump site. Eddies form in the North West Atlantic when a portion of the Gulf Stream encircles a parcel of Sargasso Sea water from its southern boundary, then breaks away from the main stream as a clockwise circling eddy of ~ 50 km in radius. These eddies move roughly along the contour of the U.S.A. continental shelf towards the southwest (Bisagni, 1976). Mizenko and Chamberlain (in press) reported that DWD-*106* was completely or partially within an eddy ~ 20% of the time during 1974–1976, and 70% of the time during 1977, which was a particularly fertile year for eddies.

When waste is dumped within an eddy, it will move in a clockwise direction at speeds of up to ~ 1 kt. It was observed that a dump of DuPont–Edge Moor waste, for instance, which was followed during 1977, moved about 60 km in 26 hr., and was 30 km east of DWD-*106*. These eddies vary in characteristics. According to Bisagni (1976), an average translational speed of an eddy is 0.2 kt., its direction southwest, and an average residence time of an eddy at DWD-*106* is ~ 3 weeks. Using the 3-week value and assuming four dumps per week, and a plume to be 45 km long and 1 km wide, it follows that 0.6% of the eddy is required for each plume. Therefore, the chance of interaction among the twelve separate plumes is low.

In the absence of an eddy, the direction of plume movement or its residence within DWD-*106* are not well known. Long-term current meter data from a site northeast of DWD-*106* give a mean flow of surface waters through the dump site of 0.2 kt. towards the southwest (Ingham et al., 1977).

Knowledge of water movements at DWD-*106* in the absence of eddies should improve, because of a recently inaugurated program using long-term drogues which can be followed from shore-side radio stations.

Biological effects

Four conclusions from chemical and physical studies bear directly upon assessing the biological consequences of waste disposal at deep-ocean sites (T. P. O'Connor and Park, 1981):

(1) At present dumping rates, concentrations of waste are 50 ppm or less within ~ 2 hr. of a dump. They may remain near these levels for more than 1 day, and for some an undetermined, longer time.

(2) Aside from being diluted, wastes as presented to the biological community, in the ocean may be altered from their original condition by precipitate formation, pH neutralization, adsorption of waste components on precipitates, or abiotic transformation of organic compounds.

(3) Waste plumes are relatively small, of the order of 50 km^2, and restricted in depth to the mixed layer.

(4) At the present frequency of dumping, each plume is a separate incident affecting a small portion of ocean and about 200 such incidents occur per year.

Biological studies at DWD-*106* site have, therefore, centered on planktonic organisms: bacteria, phytoplankton and zooplankton. The rationale is that the critical component of the marine ecosystem which could most likely suffer from the practice of dumping in the deep ocean are those planktonic

TABLE III

Average effective concentrations (EC) found to yield cell counts of *Skeletonema costatum* 50% lower than controls during 96 hr. exposures (from data supplied by industry)

Waste source	96-EC$_{50}$ (ppm)
DuPont–Edge Moor	2,010[1]
DuPont–Grasselli	970[2]
American Cyanamid	260[3]

[1] Average of results obtained with individual barge samples for May–November 1977 ($n = 7$, range $= 1100–2690$ ppm).
[2] Average of results obtained with individual barge samples for February–September 1977, excluding value of 8600 ppm from August ($n = 7$, range $= 330$ to 3100 ppm).
[3] Average of monthly results from 1976 ($n = 12$, range $= 19–750$ ppm).

organisms which occupy surface layers. Free swimming organisms can encounter plumes but unless they are particularly attracted to it, their exposure should be brief. A measure of toxicity of wastes toward the diatom *Skeletonema costatum*, as derived from information supplied to the U.S.E.P.A. by dumping permittees, is given in Table III. It would seem on the basis of these data that, except possibly for the American Cyanamid waste, there would be little effect, if any, expected on phytoplankton communities within waste plumes. However, results based on a single organism may be misleading. In tests of DuPont–Grasselli waste on growth of *S. costatum*, Murphy et al. (1980) found results to vary with the source of the organisms (Fig. 12). The most sensitive strain of *S. costatum* was one isolated from open-ocean water, whereas the same species when isolated from coastal or estuarine water was generally more resistant. Similar results were found for the coccolithophorid *Emiliania huxleyi* (Fig. 12) and the diatom *Thalassiosira pseudonanna* (not shown). In none of these cases were the tested phytoplankton affected by the DuPont–Grasselli waste at concentrations as low as those observed within plumes.

Extrapolation of laboratory results on waste effects is difficult because it is not possible to test all wastes against all possible planktonic species (Hulbert and Jones, 1977; Sherman et al., 1977). It has been demonstrated that effects upon growth of phytoplankton vary with species and even among subgroups of the same species. Therefore, it is possible that within waste plumes species composition of the phytoplankton community is altered as growth among species is differentially affected. Such an effect has been postulated by Fisher (1975) as a consequence of chlorinated hydrocarbon additions to marine systems. A shift in community structure toward smaller species is observed by W. H. Thomas and Seibert (1977) when Cu was added to controlled ecosystems moored in coastal waters. However, the community changes due to Cu additions as demonstrated by W. H. Thomas

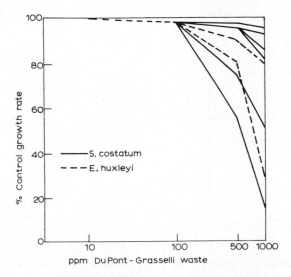

Fig. 12. Effect of DuPont—Grasselli waste on growth of *Skeletonema costatum* (six clones) and *Emiliamia huxleyi* (two clones). Each point is the average of three determinations with percentage decreases in growth among determinations varying by 0.05% or less (Murphy et al., 1980).

and Seibert (1977) were not evident until four or more days after the additions. The doubling time for the phytoplankton species grown under optimum conditions was of the order of 12 hr. (Murphy et al., 1980). If that time is relevant to field conditions and five doublings are needed to produce a measurable redistribution of species then waste would have to remain at an effective concentration for 60 hr. to yield a demonstrable effect. In more tropical waters the doubling time for phytoplankton may be only 3 hr. (Sheldon and Sutcliffe, 1978), allowing for a faster response to waste additions.

There are numerous natural restraints on phytoplankton growth such as light, temperature, nutrient concentrations and grazing pressure. These all bear upon the size and structure of communities. It is not yet known if waste dumping as presently practiced can exert an effect of significance relative to natural variations even within the relatively small scale of waste plumes.

The consequence of a change in species composition of phytoplankton is that organisms could come to dominate which are unacceptable as food to zooplankton. Concurrently, the feeding rate of zooplankton in waste plumes may be decreased. This may by itself enhance the possibility of waste affecting phytoplankton populations. Capuzzo and Lancaster (1980) have found that DuPont—Grasselli waste at low concentrations, in the 10—50 ppm range, has a slight lethal effect upon the copepod *Centropages typicus* (Fig. 13) and causes decreased feeding rates in both that organism and the copepod *Pseudocalanus* sp. (Fig. 14). Since the generation time of zooplankton organisms is longer than that of phytoplankton, it is not clear

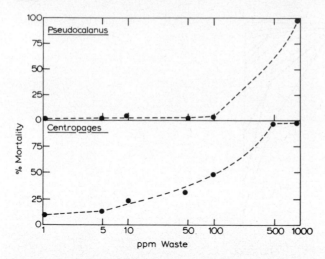

Fig. 13. Effect of DuPont–Grasselli waste on mortality of two species of zooplankton. Each point is mean of six determinations (Capuzzo and Lancaster, 1980).

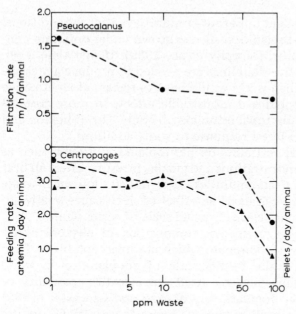

Fig. 14. Filtration rates of *Pseudocalanus* and feeding rates and fecal pellet production rates of *Centropages* exposed to DuPont–Grasselli waste for 96 hr. For filtration rates each point is the mean of 24 determinations, for feeding rates (*circles*) and fecal pellet production rates (*triangles*) points represent means of six determinations, *open symbols* are control values (Capuzzo and Lancaster, 1980).

that these small effects would lead to any measurable changes in zooplankton populations.

We pursued the studies on the effect of wastes upon marine bacteria for three reasons: (1) a decrease in bacterial activity within the mixed layer could affect the ecological balance by slowing the rate at which dead or excreted biological material is recycled; (2) a change in bacterial species composition in favor of more tolerant organisms would be evidence of a biological response to waste disposal even if the net effect upon the recycling rate were minimal; and (3) bacteria serves as a convenient class of organism with which to test chemically or physically-altered waste material.

Vaccaro and Dennett (1977) found that concentrations which lower glucose-uptake rates by 50% relative to controls were 5000 and 17,000 ppm for the DuPont–Grasselli and American Cyanamid wastes, respectively, when tested against mixed bacterial populations from DWD-*106*. This level of tolerance would seem to preclude any significant effect upon bacteria due to dumping at DWD-*106*. Also, it was found that bacterial populations within the dump site were identical to those outside the site in terms of response to waste. Sampling within the dump site was not done in conjunction with chemical analysis and did not recognize the rather discrete distribution of waste. Therefore, as was the case with early phytoplankton sampling (Ortner and Murphy, 1977), results could not be directly connected with waste

Vaccaro and Dennett (1980), using the same technique when applied to mixed populations from coastal water, found effective concentrations of 1000–2000 ppm for the DuPont–Grasselli and 100–200 ppm for the DuPont–Edge Moor wastes. For the American Cyanamid waste, R. F. Vaccaro (pers. commun., 1979) indicated a 100-ppm concentration. Relatively low concentrations of DuPont–Edge Moor waste were found, through laboratory experiments, to affect bacterial activity which may imply an effect at DWD-*106*. However, a sample taken within a visible plume of that waste did not reveal a decreased rate of glucose assimilation. The wide range of effective concentrations for the American Cyanamid waste, 100–17,000 ppm, was observed to hold for two separate waste samples. So the range is not likely due to variations in waste chemistry. The unexpected and tentative conclusion must be, though, that open-ocean bacteria are more tolerant of waste additions than are coastal communities. The effects of wastes in the ocean on the biota including bacteria are presented in some detail for various areas in this volume by Mearns (Chapter 1), Bernhard (Chapter 3), Gillespie and Vaccaro (Chapter 5) and Gunnerson (Chapter 11).

Bacterial assays with altered waste indicate that toxicity of DuPont–Grasselli waste is due to its organic content (Vaccaro and Dennett, 1980). Conversely, the toxicity of DuPont–Edge Moor waste can be decreased considerably by addition of EDTA indicating that its toxicity is due to its metal content.

DISCUSSION

Because of the immensity of the volume of water and the great depth encountered at both DWD-*106* and Puerto Rico dump sites, the main effort has been the study of the physical diffusion and advection of the waste plume at the dump site and at its subsequent trajectory regime. Some tentative conclusions have been given here on the size and concentration of plumes generated by present ocean dumping activities at DWD-*106*. The possibility of phytoplankton community being altered by the waste dumping practice is advanced. Experiments are needed to verify both the physical and biological knowledge and hypothesis.

Experimentation at sea with ocean dumping parameters is difficult. And the ability is needed to sample contaminated chemical and biological samples for longer periods of time. Because the waste plumes are relatively small, they are easily lost, and without concomitant scientific verification of the presence of waste, no correlation can be made between observations of biological community and dumping events. The integrity and reliability of field experimentation must be advanced.

As the human society looks toward the deep ocean as a possible receptacle of many wastes, including radioactive waste and manganese nodule processing waste, we must develop our competency to deal with both shallow-water and benthic-bottom-layer waste effects in the open ocean. Tools needed for the study require long planning, designing and building phases in order for them to be effective.

CHAPTER 13

AN EVALUATION OF THE EFFECTS OF MAN-DERIVED WASTES ON
THE VIABILITY OF THE GULF OF MEXICO

William M. Sackett

INTRODUCTION

The ocean has been used by man from early in recorded time as a means
of transportation and a source of food. In recent years it has been providing
chemicals such as liquid petroleum, natural gas, magnesium, bromine and
other products that are vitally needed in modern technology. The ocean
bottom offers the potential for other metal requirements of our society
through the mining of the extremely large, deep-sea deposits of metalliferous
substances such as maganese nodules. However, man is also making use of
the ocean in another less attractive way — as a dumping ground for his
garbage and chemical wastes.

Because the Gulf of Mexico receives runoff and wastes from about two-
thirds of the continental U.S.A., excluding Alaska, it is particularly sensitive
to the introduction of man-generated wastes. This presentation attempts to
give realistic appraisals of the present problems and the fate of marine life in
the Gulf of Mexico. First the circulation and mixing characteristics of the
Gulf will be discussed, and next the types and biological effects of various
man-derived chemical contaminants will be presented. Finally, an analysis of
the inputs and fates of the most serious contaminants will be combined with
the circulation and biological aspects to give a prediction of the fate of
marine life in the Gulf of Mexico.

THE SETTING: THE GULF OF MEXICO

The Gulf of Mexico is a semi-closed basin with an oceanic inflow from the
Caribbean Sea through the Yucatan Strait. This strait has a maximum or sill
depth of 1800 m. As the water masses in the ocean tend to spread horizontally and not flow up and over obstacles, only water shallower than 1800 m
in the Caribbean Sea, which in places has depths greater than 6500 m, enters
the Gulf of Mexico. As is shown in Fig. 1, most of this upper water passes
quickly through the eastern Gulf and out the Florida Straits where it becomes the Gulf Stream in the Atlantic Ocean. Generally, this current in the
Gulf intensifies through the summer months into a pronounced loop. Sometimes in late summer this loop detaches itself from the main current as a
separate eddy which then drifts westward. Adding to this westward drift are

402

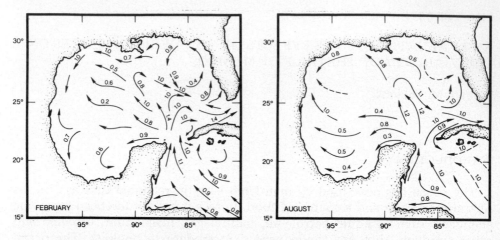

Fig. 1. Estimate average speed (knots; 1 kt. = 1.853 km hr.$^{-1}$) currents in the Gulf of Mexico from U.S. Naval Oceanographic Office Pilot Charts for February and August (Nowlin, 1971).

near-shore currents which are generally moving in a westerly direction along the northern coastline. Hence, it should be apparent that the Gulf of Mexico may be divided into two parts; an eastern part which is continually and thoroughly flushed by the loop current and a western part which is relatively isolated. It is to this western Gulf of Mexico that most of the pollutant-laden runoff and industrial wastes are being delivered; and where marine life is in the greatest danger of being damaged by man's activities.

Fig. 2 presents a W—E cross-section across the Gulf of Mexico, the Florida Straits and into the Atlantic Ocean. It shows that the Yucatan current extends down to 1800 m in the eastern Gulf, but only the upper 600 m of water can escape across the Florida Straits into the Atlantic Ocean. The Mississippi River water flow is shown spreading out into the surface water of the western Gulf. Most of the surface water down to about 55 m and intermediate water between 55 and 600 m is shown spreading to the west, with a much smaller return flow to the east. Water deeper than 600 m is shown to be trapped in the bottom of the Gulf of Mexico. However, in reality, deep water is continuously mixing upwards over the entire Gulf at a slow rate. Some studies indicate that the lifetime of a water parcel in the deep water is the order of 100 years (Mathews et al., 1973). This figure becomes important in considering the future viability of the Gulf of Mexico.

A distributive province has been defined by Moody (1967) as the area of origin and accumulation of sedimentary material and includes the rivers and streams responsible for the distribution. The Gulf of Mexico distributive province is shown in Fig. 3. It is composed of a sediment source area of $5.40 \cdot 10^6$ km^2, a distribution system of 159,890 km of rivers (primarily the Mississippi and its tributaries) and a depositional area (the Gulf of

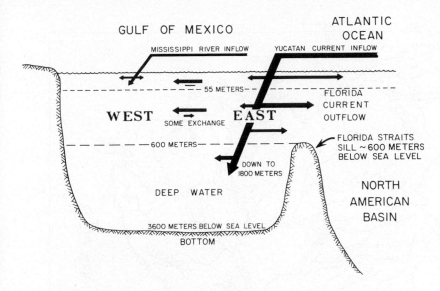

Fig. 2. West–east cross-section of the Gulf of Mexico.

Mexico) of $1.6 \cdot 10^6$ km^2. It is drenched annually on the U.S.A. continental part by ~ 420 km^3 of rain water of which one-fourth reaches the Gulf, primarily by way of the Mississippi River. The annual amount of this runoff is $\sim \frac{1}{10}$ of the volume of water on the continental shelf in the western Gulf but only $\frac{1}{800}$ of the annual volume of flow through the Yucatan Strait. For comparison purposes it would take about three years for the Yucatan current to fill the entire Gulf of Mexico basin, whereas it would take about 2400 years for the Mississippi River to accomplish this feat.

CHEMICAL CONTAMINATION

Most Earth scientists accept the findings that the Earth was formed about $4.5 \cdot 10^9$ yr. ago by the accumulation of a great number of relatively small, cold, solid bodies. An initial heating due to compressional effects and radio-active decay and the insulation enhancement of the large proto-Earth relative to small pre-Earth bodies apparently resulted in an outgassing of carbon dioxide, methane, hydrogen and other gases and especially water which con-densed to form the oceans. After much of the free hydrogen gas was lost to outer space, a buildup of molecular oxygen allowed the evolution of life as we know it. For the past two, three or four billion years, an oxygenated ocean has been the receiver of hot fluids from the Earth's interior and the runoff from the continents. These fluids contain salts resulting from the partial solution of rock materials. Over hundreds of millions of years, the ocean evolved to the point where, to a first approximation, the amounts of

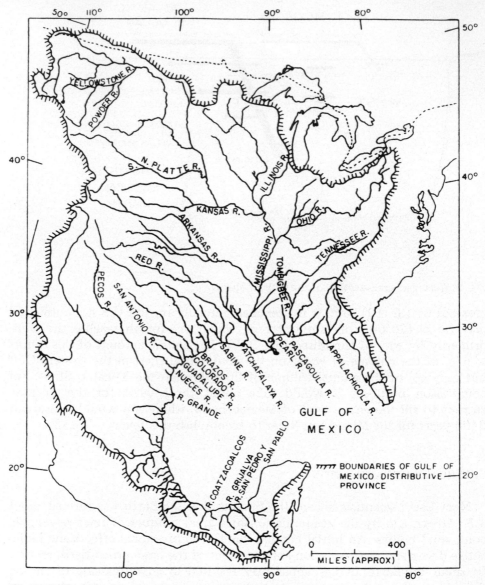

Fig. 3. Gulf of Mexico distributive province (Moody, 1967).

these salts in one form reaching the ocean each year were about equal to the amounts of these salts in another form which were leaving the ocean each year as sedimentary deposits. This steady-state situation may be described by the following equation:

$$\tau = \frac{A}{dA/dt}$$

where τ is the residence time of an element, A is the amount of the element in solution in the ocean and dA/dt is the rate of introduction or removal of the element. This discussion is presented to help the reader to accept the idea that the chemical composition of the ocean has remained about the same for approximately the last one billion years.

Man has evolved over the past several million years. Because of his relatively small numbers and his rural and primitive activities as measured by modern standards, he had little influence on the ocean until $\sim$ 1900 A.D. At about that time, industrial activity began to grow explosively. An unquenchable thirst for raw material and various forms of power, has led to a reshaping by man of much of the Earth's surface. The waste products and the products themselves have been increasingly introduced to the ocean with serious consequences to some ecosystems. It is this point in time when a chemical substance has a demonstrable effect on an ecosystem that it passes from the realm of a chemical contaminant to that of a chemical pollutant. For example, the burning of fossil fuels over the past 75 years has increased the carbon dioxide content of the atmosphere by over 5%. This CO_2 is being taken up by surface water of the ocean, but as there are no apparent effects on any marine ecosystem, it must remain in the contaminant category. On the other hand, chlorinated hydrocarbons manufactured by man, have been shown to be responsible for the extinction of several populations of marine birds. This type of chemical is clearly a hazardous pollutant.

There are large differences between the effects of various contaminants on freshwater lakes and streams and the ocean. The introduction of fertilizer-derived nitrates and also the phosphates from both fertilizers and municipal sewage systems, also containing biodegradable organic materials, has stimulated algal growth. In many instances, the accumulated organic material has, in turn, used up available oxygen in a water body and ultimately resulted in the death of all life requiring molecular oxygen. A striking example of this phenomenon (eutrophication) is Lake Erie. On the other hand, the ocean is such a large reservoir that such man-induced eutrophication is highly unlikely except in very restricted bays and estuaries. Similarly, so-called thermal pollution resulting from the discharge of waste heat by power plants has accelerated algal growth and eutrophication in inland water bodies, but there is little likelihood that this will occur in the open ocean, although a thermal-induced-shift in the types of organisms may occur in the vicinity of thermal discharges. In summary, fertilizers, sewage effluents and heat are pollutants for various inland water bodies, but only unimportant contaminants for the ocean as a whole.

Man-derived radioactive contamination of the ocean began in the early 1940's with the discharge of effluents from the Hanford atomic plant into the Pacific Ocean by way of the Columbia River. Subsequently, numerous atomic weapons were detonated with most of their radioactive ashes ultimately being transported to the ocean via the atmosphere and runoff.

TABLE I

Summary of chemical contaminants and their oceanic hazard potential

General type of contaminant and/or pollutant	Specific examples	Mode of introduction to the ocean	Oceanic hazard evaluation	Comments
Synthetic organics	chlorinated hydro-carbons	atmosphere and runoff	high on global basis	DDT has been shown to be responsible for the decline in several marine bird populations through interference with certain hormone systems
Petroleum	benzene, benzo-pyrenes	runoff; produc-tion and shipping operations	moderate on global basis	certain toxic compounds in petroleum have been shown to inhibit photo-synthesis, induce cancer in humans, etc.
Heavy metals	Hg and Pb	atmosphere and runoff	high on near-shore regional basis; low on global basis	Hg, from manufacturing operations and concen-trated in fish, has killed and maimed people in Japan
Fertilizers	nitrate and phosphate	runoff	very low	possible enhancement of primary productivity in coastal waters
Biodegrad-able sewage effluents	–	runoff	very low	possible enhancement of animal productivity in coastal waters
Thermal	–	atmosphere and runoff	very low	possible shift in types of organisms in discharge area
Radio-nuclides	^{32}P, ^{90}Sr, ^{65}Zn	atmosphere and runoff	very low	no short-term effects on organisms but long-term genetic effects may be important; ^{32}P and ^{65}Zn are magnified to near-permissible levels in oysters and seaweed near nuclear plant effluents

With the limitation in above-ground atomic testing in recent years, most present-day radioactive contributions to the sea are wastes from atomic reactors and from medical, scientific and industrial uses of radionuclides.

In the open ocean at the present time most of the radioactivity is due to naturally-occurring radioactive nuclides, particularly ^{40}K; therefore it would appear that man-produced radioactivity does not present a big hazard. However, many elements are concentrated by organisms by a thousand times or more over their levels in seawater. Thus a negligible amount of a radioactive material such as ^{32}P is concentrated by oysters and fish to levels that may be hazardous to humans. These instances are rare and presently are only being observed in the immediate vicinity of reactor waste discharges.

The potential hazard presented by long-term chronic exposure of an eco-system to radiation somewhat above natural levels is unknown. But the con-sensus by experts seems to be that above ambient levels induce proportional increases in mutations in living systems. This in turn will have dire conse-quences on future generations. Other basic problems associated with the disposal of such wastes are described in detail in this volume by Angel et al. (Chapter 10).

Various aspects of the relatively unimportant oceanic chemical contami-nants discussed above are summarized in Table I together with three classes of pollutants shown to be quite harmful to marine ecosystems. These three classes are the synthetic organic compounds, petroleum compounds and the heavy metals such as Hg and Pb.

Synthetic organic compounds

One of the major categories under synthetic organics is the group called the halogenated hydrocarbons, which includes DDT, PCB's, Freons and many other types of compounds. In recent years a great many studies have been conducted on the biological effects of these compounds, particularly the pesticide DDT [1,1,1-trichloro-2,2-bis(p-chlorophenyl) ethane]. DDT has proven invaluable in controlling the pest-borne killers (typhus, encephalitis and malaria). Over the past thirty years, millions of lives have been saved. Initially one of DDT's great advantages was its stability and persistence. However, this characteristic has become a major drawback with as much as two-thirds of the $1.5 \cdot 10^6$ t manufactured by man now distributed around the Earth via the atmosphere and ocean. As is shown in Fig. 4 for the eco-system in Long Island Sound, an additional problem is the magnification of DDT through the food chain on up to marine birds such as the cormorants and ospreys. In many areas high DDT levels have interfered with the hormonal balance in birds and affected their reproduction, with the result that some bird populations are becoming extinct because of DDT in the environment.

Other chlorinated hydrocarbons, the PCB's (polychlorinated biphenyls),

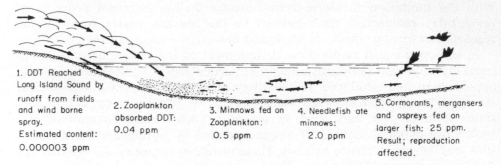

1. DDT Reached Long Island Sound by runoff from fields and wind borne spray. Estimated content: 0.000003 ppm

2. Zooplankton absorbed DDT: 0.04 ppm

3. Minnows fed on Zooplankton: 0.5 ppm

4. Needlefish ate minnows: 2.0 ppm

5. Cormorants, mergansers and ospreys fed on larger fish: 25 ppm. Result; reproduction affected.

Fig. 4. Biological magnification of DDT through the food chain in Long Island Sound (Woodwell et al., 1967).

with a high stability and desirable physical properties from a technological point of view, have been manufactured in great quantities with much escaping to the environment. Other chlorinated hydrocarbon pesticides have become very important in controlling crop and home pests. Large quantities of these toxic and persistent compounds have now been transported to the sea and are working their lethal ways on the marine ecosystem.

Petroleum compounds

Many organic compounds found in petroleum are toxic to marine life. For example, the water-soluble fraction consisting primarily of aromatic hydrocarbon compounds such as benzene and toluene at levels of $\sim$ 1 ppb interfers with the feeding behavior of some snails and crabs (Jacobson and Boylan, 1973; F. T. Takahashi and Kittredge, 1973; Johnson, 1977). At levels of 0.2 ppb, R. L. Steele (1977) showed that the fertilization of macroalgae was affected. Benzene and toluene at levels of $\sim$ 20 ppm and naphthalene at $\sim$ 1 ppm are toxic to an estuarine fish *Cyprinodon varigalis* (Petrocelli et al., 1974). Petroleum also contains multi-ring aromatic compounds (benzopyrenes) that have been found to be carcinogenic to humans. Studies on the long-term effects of below toxic levels of these compounds on marine ecosystems are underway, but years will pass before these sublethal effects are adequately evaluated.

Unlike other pollutants that are added to the ocean by atmospheric transport or runoff, most petroleum is added during transportation via ships and offshore production operations. Approximately 60% of the world oil production is transported across the sea from source areas such as the Middle East to markets in Europe and elsewhere. A significant amount of this oil is lost to the sea each year during normal handling procedures and catastrophic accidents such as the groundings of supertankers, e.g., the "Torrey Canyon", "Metula" and "Amoco Cadiz". Normal offshore production losses and blowouts such as the Santa Barbara Channel accident add to the petroleum burden of the ocean. Some critics of petroleum pollution estimate that as much as 0.1% of the $2 \cdot 10^9$ t of oil produced each year may be lost to the sea, a total of $\sim 2 \cdot 10^6$ t.

Initially much of this oil is found floating on the surface, especially in shipping lanes. For the above-mentioned near-shore accidents a good fraction of this floating oil ends up on beaches with disastrous consequences to the local marine ecosystem and especially to the birds, coated with tar, which we have all seen on TV and in pictures. Although these dramatic short-term effects are important, it is the long-term sublethal effects that most concern marine scientists. The role of petroleum hydrocarbons is also discussed in detail by Sauer and Sackett in Volume 1 (Chapter 4).

Heavy metals

The toxicity of Hg, Pb, Cd, Cu and other metals has been known for many years. For most of these toxic metals, the total amounts in the ocean are hundreds or thousands of times the amounts being brought in each year by the rivers or the atmosphere. The situation for Hg is depicted in Fig. 5. The natural and man-derived inputs approximate $12 \cdot 10^4$ t yr.$^{-1}$ whereas the total burden of Hg in solution in the ocean is $\sim 4 \cdot 10^7$ t or 330 times the input of a year. Thus it would take about three-hundred years to double the amount of Hg in the sea if no mechanism were operating for its removal from solution. If Hg were quickly distributed through the ocean, there would be no immediate pollution problem. However, this is often not the case as local man-derived sources may contaminate a rather restricted inland or oceanic-near-shore water body. The Minamata Bay, Japan, disaster is one of the most dramatic examples of this problem (Anonymous, 1972). Industrial discharges of Hg into Minamata Bay resulted in an excessive accumulation in fish and shellfish which were the main components of the diet of the local people. Many individuals subsequently died or were maimed by the poisoned fish before Hg was identified as the causitive agent.

In summary, although metals do not seem to be a pollution problem for the ocean as a whole, the oceanic levels of some of them potentially are detrimental to ecosystems in semi-restricted rather large water bodies such as the Gulf of Mexico.

FATE OF THE GULF OF MEXICO

In considering potential long-term lethal and sub-lethal effects of pollutants on life in the Gulf of Mexico, the most serious situation is for certain pollutants to be confined to the northern shelf of the western Gulf. The discussion in the first section pointed out that the Mississippi River outflow turned westward and moved across the shelf. Also the rate of exchange of

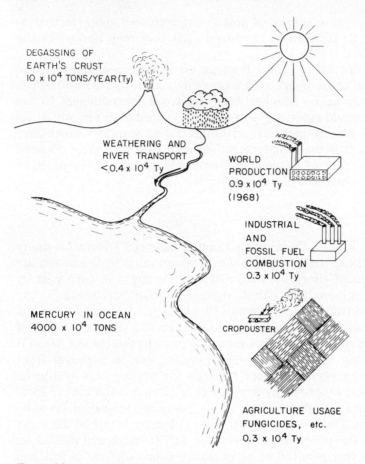

Fig. 5. Mercury pollution (Weiss et al., 1971).

water on the shelf and the open Gulf is not very rapid so that any pollutant added to the western Gulf shelf water is diluted by open Gulf water at a relatively slow rate. A rapid exchange would lead to the next most serious situation: that of the shelf water mixing into the upper 55 m of surface water over the entire western Gulf of Mexico. These two possibilities will be considered below for several important pollutants.

These two cases are the extremes and are not at all realistic; but for a presentation such as this, it is important to consider the most serious situation. If the worst cases prove unimportant then much time and money may be saved in the future by industrialists in their controversies with the environmentally concerned but often naive populace.

The present scenarios presuppose that the pollutants are stable and remain unchanged in solution in their particular boxes on the shelf or in the surface water of the western Gulf. This is, of course, not realistic. First of all, these water masses do mix downward and laterally into the eastern Gulf with a corresponding dilution effect. In addition many pollutants are adsorbed on the surfaces of particles of suspended sediment and organisms, which are transported downward by gravitational settling. This removal from solution purifies the water but, on the other hand, contaminates the bottom where other organisms such as shrimp, clams and various filter feeders may incorporate the pollutants into their tissue during feeding. Finally, most pollutants such as petroleum and the synthetic organics are not completely stable but do decompose via chemical and biochemical pathways to stable, non-toxic end-products. However, the reverse is also possible as has been shown for the conversion of toxic inorganic mercury compounds to more toxic organic mercury compounds by bacterial processes in sediments.

For the western Gulf of Mexico the principal sources of pollutants are Gulf coast rivers (primarily the Mississippi), offshore dumping of industrial wastes and offshore petroleum production operations. Atmospheric transport to the Gulf is also important but the Gulf presents such a small area for deposition relative to the remainder of the ocean that atmospheric transport is unimportant relative to the aforementioned sources.

Table II summarizes the inputs, predicted maximum levels in shelf and surface water in the western Gulf and published toxic levels for one or two types for each of the significant classes of oceanic pollutants. For the synthetic organics, DDT is the first type that is listed. Risebrough et al. (1968) estimated that 10^4 kg yr.$^{-1}$ is being introduced into the Gulf by the Mississippi. As this has been going on for 30 years, a total of $\sim 3 \cdot 10^5$ kg or ~ 0.3 t is the estimated input. This is probably a minimum figure as much has been brought into the Gulf via other rivers draining agricultural lands, particularly cotton fields, where much DDT has been used. Using the figures presented in Table II, predicted maximum levels of DDT for shelf and surface water in the western Gulf of Mexico are 0.05 and 0.01 ppb, respectively. These levels are more than an order of magnitude below the toxic levels reported by Wurster (1968) but on the basis of the 10^7-fold magnification of DDT presented in Fig. 4 (Woodwell et al., 1967), maximum DDT levels in Gulf coast marine birds may approach 500 ppm. This is two orders of magnitude higher than necessary for interference with their reproductive success. These simple calculations suggest that DDT levels are critical in the Gulf of Mexico and are probably responsible for phenomena such as the near overnight disappearance of the brown pelican, a symbol for the entire Gulf region (Jehl, 1969).

For over twenty years industrial wastes have been dumped offshore Texas. One type of waste is the chlorinated hydrocarbons, such as dichloropropane, trichloropropane, dichlorobutane and trichloropropane: ~ 0.6 t day^{-1} have been dumped (N.A.S., 1975). For twenty years this could build up a concentration of 60 and 10 ppb in western shelf and open Gulf surface

TABLE II

Summary of inputs, maximum concentrations and toxic levels for various pollutants

Contaminant and/or pollutant	Principal inputs			Estimated or maximum concentrations (ppb)			Toxic levels	
	pathway	period	amounts (10^3 t)	pre-man	NW shelf water*1	W surface water*2	concentration (ppb)	reference
Synthetic organics:								
DDT	Mississippi River runoff	1944–1974	0.3	0	0.05	0.01	10	Wurster (1968)
Chlorinated hydrocarbons; industrial wastes	offshore dumping	1953–1973	400	0	60	10	19	N.A.S. (1975)
Petroleum:								
C_5–C_{10} hydrocarbons	transportation and offshore production operations	1964–1974	4.5	~ 0	0.6	0.09	100 / 100 / 0.2	bacteria; F. Walsh and Mitchell (1973) snails and crabs; Johnson (1977) macroalgae; R. L. Steele (1977)
Heavy metals:								
Hg	Mississippi River runoff	1944–1974	1.2	0.03	0.1	0.03	~ 10	Petrocelli et al. (1974)
Pb	Mississippi River runoff	1944–1974	2.5	0.03	0.3	0.06	~ 10	estimate based on Hg

*1 Volume taken as $6.4 \cdot 10^{15}$ l; *2 volume taken as $4.5 \cdot 10^{16}$ l.

waters, respectively; concentrations which are in the range of the threshold toxicity levels for these compounds (N.A.S., 1975). These higher maximum levels may be balanced by greater decomposition rates for these simpler chlorinated hydrocarbons as compared to DDT-type compounds. Only extensive further studies of these and other synthetic organics such as the PCB's, Freons, cleaning solvents, etc., will determine the effects of contemporary levels of these compounds on the marine ecosystem of the Gulf of Mexico.

The effect of PCB's and DDT on the Gulf of Mexico is also discussed in great detail in this volume by Trefry and Shokes (Chapter 4), and on the Mediterranean Sea by Bernhard (Chapter 3), as well as for the west and east coasts of the U.S.A. by Mearns (Chapter 1) and Gunnerson (Chapter 11), respectively.

For the major class designated as petroleum, only the group of compounds having from six to fourteen atoms of carbon is listed. It is this group which is easiest to identify, in so far as types and amounts, and for which it is easiest to measure its lethal effects on marine organisms. The higher-molecular-weight compounds, comprising by far the major fraction of liquid petroleum, are those that are most visible in the form of tars on beaches, floating on the sea surface, and coating and killing marine organisms and birds. It is also this latter fraction which contains the benzopyrenes and other compounds which might have long-term but extremely difficult to measure major sub-lethal effects on ecosystems. Although years of tedious studies will have to be conducted to understand the effects of the high-molecular-weight components, good progress has been made in the past few years on the biological effects of C_1-C_{14} hydrocarbons on various organisms. Some of these were reported in Chapter 4 by Sauer and Sackett in Volume 1.

According to a study by J. M. Brooks et al. (1977) the input of C_5-C_{10} hydrocarbons into the Gulf is $\sim 450 \, t \, yr.^{-1}$ or for a ten-year period, a total of $\sim 4.5 \cdot 10^9$ g. This input is considerably higher than expected based on normal operating loses because of two oil-industry practices; the underwater venting of non-commercial amounts of gases issuing from gas–liquid separators located on offshore platforms and the discharging of brines with high concentrations of the relatively water-soluble components of petroleum. The predicted maximum levels in western shelf and open Gulf surface water are 0.6 and 0.09 ppb, respectively. These values are in the range observed to inhibit primary productivity (J. M. Brooks, 1975). On the other hand, these hydrocarbons are relatively volatile and are probably lost to the atmosphere at a rapid rate and the concentrations in the Gulf, in all probability, remain substantially below experimentally determined toxic levels.

Hg and Pb are usually considered as the most serious metal pollutants in the environments, with Hg being responsible for the Minimata disaster and Pb possibly in part, for the mental, emotional and other health problems of some urban dwellers (Patterson, 1965). Because of the volatility of elemental

414

Hg and the lead alkyls used as antiknock additives in gasoline, these metals are transported through the atmosphere over the entire globe. Deposition into the Gulf from the atmosphere is probably in proportion to the area of the Gulf compared to the entire ocean — a small fraction.

As indicated in Fig. 5, the main input of Hg is the outgassing of the Earth's crust with an estimated annual input six times more than man-associated production and losses. These are in turn 300 times less than all the Hg in the ocean, which means that it would take 300 years to double the Hg concentration. Actually the time required would be much longer as most Hg entering the ocean combines with sediments and is removed from further interaction with the open system. N.A.S. (1975) reported that the flows of mercury involving man are roughly equal to natural flows via rivers. Referring to Table II, the best estimates of the concentration of Hg in the Mississippi River lead to an input which over a 30-yr. period could increase the burden in the western shelf and open Gulf surface water to 0.1 and 0.3 ppb, respectively. These maximum estimated levels are considerably below toxic levels measured by Petrocelli et al. (1974). In a similar manner for Pb, input via the Mississippi for thirty years could not significantly raise the Pb burden in the Gulf of Mexico. Additional discussions on the effects of heavy metals in the Gulf of Mexico in the volume is provided by Trefry and Shokes (Chapter 4), in the Mediterranean by Bernhard (Chapter 3), and off the west and east coasts of the U.S.A. by Mearns (Chapter 1) and Gunnerson (Chapter 11), respectively.

In summary, DDT continues to be the most serious pollutant in the Gulf of Mexico, primarily because of its characteristic magnification through the food chain. C_6–C_{14} hydrocarbons may be important in the areas where they are discharged but, because of their volatility and only moderate stability, they are probably not important on a long-term basis. High-molecular-weight petroleum compounds may have important long-term, but as yet undiscovered sub-lethal effects. Heavy metals such as Hg and Pb do not seem to be serious pollutants in so far as the open Gulf of Mexico is concerned but may be important in the vicinity of industrial wastes and sewage outfalls.

ACKNOWLEDGEMENT

Preparation of this manuscript was made possible by NSF Grant OCE-80-02455.

CHAPTER 14

ENVIRONMENTAL IMPACTS OF DEEP-OCEAN MINING —
THE IMPORTANCE OF MANGANESE

Charles L. Morgan

INTRODUCTION

The mining and processing of manganese nodules involve the removal of
rather large quantities of Mn-oxides and other materials from the deep
ocean. When investigating the potential environmental impacts of these
activities, it is critical to understand as much as possible about the natural
processes which control the distribution of these materials. In particular,
the agents of transfer which are capable of moving more than $5 \cdot 10^5$ t of
dissolved and fine colloidal Mn into the North Eastern Tropical Pacific
(NETP) (lat. 5–20°N, long. 110–150°W) each year (Elderfield, 1976) must
be examined to identify and quantify the transport of this material through
the potentially affected ecosystems. For example, to determine how the
nodule pick-up operation will affect Pacific ecosystems (Fig. 1A), one must
identify the mechanisms which remove dissolved and particulate Mn from
the water column and the rates at which they operate. To establish the
optimum processing rejects disposal scheme for any particular process plant
site (see Fig. 13), one must examine the ways in which Mn is carried through
the relevant natural systems. One important class of ocean mining environmental impacts which is not considered in this approach is the direct and
indirect benthic impacts which result from the pick-up of manganese nodules
and the accompanying dispersal of sediments brought about by the pick-up
operation. These impacts are discussed in detail in this volume by Ozturgut
et al. in the next chapter (Ch. 15).

The particular processes which move Mn from place to place in aquatic
and marine environments have been the objects of research by limnologists,
oceanographers, Earth scientists, and others for many years (e.g., J. Murray
and Irvine, 1895; Kindle, 1932), and in the last ten to fifteen years we have
seen the production of a wealth of studies dealing with nearly all pertinent
aspects of Mn in natural systems (cf. Hirota, 1977a). Reasons for this long-
term and recently intense scientific concern are not difficult to deduce.
Transition metals play decisive though frequently complex roles in geological, oceanographic and ecological problems. Mn, at an approximate
concentration of 1000 ppm in the Earth's crust, is the third most common
transition metal (Mason, 1958). Moreover, the abundance and the oxidation–
reduction and oxide surface chemistry of this element make it potentially

capable of controlling the behavior of several other metals in natural aqueous inorganic systems. Since the recent emergence of widespread public concern about the ecological welfare of our lakes, rivers and oceans, these natural systems have been scrutinized in great detail. The results of some of these studies provide much useful information relevant to an evaluation of the environmental impact of ocean mining. The following discussion will explore several of the latest such studies in an attempt to identify the key natural processes which will be involved and to estimate the relative effects of mining activities upon the operation of these processes.

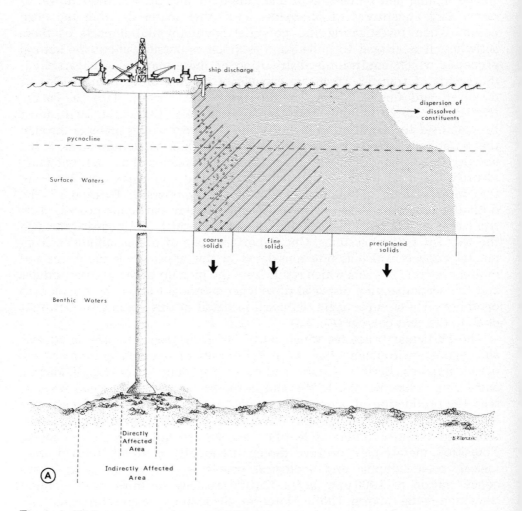

Fig. 1. A. Mining system environmental impacts.

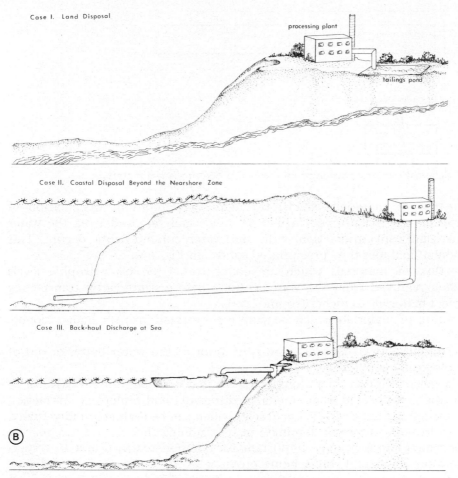

Case I. Land Disposal

processing plant

tailings pond

Case II. Coastal Disposal Beyond the Nearshore Zone

Case III. Back-haul Discharge at Sea

(B)

Fig. 1. B. Disposal options for process reject materials.

MASS BALANCE MODEL FOR REJECT MATERIALS

Problem definition

The natural solid and solution fluxes which remove reject products from mining and processing discharge sites (Fig. 1A and B, respectively) can be simplistically described for a steady-state condition by the six components depicted in Fig. 2. These components can be rigorously defined for any particular discharge location, and are generally defined as follows:

I = input flux of solid and dissolved reject materials.

A = flux of materials which rapidly settle out of the water column, interacting significantly with the receiving waters only through physical dispersion. This component includes the "coarse" and "fine" solids depicted in Fig. 1A.

418

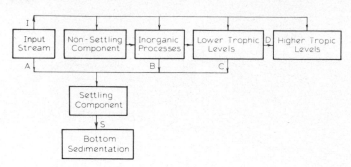

Fig. 2. Schematic representation of mining and processing reject materials.

B = flux of materials which, through coprecipitation, flocculation or adsorption, are transformed into inert solids and removed from the water before significant interaction with the water column biota occurs. This component includes the "precipitated solids" in Fig. 1A.

C = flux of materials which are concentrated by lower trophic levels and then removed from the water column before significantly transferring the reject materials to higher trophic levels.

D = flux of materials which become incorporated into the higher trophic levels.

S = output flux of materials onto the floor of the water body, or out of the biologically active zone.

This scheme has an operational value in that it enables the investigator to divide the complicated phenomena of dispersion and biological interaction into overlapping but distinct time frames which can be evaluated individually. Thus short-term processes dominate in determining flux A, and successively longer time periods become important for components B, C and D, respectively. More important, the scheme provides a very useful conceptual framework for dealing with these processes, by forcing the investigator to concentrate upon answering the key questions in proper sequence. Before the extent of biological interaction can be evaluated for any discharge (fluxes C and D), components A and B must be determined to arrive at an estimate of the amount of materials that will be available for subsequent interactions. All too frequently the task of environmental assessment is initially approached with a set of experiments that measure how various arbitrarily defined levels of discharge materials affect certain biota. Such experiments are premature before the appropriate concentration levels and chemical states of the rejects have been at least tentatively determined.

Computer model

As a first step in this approach, one can use a computer model to estimate the dispersal of solids and dissolved constituents through various well-defined systems. One such model has been developed by Brandsma and Divoky (1976) for the U.S. Army Corps of Engineers Dredged Materials Research Program, and is based upon the work of Koh and Chang (1973). This model includes provision for the settling out of particulate matter and is designed to allow generality of application to a wide variety of discharge situations. The model includes dynamics to describe discontinuous or continuous discharges and is divided into three successive phases:

(1) A jet (continuous) or instantaneous discharge (discontinuous) phase where momentum and buoyancy forces dominate the dispersal of materials.

(2) A dynamic collapse phase where neutral buoyancy is achieved and horizontal spreading dominates.

(3) A passive dispersion phase where the dispersion is calculated on the basis of a user-defined length (the effective turbulent diffusion coefficient is proportional to the $\frac{4}{3}$ power of this length (cf. Brandsma and Divoky, 1976, p. 18).

It must be emphasized that, at the present stage of development, such models cannot be used as accurate predictive tools. Calculation of such critical parameters as horizontal and vertical diffusion coefficients depend upon rather arbitrarily defined constant coefficients, and the water current and density structures are in most cases only crudely defined. Rather, the key value of such modelling is in providing guidelines for the detailed design of laboratory and field tests. The data obtained from such tests can then be used to greatly improve the validity of subsequent model calculations by providing ground truth data. An example of the usefulness of this approach is presented in Appendix A, which includes the description of a simulation of the Deep Ocean Mining Environmental Study (D.O.M.E.S.) monitoring of the "SEDCO 445" ocean mining test in 1978 (discussed in Chapter 15 by Ozturgut et al.). Though limited by the fact that some of the data necessary for direct comparison of the actual test with the simulation were not available, the exercise does show that the dispersion measured by the government scientists appears to be much greater than that predicted by the model (see Appendix A, Table A-II).

Important outputs of such simulations are the predicted concentration gradients for solid and dissolved rejects, when scaled up to the size of a full-scale system (see Appendix A, Fig. A-1). These estimates give a first indication of the expected ranges over which the various discharge components can exert any influence in the surrounding waters, and thus define the specific parameters which need to be subjected to further field and laboratory analysis. In this case, the need to evaluate the fate of Mn in the discharges from mining and processing operations is apparent from these initial estimates. The short-term dilutions of more than $1:10^3$ which are predicted will reduce the concentrations of most constituents of mining or processing reject materials to rather low levels soon after discharge. Since Mn is present in nodules and in most process rejects at levels which are more than an order of magnitude higher than most other constituents, it will be present in significant concentrations longer and at much greater distances than other components. Moreover, as discussed below, the presence of Mn-oxides in the reject materials may have significant effects upon the fates of other components (especially process rejects) and should thus be considered as a prerequisite for the adequate description of these discharges.

These simulations also emphasize the need to understand how the very fine-grained discharge solids behave in marine waters. Since the settling velocities of fine-grained solids are vanishingly small, there is presently no mechanism (other than dispersion) provided in the model to account for the fate of this component. As discussed subsequently (p. 428), there appear to be effective natural mechanisms which remove fine-grained solids from marine surface waters, but they have yet to be quantified to the point where they can be used in these modelling exercises. Subsequent sections in this chapter will examine these and other problems within the context of the representation given in Fig. 2, and will focus upon the role of manganese in determining components A, B and C for various discharge situations. The use of models for disperion studies in the respect to disposal of wastes on the east coast of the U.S.A. is discussed in this volume by Gunnerson (Chapter 11) and by Simpson et al. (Chapter 12). The basic principles of the role of particulates in the ocean is presented by Eisma (Chapter 9).

RELEVANT ASPECTS OF MANGANESE CHEMISTRY

Current popular literature which deals with environmental issues frequently contains terms such as "heavy metals" or "trace metals" to refer to any and all of the other fifty transition elements (lanthanides and actinides). This terminology is effective in lumping into one category a most diverse set, and is misleading in that it associates essential nutrients with universal toxins and common, inert materials with noxious pollutants. A review of some peculiarities of manganese chemistry is useful to distinguish this element from the other "heavy" metals and, more important, to provide guidelines for use in the development of accurate models of the natural transfer processes. Three aspects of aqueous manganese chemistry are very important: (1) oxidation–reduction behavior; (2) the adsorption chemistry of Mn(IV)-oxides in aqueous solutions containing other metals; and (3) the effects of dissolved organic species in the above reactions. These three are discussed in sections on relevant aspects of manganese chemistry.

Oxidation–reduction processes

In nature, Mn is usually found in the divalent, trivalent and tetravalent states. J. J. Morgan and Stumm (1970) have shown that in most natural systems, dissolved, trivalent manganese, Mn^{3+}, is not stable, and spontaneously reacts to form solid or dissolved divalent and tetravalent manganese species. Mn-oxides form from various combinations of these oxidation states, and are not generally stochiometric. In the laboratory, Bricker (1965) has shown that several distinct oxide phases can form from dissolved manganese, depending upon the specific conditions of pH and oxidation potential. These oxides are thermodynamically stable with respect to most dissolved forms, but they precipitate quickly only at pH-values of 8.5 or higher. Of particular importance here is the fact that oxidation is autocatalytic, i.e. the presence of the oxide surface greatly increases the precipitation rate from solution (J. J. Morgan and Stumm, 1970, p. 534). Hem (1963) has discovered that the oxidation is also catalyzed by other solids, such as silicate sand, though the autocatalytic precipitation is generally more effective.

Adsorption chemistry

The ability of Mn-oxides to remove transition metals from solution by means of surface adsorption is well documented and is used as the basis of several mechanisms which have been proposed to account for the metal enrichment observed in manganese nodule deposits (Goldberg, 1954; Burns and Brown, 1972; see also the section "Studies in natural systems"). This adsorption capacity is very important to a consideration of process reject disposal, since process rejects may contain significant concentrations of dissolved metals.

When a solid surface interacts with simple dissolved, complexed, or colloidal metal species in seawater the principal factor which determines the relative stabilities of the bonded and unbonded species *initially*, is the degree to which the surface-to-metal species bond can displace the surface-to-water bond. The surface can be defined one way using an electrostatic model such as the Gouy theory or the Stern model (W. J. Moore, 1962, p. 75), which balances surface electric charge against molecular diffusion and defines the chemically enhanced zone (the diffuse double layer) around the surface as the region where there is a majority of either positive or negative charges. The surface can also be defined on a finer scale (since these theories do not apply to the actual surface–solution interface but only to the adjacent volume), using molecular models of the solid bonding to water or hydroxyl groups and evaluating the process as one of ligand exchange by the adsorbed metal. It is necessary to employ both molecular and electrostatic models to derive a preliminary picture of how surface adsorption must proceed on the natural surfaces. Also, since natural processes often are controlled by the kinetics of surface reactions rather than the thermodynamic states of the surface, the predictions from these equilibrium models must be applied to natural systems only with great care.

Several laboratory experiments have been performed to measure the effectiveness of manganese oxides in removing various metals from solution.

D. J. Murray et al. (1967) examined the adsorption characteristics of hydrous Mn-oxides with respect to Ni^{2+}, Cu^{2+} and Co^{2+}. They have concluded the following from these analyses:

(1) At low concentrations these cations are adsorbed into the diffuse double layer, replacing protons.

(2) At higher concentrations, pH-independent sorption takes place, indicating that the cations are in some way being incorporated in the MnO_2 matrix.

(3) Co^{2+} is adsorbed to a higher degree than the other two cations. J. W. Murray (1973, 1974, 1975) reported that the sorption capacity of Fe-hydroxides for Co^{2+} is much lower than that of the Mn-oxides analyzed by D. J. Murray et al. (1967).

It is clear from these and other studies (e.g., Parks, 1967) that, in most natural aquatic systems, Mn-oxides will be active in removing dissolved metals from solution at significant rates. However, the net effectiveness of this factor in determining the over-all characteristics of any discharge must be measured separately for every specific situation, since this adsorptive capacity is dependent upon the highly variable ambient concentrations of all species in solution as well as the states of the oxide surfaces.

Biological interactions

As discussed below, the potential interactions between dissolved and solid manganese phases and organic compounds are extremely important to the issues addressed here. Recent research makes it increasingly clear that interactions between transition metals and organic matter play a fundamental role in natural transport processes. For manganese, the following factors are relevant:

(a) The affinity of Mn^{2+} for organic ligands containing nitrogen is summarized by the well-known Irving–Williams order for divalent metal ions: $Mn^{2+} < Fe^{2+} < Co^{2+} < Ni^{2+} < Cu^{2+} < Zn^{2+}$ (cf. Cotton and Wilkinson, 1972, p. 576). In addition, equilibrium calculations performed by Morel and Morgan (1968) upon a system of nine metals and nine organic ligands show that Mn^{2+} is less susceptible to organic chelation by a variety of ligands than any of the other transition metals in the study and behaves in a manner similar to Ca, Mg and Sr. Table I, computed from the results of this computer study, summarizes these conclusions.

(b) The rates of manganese oxidation are lowered by the presence of certain organic ligands such as EDTA or pyrophosphate (J. J. Morgan and Stumm, 1970, p. 294). The causes of this reaction inhibition are not known.

(c) The rates of manganese oxidation may be greatly accelerated in the proximity of photosynthesizing organisms. As discussed above, this reaction is extremely pH-dependent. The photosynthetic reaction, when operating

TABLE I

Net susceptibility to organic ligands as (uncomplexed metal concentration)/(total metal concentration) ratios [computed from Morel and Morgan (1968)]

Ca	0.933	Mn(II)	0.933	Cu(II)	0.006
Mg	0.933	Fe(III)	0	Zn(II)	0.501
Sr	0.933	Ni(II)	0.158	Cd(II)	0.692

in waters which are buffered by carbonates, consumes protons as follows (Brock, 1966, pp. 10 and 177):

$$H^+ + HCO_3^- \rightarrow CO_{2\,(dissolved)} + H_2O$$

$$6CO_{2\,(dissolved)} + 6H_2O \rightarrow C_6H_{12}O + 6O_2$$

Thus, in the solution immediately surrounding photosynthesizing organisms, a local volume of relatively high pH can be expected; Mn^{2+} which comes into this zone will be much more likely to oxidize and provide a substrate for subsequent autocatalytic oxidation.

Potential relevance to natural systems

Much of the chemistry discussed above is strictly valid only for well-defined laboratory systems, but it does suggest the following guidelines potentially applicable to the natural systems:

(a) In most aquatic and marine environments, the most stable forms of manganese are the Mn(IV)-oxides. Dissolved Mn^{2+} can persist for long periods of time, but in oxygenated waters where solid surfaces are present (especially Mn(IV)-oxide surfaces) or where photosynthesis occurs, Mn^{2+} may precipitate quite rapidly. Conversely, once formed, these oxides probably would not be capable of dissolving back into natural waters unless anaerobic conditions occur.

(b) Mn^{2+} has a low affinity for most organic ligands, and its oxidation is inhibited by their presence. One would, therefore, not expect Mn to be concentrated in living organisms like metals such as Hg or Cd (i.e. in Fig. 2, fluxes C and D should be very small).

(c) The adsorptive capability of Mn(IV)-oxides is significant, and in natural systems may be responsible for removing many other metals. Thus, the dispersal of manganese nodule particles in natural waters may result in a net decrease in the dissolved metal content in these waters rather than an increase (i.e. in Fig. 2, flux components A and B should be dominant).

These predictions and others have been under investigation for several years by oceanographers, limnologists and others, and the results of a number of these studies, discussed below, generally support these laboratory predictions with actual field observations.

PROCESS REJECT DISPOSAL ON LAND

Assessment of the environmental impact of processing plants can only be done on a case-by-case basis, as the impact is entirely dependent upon the particular site chosen and the particular discharge composition which will be produced. However, freshwater systems will be involved in some manner whenever the material is not discharged directly into ocean waters (Case I, Fig. 1B). For this situation some relevant observations can be made which are independent of the actual site chosen. These observations are derived from the general composition of the tailings material and the previously noted chemical characteristics of Mn.

Studies in natural systems

Major constituents of all process rejects will be sea salts (NaCl, $MgCl_2$, etc.), deep-ocean silicate sediments, and variable amounts of inert and soluble species including possibly some process reagents (cf. Dames & Moore, Inc. and E.I.C. Corp., 1977). If only Ni, Cu and Co are extracted from the ore (as is the case for most of the proposed processes), then large amounts of Mn will comprise the majority of the non-aqueous discharge. The particular manganese phases present will vary, but Mn(IV)-oxides will probably be the major solid component for all three-metal process types. If the pH of the reject solid–solution mixture is adjusted to normal freshwater levels (generally $6 < pH < 9$), then, as discussed above, the well-known sorptive capacity of Mn-oxides will be available to remove remaining Mn and other metals from solution.

The effectiveness of Mn-oxides in the removal of several metals in natural systems has been recently confirmed by several workers. Callender and Bowser (1976) show that in many freshwater deposits relatively high concentrations of Cu, Ni, Co, Zn and other metals are clearly associated with authigenic Mn-oxides on the surfaces of several lake bottoms. These deposits form whenever natural waters flow through the appropriate gradients of pH and oxidation potential, such as when relatively acidic river waters flow into a relatively alkaline lake or when aquifer-driven or compaction-driven flows of oxygen-depleted groundwaters surface through the sediments into a highly oxidizing body of water. These and other similar situations all result in extensive removal of metal cations from natural waters and onto Mn-oxides and suggest that an analogous process would occur when process rejects are exposed to similar conditions. Means et al. (1978) have demonstrated a clear association of Mn-oxides with Co and also with various actinides in soils and river sediments. These researchers suggest that mixing of Mn-oxides with independent discharges of toxic metals may result in the effective removal of the toxic elements from potential interactions with biota.

Recent U.S. government-sponsored studies

Four recent studies have been produced which are directly relevant to the issue of disposal of process reject materials on land. One, conducted for the National Oceanographic and Atmospheric Administration (N.O.A.A.) by the Dames & Moore, Inc. and E.I.C. Corp. (1977), discusses six potential processes and, on the basis of these, attempts to characterize the land, resource, personnel requirements and the reject materials' discharge characteristics for the different process types. One key conclusion from this study is: if land disposal is chosen, then complete containment of the discharge solution and solids would be required in many locations. This conclusion is based in part upon the significant amounts of sea salts produced, which could clearly do harm to freshwater rivers and lakes if discharged in sufficent quantity to significantly affect the natural salinities,

Another study (Stephen-Hassard, 1978) discusses the specific effects of locating a processing plant on the Island of Hawaii. This effort evaluates the possible beneficial uses for the process rejects in re-claiming barren lava fields for agricultural and other uses, and concludes that these uses may indeed provide a positive and productive alternative to other possible methods of disposal. In this chapter, the key emphasis is placed upon the removal of toxic metals such as Co, Cu and Zn from the dissolved or "reactive" component of the rejects as the main problem to solve. Two other geographically specific workshop studies of process plant siting have been sponsored by N.O.A.A. and assembled for the Gulf of Mexico coast (Bragg, 1978) and for the west coast (Neal, 1977) of the U.S.A. Both of these studies concur with the Dames & Moore, Inc. and E.I.C. report by finding that land disposal would probably necessitate either extensive treatment or "complete containment" where solutions are removed only by evaporation. However, the Gulf coast study does identify a number of potential coastal sites for such activities, should the above-mentioned problems be solved.

Summary — Land disposal of process rejects

An adequate determination of the environmental impact of land disposal of process rejects will be possible only in most specific terms, and will be unique for each combination of process site and process technique. Two components which are common to the rejects from all three-metal (Cu, Co and Ni) processes and which must be considered in an environmental impact analysis are: (1) soluble sea salts such as NaCl and $MgCl_2$ which, at some state of the process, must be discharged in significant quantities; and (2) toxic metals (e.g., Cu, Hg, Co, or Ni), which must be effectively removed from any potential reactions with biota.

Disposal of soluble sea salts could constitute a problem if ocean disposal is prohibited, and would require complete isolation from freshwater systems. Toxic elements are found at low levels in manganese nodules and would be brought into solution by most process methods. Removal of these elements from solution could be accomplished through treatment with a base to bring the pH to within natural levels. Subsequent to this initial treatment, the metals would probably be inert in normal oxic environments. Due to the strong adsorptive characteristics of Mn-oxides in the reject materials could fill a beneficial role by removing other inputs of toxic metals.

EFFECTS UPON MARINE ECOSYSTEMS

Ocean mining can potentially interact with marine ecosystems through the mining operations (Fig. 1A) and, if ocean disposal of process rejects is considered either as a method employed in land processing (Fig. 1B, Cases II and III), or from an at-sea process, through discharge of reject materials. Both types of activities involve the transport of large quantities of Mn through the water column (except processes which remove Mn), and thus a consideration of the natural transport processes which control the distribution of this metal in the marine environment is a necessary first step in the attempt to predict potential environmental impacts. Such a treatment, presented below, is useful in putting the problems involved into a rational perspective and in identifying the key mechanisms which should be evaluated in detail for future environmental impact assessment. The manganese nodule province which will be mined initially is located in the heart of the world's largest ocean, the Pacific. Comparisons of mining activities to nearly any of the natural processes occurring in the NETP (Welling, 1972) quickly show that mining operations will be extremely small within the realms of this enormous expanse. (See Appendix B for examples.)

Arguments such as these set limits upon the temporal and spatial perturbations which can be expected, and give one a high confidence that environmental impacts will be insignificant or easily avoided. Nevertheless, the nearly global extent of this environment also makes it crucial that all potential impacts be identified and monitored. This effort must be continuous throughout the development and operation phases of a mining enterprise, and must be directed toward: (1) the establishment of a complete description of mass flows through the natural systems; and (2) basic understanding of all biological interactions. The present state of competence in oceanographic measurement techniques gives us the ability to determine a host of physical, chemical and biological parameters with unprecedented accuracy; application of these and future scientific tools to this effort will generate a vast amount of information relevant to nearly all aspects of deep-ocean environments. The role of Mn is discussed below in an attempt to further clarify the magnitudes of mining and environmental impacts and to identify an initial focus for environmental monitoring efforts.

Geochemical cycling of dissolved and fine-grained manganese through the oceans

A comprehensive mass balance model for describing the distribution of 65 elements among the sediments and sedimentary rocks of the Earth has been developed by Horn and Adams (1966). This model equates the sum of all sediment components to a certain mass of igneous rock for each element. Published data for the compositions and abundances of various rock types are used in an interative least-squares computer calculation sequence to determine the most consistent set of elemental compositions and abundances and the mass of igneous rocks which have been converted to sediments (estimated $2 \cdot 10^{18}$ t). Using this technique, Horn and Adams were able to balance nearly all elements between the igneous and sedimentary categories. Mn, however, exhibited a fairly large surplus in sedimentary rocks, and could not be brought into accord with the over-all balance. From Table II, it appears that almost all of the Mn has been accumulating in oceanic sediments, and is predominantly found in the hemipelagic or shallow marine deposits.

One explanation to account for this surplus of Mn ($22.9 \cdot 10^{14}$ t) can be inferred from the work of Lynn and Bonatti (1965). This study demonstrates the possibility of diagenetic (compaction-driven) re-mobilization of Mn in sediments where anoxic conditions persist, leading to a net depletion of Mn in buried sediments and a net enrichment of this metal at the sediment surface (when oxygenated bottom water is present). This

TABLE II

Model of manganese distribution in various rock types (after Horn and Adams, 1966)

Rock type	Continental shield	Mobile belt shelf	Hemi-pelagic	Pelagic	Weathered igneous rock	$A + B + C + D - E$
Total mass manganese $(10^{-20} g)$	0.573 (A)	1.44 (B)	27.2 (C)	16.1 (D)	22.4 (E)	22.9
Abundance of manganese (ppm)	550	497	6,040	5,820	937	

process could lead to a large overestimate of Mn abundance if the composition of the entire sediment mass is derived from analyses of surface samples (as is the case for pelagic and hemipelagic sediments). Bender (1971) has shown that diagenetic migration of Mn cannot occur in most pelagic sediments, but Bender's arguments do not apply to many hemipelagic sediments.

Elderfield (1976) identifies several independent lines of reasoning which also predict that the dissolved and colloidal Mn present in oceanic deposits is more than can be explained by weathering of igneous rocks, and concludes that submarine volcanism must contribute significant amounts of Mn to these sediments. Boudreau and Scott (1978) show that, based upon Elderfield's estimates of the over-all mean flux rate of Mn to oceanic sediments ($\sim 10^{-6} \, g \, cm^{-2} \, yr.^{-1}$), manganese nodules are growing at rates which are limited only by the rate of diffusion of Mn^{2+} to the growing surfaces. These workers demonstrate further that these diffusion-limited growth rates agree quite well with published rates calculated using radiometric dating techniques.

Though the quantitative description of natural manganese transport through the oceans is by no means complete, these studies support the following conclusions:

(1) The natural flux of dissolved and colloidal manganese into marine sediments in the North Pacific is $10^{-6} \, g \, cm^{-2} \, yr.^{-1}$.

(2) Much of this Mn comes from river water, but a significant amount must also come from submarine volcanic sources.

(3) The surplus of Mn in marine sediments (indicated by the model of Horn and Adams) can be explained in part by diagenetic enrichment of Mn in the surficial sediments of hemipelagic deposits and partly by submarine volcanism.

Mechanisms which remove manganese from seawater

The above geochemical studies give general guidelines regarding the overall transport rates through and sources of Mn in the oceans; they do not identify the mechanisms through which these transport processes operate. For deep-water deposits at least, most workers currently support the contention that Mn, as well as several other metals, is removed from the water column by living planktonic organisms. Some recent evidence for this proposal is described as follows. For many years it has been known that the surficial sediments in the deep Pacific accurately reflect the biological activity which occurs in the surface waters above them, and that zones of high surface productivity are immediately above sediments with components of biogenic material of the same species composition (Menard, 1964, pp. 153–169). These well-defined zones are somewhat puzzling in one respect. The biological matter which makes up these sediments is generally composed of rather small organisms, including significant and sometimes dominant percentages of organisms which are smaller than $40\,\mu m$ (J. P. Riley and Chester, 1971, p. 263). According to Stokes' law of settling, particles this small should take months to sink the thousands of meters to the ocean floor, and should thus be widely dispersed before coming to rest.

One mechanism which can account for this apparent discrepancy involves the concentration of the small particles into relatively large fecal pellets (several hundred micrometers) by zooplankton and other larger consumers prior to settling out. Boström et al. (1978) using a computer model similar to the one developed by Horn and Adams but designed specifically to examine the various components of pelagic sediments, show that virtually all of the Cu, Ni and Fe in authigenic deposits of the NETP are clearly associated with the accumulation of biological matter. Thus, biological processes in the deep ocean appear to act as a very efficient filtering system which removes fine solid material from the water column by concentration into large fecal pellets. This mechanism can account for virtually all authigenic deposits of certain key metals in the NETP.

CONCLUSIONS

On the basis of the above considerations, it is possible to make some general conclusions about the environmental effects of ocean mining. These conclusions are summarized below for freshwater and marine systems. They are organized in accordance with the mass balance model described in the section discussing mass balance models of reject materials.

Freshwater systems

I: For three-metal processes, the input flux will consist of Mn-oxides, deep-sea clay, and various metals and soluble sea salts in solution.

$A + B + C = S$ (Fig. 2): Virtually all solids and dissolved metals can be removed from the rejects stream by conventional treatments with alkaline materials such as lime, followed by a residence time in settling ponds. The strong sorptive characteristics of Mn-oxides will act to insure that toxic metals will not be released during the initial solid–solution separation or after the settled solids are returned for other land uses. In fact, if these solid rejects are used in oxic soils, they may act to actually reduce the metal content of waters which come into contact with them.

D: A principal component not removed from solution by lime-type treatments will consist of soluble sea salts. If ocean dumping is not permitted, special precautions must be employed to insure that these materials do not endanger freshwater ecosystems.

Marine systems — Mining operations

I: Nodule pick-up operations will discharge an undetermined amount of deep-sea sediment and fine-grained manganese nodule particles along with benthic water brought up in the operation.

A: On the basis of current government monitoring of test mining operations (see Chapter 15 by Ozturgut et al.) it is apparent that virtually all the input solids settle out of the surface waters within a few hours. Since these materials are in complete equilibrium with benthic waters, we expect them to be completely inert after they are re-deposited upon the ocean floor.

B: This component will be extremely small since the discharge materials are essentially in equilibrium with seawater. However, it is possible that a local depletion in metals will result in the nearby waters, due to absorption onto Mn-oxides.

C: Extremely fine-grained solids will remain in the water column for long periods of time unless they become physically concentrated by some biological mechanism. When this happens, the relative lack of affinity for organic ligands which is characteristic of Mn suggests that this material will be quickly removed from the water column in the same ways that naturally-derived Mn is removed.

D: Since most chemical and physical processes predict the rapid removal of these materials from biological systems, it is extremely unlikely that upper trophic levels will be affected by the discharge to any meaningful extent.

S: Based upon the known characteristics of the discharge materials and preliminary monitoring of test mining operations, we expect virtually all of the discharge materials to settle out of the receiving waters before any significant biological interactions occur, that is,

A + B = S: It is possible that discharged solids will be slowed down at the pycnocline (Fig. 1A) on the trip to the ocean floor. If the resulting accumulation at the pycnocline is proved to be detrimental, it will be possible to discharge these materials below this level. However, it is important to clearly prove that discharge at a depth below the pycnocline is necessary, since a deep-water discharge will not be able to take advantage of the high mixing rates and biological filtering mechanisms which are available in surface waters.

Marine systems — Process reject materials discharge

I: A full-scale mining operation ($5000\,t\,day^{-1}$) will remove $\sim 10^{12}\,g$ Mn from the ocean floor each year. In three-metal processes this Mn will be the largest component of the reject materials (treatment of the rejects with a base is assumed to consume excess acid before discharge). This Mn will probably be in the form of $Mn(IV)$-oxides. In addition, $2 \cdot 10^{10}\,g$ of soluble salts, and variable amounts of soluble process reagents will be produced.

A: Since the particle-size distribution of reject solids will be variable from one process to the next, it is not possible to generally predict the percentage of material which will rapidly settle out. However, since solid—liquid separations are required in most processes, it will probably be to the advantage of the processor to minimize the fine-grained component of the solids. Since the ocean sediments contain nearly all of the Mn in the Earth's sedimentary environments (Table II), the most appropriate disposal site may indeed be in this location. For a coastal slurry discharge (Fig. 1B, Case II), it is possible that the discharged Mn may participate in subsequent diagenetic activity along with the natural Mn. For a deep-water discharge (Case III), the solids will probably remain inert.

B: Flocculation, coprecipitation, and specific chemical adsorption will probably make this a significant component of the output flow. Dissolved Mn and other metals will rapidly be removed by these mechanisms in all oxygenated marine waters. The specific rates at which these reactions occur will be functions of the particle-size distribution, the turbulence, dispersion rate and other parameters, and will be determined only after extensive laboratory and field testing.

C: Fine collodial and dissolved metals which are not immediately brought out of solution during the initial mixture with seawater will be suceptible to: (1) oxidation in the vicinity of photosynthesizing organisms; and (2) concentration and subsequence precipitation in fecal pellets. Since these processes in turn lead to additional (surface-active) adsorption, this component may be self-limiting in its dispersion.

D: As with mining reject materials, the known natural processes will act to remove reject materials rather than to pass them on to higher trophic levels. Though preliminary estimates indicate that this component will be insignificant and unmeasurable, it is important to include tests to measure the concentrations of metals in the higher trophic levels as a part of the monitoring program during test and development phases.

S: Components *A*, *B* and *C* are expected to settle rapidly out of surface waters through a combination of simple settling and rapid chemical and biological concentration followed by settling. To optimize the chemical and biological removal processes for deep-water (Case III) disposal, discharge in surface waters (above the pycnocline) is preferable. However, as stated above, the discharge depth can be altered if it becomes more advantageous to do so. For Case II discharges, the optimum discharge depth will depend upon the specific site characteristics.

Pertinent detailed information is further presented in this volume by Eisma (Chapter 9); and comparable data are applied to dredge spoils discussed by Herbich (Chapters 6 and 7). Similarly, the results of model studies (see Appendix A) on dispersal of wastes are described by Gunnerson (Chapter 11) and by Simpson et al. (Chapter 12).

432

ACKNOWLEDGEMENTS

The following individuals have been indispensable in the formulation of this chapter: Mr. Conrad G. Welling, for providing the inspiration and management genius behind every aspect of this ocean mining program; Mr. Don Hamm, for surmounting the convolutions of the U.S. Army Corps of Engineers Waterways Experiment Station computer model and modifying it for our application; and Ms. Barbara Pilarcyzk, for giving freely of her time to produce useful and well-constructed illustrations.

The American Mining Congress Undersea Mineral Resources Committee Ad Hoc Working Group on Environmental Matters, and specifically the Working Group Chairman Mr. Ray Kaufman, have been extremely instrumental in the development and review of far-sighted and rigorous programs dedicated to the understanding and control of the environmental impacts of ocean mining. Many of the ideas and needed support in this effort have come from this group of individuals.

APPENDIX A — CASE *1*: A SIMULATION OF THE OCEAN MANAGEMENT, Inc., "SEDCO *445*" OCEAN MINING TEST, APRIL—MAY 1978

(A) Description of the model inputs

(1) Physical oceanography. The density structure (Table A-I) is taken from the D.O.M.E.S. data from the April–May monitoring of the "SEDCO *445*" nodule test. The surface (above the pycnocline) current of 8 cm s^{-1} and subsurface (below the pycnocline) current of 4 cm s^{-1} are estimated from D.O.M.E.S. data given in Halpern (1979).

TABLE A-I

Model surface water density structure

Depth (m)	Density $(g \, cm^{-3})$	Depth (m)	Density $(g \, cm^{-3})$
0–65	1.021	80–130	1.024
65–75	1.022	130–200	1.025
75–80	1.023		

(2) Mining rejects composition. In the absence of any reliable estimates of the fall velocity profile of the solids from the "SEDCO *445*" tests, a range of three was picked to represent 1.0-μm ($9.08 \cdot 10^{-5} \text{ cm s}^{-1}$), 25-$\mu$m ($5.68 \cdot 10^{-2} \text{ cm s}^{-1}$) and 100-$\mu$m ($0.98 \text{ cm s}^{-1}$) diameter sizes. The concentrations of these three (arbitrarily chosen to sum to the "SEDCO *445*" high estimate of 22 g l^{-1}) are 1.10, 12.2 and 8.6 g l^{-1}, respectively. Only the results for the 1.0-μm size components are discussed here since no data relevant to the rapidly-settling fractions were collected in the monitoring study.

(3) Discharge configuration. The discharge vessel is stationary, positioned in the center of

the upstream side of the computing grid. The discharge pipe has a diameter of $\sim 46\,cm$ (1.5 ft.), and is directed downward, discharging continuously throughout the simulation, at a rate of $159.41\,s^{-1}$ ($5.63\,ft.^3\,s^{-1}$) (also from the "SEDCO 445" data).

(4) *Calculation coefficients.* These are, with one exception, the default values provided with the model. The horizontal dispersion parameter, A_λ has a default value of 0.005, supposedly characteristic of turbulent estuarine conditions. For this simulation, a value of 0.002 was chosen (Brandsma and Divoky, 1976, p. 18).

(B) *Description of model outputs — Dispersion of fine particulates*

Table A-II shows how the dispersion of the 1.0-μm component of the simulation compares with two of the surface transects measured by the D.O.M.E.S. scientists. The results are normalized to a peak value of one in all cases. Absolute comparisons are not possible without the collection of the additional data necessary to completely characterize the actual discharge flux from the test vessel. The key difference in both comparisons is that the field tests indicate that a significantly greater dispersion is occurring in the field than is predicted by the simulation. This result is the usual one which is observed in most model calculations, and has not yet been satisfactorily explained. Increasing the value of the dispersion parameter, A_λ, is not an adequate solution to this problem, because an increase which is sufficient to reconcile the model with the field data for the 2.1-hr. transect, is too large an increase to account for the shape of the 4.8-hr. transect. The disagreement may be resolved by modification of the jet and dynamic collapse phases as well as the passive dispersion phase, but such modifications require data which describe the very short-term (1–200 s after discharge) characteristics of the discharge plume. On the other hand, the linearity of the response of the nephelometer with rising particle concentrations is not well established, and possible nonlinearities could also explain the different results.

TABLE A-II

Discharge plume description after 2.1 hr. for Cases *1* and *2*

Perpendicular distance from plume axis (m)	Relative model concentration	Relative nephelometer concentration[*]
Case 1:		
+ 400	0	0
+ 200	0.34	0.33
0	1.0	1.0
− 200	0.34	0.67
− 400	0	0.05
Case 2:		
+ 400	0.01	0
+ 200	0.17	0.67
0	1.0	1.0
− 200	0.17	0.67
− 400	0.01	0

[*] Measured from Burns et al. (1980, fig. 4.5, transect H).

Thus, this application of the model has resulted in at least three valuable conclusions:

(1) Presently the model outputs are apparently low in predicting the actual dispersion of suspended fine-grained particulate matter.

(2) An adequate understanding of the processes which operate in dispersing materials in open waters can be achieved only after the dynamics of the very short-term, momentum-driven phases of dispersion are measured in field tests.

(3) Accurate nephelometer calibration is an essential element of any field plan using this device to track suspended solids.

CASE 2: SIMULATION OF A 5000-t day^{-1} DISCHARGE OF PROCESS REJECTS

(A) Description of the model inputs

The oceanographic parameters and the dispersion parameter, A_λ, are the same as in Case 1. The inputs for the discharge have been raised to the level of production of a full-scale mining operation, ~ 5000 t day^{-1}, to represent the baseline case of the effluent from a one-mining system processing plant. The discharge configuration is similarly scaled up to a flow rate of 1160 l s^{-1} (41 ft.^{3} s^{-1}) through a $\sim$ 91.5-cm (3-ft.) diameter pipe.

(B) Description of model outputs — Dilution of fluids

A key output from this particular simulation is an estimate of how the dilution of fluids will reduce the concentration of dissolved materials in the discharge. This information is summarized in Fig. A-1.

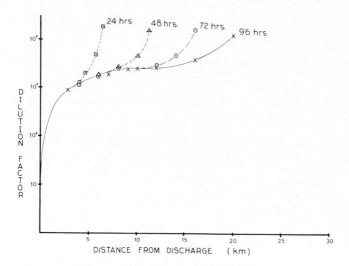

Fig. A-1. Dilution of reject fluids for Case 2.

Of prime importance here is the fact that the jet phase of the discharge provides an almost instantaneous dilution of more than 1000, regardless of how long the discharge has been operating. This "instantaneous" dilution must be scrutinized carefully for any particular discharge to optimize the actual near-term gradients. The jet phase dilution can be modified by the geometry of the outfall (pipe diameter and discharge angle) and the characteristics of the discharge itself (such as density and solids content). The optimum configuration for any particular discharge will not necessarily be the one which maximizes dilution, since processes which are often environmentally desirable, such as coagulation of particles, adsorption of dissolved species, and rapid sinkage of dense fluid plumes, are favored by relatively high concentrations of reject components.

Thus, this large-scale simulation reinforces the need to focus upon the near-field, short-term processes which occur when the discharge is introduced to the receiving waters. That is, flux components A and B (Fig. 2) must be clearly understood in order to adequately predict the environmental consequences of the discharge.

APPENDIX B — A COMPARISON OF OCEAN MINING vs. NATURAL EFFECTS ON THE MARINE ENVIRONMENT

The following set of comparisons illustrates the scale of the impacts on natural systems which may be caused by ocean mining. The Clarion–Clipperton zone of active mining is defined as the area between lat. $5°$ and $20°N$, and long. $110°$ and $160°W$. One mining operation is assumed to discharge 5000 t of material into the marine environment (roughly equivalent to the sum of mining and process rejects). Calculations and references are listed underneath each comparison. The object is to put the environmental impact of ocean mining into a perspective which can be readily understood by someone who is not familiar with the sizes and rates which typify oceanographic processes.

(a) Between latitudes $5°N$ and $20°N$, the Pacific Equatorial Counter Current transports about $3 \cdot 10^{16}$ m^3 of water per day to the east. By comparison, all the rivers of Asia transport about $3 \cdot 10^{10}$ m^3 day^{-1}. Thus discharging 5000 tons of materials per day into the Pacific Counter Current is analogous to discharging 5 kg (11 lb.) of materials per day, distributed through all the rivers of Asia.

Calculation: $5000 \times 3 \cdot 10^{10} \times 1000$ (kg t^{-1}) $: 3 \cdot 10^{16} = 5$ (kg).
References: Kendall (1970) and D. F. Martin (1970).

(b) The clearest water in the Atlantic Ocean is found about 965 km seaward from the mouth of the Amazon River. Visibly turbid water can be found ~ 80 km from shore. The Amazon carries a load of fine material which averages $\sim 3 \cdot 10^6$ t day^{-1}. By analogy, a mining system which discharges ~ 5000 t day^{-1} into the sea would be undetectable by any means ~ 1.5 km from the discharge, and would be invisible 150 m from the discharge.

Calculations: $600 \times 5000 : 3 \cdot 10^6 = 1$ (mi.) $= \sim 1.5$ (km); and
$\quad\quad\quad 50 \times 5000 : 3 \cdot 10^6 - 0.08$ (mi.) $= 500$ (ft.) $= 150$ (m).
Reference: Gibbs (1974).

(c) Mn is accumulating naturally in the sediments of the Clarion–Clipperton zone at a rate of $\sim 6 \cdot 10^5$ t day^{-1}. Twenty mining operations working simultaneously can displace $\sim 5\%$ of this amount of Mn.

Calculation: $5000 \times 0.3 \times 20 : 6 \cdot 10^5 = 5$ (%).

(d) Twenty 5000-t-day^{-1} mining operations working constantly for twenty years over nodule deposits with abundances of 10 kg m^{-2}, will mine an area which covers between 0.6 and 3% of the Clarion–Clipperton zone, depending upon the mining efficiency.

Calculation: 5000×1000 (kg t^{-1}) $\times 10^{-1}$ (m^2 kg^{-1}) $\times 365 \times 10^{-6}$ (km^2 m^{-2}) $\times 20 :$
$\quad\quad\quad 12 \cdot 10^6$ (km^2) $= 0.6$ (%) with 100% efficiency, 3 (%) with 20% efficiency.

Reference: Ryan and Heezen (1976).

(e) The rivers of the world discharge $\sim 10^7$ t day^{-1} of dissolved metals into the world's oceans. Assuming that all the discharge solids (5000 t day^{-1}) can be put into solution it would take <u>2000</u> mining operations to put an equivalent amount of metals into the sea.

Calculation: $10^7 : 5000 = 2000$.
Reference: D. F. Martin (1970).

(f) Atmospheric fallout of Hg into the oceans occurs normally at a rate of ~ 168 t day^{-1}. Twenty operations combined will be discharging a maximum of <u>5</u> kg Hg day^{-1}.

Calculation: $20 \times 5000 \times 5 \cdot 10^{-8}$ (g Hg per g ore) $\times 10^3$ (kg t^{-1}) $= 5$ (kg).
References: Goldberg (1971) and Toth (1977).

CHAPTER 15

IMPACTS OF MANGANESE NODULE MINING ON THE ENVIRONMENT:
RESULTS FROM PILOT-SCALE MINING TESTS IN THE NORTH
EQUATORIAL PACIFIC

E. Ozturgut, J. W. Lavelle and R. E. Burns

INTRODUCTION

Commercial deep-sea mining of manganese nodules in the North Equatorial
Pacific is scheduled to begin in the 1980's. The mining system involved will
gather nodules from the sea floor with towed or self-propelled collection
devices, and will draw them through a pipe to the surface. On board the
mining vessel, the material brought from the depths will be winnowed of
nodules; the nodules will be conveyed to a transport vessel, and the residual
mixture of bottom water, pelagic silts and clays, nodule fragments and
benthic biota will be discharged at or near-surface depths.

On the ocean floor, the mining collector will scour the surficial layer of
sediment along the track. Fine bottom sediments will be suspended, and a
benthic plume is created. At the surface, the discharge mixture will create
both a dissolved and a particulate plume. Since both the benthic and surface
plumes may affect the biota, the extent and effect of mining plumes will
require assessment.

The Deep Ocean Mining Environmental Study (D.O.M.E.S.) was initiated
in 1975 to address environmental impact questions so that environmental
guidelines for mining could be developed before full-scale mining began. The
efforts have drawn on the experiences and problem identification of Roels et
al. (1973), who had earlier looked at the initial prototype mining activity of
Deepsea Ventures, Inc. on the Blake Plateau. In the first three years of the
D.O.M.E.S. program, the efforts concentrated on: determining pre-mining
environmental conditions, to which mining perturbations could later be compared; and on forecasting, with the limited data available, both the fate and
effects of the benthic and surface plumes. The results of the site surveys, in
terms of the spatial and temporal variability of oceanographic conditions,
are found in Ohman et al. (1979); the identification of specific processes
leading to impact and conceptual and mathematical models of the fate and
effects of mining plumes are discussed by Ozturgut et al. (1978).

In 1978, two successful mining tests of engineering feasibility were conducted and completed in regions where commercial scale mining is likely to
occur first. Those tests provided the first opportunity to observe mining
operations such as those envisioned for the next decade, and they allowed

438

comparisons of earlier estimates of mining perturbations with actuality. Each test saw hundreds of metric tons of manganese nodules brought to the surface from a depth of ~ 5000 m, which established the engineering feasibility of deep-sea mining and provided data on collector design and nodule pumping capabilities. Both tests were monitored to provide data useful in assessing future mining environmental perturbations and possible environmental effects.

This chapter summarizes the results of the measurements taken during those tests and during related laboratory experiments and updates estimates of potential mining impact on the region based on those findings. For completeness, when new findings were not available on specific effects, related studies and earlier information have been summarized. Because of the large scope of the environmental fate and effect question, however, details available in earlier reports have been necessarily omitted (Burns et al., 1980; Ozturgut et al., 1980).

Two caveats need be given for the impact assessment aspect of this chapter: (1) since the prototype mining tests were scaled versions of future commercial systems, these results need be cautiously extrapolated to the conditions during a full-scale mining operation; and (2) since tests monitored were of short duration and intermittent, only short-term effects are addressed here. This chapter while limited in these ways, is intended to present the best available knowledge on mining discharge fate and effects. These results, after completion of ongoing studies, will be used to provide better estimates of the effects of commercial scale mining.

MINING TESTS

Ocean Management, Inc. (O.M.I.) conducted successful deep-sea mining tests during the period March–May, 1978, near 9°N and 150°W (Fig. 1, site A). Using the converted drilling ship "SEDCO 445", O.M.I. mined ~ 900 t

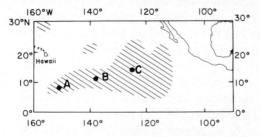

Fig. 1. The mining tests reported here took place near sites A and C. Sites A, B and C were also the focal points of Deep Ocean Mining Environmental Studies (D.O.M.E.S.) investigations (Ozturgut et al., 1978). *Hatched area* indicates the region where Ni content of manganese nodules is 1–2 wt.% (data form Horne et al., 1973).

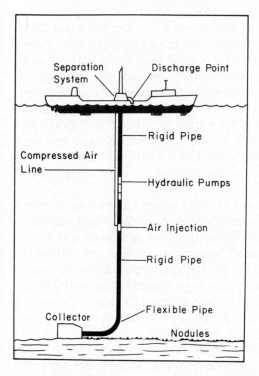

Fig. 2. A schematic diagram of a mining system. Both hydraulic pumping and compressed air lifting have been separately used in mining tests.

of nodules from a depth of 5100 m during 102 hr. of actual collector operations. Mining operations, by virtue of their engineering test nature, were intermittent and took place over three intervals: March 28 (15 hr.), April 6—8 (54 hr.), and May 1—4 (33 hr.).

Ocean Mining Associates (O.M.A.) were also successful in deep-sea mining tests in October and November 1978 at 15°N and 126°W (Fig. 1, site C). In those tests, ~ 500 t of nodules from 4300 m were taken on board the converted ore carrier "Deepsea Miner II" during an 18 hr. mining period (November 10).

In both mining tests, nodules were gathered by collectors which were towed across the bottom (Fig. 2). The sled-mounted collectors were attached by a flexible rubber pipe to a rigid drill pipe which served as nodule conduit. The drill pipe inner diameter was on the order of 22 cm in the O.M.I. and 15 cm in the O.M.A. tests. Collector size in both cases was ~ 5 m long, 3 m wide and 2 m high. In both tests, the ground speed of the collector averaged ~ 0.5 kt.

The collectors were designed to remove nodules from the sea floor, reject large nodules, winnow much of the fine sediment, and pass nodules to the lifting pipe intake. Although nodules in the collector were subject to washing by the same high-velocity water stream that lifted them from the sea floor, some bottom sediment in addition to bottom water and nodules were drawn into the pipe. In the pipe, the lifting force was of two types. In one case, hydraulic pumps located below the sea surface caused the mixture of nodules, water and bottom sediment to ascend to the surface. At other times, compressed air was injected at depths of 1500–2000 m along the lift pipe to create a positively bouyant mixture. The hydraulic lifting technique was used by O.M.I. in March and April 1978, and the air-lift technique was used by O.M.I. in May and by O.M.A. in November 1978. Distinction between lift techniques is made in the description of the discharge; only air-lifted discharge was monitored as a surface plume.

The mixture of nodules, sediment and bottom water pumped or air lifted to the surface was passed through a separation system which was designed to winnow nodules and nodule fragments from the fine sediment and water. The extent of the separation determined the characteristics and ultimately the fate of the discharged effluent. In the O.M.I. tests, material passed through a centrifugal separator, a shaker screen to remove stray nodules and large nodule fragments, and a settling tank which was effective in removing all but the smallest sized particles ($< 100 \mu m$). In the O.M.A. tests, centrifugal separation alone was employed.

Both tests took place in an open-ocean environment; a description of each site has been given in Ozturgut et al. (1978). Surface currents are typically $25 \, \mathrm{cm \, s^{-1}}$, and surface winds of $10 \, \mathrm{m \, s^{-1}}$ are common. At both sites, the upper oceanic surface layer consists of a mixed layer some 45–70 m thick with a mean salinity value of $\sim 34\%_{00}$, and a mean temperature of $\sim 26°C$ (Halpern, 1979). The pycnocline, the region of high density gradients beginning below the mixed layer, is typically $\sim 80 \, \mathrm{m}$ thick. Ambient particulate concentration levels in the mixed layer average about $40 \, \mu \mathrm{g} \, \mathrm{l}^{-1}$, increasing by no more than $\sim 20\%$ in the pycnocline (Baker et al., 1979). The bottom sediment is principally a siliceous clay at the O.M.I. site and a siliceous ooze at the O.M.A. site (Hein et al., 1976; Hein and Jones, 1977; F. Simpson et al., 1977). The dispersed median diameters of bottom sediment at both sites are from 2 to $4 \, \mu \mathrm{m}$.

Monitoring studies of the mining tests were conducted from the N.O.A.A. ship "Oceanographer" and the mining vessels. Prior to each test, ambient measurements of physical, chemical and biological parameters were made in order to establish a baseline against which mining-introduced perturbations could be compared. During the mining activities, the plume generated by the mining discharge was investigated and further measurements of environmental parameters were made. Both benthic plume and surface plume studies were conducted during the O.M.I. tests; only surface plume was monitored during the O.M.A. tests. The measurement techniques and detailed analysis of these monitoring studies have been reported by Burns et al. (1980) and Ozturgut et al. (1980).

Since these mining tests were designed as tests of the mining systems, discontinuous operation and variable production rates were common during both O.M.I. and O.M.A. tests. The short duration and intermittence of mining activities did not allow all planned environmental studies to be completed.

SURFACE DISCHARGE

As in any potential pollution investigation, identification of the source characteristics is essential in estimating the potential impacts of nodule mining. The surface discharge from the mining ship consisted of bottom and interstitial water, bottom sediments and nodule fragments. Macerated biota, though not a significant fraction of the volume discharge, was observed from time to time (L. Kimrey, pers. commun., 1978). Bulk properties of the discharge are found in Table I. Comparable discharge rates of nearly $100 \, l \, s^{-1}$ Table I) were attained in both air-lift tests. The rates indicate little variation between either type of lift mechanics or between mining consortia.

Total particulate concentrations of the discharge are based on gravimetric analysis of samples taken aboard the mining vessels at points just prior to the effluent entering the water. The large confidence limits on each test con-

TABLE I

Discharge characteristics averaged over the mining period for mining tests conducted by Ocean Management, Inc. (O.M.I.) and Ocean Mining Associates (O.M.A.)

	O.M.I.		O.M.A.
	pump	air lift	air lift
Average flow rate of discharge[1] $(l \, s^{-1})$	160	100	95
Total particulate concentration $(g \, l^{-1})$	12.7 ± 9.56	6.8 ± 5.61	5.8 ± 3.91
Solid discharge rate $(g \, s^{-1})$	2,030	680	550
Solids in the discharge (vol.%)	0.55	0.30	0.25
Temperature of discharge (°C)	8.5	7–10	4.4–5.2
Bulk density of discharge[2] $(g \, cm^{-3})$	1.034	1.030	1.031
Nodules collected[1] $(10^6 \, t \, hr.^{-1})$	11	5	27

[1] Data supplied by the mining companies.
[2] Bulk densities calculated using a salinity value of 34.70⁰/oo.

centration mean reflect the variability in the efficiency of the mining activities during the tests. The mean concentration of the discharge for all tests is $8.4\,g\,l^{-1}$, while the mean solid discharge rate is $1.1\,kg\,s^{-1}$. The volume fraction concentration of the solids in the discharge ranging from 0.25% to 0.55% is computed from the total particulate concentration values, and average flow rates of the discharge by assuming a specific gravity of non-nodule solids of $2.7\,g\,cm^{-3}$ (Richards, 1977).

The temperature of the discharge was measured as the discharge passed overboard, and was from $3°$ to $9°C$ higher than bottom-water temperature ($1.44°C$ at both sites) and from $16°$ to $22°C$ lower than the sea-surface temperature of $26°C$. The temperature change is in part due to heat conduction through the lift pipe as colder bottom water passes through the warm near-surface region; and may also be partly due to frictional heating within the pipe. The higher discharge temperatures in the O.M.I. test may also reflect longer time spent in shipboard separation systems at ambient air temperature. The salinity of the discharge was not measured but is assumed to be very near the value of $34.68°/_{00}$ (O.M.A.) and $34.72°/_{00}$ (O.M.I.) measured near the bottom during earlier studies of both areas (Ozturgut et al., 1978).

Estimates of bulk density were derived from the values of temperature, salinity and particulate load (Table I). The resulting bulk densities, ranging from 1.030 to $1.034\,g\,cm^{-3}$, are greater than ambient bulk density ($1.022\,g\,cm^{-3}$) by 0.8—1.2%. The magnitude of the bulk density difference between the discharge and ambient water, the volume discharged, and the manner with which the discharge enters the water control the initial behavior of the plume were also studied.

Discharge samples from both tests were frozen and returned to the laboratory for analysis. Samples were thawed, shaken and washed through a 64-μm sieve with artificial seawater. An average of $7\% \pm 5\%$ of the mass of each O.M.I. sample and $44\% \pm 21\%$ of the mass of each O.M.A. sample failed to pass through the sieve. The material retained on the screens was primarily black manganese nodule fragments resulting from the fracturing and abrasion of nodules during collection and lift.

Material retained on the screen was subjected to settling velocity analysis. These analyses indicate that the settling velocity distributions of the coarse fraction of samples from the O.M.I. test are skewed toward the finer end of the spectrum (quartz spherical equivalent diameter of 64 μm). The average median settling velocity of three O.M.I. distributions is $1.43 \pm 0.79\,cm\,s^{-1}$. In contrast, the settling velocity distributions from the O.M.A. test sample coarse fraction are skewed toward the coarse end of the spectrum (quartz spherical equivalent diameter of 1000 μm). The average median settling velocity of nine O.M.A. distributions is $11.9 \pm 11.3\,cm\,s^{-1}$, corresponding to a size of $\sim 400\,\mu$m (quartz spherical equivalent diameter). The difference in settling distributions is indicative of different separation techniques employed by O.M.I. and O.M.A. In addition, the differences in distributions point to the need to analyze each mining test individually and draw common conclusions with care.

The fraction smaller than $64\,\mu$m contained both nodule fragments and normal sediment, but the ratio of nodule fragments to sediment in the O.M.A. samples was considerably greater than that of the O.M.I. samples. Laboratory analyses of the fine fraction of the discharge ($< 64\,\mu$m) show that the rate of settling is dependent primarily on the initial concentration of particulates within the settling tube, suggesting that flocculation is important in controlling the rapidity at which fine clays and silts from mining settle (Lavelle et al., 1978).

Laboratory settling experiments were not possible of the fine fraction ($< 64\,\mu$m) of discharge in a saline environment at concentrations as low as $10\,\mathrm{mg\,l^{-1}}$, corresponding to initial post-discharge concentrations. Vertical profiles and horizontal nephelometer transects in and through the O.M.I. plume have, however, been used to infer a discharge average effective settling velocity in the range of 0.05 to $0.07\,\mathrm{cm\,s^{-1}}$, corresponding to a 10 to 23—27-μm quartz spherical equivalent diameter particle (Lavelle et al., 1978).

SURFACE PLUME

In both tests, the mining effluent passed over the side and fell to the sea surface (Fig. 3A), a distance of ~ 3 m (O.M.I.) or 5 m (O.M.A.). The water directly below the discharge point was light brown in color and, in the case of the air-lift systems, was characterized by frothiness. Orientation of the vessel with surface winds and surface currents occasionally caused pooling of discharge along the mining vessel. A sharp front between discharge and ambient water near the discharge point was common (Fig. 3B) as the plume moved away from the ship. The color changed rapidly appearing on the surface as a light-green discoloration at a distance of several ship lengths (~ 150 m) from the mining vessel (Fig. 4A and B).

Particulates

Extent of the surface plume was tracked with a system which pumped water from within 5 m of the surface through a nephelometer aboard the "Oceanographer". During each crossing, water samples were drawn from the pumping line for nutrient and particulate analysis. Particulate concentrations recorded as light scattering intensities in several transects across the O.M.I. and O.M.A. plume can be found in Figs. 5 and 6.

Transects during the O.M.I. tests (Fig. 5A) were made relative to a drogued marker buoy deployed in the plume at early plume age. This procedure allowed the same parcel of water to be tracked in time. A planar view of the plume was reconstructed from the transects (Fig. 5B). The plume extended beyond 5 km along the axis and is nearly 1 km wide over much of its length.

The O.M.A. transects (Fig. 6), in general, display a similar sequence to those of Fig. 5. However, transects had to be made in a way which precludes compensation for variable discharge rates so that a planar view cannot be given. Variable discharge is thought to be the reason for the broad profile at 6.5 hr. Maximum concentration on the transect closest to the mining vessel (0.25 hr.) was $\sim 900\,\mu g\,l^{-1}$. The last transect farthest from the mining ship (11.75 hr.) had a maximum particulate concentration of $\sim 40\,\mu g\,l^{-1}$ above the ambient concentration of $30\,\mu g\,l^{-1}$.

During the O.M.A. test, a beam transmissometer was also used simultaneously with the flow-through nephelometer. The beam transmissometer indicated a higher concentration variability and a narrower plume than the nephelometer in the early plume ages. These differences were due to the configuration of the flow-through nephelometer system: the stream path length ($\sim 15\,m$) and a mixing chamber which smoothed the small-scale variability.

Fig. 3. A. The stern of the mining vessel, R/V "Deepsea Miner II", and the discharge pipe near amidships. The discharge pipe was 40 cm in diameter and locaed ∼ 5 m above the sea level. The outfall of discharge is frothy due to the release of entrained air.

B. Air-lift discharge plume during the O.M.I. test. The discharge pipe had a diameter of 30 cm and was positioned 3 m above the sea surface.

The short duration of mining prevented a definition of the plume depth at the point of discharge. The only information available comes from samples pumped on board the "Oceanographer" during transects through the O.M.I. plume at a distance of ~ 700 m from the mining ship. Concentrations as large as $700\,\mu g\,l^{-1}$ occurred in samples pumped from a depth of 17 m; samples below the 17 m level were not taken. The occurrence of mining particulates at 17 m depth, 700 m from the discharge point, may indicate for an initial distribution only a few tens of meters deep. Vertical profiles taken in the O.M.A. plume (Lavelle et al., 1978) support that suggestion by showing that the initial distribution of fine particulates did not extend beyond 30 m. Based on a 20-m initial intrusion depth and the observed width of the plume at the discharge point, a fair estimate of the initial dilution factor for the discharge is 10^3. Concentrations after discharge, when passive dispersion begins, are therefore likely to be on the order of $10\,mg\,l^{-1}$.

The fine fraction of the discharge is presumed to be most long-lived in the mixed layer. Information of relative concentrations of fine particulates was sought through measurement of size distributions. Particle sizes, expressed in terms of equivalent spherical diameters, were measured by Coulter counter, a device which computes the displacement volume of particles passing through an orifice of known size.

Plume sample distributions had relatively more particulates at the smaller size than did ambient samples (Fig. 7). The presence of mining discharge particulates in 1 hr. old plume (O.M.A.) water is suggested by the higher concentration of smaller size particles at 5, 15 and 25 m depths than at

Fig. 4. A. The mining vessel, R/V "Deepsea Miner II", steaming south with a speed of
~ 0.5 kt. and the winds from NE. The plume is seen astern.
B. The O.M.I. mining vessel, "SEDCO 445", and the surface plume.

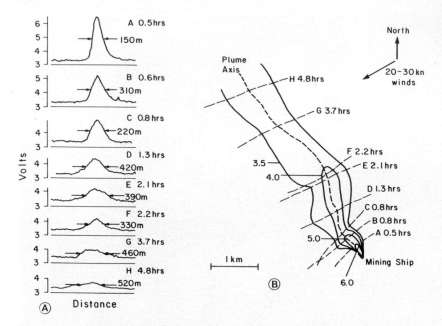

Fig. 5. A. A sequence of nephelometer traces through the O.M.I. plume at increasing plume ages. Plume width at half amplitude is shown. Calibration data for nephelometer is not available.

B. Planar reconstruction is the steady-state O.M.I. mining plume. Contour levels are in volts; background voltage level was 3.3 V.

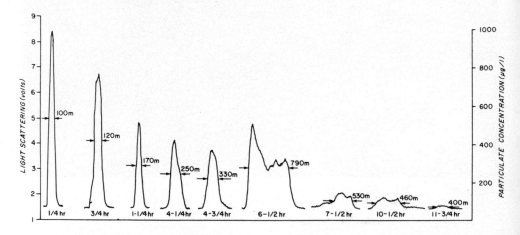

Fig. 6. Light scattering values (in volts) from the flow-through nephelometer on the left and computed particulate concentrations on the right at different ages of the plume during the O.M.A. test. The width at half amplitude indicated accounts for variable ship speed and oblique crossing angles. Nephelometer system was calibrated with sediments from the mining site.

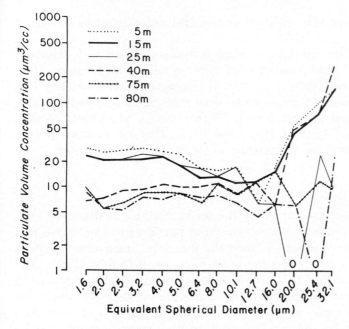

Fig. 7. Size distribution of particulates in 1 hr. old plume water during the O.M.A. test. The samples from 5, 15 and 25 m depths had high concentrations of mining discharge particulates.

depths below. Size distributions therefore provided a field signature for the plume.

Dissolved materials

More than 99% of the volume of the surface discharge is bottom water, which is richer in nutrient content than mixed layer water. Addition of nutrients into the nutrient poor mixed layer could enhance primary production. Earlier D.O.M.E.S. estimates indicated that nutrients added by the discharge would be undetectable in the plume because of the small volumes discharged and the mixing of the discharge into surface water.

During the O.M.I. test, the hypothesis that nutrients in the plume would be undetectable was tested. Transects were made as close to the mining ship as possible, during which water from two depths was pumped aboard the "Oceanographer" for analyses. Separation distance between ships during the transects was ~ 700 m, at which point the plume was ~ 100 m wide. Nitrate and silicate concentrations were used as indicators of nutrient enrichment, since concentrations of these nutrients in bottom water are nearly 25 and 50 times, respectively, those of the surface mixed layer. Statistical comparisons of ambient and the plume water samples taken on those transects indicate that there were no appreciable nutrient concentration differences (Burns et al., 1980).

In an air-lift system, the mining discharge is supersaturated with dissolved gases, and it is known that exposure of fish to such water could affect mortalities (Fickeisen and Schneider, 1976). However, fish mortality was not expected during mining, because small flow rates and subsequent mixing should not allow unusual oxygen levels. Measurements of oxygen profiles made during both tests verified this hypothesis. Oxygen profiles in the plume were not different from the profiles made in the ambient water.

Trace metals

Trace metals could be introduced to the water column as dissolved constituents from bottom water or via release from particulates (sediments and nodule fragments) in the discharge. Since the dissolved trace-metal content of the bottom water and surface layer are not too different (Healy and Richards, 1976), there would be no detectable trace-metal increases due to the discharge of bottom water. On the other hand, deep-sea clays are particularly enriched in many heavy metals, and concern has been expressed that these metals may be directly released to the water column and be taken up by plankton. In addition, nodule fragments could also provide trace metals.

Rosenbauer and Bischoff (1978) conducted experiments with resuspended bottom sediments from mining sites in oxygenated and deoxygenated seawater. In these experiments, samples were analyzed for Ni, Zn, Cr, Fe, Cu, Mn and Al. Aside from a few cases of suspected contamination, all concentrations were below detection limits in the oxygenated experiments. The deoxygenated experiments showed that only Mn and Ni were released in very small quantities (0.03 ppm Ni, 0.03–2 ppm Mn) at the end of six months. Rosenbauer and Bischoff concluded that the potential of sediments to release heavy metals is insignificantly small in both oxygenated and deoxygenated environments.

Robertson and Rancitelli (1973) found that crushed nodules in seawater released Fe, Zn and Co in significant amounts in several days. However, according to Rosenbauer and Bischoff (1978), contamination may have affected these results. Because neither set of experiments on nodules included analyses for the toxic metals Hg, Cd and Ag, and because toxic response can occur at metal ion concentrations well below the normal detection limit, these experiments — using a concentrating technique — are being undertaken again on both crushed nodules and with discharge material (M. Benjamin, pers. commun., 1978).

Pycnocline accumulation

The possibility of spreading and accumulation of fine mining particulates at and along the pycnocline is a frequently expressed concern. The concern stems from other observations of non-living particulate maxima in layers where large gradients of water mass characteristics exist. Changes in density, temperature, or vertical turbulent mixing within the settling column can result in a retardation or cessation of particulate settling.

Data bearing on the question of pycnocline accumulation were taken during the O.M.I. tests in two forms: (1) nephelometer casts through the pycnocline; and (2) water samples above and in the pycnocline. Although both data sets suggest a small increase in particulate matter at the top of the pycnocline in comparison to pre-mining samples (Burns et al., 1980), the differences are within expectations of natural variability. Moreover, chemical analysis of particulate samples, a more sensitive indicator of plume presence, did not indicate the presence of mining particulates in the pycnocline.

A model which incorporates the difference between mixed layer and pycnocline vertical diffusivity (Ichiye and Carnes, 1977) also indicates that no particulate accumulation layer should be observed at the base of the mixed layer. For very fine particulates ($\gtrsim 10 \mu m$) and at a typical value of $10^{-2} m^2 s^{-1}$ for vertical turbulent diffusivity in the mixed layer, and no vertical diffusivity below, the effect of the pycnocline is to increase residence time and make the vertical distribution of particulates uniform in the mixed layer at times larger than 12 hr.

Light attenuation

Mining particulates reduce the amount of solar radiation at depth and thereby directly affect the primary production in the mining area. Irradiance measurements were made during the O.M.A. test to quantify the solar attenuation—particulate concentration relationship. Light penetration profiles were made with a quantum meter, sensitive over the photosynthetically active band (400—700 nm), and with a blue light (475 nm) sensitive photometer.

Light profiles and particulate samples were simultaneously obtained in 1 hr. old plume water. The distribution of particulate concentrations in the water column indicated that the mining particulates were confined to the upper 25 m layer with an average concentration of $440 \mu g l^{-1}$. Light profiles taken in ambient and plume waters are compared in Fig. 8. Only one set of light profiles could be taken because of the limited nature and timing of the mining tests.

The photosynthetically active radiation (PAR) measurements make clear the presence of the plume in the upper 25 m. At a depth of 20 m, the PAR is

452

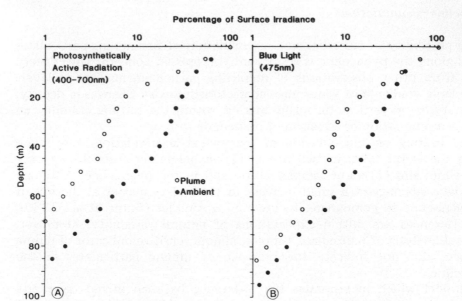

Fig. 8. Profiles of photosynthetically active radiation (A) and blue light (B) in ambient and 1 hr. old plume water. The increased attenuation in plume water is due to the presence of mining discharge particulates in the upper 25 m. The ambient light profiles were taken immediately after the plume profiles.

seen to be reduced from 28% to 6%. The slopes of the irradiance curves, and the attenuation coefficient, have values in the upper 25 m of 0.034 and 0.14 m^{-1} for ambient and plume water, respectively. The break in the slope at the base of the mixed layer (45 m) is a general feature of the area and is a result of the chlorophyll a maximum located there. Blue-light profiles have features similar to the PAR profiles. The attenuation coefficient for blue light in the upper 25 m, where the average mass concentration was approximately 440 μg l^{-1}, is 0.105 m^{-1}. These measurements and a blue-light clear water attenuance value (Jerlov, 1976) allow the calculations of blue-light levels in plume areas of differing particulate concentration.

BACTERIAL GROWTH AND OXYGEN DEMAND

The growth of bacteria is stimulated by increases in substrates, and particulates from the discharge could provide that increased surface area. Increases in bacterial biomass and oxygen demand could then result. Since bacteria are digested by zooplankton, however, bacterial growth would add nutritive value to the mining particulates.

Both increased biomass and oxygen demand from increased substrates have been documented in shipboard experiments using fresh sediment and mining discharge (Ozretich and Ozretich, 1978; Ozretich, 1979). In addition, measurements fron the O.M.I. plume showed that a higher biomass was present near the bottom of the mixed layer than in ambient water from the same depth zone. Ozretich (1979) concluded that some of this excess biomass was the result of bacterial growth in the plume.

The effect of bacterial growth through increased oxygen demand in the mixed layer would be inconsequential due to oxygen mixing across the air–sea interface and production of oxygen by the phytoplankton. However, an increase in oxygen demand could have a deleterious effect below the pycnocline at depths between 150 and 600 m where an oxygen minimum zone occurs. This oxygen minimum zone is most pronounced, i.e. almost depleted in oxygen, toward the eastern end of the D.O.M.E.S. region. Further removal of existing oxygen could create anaerobic conditions in this zone, i.e. production of hydrogen sulfide after the depletion of dissolved nitrate ions. J. J. Anderson (1978), however, concluded that the hydrogen sulfide production stage would likely never be reached. Long-term experiments of bacteria—sediment interactions (Ozretich and Ozretich, 1978; Rosenbauer and Bischoff, 1978) further invalidate the hydrogen production hypothesis. Increased oxygen demand due to mining discharge should have no adverse effect in the mining region.

EFFECT ON PHYTOPLANKTON

Possible effects on the marine ecosystem at the phytoplankton level from the introduction of surface discharge include: (1) reduction in primary production due to increased light attenuation; (2) enhancement in primary production due to addition of nutrient-rich bottom water; (3) change in species composition of the surface phytoplankton populations; (4) trace-metal uptake by phytoplankton and inhibition of primary production.

Decreased solar irradiance

Phytoplankton photosynthesis, the assimilation of inorganic carbon into plant biomass using sunlight as the energy source, forms the base of the marine food chain. The rate of photosynthesis is affected by light quality, intensity and duration. In the sea, the intensity of radiation diminishes with depth due to the scattering absorption of the light by the water itself and by the particulates in it. In the open ocean, attenuation greatly depends on the concentration of particulates (see Eisma, Chapter 9 in this volume). Mining particulates will increase light attenuation and thereby directly affect the primary production in the mining area. Experiments were conducted during both mining tests to obtain a quantitative estimate of this effect (Chan et al., 1980).

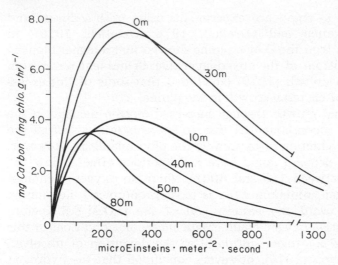

Fig. 9. Photosynthesis–irradiance curves for site C ambient samples from depths indicated (courtesy of A. T. Chan).

During the O.M.A. tests, primary production in samples taken at six depths was measured in 3 hr. (midday) onboard incubation experiments. The results from each sample depth were fit with a photosynthesis — irradiance (P vs. I) equation (Denman and Platt, 1975). Continuous $P-I$ curves were then constructed from the equation and best-fit parameters (Fig. 9).

In the D.O.M.E.S. region, the bulk of the phytoplankton population is concentrated at depths just below the mixed layer corresponding to 10–20% light levels where more nutrients are available than the mixed layer (El-Sayed et al., 1977). Reduced light levels due to daily variation in the incoming radiation or due to mining particulates would shift the optimum light levels into the nutrient-poor region of the water column and reduce the total production. From the light adaptation characteristics, i.e. the $P-I$ relationship, and from reduced light levels measured in the mining plume (Fig. 8), a photosynthetic rate within the plume was calculated as a percentage of the ambient photosynthetic rate for each depth (Table II).

Fig. 10 shows a comparison of ambient production rates measured in situ with the estimated production rate at a point along the axis of the plume representing discharge 1 hr. old. The total reduction in productivity over the entire euphotic zone at this point amounts to 40%. This change in primary production is of the same order of magnitude as the natural variability which results from the day-to-day variation in solar irradiance. On November 6, 1978, a dark day, the daily PAR received was $\sim \frac{1}{3}$ the amount received during other days of in situ primary production measurements (Table II); the

TABLE II

In situ rate of primary production in ambient waters (real measurements) and in 1 hr. old discharge plume (estimated)

Depth (m)	Primary production (mg C m^{-3} day^{-1})					
	Nov. 5, 1978		Nov. 6, 1978		Nov. 7, 1978	
	ambient	plume	ambient	plume	ambient	plume
0	3.40	3.40	7.28	7.28	7.54	7.54
5	0.39	0.42	–	–	–	–
10	0.64	0.64	1.00	0.82	1.50	1.50
20	0.40	0.32	0.93	0.34	1.50	1.22
30	3.52	1.44	1.08	0.36	1.27	0.52
40	5.58	3.01	0.61	0.13	7.00	3.77
50	–	–	0.54	0.14	1.76	1.23
60	1.15	0.58	1.26	0.34	1.05	0.53
70	2.71	1.30	0.66	0.16	1.81	0.87
80	0.31	0.15	0.78	0.19	0.77	0.38
90	–	–	–	–	0.47	0.23
100	0.01	0.01	0.15	0.04	–	–

Daily PAR *photosynthetically active reduction* (Ein m^{-2} day^{-1}):

20.2		7.1		22.7	

Integrated production rate (mg C m^{-2} day^{-1}):

187.2	102.4	110.6	68.6	206.4	139.4

Reduction in primary production in the plume:

45%		38%		33%	

integrated primary production for this day is on the average 45% less than the brighter days.

For a given parcel of water-bearing phytoplankton and mining particulates, the shading effect of mining particulates is only temporary as most of the mining particulates settle within a period less than a day. Since the adaptation of the phytoplankton to a new light regime takes 2–3 days (Steemann Nielsen, 1975), the short-term shading effect of particulates is not likely to affect the light-adaptation characteristics of the phytoplankton.

Nutrient enrichment

The bottom water in the D.O.M.E.S. region is rich in nutrients, and the introduction of bottom water with the discharge will increase the nutrient concentration in the ambient surface water. Since the surface mixed layer is poor in nutrients, it has been frequently suggested that the addition of discharge water would stimulate phytoplankton growth.

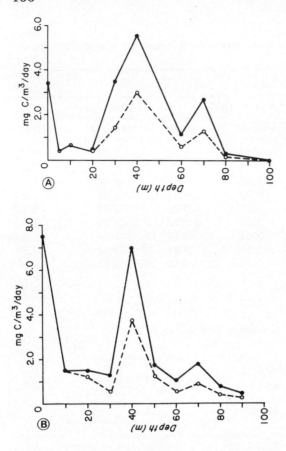

Fig. 10. Measured production rates in the ambient water (*solid circles*) and estimated production rates at the same station as a result of reducing photosynthetically active radiation (PAR) by the amounts corresponding to the reduction of PAR in 1 hr. old plume water (*open circles*).

Experiments were conducted during both mining tests to determine combined stimulatory—inhibitory effects. Ambient water samples were inoculated with different concentrations of mining discharge which had been allowed to settle overnight to remove most of the particulates. Concentrations were chosen to approximate conditions in the mixed layer near the discharge point. The photosynthesis—irradiance (*P–I*) relationship of the treated samples was studied and compared to controls (Fig. 11). No significant difference was observed between the experimental and the control samples, indicating that discharge does not have any experimentally demonstrable effect on productivity (Chan et al., 1980). Earlier experiments conducted with bottom water and sediments from the D.O.M.E.S. region produced similar results (El-Sayed et al., 1977).

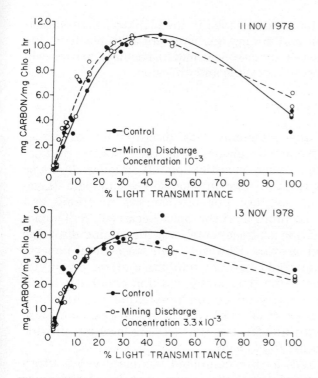

Fig. 11. Photosynthesis–irradiance curves for ambient surface water (control) and for ambient water inoculated with mining discharge at two different concentrations (Chan et al., 1980).

Species composition

There has been a concern that the addition of bottom water would affect the species composition of the surface phytoplankton populations by changing the nutrient content of the water or by introducing algal spores from the sea-floor sediments. Incubation experiments conducted with bottom sediments from D.O.M.E.S. sites A, B and C showed that, at the concentration of sediments estimated at the discharge point in pilot mining tests ($10\,\text{mg}\,\text{l}^{-1}$), the species composition after 72 hr. was similar to the initial and control samples (El-Sayed et al., 1977). No cells which could be identified as having littoral or benthic origin made a significant contribution to the standing stock of phytoplankton in these experiments.

In commercial mining operations, no significant changes in species composition of phytoplankton of the mining area are anticipated. Changes in species composition could occur where there are pronounced and enduring changes in physical environmental factors, so that species better adapted to the new environment have a chance to replace previously dominant species.

However, permanent or even long-term change in the plankton environment should not occur. As plankton and mining discharge are advected from the mining ship, concentration levels of dissolved and particulate matter will return to ambient levels as a result of settling and diffusion.

Trace-metal uptake by phytoplankton

Trace-metal release from the sediments was discussed in an earlier section; the potential of trace-metal release from the abraded nodules in the discharge is currently under investigation. If mining discharge releases trace metals, phytoplankton may uptake and accumulate them. Some of the trace metals are essential for biochemical functions, and some have been suggested to be limiting in the open ocean (W. H. Thomas et al., 1974); in this case, addition of trace metals via the mining discharge could enhance phytoplankton growth. However, high trace-metal concentrations or otxic metal release could have an inhibitory effect on phytoplankton growth. In some cases, the trace-metal addition could also affect the species composition of the phytoplankton population as shown by Menzel (1977) and W. H. Thomas et al. (1978a). Phytoplankton could also take up toxic trace metals with no effect at that trophic level but provide an entry for metal into the marine food chain.

Experiments designed to look at the combined stimulatory–inhibitory effects of discharge were described earlier. Because no significant difference was observed between experimental and control samples in those experiments, mining discharge is judged to have no inhibitory effect on phytoplankton growth exclusive of shading.

EFFECT ON MACROZOOPLANKTON

The potential effects of mining-related particulate matter on the vital activities and species composition of macrozooplankton have been described by Hirota (1977b, 1978). The directly observable or measurable effects of discharged particulates and their inferred ecological consequences can be summarized in the following four subject areas: (1) direct morality over a shot time period of 1–2 days (approximating the duration organisms may reside in the plume), excluding longer-term delayed mortality and sublethal physiological stress; (2) ingestion of discharge particulates, changes in ingestion rates of suspended particulate matter, changes in the production and/or settling rates of fecal pellets; (3) ingestion and external adsorption of the discharge particulates by macrozooplankton, including changes of elemental composition in whole organisms and their fecal matter; and (4) short-term differences in the spatial distribution, abundance and species composition of zooplankton in a discharge plume vs. ambient seawater.

Mortality

The direct effects of mining particulates on short-term mortality of a number of marine, tropical copepod species were evaluated with two different approaches: (1) controlled, "sediment challenge" experiments with direct microscope counts of live and dead animals after incubations in seawater with or without added particulates (discharge or sediments from D.O.M.E.S. site *C*); and (2) net tow collections at sea with short-term shipboard incubations of zooplankton species, using a vital staining technique to evaluate percent mortality.

The controlled laboratory "sediment challenge" experiments were carried out at sea and in Hawaii with oceanic species at low total particulate concentrations ($\sim 0.1-1.0 \, \text{mg} \, \text{l}^{-1}$). Mean values of mortality as the difference between experimental treatments and corresponding controls are less than 5% for most species after 24 hr. of incubation and generally higher for 48-hr. periods (Table III). For species of *Undinula* and *Labidocera* with many observations, when all data are grouped together per species for experimental treatments vs. controls, there are no statistically significant differences in these grouped mortalities. Thus, it is concluded that neither suspensions of bottom sediment from site *C* nor O.M.A. particulate matter discharge appear to contain toxic materials or cause great increases in mortality for short periods (1–2 days), at least for the species tested.

Net tows in the plume and in ambient water were also carried out aboard the "Oceanographer" during mining tests for determinations of percent mortality of surface-living copepod species, using a modified procedure of the neutral red vital staining technique of Dressel et al. (1972). In the O.M.I. test, the effect of the plume (at an age of 8–10 hr.) on mortality was negligible compared to values in ambient water (Hirota, 1978). O.M.A. test results are not yet available.

Ingestion

The fine fraction of the discharge, remaining longer in the upper layer than coarser discharge particulates, is more likely to be ingested by filter-feeding zooplankton. Such ingestion of non-living particulates have been widely reported (P. G. Moore, 1978). The result is aggregations of fine particles, i.e. fecal pellets which rapidly sink. Calculations show that zooplankton are not likely to contribute to the removal of mining particulates from the upper layer on a short-term basis. However, over periods of months to years, pelletization can be an important mechanism for clearing the upper layer of fine mining particulates.

Laboratory ingestion experiments using both discharge and suspensions of bottom sediments were conducted to verify intake of mining discharge particulates by macrozooplankton (J. Hirota, pers. commun., 1978). In these

TABLE III

Percent mortality difference between treatments (plus sediment) and controls (natural seawater only) for incubations of species for 24 and 48 hr. durations

Species	Percent of mortality of treatment minus mortality of control					
	24 hr. duration			48 hr. duration		
	N	$\bar{X}$ (%)	S.D. (%)	N	$\bar{X}$ (%)	S.D. (%)
Undinula vulgaris[1] (Kaneohe Bay)	33	0.14	1.74	33	0.62	7.92
Undinula darwinii[1] (open ocean)	4	0	0	2	28.12	30.94
Labidocera madurae[1] (Kaneohe Bay)	12	5.26	16.07	6	6.64	16.89
Labidocera acutifrons (Kaneohe Bay)	1	8.3	–	1	27.27	–
Labidocera acutifrons (open ocean)	12	4.40	7.19	10	4.83	18.98
Acrocalanus inermis (Kaneohe Bay)	6	0.45	1.13	6	− 3.14	3.34
Acartia hamata (Kaneohe Bay + Hanauma Bay)	6	− 2.60	3.43	6	3.66	6.20
Anisomysis sp.[2] (Hanauma Bay)	2	0	14.14	2	12.5	17.68
Centropages gracilis (open ocean)	6	− 0.88	6.28	6	− 2.89	6.82
Euchaeta rimana (open ocean)	4	0	2.89	4	2.22	8.125
Oncaea venusta (Hanauma Bay)	1	0	–	1	0	–
Macrosetella gracilis (open ocean)	2	0	0	2	9.09	0

Additions of sediment are as mining effluent collected during the O.M.A. test at site C, except for species of Undinula and L. madurae as indicated. Symbols N and $\bar{X}$ are the number of observations per species and the mean of each group, respectively; S.D. = standard deviation.
[1] Data from incubations using red clay particulates from box cores at site C.
[2] A marine semi-benthic mysid species.

experiments, the rate of fecal pellet production is used as an estimate of ingestion rate. Preliminary analysis of data show both an inter-species variation of pellet production rates and dependency on the duration of the experiments (viz. 24 vs. 48 hr.). The second dependency is confounded by higher natural mortality of animals during the second 24 hr. of incubation. In general, the trend is toward a 3–4-fold increase in pellet production rates above those for controls for sediment concentrations of $\sim 0.1-1.0\,\mathrm{mg\,l^{-1}}$. However, more work is needed before definitive conclusions of ingestion rates of mining particulates by zooplankton can be made.

The vertical flux of pelletized fine mining particulates is dependent on the settling velocity of pellets as well as on their production rate. Fecal pellets were produced in the laboratory by incubation of various species of copepods in ambient and sediment laden water. Settling velocity of pellets was subsequently measured. Results show generally that ambient pellets from controls sink with greater variability than ones produced in the presence of sediment, that sinking speeds are less for ambient pellets of similar volume, and that speeds range from ~ 20 to $220\,\mathrm{m\,day^{-1}}$. The latter speeds are for larger pellets of *Undinula vulgaris*.

Abundance and species composition

During the mining tests, neuston (surface) net collections were made in both ambient and plume water for determinations of standing stocks and occurrences of major macrozooplankton taxa. In the O.M.I. test, the neuston dry weight, ash percentage, and species count data from tows taken during premining conditions and in the mining plume could not yield conclusive cause–effect results since there were six days between these two sets of collections (Hirota, 1978). In the O.M.A. test, neuston tows were made across the 0.75, 1.25, 4.25 hr. old plumes and also in the ambient water shortly after each crossing of the plume. During the plume crossings, the concentrations of particulates in the upper few meters were less than $1\,\mathrm{mg\,l^{-1}}$ (Fig. 6). The results from these two sets of three replicate tows indicate, that while there were higher catches in the plume by about two-fold, no apparent differences in the ratios of analytical chemistry constituents of the macrozooplankton between the ambient and plume samples (Table IV) were observed. These results imply that there was at least no major decrease in the abundance of neustonic macrozooplankton at plume concentrations less than $1\,\mathrm{mg\,l^{-1}}$; and that zooplankters captured by 0.18-mm-mesh nets had not eaten or adsorbed sufficient amounts of mining particulates to cause alteration of their measured chemical composition (Table IV). The enumeration of zooplankton taxa from these samples for evaluating differences in species composition is not yet completed.

Trace elements

As discussed in a previous section above, the potential of trace-metal release from mining discharge is currently under investigation. If mining discharge releases trace metals, zooplankton could accumulate them by assimilation of metals from food and detrital particles ingested by adsorption

TABLE IV

Macrozooplankton chemistry data for sets of three replicate neuston net tows were taken in ambient and in plume water (0.75, 1.25 and 4.25 hr. old) on November 10, 1978

Local time	(mg/100 m^2)			Ratios				
	DW	C	N	C/N	A/DW	C/DW	N/DW	C/AFDW
Ambient:								
20^h47^m–20^h56^m	160.02	56.8	14.3	3.98	0.2612	0.3548	0.0892	0.4081
21^h42^m–21^h52^m	123.65	37.2	9.6	3.89	0.3441	0.3005	0.0773	0.4582
22^h39^m–22^h49^m	118.15	40.7	10.6	3.86	0.2828	0.3449	0.0894	0.4808
Plume:								
21^h09^m–21^h17^m	282.89	99.6	26.3	3.78	0.2814	0.3521	0.0931	0.4901
21^h32^m–21^h37^m	271.16	95.5	23.9	4.00	0.2787	0.3520	0.0880	0.4881
22^h26^m–22^h33^m	223.45	66.2	16.8	3.94	0.3284	0.2962	0.0751	0.4733

Plume water was determined by nephelometry readings.
DW = dry weight; C = carbon; N = nitrogen; AFDW = ash-free dry weight).

onto their body surfaces and also from the water passing over respiratory surfaces (A. G. Davies, 1978). Upper water column organisms were collected during the D.O.M.E.S. baseline studies and analyzed for trace metal (Hall et al., 1977).

It is known that zooplankton metabolism plays an important role in the removal of trace elements from the surface layer. The particular importance of fecal pellets in this process has been noted by several workers (e.g., Fowler, 1977). The relative elemental composition of selected copepod pellets produced in the presence of sediment vs. those pellets obtained from ambient water are still being investigated. The preliminary results indicate that Al/Fe, Si/Fe and Si/Al ratios, and ratios with Mn and Ca appear to be useful indicators of the ingestion of mining sediment. The elemental composition of mining discharge particulates in fecal pellets shows great enrichment in Mn, Al and Si, a slight increase of Fe, and decrease of Ca relative to the composition of pellets from seawater (Hirota, 1978).

EFFECTS OF SURFACE DISCHARGE ON FISH

The surface discharge of mining particulates and consequent turbidity in the upper layer may affect the fishes directly or indirectly. Direct effects are those factors that influence physiological activities, and indirect effects follow from the consequences of the modified light regime on lower trophic levels.

A comprehensive review of existing information on fishes of the upper waters (to $\sim 1200\,\mathrm{m}$) of the D.O.M.E.S. area has been completed by Blackburn (1976). This study included a literature survey to describe the distribution, abundance, ecology, migration, and general biology of upper and mid-water fishes, including eggs and larvae. The D.O.M.E.S. area is in the feeding and spawning grounds of yellowfin, big eye, and skipjack tuna; and generally, the highest abundance of these species are found south of 10–12°N and to the west. The northeast sector of the D.O.M.E.S. area has a paucity of tuna.

Direct effects

There is a considerable body of information on the effect of turbidity and suspended particulates on freshwater fish; most of the literature has been reviewed and the recent studies have been summarized by P. G. Moore (1978) and Stern and Stickle (1978). Many of the studies reviewed by Stern and Stickle (1978) indicate that a wide variety of species can tolerate increased concentrations of suspended material such as those encountered during the disposal of dredged material, and that most physiological changes which do occur are reversible. Wallen (1951) investigated the direct effects of turbidity in the laboratory on sixteen adult species of freshwater fish and concluded that concentrations below $55\,\mathrm{mg\,l^{-1}}$ were not harmful. Alabaster (1977) thought harm unlikely for freshwater species at concentrations below $25\,\mathrm{mg\,l^{-1}}$. Dose–mortality studies for fish exposed to suspensions of natural sediments and pollutants is summarized in P. G. Moore (1978). Those studies indicate both concentration and exposure time determine lethality. The dependence of lethality on both variables makes reasonable predictions of mortality from mining plumes beyond the state of the art, but mortality from mining is probably insignificant.

Field data relating to purely marine species are very limited. Ritchie (1970) detected no effects of overboard dredge spoil disposal on 44 species of fish in upper Chesapeake Bay. Abundant flora and fauna and an important fishery, including albacore and salmonids, also flourish offshore from the mouth of the Columbia River, in water containing particulate concentrations of between 10^4 and $10^5\,\mu\mathrm{g\,l^{-1}}$ (Pruter and Alverson, 1972).

The effects of sediments from the D.O.M.E.S. site and consequent turbidity on tuna behavior were investigated for D.O.M.E.S. by Barry (1978). In these experiments, yellowfin tuna and kawakawa were placed in tanks with continuous flow of seawater. No ill effects in tuna were observed as a result of short-term exposure in concentrations ranging from 1 to $10\,\mathrm{mg\,l^{-1}}$. Behavioral responses of tuna to turbidity were mixed. Tuna were sometimes observed to avoid turbid areas, the avoidance appearing to be visually mediate. On the other hand, at particulate concentrations up to $10\,\mathrm{mg\,l^{-1}}$, tuna would feed within the turbid region. During the mining tests, no special study was conducted to investigate fish behavior.

Indirect effects

All epipelagic and mesopelagic fish in the D.O.M.E.S. area eat zooplankton or other animals that are ultimately dependent for their food upon zooplankton (Blackburn, 1976). A permanent change in the abundance and productivity of phytoplankton or zooplankton communities would be expected to result in subsequent changes in the biomass of fish as shown by Blackburn (1973) for the eastern part of the D.O.M.E.S. area. However, earlier presented evidence implies that the reduction in primary production due to pilot-scale mining is within natural variability. Moreover, no short-term effects of mining on zooplankton were observed during both tests. Because changes in the lower trophic levels are small and localized, a subsequent permanent change in the biomass of fish resulting from mining would not be expected.

BENTHIC IMPACT

The collector tracks, benthic plume and biologic effects were examined at the O.M.I. test site. Benthic biota were sampled several months before mining by extensive box coring and bottom photography to establish the benthic population characteristics. Just prior to mining, four arrays equipped with near-bottom recording nephelometers and current meters were moored in the immediate vicinity of the mining site. During the mining test, the benthic plume was also monitored from the "Oceanographer" with nephelometers and with water samples for particulate mass analyses. Following mining, the mining site was sampled with box cores, and photographic transects were made across collector tracks and through the mining area. Details of methodology and results are given in Burns et al. (1980).

COLLECTOR DISTURBANCE

As the collector moves along the sea floor gathering nodules, it removes bottom material to a depth of several centimeters over a track width of a few meters in prototype mining or 10–15 m in full-scale operation (Welling, 1979). Based on the mass of sediment disturbed by the collector per unit time and on the measured particulate mass discharge rate in the mining tests, the bulk (96–98%) of non-nodule bottom material entering the collector is separated from the nodules and is left near the sea floor. In addition to pelagic silt and clay, this includes micronodules, microfauna, and nodule and macrofaunal fragments.

Bottom photographs indicate that the collector pushes bottom material aside on the outside of the collector sled edges (Fig. 12A) with heavy redeposition occurring near the tracks, this results from the very rapid settling of most of the material initially resuspended by the collector. Five to ten meters away from the collector track (Fig. 12B), the sediment blanket can be estimated to be a few centimeters thick since the photograph indicates heavy draping by sediment; and nodules in this area are typically 2–6 cm in diameter. On other photographs, resedimentation is still discernible to an average distance of 100 m on either side of the tracks (L. Parsons, pers. commun., 1978).

The collector was monitored continuously with closed-circuit television on the O.M.I. mining vessel. During the movement of the collector along the bottom, the resuspension of sediment was observed. Although orientation and range of the video monitor was limited, the resuspended material was observed initially as a very heavy cloud, most of which was redeposited almost in proximity to the collector track. The thickness of the redeposited sediment at large distances from the collector could not be observed; however, estimates of this thickness are being made. Such predictions are essential for evaluating impact on the benthic organisms in the mining area.

Benthic plume

The benthic plume, that fraction of the resuspended sediment mass which travels farther than a few meters from the collector track, was assessed principally by examining local changes in particulate concentrations near the sea floor (Table V). Nephelometer 30-l Niskin bottle deep casts to within 5 m of the bottom were taken before, during and after mining at stations located around the perimeter of the mining area. The benthic plume was easily identified in nephelometer profiles; plume profiles sharply contrasted with ambient profiles which were nearly constant in the bottom 200 m (Fig. 13). These data and analyses of water samples suggest the benthic plume is characterized by a thickness of the order of a few tens of meters within the mining area and particulate concentrations of 15 to 150 $\mu g\,l^{-1}$. Mean ambient concentrations were approximately $5\,\mu g\,l^{-1}$.

The fraction of the benthic plume not settling quickly to the bottom will be transported by the bottom currents, resulting in a thinner and thinner sediment depositional layer at increasing distances away from the source. In pre-mining studies, six-month mean speeds of currents 30 m above the bottom at D.O.M.E.S. sites were: 2.1, 5.2 and 4.9 cm s^{-1} at sites A, B and C, respectively (Hayes, 1979, 1980). Such speeds support a daily advection distance of 2–5 km. Higher instantaneous speeds also occur; current speeds as much as 24 cm s^{-1} at 6 m off the bottom were measured at site A (Hayes, 1979).

Fig. 14 shows the local change of particulate concentration recorded at two nephelometers moored 10 km downstream on the mining site and 4 m above the sea floor (Steffin et al., 1979). The instruments indicated an increase in particulate concentration from 6 to 7 days after the start of the mining test. Presence of the benthic plume was confirmed by data from a deep cast between the two moorings (Fig. 13, profile *4*) the suspended particulate load was 9 and 14 $\mu g \, l^{-1}$ at 5 m, and 10 and 13 $\mu g \, l^{-1}$ at 50 m.

The plume arrival time and the nephelometer locations relative to mining suggest plume advection velocity of 3–4 cm s^{-1} to the E-NE. The mean speed and the direction of the currents measured by instruments 4 m off the bottom during the same period were reasonably compatible at 1–3 cm s^{-1} to

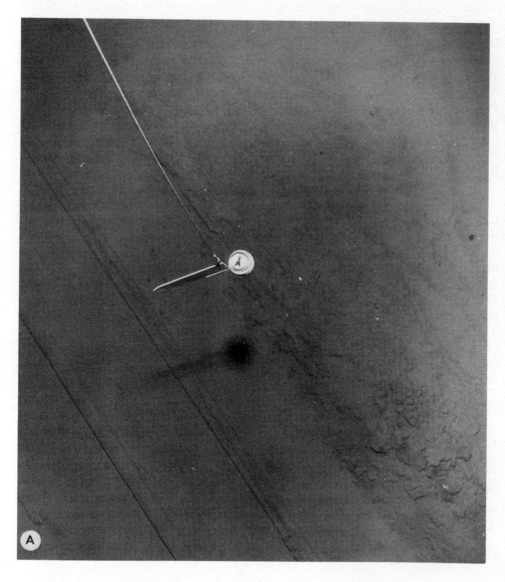

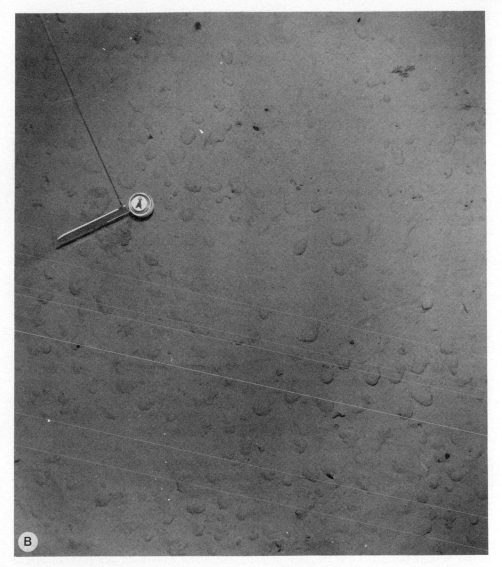

Fig. 12. A. Collector tract and neighboring region where material has been pushed by the collector runner. The absence of nodules on the trace indicate high collector efficiency. Raised areas on the trace occur between collector runners. Compass is 8 cm in diameter. B. An area within 5 m of the collector trace showing the heavy resedimentation resulting from the mining. Normal color contrast between nodules and sediment is no longer seen.

the N-NE (Steffin et al., 1979). These data in toto suggest that the benthic mining plume, with particulate concentrations twice that of the ambient within 50 m of the sea floor, will persist for at least the order of days and move distances of tens of kilometers away from the source.

468

TABLE V

Particulate matter concentrations in the benthic layer at stations taken after mining on the perimeter of the mining site and ambient particulate concentrations at one of the stations

Depth above bottom (m)	Particulate concentration (μg l^{-1})				
	post-mining station No.				average of three pre-mining stations
	1	2	3	4	
5	36.4	143.3	18.2	9.0/14.2*	6.9
15	8.9	125.3	–	–	11.0
30	7.5	19.2	–	–	6.0
50	5.5	7.2	28.0	13.4/10.6*	6.0

* Replicate values.

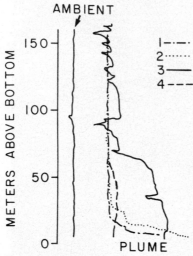

Fig. 13. Vertical profiles of light scattering value near the sea floor. Increases in nephelometer voltage indicated the presence of particulates and were used as a guide obtaining samples of the benthic plume. Ambient and plume profiles are offset; plume voltage means above 100 m were identical with ambient voltage levels.

CHEMISTRY AND MICROBIOLOGICAL ACTIVITY

The chemistry of the interstitial water of the sediments differ very little from that of bottom water, indicating that the sediments are not rapidly undergoing diagenesis (Callender and Bischoff, 1976). Therefore, no significant chemical changes in bottom water can be expected from mining.

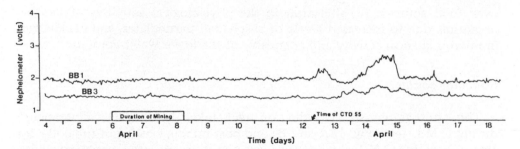

Fig. 14. Variations in light scattering values from nephelometers moored 10 km from the mining site. The 0.5-V separation between the two records is a reflection of instrumented differences. The pointer on April 12 (CTD 55) indicates the time at which a bottom cast (Fig. 13, profile 4) was made.

NH_4^+ is the only significant exception; it is enriched by a factor of 13, suggesting bacterial conversion of small concentrations of solid organic matter in the sediment. However, the amount of interstitial water released by mining is very small and the NH_4^+ would be quickly diluted in the bottom water.

The benthic plume can be expected to increase oxygen demand in the lower water column for several reasons. First, particulate organic matter from dead benthic biota (estimated below) will be suspended within the water column, and oxygen will be consumed as the organic matter undergoes bacterial decomposition or ingestion by larger invertebrates. Assuming that the oxidation will take place within 5 m of the sea floor, the oxygen demand in the lower water column from decomposition of this material would be exceedingly small $(0.1 \mu gAt. O_2 l^{-1}$, compared to ambient levels of $350 \mu gAt. O_2 l^{-1})$ and need not be considered further. The introduction of particulate organic matter from bottom sediments could provide another source of oxygen demand. The organic C content of the sediments is on the average 0.24 wt.% (Bischoff and Piper, 1976); however, a very large fraction of the sediment particulate organic matter would be unreactive and thus not subject to breakdown by bacteria (J. Hedges, pers. commun., 1978). Also, suspended particles, either organic or inorganic, stimulate bacterial growth and oxygen consumption merely by providing sites for attachment. In any case, bottom waters are well oxygenated and would withstand relatively large increases in oxygen demand.

EFFECT ON BENTHIC FAUNA

Possible effects of mining on the benthic community include: (1) destruction of benthos in the path of the collector; (2) increased mortality of benthic organisms due to smothering or rapid decomposition of sediments over food sources; (3) alteration in the physiological activities of benthic organisms due to increased levels of suspended particulates; and (4) changes in microbiological activity and chemistry of the lower water column.

Collector contact

The photographs of the collector track taken by the "Oceanographer" during O.M.I. test (Fig. 13A and B) and post-mining viewing of the collector tracks from the O.M.I. video tapes suggest that, as expected, organisms living on the surface of the sea floor, the epibenthos and near-surface fauna on the path of the collector are destoryed. The benthic population in the immediate vicinity of the collector also appears to be buried under 10–20 cm of sediment. At distances of some 3–4 m from the track, the thickness of the sediment blanketing the sea floor diminishes to the order of a few centimeters. Using a disturbance width of 10 m and a collector ground speed of 0.5 kt., it is estimated that $2.4 \cdot 10^5$ m^2 day^{-1} were affected during the mining tests. The biomass of the benthos involved, some 6000 epibenthic and 25 million macro-faunal organisms per square meter (0.33 g m^{-2}, wet weight; Hecker and Paul, 1979) is estimated to be $\sim$ 75 kg daily.

The significance of this loss rate or the larger loss rate under full-scale commercial operations for the enite benthic community is yet to be evaluated. The picture is made compled by the fact that a mining system does not strip the entire mine site of benthos: the collector cannot be made to sweep in perfectly overlapping swaths (estimated sweep efficiency of the collector 40–75%) and some unfavorable bathymetric features (unminable portion in a mining site 10–40%) will not be mined (U.N.O.E.T.O., 1979). Recolonization also has not been addressed. Fig. 15, however, does show an epibenthic organism in an O.M.I. collector track one year after the cessation of mining.

Resedimentation

Most deep-sea organisms live on or within the top centimeter of the sediment, and the study of their morphologies suggests that they are not adapted for strenuous burrowing (R. Hessler and P. Jumars, pers. commun., 1978). Since the natural sedimentation rate in the deep sea is on the order of millimeters per thousand years, in the region of heavy burial resulting from mining, most of the fauna are likely to be buried and killed.

The sediment blanketing will also affect the food supply in the mining region. The little food that does reach the deep-sea floor almost surely exists as a minutely thin layer on the sediment surface (Hessler and Jumars, 1974). This is the pattern of food distribution to which most deep benthic species are adapted. Since most such animals are small, less than 0.5 mm in length, even a slight alteration from this pattern would be a significant deviation

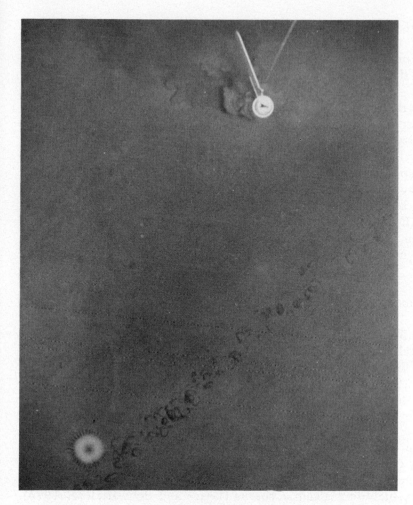

Fig. 15. The O.M.I. trace one year after mining showing recolonization by a benthic organism. Strip of unmined nodules is also seen.

from the normal conditions. Thus, what might be regarded as an insignificantly thin layer of resedimented material due to mining might actually be lethal; it diminishes the already meager food resource. For this reason, the zone of significant mortality may extend far beyond that where mortality is caused by mechanical damage or deep entombment.

There is no precedent to the impact of deep-sea mining on benthic organisms. An analogy may be made between the impact on benthic organisms from severe natural events and the impact from the disturbances caused by deep-ocean mining. This analogy has been studied in some detail by Ryan and Heezen (1976) who assessed the effects of burial of bottom life by consideration of episodic events such as ash falls from volcanic eruptions and turbidity currents. A review of these phenomena indicates that the thickness and rate of accumulation of the deposits which entraps and smothers the benthic community are important. According to Jumars (1977), what clearly precludes a valid analogy is the environmental instability characteristics of deep-sea regions subject to turbidity flows. Such flows occur most frequently in areas of high sedimentation and/or tectonic activity. The species found in these areas are precisely the ones which have evolved to survive turbidity flows (Jumars and Hessler, 1976). Thus, their response to any given turbidity deposit is not the likely response of a community which has never experienced a similar calamity.

A series of box cores and bottom photographs were taken prior to and after the O.M.I. mining test to document and quantify the impact of mining. Analysis of the results, revealing the impact of the pilot-scale mining on the macrobenthic community is expected shortly (P. Jumars, pers. commun., 1980).

Increased particulate levels on benthic organisms

The observations of the benthic plume during the O.M.I. test indicate that particulate levels within 50 m of the bottom were many times the ambient (Fig. 13) and that these increased levels could last at least a week. Suspension feeders, 19% of the total macrofauna (Hecker and Paul, 1979), and other benthic fauna in the vicinity of the mining area will have to endure these higher levels. There is no firm basis for prediction of the impacts of such events.

The bottom disturbances are expected to have a marked effect on the bottom living community of active swimmers. Bottom-mounted cameras have shown that bottom disturbances can attract a large number of deep-sea scavengers, including gigantic sharks (Isaacs and Schwartzlose, 1975). The free-fall baited trap samples obtained during the D.O.M.E.S. baseline studies indicated that a large population of active scavengers exist in the bottom waters (Hecker, 1977). Frequently, whole fish used as bait were totally consumed leaving only cartilage and bones. In several cases, fish caught in the traps were apparently attacked by the amphipods, demonstrating the ability of the latter to utilize rapidly any large parcel of food. Thus, avid scavengers are likely to make short work of macrofauna exposed or injured by the collector.

CONCLUSION

Field studies and laboratory experiments have been conducted in conjunction with manganese nodule mining tests to provide better estimates of mining parameters and the effect of mining on the environment. Results of this study include:

The mining system collector is efficient in rejecting the unwanted sediment at the sea floor; from the mining parameters, this efficiency is estimated to be 96–98%. On the average, the sediment to nodule ratio in the pipe is 1:4, and solids lifted in the pipe constitute $\sim$ 3% of the mass lifted. The average bulk density of bottom water and particulates discharged at the surface is $\sim$ 1% greater than the ambient density.

The discharge rates of particulates were highly variable. Discharged material larger than 64 μm was principally nodule fragments. In the O.M.A. test, where the separation system allowed less chance for coarser fragments to be winnowed from the discharge, material greater than 64 μm constituted 44% to the mass of the discharge on the average. It follows that abraded or fractured nodule fragments or micronodules with diameters ranging from 64 to 1000 μm constituted at least 2.5% of the nodule material brought up to the surface.

Results of laboratory tests on the fine fractions ($<$ 64 μm) of the discharge suggest that flocculation may be important in determining the settling behavior of fine mining particulates. Analysis of O.M.I. vertical profiles and nephelometer transects in and through the plume, using models of particulate dispersion, suggest that the effective mean settling velocity is in the range of 0.05–0.07 cm s^{-1}, substantially higher than has been measured with dispersed sediments collected at the mining site (Lavelle et al., 1978). Settling velocity distributional properties for the fine fraction, beyond an estimate of mean, are not possible with the available data. Because the mean effective settling velocity is higher than that of dispersed sediments, the dimensions of the surface plume in the mining tests were smaller (3–4 km^2) than previously predicted.

At the sea floor, most of the sediment resuspended by the collector return to the bottom within tens of meters of the collector track. The benthic macrofauna on the path of the collector is destroyed. However, all benthic macrofauna in the mining area will not be destroyed by collector contact because part of the area will remain unmined due to the occurrence of undesirable bathymetric features and sweeping efficiency of the collector (40–75%). The fine fraction of the resuspended sediments can remain suspended for at least a week or longer. For typical bottom currents in the manganese nodule province, resedimentation can occur over a distance of tens of kilometers from the mining area.

The effect of mining discharge on phytoplankton is limited to that caused by increased turbidity. Increased light attenuation due to increased turbidity reduces the primary production rate within the plume. The overall reduction in the primary production rate, along the plume axis at a distance of $\sim$ 500 m from the mining vessel, was within that caused by natural variability of the incident solar radiation. The short-term shading in the plume is not likely to affect the light-adaption characteristics of the phytoplankton.

474

Because the concentration levels of particulates and dissolved matter in the plume will return to ambient levels within a few days and because species composition changes occur over longer time scales, it is believed that species changes of plankton will not take place. Based on both mining tests and laboratory experiments, it is concluded that the abundance and mortality of macrozooplankton will also not be affected by the mining plume.

Uncertainties in the mining test schedules and the intermittent and short-term nature of the mining activities have made the monitoring efforts difficult and left several questions unanswered. We recommend that a longer duration, sustained mining operations be monitored to permit further observations toward determining in situ settling velocities of mining particulates; and toward the assessment of possible longer-term biological effects.

ACKNOWLEDGEMENTS

We are grateful to many friends and colleagues who shared their ideas and data with us: G. C. Anderson, E. T. Baker, A. T. Chan, J. Hirota, P. A. Jumars, R. J. Ozretich, J. W. Padan and S. A. Swift. We thank all D.O.M.E.S. personnel and the officers and the crew of the N.O.A.A. ship "Ocean-ographer" who participated in the field work or provided help. We also thank the mining consortia for their cooperation throughout this study. This investigation was funded by the Marine Ecosystems Analysis Program of N.O.A.A.

CHAPTER 16

SEDIMENT DISPERSION AND OTHER ENVIRONMENTAL IMPACTS OF DEEP-OCEAN MINING IN THE EASTERN TROPICAL PACIFIC OCEAN

Takashi Ichiye and Michael Carnes

INTRODUCTION

Manganese nodules were discovered to be distributed on the deep-sea floor of the world oceans during the famous scientific cruise of the H.M.S. "Challenger" about a century ago and have been the object of scientific curiosity since then. However, developments in deep-ocean engineering and awareness of the scarcity of land deposits have recently stimulated various governments and industries to consider retrieving the nodules as mineral resources, since these nodules are found to contain Ni, Cu and Co besides Mn. It has been assessed by many scientific cruises that nodules with Ni content above 1 wt.% are distributed within a narrow belt of the eastern tropical Pacific Ocean as indicated in Fig. 1 (Menard, 1976). It is reported that each of the several mining firms and international mining consortia operating in this area plan to mine from 1 to 3 Mt in dry weight (DW) of nodules annually. A mining ship may recover 5000—10,000 t of nodules per day and operate about 300 days per year. It is estimated that each ship will discharge ~ 1000 t DW of bottom sediments and $2 \cdot 10^4$ m^3 of bottom and interstitial water at the surface as waste products in order to recover 5000 t day^{-1} of dry nodules (N.O.A.A., 1976).

Discharged sediments may cause environmental changes in the ocean upper layer by interacting with planktonic organisms directly and by changing near-surface turbidity and thus absorption of solar radiation. Also, introduction of bottom and interstitial water in the upper layer and churning of the upper-layer water by the mining operation may change properties of the upper water in the mining area. It is out of the scope in this chapter to discuss the biological as well as the climatological effects caused by deep-ocean mining, because the processes involved in these effects are so complicated that special and completely novel approaches are necessary. Therefore, this chapter will be concerned only with an assessment of the behaviour of the waste sediments discharged from a mining ship, the physical effects of the wastewater discharged, and the mining operation from the point of view of physical oceanographers. It is also useful to describe the physical oceanography of the eastern tropical Pacific Ocean, particularly a rectangular area from 5° to 20° N latitude and from 110° to 180° W longitude, where intensive mining is expected in the early 1980's.

CIRCULATION IN THE MINING AREA

The mean subsurface circulation can be represented by the topography of a constant potential density surface depicted here in a chart modified from Tsuchiya (1968), derived from hydrographic data collected up to 1967. Fig. 1 indicates the depth contour of the 160-cl per ton specific volume anomaly surface. The area surrounded by the broken line is expected to be mined extensively and is called the D.O.M.E.S. (Deep Ocean Mining Ecosystem Study) area by the N.O.A.A. (1976) Environmental Research Laboratory. Three currents are distinguished in the D.O.M.E.S. area: the North Equatorial Current (NEC), the Equatorial Counter Current (ECC) and the South Equatorial Current (SEC). The NEC is the southern component of the counter-clockwise gyre mainly generated by the anticyclonic atmospheric circulation in the North Pacific Ocean, and the SEC is the counter part in the South Pacific Ocean. Because of the asymmetry of the atmospheric circulation in the North Pacific Ocean, and the SEC is the counter-part in the South Pacific Ocean. Because of the asymmetry of the atmospheric circulation about the equator, the SEC is found ~ 4—5° north of water against the western boundary of the ocean. These three currents are distinguished from the mid-latitude currents by their almost zonal pattern, although there are some indications suggesting meridional components due to climatological gyres (Wyrtki, 1966; Tsuchiya, 1968) discussed later.

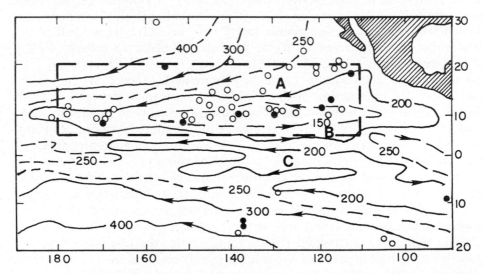

Fig. 1. Depths (m) of the 160-cl-t^{-1} specific volume anomaly. *A*, *B* and *C* indicate the North Equatorial Current (NEC), North Equatorial Counter Current (NECC) and South Equatorial Current (SEC), respectively. Nodule locations are shown in *open circles* (nickel content is 1–1.5 wt.%) and in *closed circles* (> 1.5 wt.% Ni).

Since waste sediment and water are to be discharged within the upper 10 m from a mining ship, the surface current seems to be more important than the current in the upper few hundred meters which is shown in Fig. 1, although the latter represents a more stable climatological circulation pattern. Therefore, the mean surface currents are determined from the National Oceanographic Data Center (N.O.D.C.) ship drift data (from 1900 to the present but mainly after 1945) in each season for each rectangular area of $2°$ latitude by $10°$ longitude in the Eastern Pacific from the equator to $30°N$. These are plotted in Fig. 2 (Horne, 1980).

The surface current pattern also indicates zonal structure and the three distinct currents. The northern part of the SEC is stronger, with maximum speed reaching 1.5 kt. (0.75 m s^{-1}), and more zonally consistent in fall and winter than in spring and summer. It reaches at least $4°N$ in all seasons and is stronger in the eastern part. The ECC between $6°$ and $10°N$ is strongest and best defined in fall and weakest in spring. The NEC is less variable seasonally in speed than the other two currents, although in spring it reaches farthest south down to $8°N$. It is strongest between $10°$ and $16°N$ but always extends as far north as $24°N$ except in winter. It is important that the northern SEC is always stronger than the NEC, since this indicates that the surface currents are not only driven by the surface winds, but also by the geostrophic process which causes stronger current near the equator for the same pressure gradient. The seasonal change of the ECC may reflect the atmospheric circulation which shows the well-defined doldrum in fall.

In order to show some synoptic features of the surface currents, Fig. 3 indicates charts of geopotential anomaly in joule per kilogram at the surface relative to the 500-dbar level from EASTROPAC data (Tsuchiya, 1974). These charts indicate the difference of the surface current from the current at ~ 500 m, where the flow is considered negligibly small. They show not only the three currents described above but also several large gyres embedded in these almost zonal flows. Particularly, the ECC is not a continuous eastward flow but flows around gyres located in an almost zonal row, as a belt around wheels turning clockwise or counter-clockwise. Some of these gyres are recognized to persist from season to season with some displacements and change in shape, as for example, the anticyclonic gyres with centers at $5°N,122°W$ and $14°N,98°W$ ($104°W$ in August—September) and a cyclonic gyre off Central America. The latter, in particular, is called the Costa Rica Dome. It is unknown how these gyres move or change their shape and extent, but they may be called climatological gyres since they seem to be preserved from year to year.

As discussed later, the thermocline depth is important for estimating the dispersion of the waste sediments or water in the upper layer. Therefore, the depth of the thermocline determined from 29 years of bathythermograph data is shown for February and July in Fig. 4A and B, respectively (M. K. Robinson and Bauer, 1971). The thermocline depth is defined as the depth at which the temperature is $\sim 1.7^\circ$C (2°F) less than the surface temperature. It is noted that the D.O.M.E.S. area shows a shallow thermocline depth compared to regions to the north and west both in winter and summer. Particularly, the shallow thermocline zone exists along latitude 10°N almost the whole year due to Ekman upwelling in this zone as discussed later. The thermocline depth is, in general, deeper in winter than in summer. Also, the depth follows the geostrophic flow pattern more faithfully in winter than in summer. The February thermocline topography (Fig. 4A) corresponds approximately to the mean surface current in Fig. 2D, whereas in the July chart (Fig. 4B) the smaller depth between 5° and 15°N and to the east of 160°W may be due to higher surface temperatures caused by surface heating.

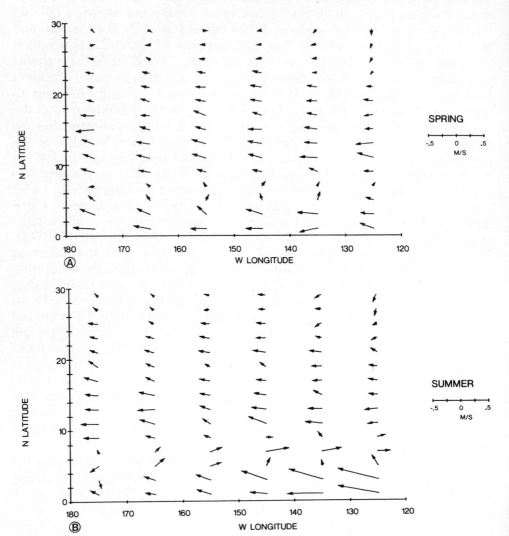

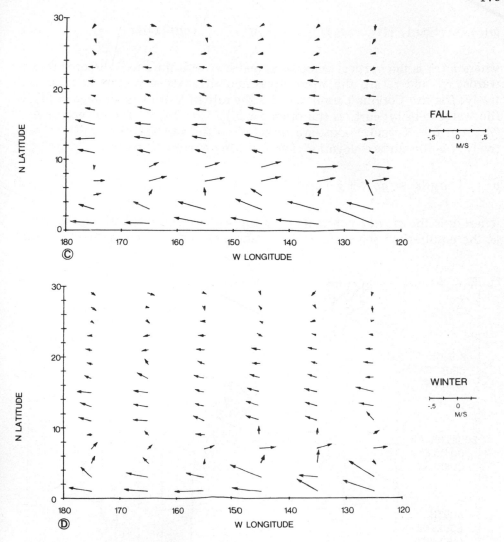

Fig. 2. The seasonal mean current in the tropical Eastern Pacific Ocean determined from ship drift data (Horne, 1980).

VERTICAL CURRENT

The vertical current is particularly important in dispersion and transport of the waste sediments. However, it is impossible to measure directly the vertical velocity in a wide area of the ocean. Therefore, the vertical velocity is estimated indirectly by use of dynamic relations. One such formula is given by:

480

$$w(z) = -\text{curl}_z\left[(\vec{\tau}_w - \vec{\tau}_z)f^{-1}\right] + \beta(\rho f^2)^{-1} \int_{-\zeta}^{z} (\partial p/\partial x)\,dz \qquad (1)$$

where $w(z)$ is the vertical velocity at depth z, with the z-axis directed downwards; $\vec{\tau}_w$ and $\vec{\tau}_z$ are the wind stress and shear stress vectors at z, respectively; f is the Coriolis parameter; β is the rate of latitudinal change in f; ρ is the water density; curl_z is the operator $\partial A_y/\partial_x - \partial A_x/\partial_y$ for the vector $\vec{A} = (A_x, A_y)$, the X- and Y-axes are directed to the east and the north, respectively; ζ is the surface elevation; and p is the pressure. The latter is given by:

$$p = \int_{-\zeta}^{z} \rho g\,dz \approx g\rho_s\zeta + g \int_{0}^{z} \rho\,dz \qquad (2)$$

where g is the gravity constant; ρ_s is the surface density; and $z = 0$ is taken at the equilibrium sea surface. Eq. 1 can be obtained by cross-differentiating

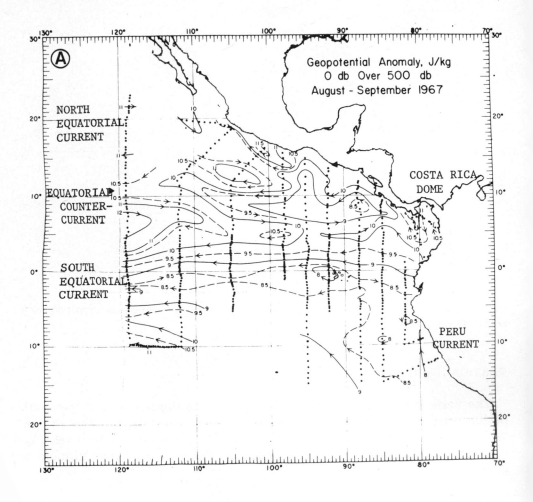

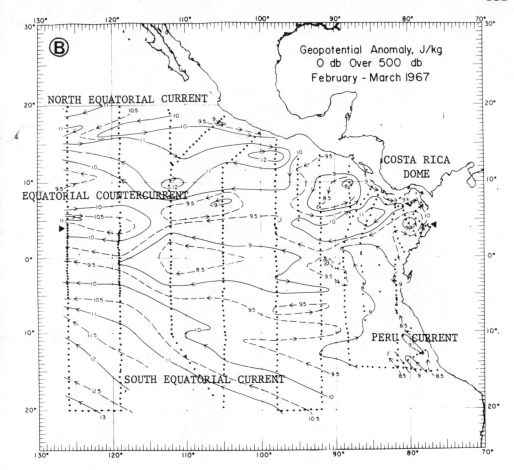

Fig. 3. Geopotential anomaly at the surface relative to 500 dbar from EASTROPAC data (Tsuchiya, 1974).

the stationary, quasi-geostrophic momentum equations and substituting the result into the continuity equation with the boundary condition of vanishing vertical velocity at the surface (Ichiye, 1958; Overstreet and Rattray, 1969).

The last term of the r.h.s. of eq. 1 represents the effect of planetary divergence caused by a change in magnitude of the meridional component of a geostrophic flow (v_g) and can be expressed by:

$$w_p = \beta f^{-1} \int_{-\zeta}^{z} v_g \, dz \qquad (3)$$

The magnitude of w_p becomes significant near the equator and also at great

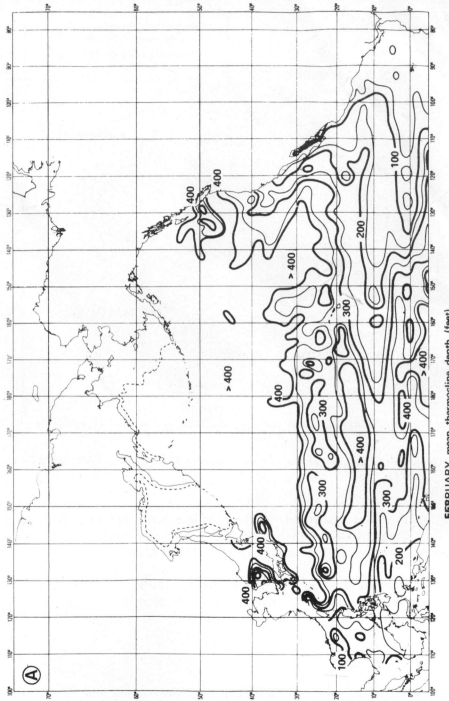

FEBRUARY mean thermocline depth (feet).

Ⓐ

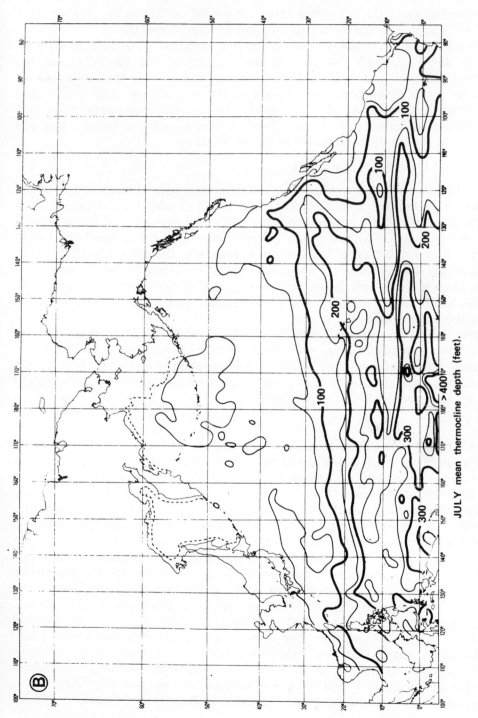

JULY mean thermocline depth (feet).

Fig. 4. The mean thermocline depth in feet (1 ft. = 0.3048 m) (M. K. Robinson and Bauer, 1971).

depths. For example, for $v_g = 1\,\mathrm{cm\,s^{-1}}$ at the $10°$ latitudes, w_p reaches 10^{-4} and $10^{-3}\,\mathrm{cm\,s^{-1}}$ at the depths of 100 and 1000 m, respectively.

Munk (1966) postulated that there is a constant upwelling velocity of $1.2\,\mathrm{cm\,day^{-1}}$ $(1.4 \cdot 10^{-5}\,\mathrm{cm\,s^{-1}})$ in the interior of the Pacific Ocean excluding the top and bottom kilometer. He argued that this is caused by bottom water formation in the Pacific at the rate of $\sim 4 \cdot 10^{20}\,\mathrm{g\,yr.^{-1}}$ due to annual freezing of Antarctic pack ice. This was based on the evidence that the vertical distributions of conservative properties such as temperature and salinity in the interior can be fitted to the solution:

$$S = \tfrac{1}{2} S_0 [1 + \exp(-K_z z\,w^{-1})] \tag{4}$$

of a diffusion equation, where K_z is the vertical eddy diffusivity; S is temperature or salinity; and S_0 is the value at $z = 0$ taken at a depth of 1 km from the surface. He conjectured that a K_z of $1\,\mathrm{cm^2\,s^{-1}}$ can be accounted for by turbulence in the ocean interior, generated by internal tides with energy flux $4 \cdot 10^{-6}\,\mathrm{erg\,g^{-1}\,s^{-1}}$, about one-sixth of the total tidal dissipation. Therefore, the values of w_p estimated above are larger by orders of magnitude than the one postulated by Munk (1966) if $v_g = 1\,\mathrm{cm\,s^{-1}}$. However, it is doubtful that this magnitude of the mean meridional velocity exists all over the Pacific interior.

The shear stress vector $\vec{\tau}_z$ can be expressed by $K_z(\partial \vec{u}/\partial z)$, where $\vec{u}$ is the horizontal velocity vector. The shear stress becomes much smaller in the thermocline where the static stability reduces the vertical eddy viscosity K_z. Thus, the vertical velocity in the thermocline or just below the upper mixed layer can approximately be given by:

$$\vec{w}_m = -f^{-1}\,\mathrm{curl}_z\,\vec{\tau}_w \tag{5}$$

This is called Ekman pumping.

The r.h.s. of eq. 5 is calculated by use of the monthly mean wind stresses averaged over a rectangle covering $2°$ of latitude and $10°$ of longitude based on the wind data obtained by ships over about 29 years (Wyrtki and Meyers, 1975). The vertical velocity in $10^{-5}\,\mathrm{cm\,s^{-1}}$ is plotted in Fig. 5 in bimonthly charts. The values are not valid within $2°$ of latitude from the equator as the Coriolis parameter vanishes at the equator.

The upwelling zone which becomes important in dispersion of the waste sediments exists throughout the year in the D.O.M.E.S. area. The magnitude of the vertical speed is, in general, larger in the upwelling area than in the downwelling area. Maximum values of both areas reach above $10^{-4}\,\mathrm{cm\,s^{-1}}$, thus justifying the neglect of w_p compared to w_m for the vertical velocity in the thermocline. The upwelling zone extends eastwards in latitudes lower than $10°\mathrm{N}$ from November to June with a maximum intensity reaching more than $10^{-3}\,\mathrm{cm\,s^{-1}}$ in March—April. The upwelling zone moves northwards near $10°\mathrm{N}$ and reduces in area and intensity in July—August and then again increases in area and intensity in September—October. The increase in downwelling speed and area south of 10°N in July—August does not reflect on the depth of the thermocline in summer as is indicated in Fig. 4B, because the solar heating increases the sea surface temperature, and thus decreases the depth of the thermocline in spite of the increased down-welling.

In and near the equator, eq. 1 fails because the Coriolis factor f vanishes at the equator. Instead, the vertical velocity can be computed by use of the balance among the advective terms including the product of vertical velocity and the vertical shear of the zonal current, frictional forces, zonal component of the Coriolis force and the pressure gradient (Ichiye, 1966). The zonal average of the vertical velocity obtained at the equator thus reaches a maximum of $6 \cdot 10^{-3}$ cm s^{-1} (upward) at ~ 200 m. Further, at certain depths the vertical speed reaches 10^{-1} cm s^{-1} at two or three locations between 90° and 140°W. Chekotillo (1970) obtained similar high speeds of 10^{-2} cm s^{-1} as an average and of 10^{-1} cm s^{-1} occasionally in the equatorial Western Pacific Ocean.

BASIC PHYSICAL PARAMETERS FOR SEDIMENT DIFFUSION

Besides the currents and their fluctuations, there are several physical parameters which affect the diffusion and transportation of the sediments discharged at the surface. One such is related to the physical nature of sediments and does not depend on the dynamics of the ocean water. This is the settling velocity, w_s, of a sediment particle and depends on the density, size and shape of the particle. For a spherical particle, w_s is given by Stokes' law (Batchelor, 1967). For particles of other shapes, w_s is determined both from hydrodynamic formulas and experiments (Munk and Riley, 1952). The results are plotted in Fig. 6 for: a sphere; a cylinder with diameter, d, falling laterally; and a plate with thickness $\frac{1}{3} d$ (d is the short-side length) falling along d (with a 10° angle from the horizontal direction), all with specific gravity 1.2 g cm^{-3}. The predominant sediments of the D.O.M.E.S. area are red clays with diameters from 0.5 to 2.5 μm and wet specific gravity of 1.2—1.7 g cm^{-3}. Turbidites located in isolated areas have a size ranging 0.8—243 μm and wet specific gravity ranging from 1.2 to 2.4 g cm^{-3} (Horne et al., 1974). Theoretically, therefore, the sediment particles from mining operations have settling velocities from 10^{-4} m s^{-1} for coarse particles of spherical shape 40 μm in diameter to 10^{-5} m s^{-1} for fine sediments of spherical shape and 3 μm in diameter. However, measurements of discharged sediments by N.O.A.A. researchers for Ocean Mining Associates indicate 44% ± 21% of the mass failed to pass through a 64-μm mesh sieve (see the previous chapter, Chapter 15), probably due to the special separation technique employed by them. Also, in actual situations flocculation and coagulation may increase the average settling velocity. Therefore, a settling velocity of 10^{-4} m s^{-1} seems to represent the average value for the sediments discharged.

The eddy diffusion coefficients introduced in eq. 4 are other parameters characterizing dispersing effects of the water motion. These coefficients are different in the vertical and horizontal direction by orders of magnitude.

The vertical eddy diffusivity depends mainly on vertical velocity shear and stability and to a lesser degree on the type of diffusant. For suspended matter, Ichiye et al. (1972) obtained a vertical eddy diffusivity of 10^{-4}—$8 \cdot 10^{-4}$ $m^2 s^{-1}$, whereas the zonal average eddy viscosity at the core of the Equatorial Undercurrent was determined to be $4.6 \cdot 10^{-4}$ $m^2 s^{-1}$ (Ichiye, 1966). The classical study of the eddy diffusivity and viscosity in

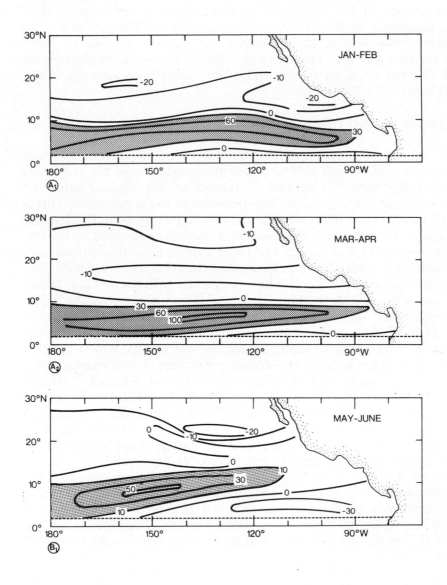

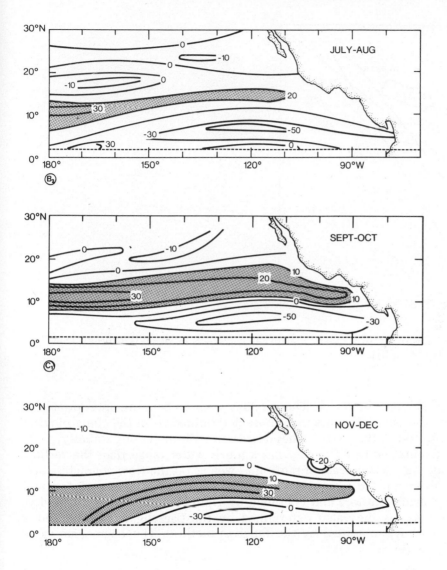

Fig. 5. The vertical current at the bottom of a mixed layer in 10^{-5} cm s^{-1} computed from the curl of the mean wind stress. The positive and negative values indicate upwelling and downwelling, respectively.

the upper 100 m of the world ocean gives values ranging from 10^{-4} to 10^{-1} m^2 s^{-1} (Defant, 1961). Carnes (1975) determined the eddy diffusivity in the tropical Atlantic Ocean at about 13°N and 54°W east of Barbados. He used time series data obtained four times a day for temperature and salinity, and obtained different values for temperature and salinity, ranging from $4 \cdot 10^{-4}$ to $6.5 \cdot 10^{-3}$ m^2 s^{-1} for the former and from 10^{-4} to

488

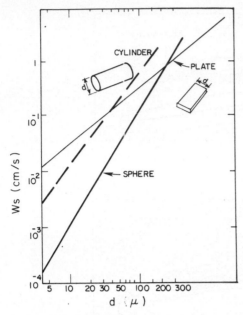

Fig. 6. Settling velocity of a sphere, a plate and a cylinder vs. linear dimensions [based on Munk and Riley (1952)].

$4 \cdot 10^{-3}$ m² s⁻¹ for the latter. These values seem to be more appropriate for dispersion of waste sediments than those determined from the climatological data. He also found that both values decreased with increasing stability.

The horizontal eddy diffusivity has a much wider range than the vertical one and depends on the scale of motion as well as on the flow characteristics. The diffusivity in the ocean was determined from diffusion of dye and drifters by many authors and compiled by Okubo (1971). This diffusivity, denoted by A_h, is the value determined by the Lagrangian method and is different from K_h obtained by the Eulerian method. The Lagrangian value A_h is given by (in m² s⁻¹):

$$A_h = 5.2 \cdot 10^{-5} l^{1.15} \tag{6}$$

where l is size of the diffusant cloud in meters. The power of l is slightly different from the $\frac{4}{3}$ postulated by the dimensional argument of turbulence theory (Csanady, 1973). The actual plumes of waste sediment in the pilot-scale mining operation (Chapter 15) had widths from 100 m 30 min. after discharge to 1 km several hours later and reached a length of more than 5 km at that time. Therefore, the Okubo diagram predicts that the lateral eddy diffusivity ranges from 0.5 to 5 m² s⁻¹ and that the longitudinal diffusivity is more than 50 m² s⁻¹ within several hours after the start of the mining operation.

In order to discuss the long-term effects for a duration longer than a few days, this kind of horizontal eddy diffusivity is inadequate. Ichiye and Carnes (1977) obtained the Lagrangian eddy diffusivities A_u and A_v, in the zonal and meridional directions, respectively as $A_u = 0.7 \cdot 10^2$, $8.32 \cdot 10^3$ $m^2 \, s^{-1}$ and $A_v = 2.1 \cdot 10^3$, $3.5 \cdot 10^2 \, m^2 \, s^{-1}$ from drifters tracked for about a month in the ECC between 125° and 155°W and 4° and 10°N in 1975 (Kerut, 1976). Apparently, these values are higher by orders of magnitude than those estimated for the mining plumes.

More appropriate for dispersion on the climatological scale are the variances of the drift of the ships, of which the average values are plotted in Fig. 4. The variances $\sigma_u^2 \, (= \overline{u^2})$ and $\sigma_v^2 \, (= \overline{v^2})$ for the zonal and meridional components, respectively, and their covariance $\sigma_{uv}^2 \, (= \overline{uv})$ are computed from the same data source as in Fig. 4. The variance ellipse is computed from the variances and covariance for each rectangular area, of 2° latitude by 10° longitude, where the direction of the major axis, θ, is measured counter-clockwise from the east and is expressed by:

$$\tan(2\theta) = \overline{uv}/(\overline{v^2} - \overline{u^2}) \tag{7}$$

The half-lengths of the major and minor axes of this variance ellipse correspond to σ_{u_0} and σ_{v_0}, respectively, where $\sigma_{u_0}^2$ and $\sigma_{v_0}^2$ are the variances of the velocity components in the direction of the major and minor axes, respectively (Horne, 1980).

The variance ellipses are plotted for 0° to 20°N and 120° to 180°W in Fig. 7. In general, the area of the ellipses is larger to the south of 10°N than to the north of that latitude. They are also larger in the western part than in the eastern part except near the equator. It is noticeable that the ellipse is large in the zone of the ECC where the mean flow is rather weak or its direction is not consistent. In general, the major axis of the ellipse is parallel to the mean flow again except in the ECC.

The variances of positions of drifters or particles can be expressed by the variances of the speed in the direction of the major and minor axes of the velocity variance ellipse, if the Lagrangian correlation coefficients of the velocity u and v are known as (Pasquill, 1974):

$$\sigma_X^2 = \sigma_{u_0}^2 \int_0^t R_{u_0}(s)(t-s)\,ds$$

$$\sigma_Y^2 = \sigma_{v_0}^2 \int_0^t R_{v_0}(s)(t-s)\,ds \tag{8}$$

respectively, where X and Y are the major and minor axes of the ellipse, respectively; and $R_{u_0}(t)$ and $R_{v_0}(t)$ are the autocorrelations of the velocity components u_0 and v_0, respectively. There are no data available for determining these autocorrelations over the entire D.O.M.E.S. area. However, the average values of R_{u_0} and R_{v_0} computed from the zonal and meridional

490

SPRING

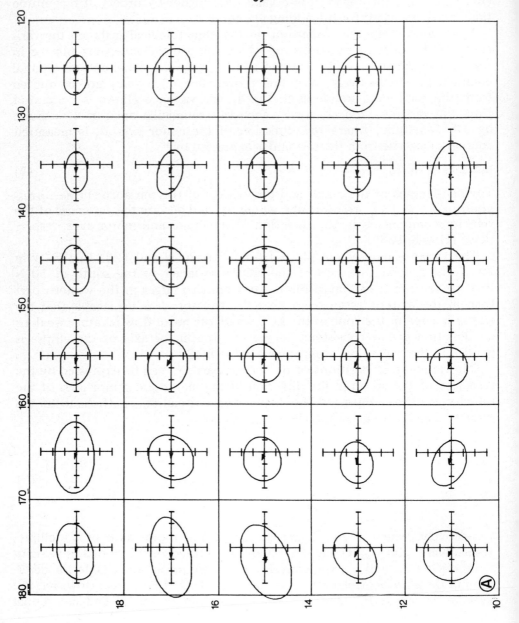

491

SPRING

0 M/S 1

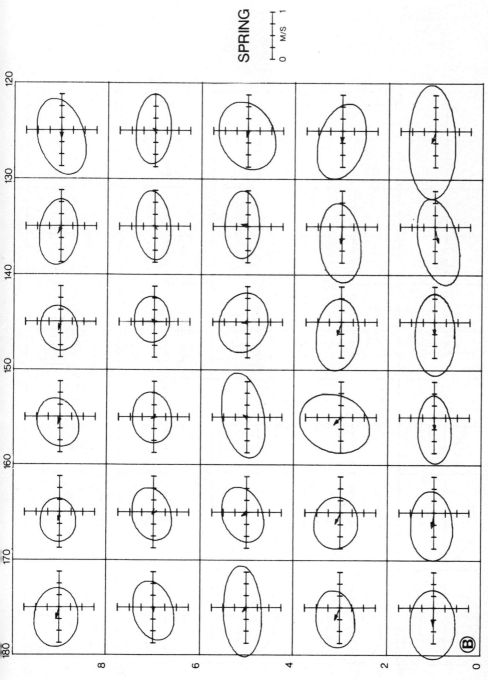

Fig. 7A and B. For caption see p. 497.

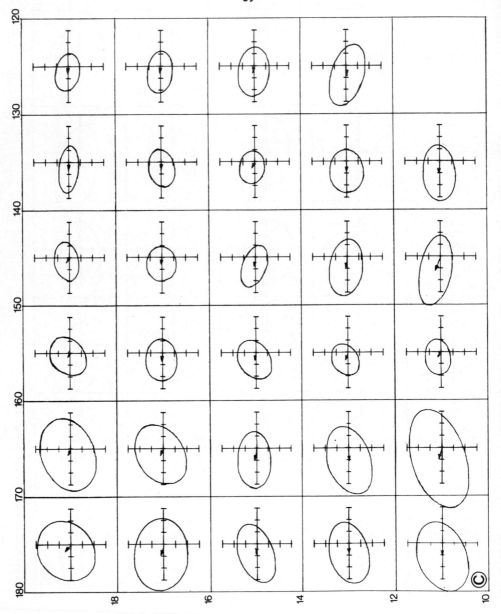

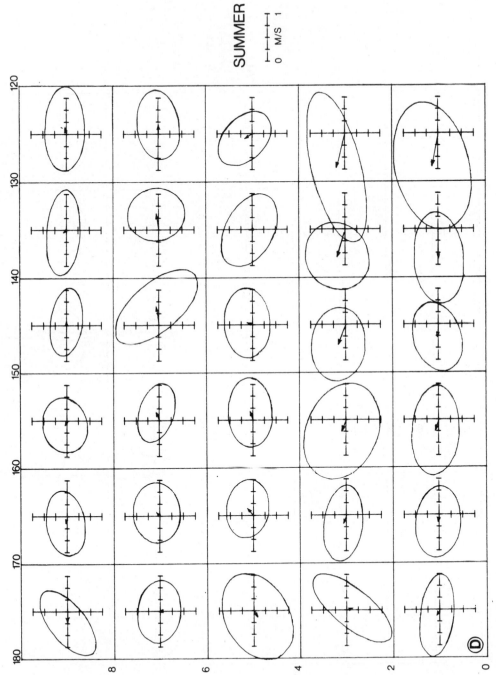

SUMMER

Fig. 7C and D. For caption see p. 497.

494

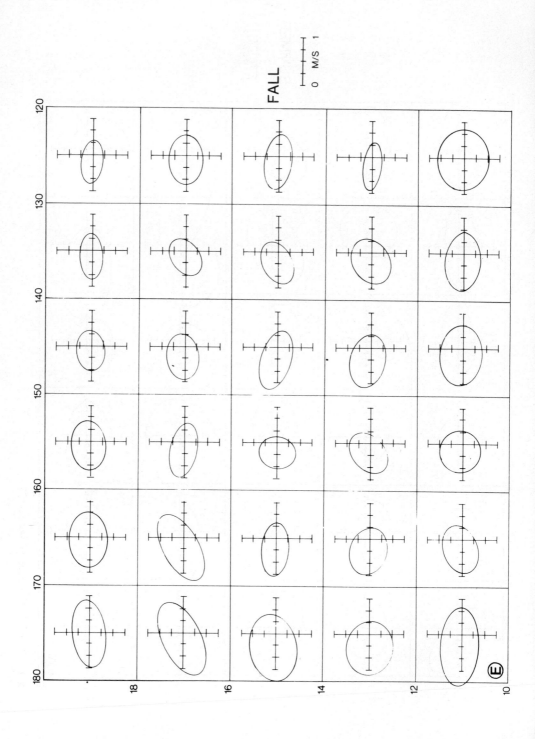

FALL

0 M/S 1

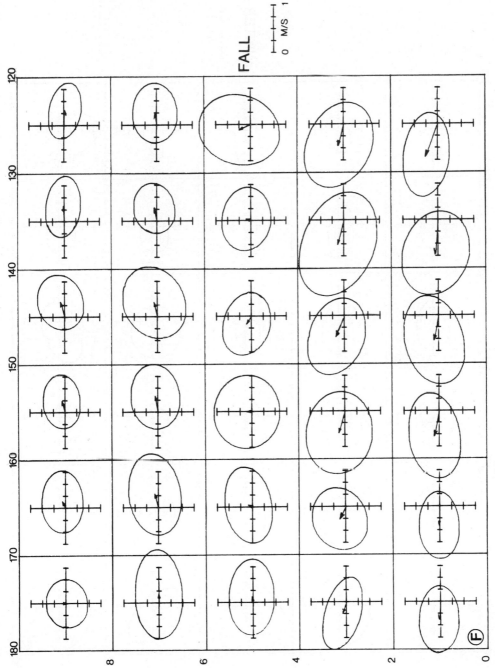

FALL

Fig. 7E and F. For caption see p. 497.

496

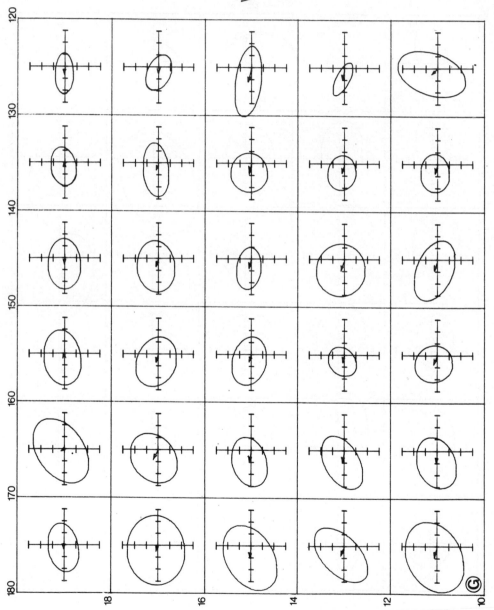

WINTER

0 M/S 1

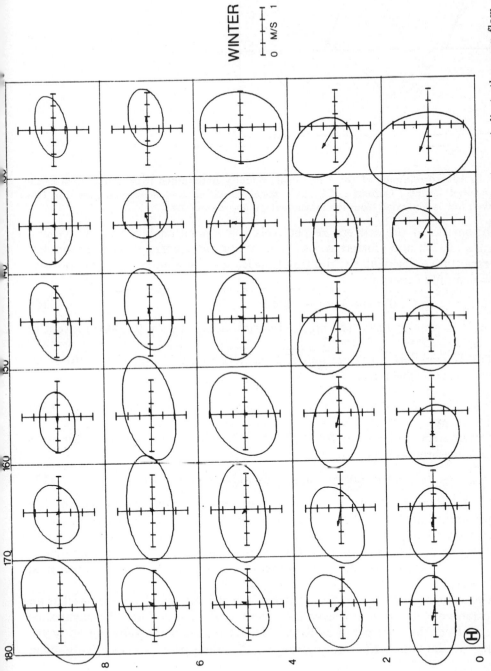

WINTER

0 M/S 1

Fig. 7. Ellipses of the current variances in each season determined from ship drift data. *Arrows* indicate the mean flow. The ellipse center is at the end-point of each *arrow* (Horne, 1980).

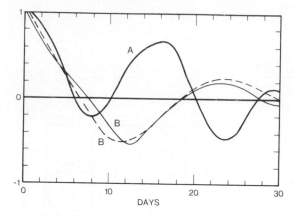

Fig. 8. Autocorrelations of the Lagrangian velocity determined from drifter data in the Equatorial Counter Current (ECC). A = zonal direction; B = meridional direction (the *broken line* is generated from the mathematical approximation).

components of the drifters, respectively, have been analyzed by Ichiye and Carnes (1977) and plotted in Fig. 8, assuming these correlation functions are representative. Particularly, the function R_{v_0} can be approximately expressed with a simple function:

$$R_{v_0}(t) = \exp(-0.0544t) \cos(0.254t) \tag{9}$$

where t is time in days. The approximation of R_{v_0} by eq. 9 is also plotted in Fig. 8.

   The Lagrangian horizontal eddy diffusivity in the X- and Y-directions can respectively be given by:

$$A_X = \sigma_{u_0}^2 \int_0^\infty R_{u_0}(t)\,dt \quad \text{and} \quad A_Y = \sigma_{v_0}^2 \int_0^\infty R_{v_0}(t)\,dt \tag{10}$$

The numerical integration for A_X and the analytical integration with eq. 9 for A_Y lead respectively to:

$$A_X/\sigma_{u_0}^2 = 4.77 \cdot 10^5 \text{ (s)} \quad \text{and} \quad A_Y/\sigma_{v_0}^2 = 6.97 \cdot 10^4 \text{ (s)} \tag{11}$$

Applying these relations, σ_{u_0} and σ_{v_0} can be determined from half of the major and minor axes of the variance ellipse in Fig. 7. The range of the eddy diffusivity so determined is from $4 \cdot 10^3$ to $5 \cdot 10^5$ m^2 s^{-1} and is much larger than those determined from the Okubo diagram for plumes and from drifter data. Therefore, these values should be interpreted as valid for climatological purposes.

   The velocity determined from the drift of the ship can be considered as an Eulerian velocity when it is averaged over a limited area. Then the covariance $\sigma_{uv}^2 = \overline{uv}$ can be considered as Reynolds' stress. Then the Eulerian eddy viscosity in the zonal, K_x, and meridional, K_y, directions is respectively given by:

$$K_x = -\overline{uv}\,(\partial V/\partial x)^{-1} \qquad \text{and} \qquad K_y = -\overline{uv}\,(\partial U/\partial y)^{-1} \qquad (12)$$

where U and V are the average zonal and meridional velocity, respectively. The mean current and the covariance are determined for rectangles of $10°$ longitude by $2°$ latitude. Thus, the value of K_y is more reliable because of a more accurate $\partial U/\partial y$ gradient. The results indicate that the magnitude of K_y ranges from 10^5 to 10^3 m^2 s^{-1}, which is almost similar to those of A_X and A_Y (Horne, 1980). However, K_y becomes negative in some areas. Actually, positive and negative areas are almost zonal, suggesting that positive values represent dissipation of the mean zonal flow energy by turbulence, whereas the negative values indicate that the energy of the mean current is supplied by eddies. The dissipation and supply of the mean flow from turbulence can be compared with the mean flow of Fig. 7. In spring the ECC is dissipated whereas the NEC from $10°$ to $15°$N in the west and to $20°$N in the east is supplied energy by eddies. In summer the area east of $160°$W and south of $12°$N is supplied by eddies, whereas again the NEC between $8°$ and $14°$N in the west and between $12°$ and $18°$N in the east is dissipated by turbulence. In fall the ECC between $6°$ and $10°$N is consistently supplied energy by eddies, showing intensification, whereas the rest of the current system shows dissipation. In winter, the eastern part of the SEC and the NEC from $12°$ to $18°$N in the eastern part show dissipation, but the other part, particularly the ECC is supplied energy by eddies.

DISPERSION FROM A POINT-SOURCE

If the waste sediment is discharged from a mining ship of draft 10 m and width 10 m, it is expected to fill the vertical cross-section of 10^2 m with a uniform concentration within a few minutes after discharge. Therefore, when the concentration of the sediment is studied for the domain of a few kilometers distance from the ship, the source can be considered almost as a point-source. The prediction models can then be based on the theory of Lagrangian diffusion from a point-source. In such models the solution can be generalized to non-stationary and variable currents and eddy diffusivities by use of numerical integration without solving differential equations (Frenkiel, 1953).

The concentration of sediment discharged from a point-source at the surface can be expressed in non-dimensional form as:

$$S = S_2 \int_0^t (2\pi s)^{-3/2} \exp\left[-\{(x-\mu s)^2 + y^2\}(4s)^{-1}\right]$$

$$\times \left[\exp\{-(z-\omega s)^2(4s)^{-1}\} - (\omega/2)\int_0^\infty \exp\{-(z+\eta-\omega s)(4s)^{-1}-\eta\}\,du\right]\,ds$$

$$(13)$$

This relation is obtained by integrating the Gaussian distribution for the sediment which is subjected to horizontal and vertical diffusion and a mean flow U in the x-direction. The distribution satisfies the boundary condition that there is no flux of sediment at the surface. In eq. 13, all the quantities are non-dimensionalized. The concentration S is scaled with S_2; the time t with T_2; and distances x, y and z are scaled with L_{2x}, L_{2y} and L_{2z}, respectively. The characteristic concentration S_2, time T_2 and lengths L_{2x}, L_{2y} and L_{2z} are respectively defined by:

$$S_2 = RUA_x^{-1}(2A_y A_z)^{-1/2} \quad \text{and} \quad T_2 = A_x U^{-2} \tag{14}$$

$$L_{2x} = A_x U^{-1}, \quad L_{2y} = (A_y A_z)^{1/2} U^{-1} \quad \text{and}$$

$$L_{2x} = (A_z A_x)^{1/2} U^{-1} \tag{15}$$

In these relations, R is the discharge rate. The parameter ω represents the settling velocity w_s which is non-dimensionalized by:

$$\omega = w_s U^{-1} (A_x / A_z)^{1/2} \tag{16}$$

The parameter μ is a scaled velocity, and may be variable with y, z and t but equals unity when the advective current is constant.

The scaling from eqs. 14—16 is based on the horizontal velocity and eddy diffusivities and is suitable for concentration in the domain at a distance less than 10 km from the source and at a short time, say less than a few days, after the start of the discharge. To show magnitudes of these scaling parameters, Table I indicates typical values of dimensional parameters suitable for a deep-ocean mining operation and its operational area. For these characteristic values, the scaling parameters are given by:

$$T_2 = 2 \cdot 10^4 \text{ s}; \quad L_{2x} = L_{2y} = 2 \text{ km}; \quad L_{2z} = 6.32 \text{ m};$$

$$\omega = 0.016; \quad \text{and} \quad S_2 = 25.9 \text{ mg l}^{-1}$$

The scaling parameters for other characteristic values are listed in Table II. In the case of $\omega \ll 1$, which is suitable for finer particles, eq. 13 can be simplified to:

$$S/S_2 \approx 2^{-1/2}(2r\pi)^{-1} \exp\left[\tfrac{1}{2}(x + z\omega)\right]\left[e^{-2r} \text{ erfc } \{(r-1)t^{-1/2}\}\right.$$
$$\left. + e^{2r} \text{ erfc } \{(r+1)t^{-1/2}\}\right] \tag{17}$$

where r is the distance from the source.

TABLE I

Characteristic values of physical parameters suitable for a deep-ocean mining operation

A_x, A_y, K_x, K_y (m² s⁻¹)	10^2	A_z, K_z (m² s⁻¹)	10^{-3}
h (m)	25	U (m s⁻¹)	0.2
R (t day⁻¹, g s⁻¹)	10^3, 11.57	w_s (m s⁻¹)	10^{-5}

TABLE II

Scaling parameters for the near-source region in the Lagrangian diffusion from a point-source

Time and space scaling:

A_x (m²s⁻¹)	T_2 (s) U(ms⁻¹)=10^{-2}	10^{-1}	L_{2x} (km) 10^{-2}	10^{-1}	L_{2z} (m) ($A_z = 10^{-4}$ m²s⁻¹) 10^{-2}	10^{-1}	($A_z = 10^{-3}$ m²s⁻¹) 10^{-2}	10^{-1}
10^2	10^6	10^4	10	1	10	1	31.6	3.2
10^3	10^7	10^5	10^2	10	31.6	3.2	10^2	10

Characteristic concentration, S_2 (μg l⁻¹):

A_z (m²s⁻¹)	($A_x = A_y = 10^2$ m²s⁻¹) U(ms⁻¹)=10^{-2}	10^{-1}	($A_y = 10^3$ m²s⁻¹ = $10A_x$) 10^{-2}	10^{-1}
10^{-4}	8,183	81,828	2,588	25,879
10^{-3}	2,588	25,879	259	2,588

Parameter, ω

A_x (m²s⁻¹)	w_s (ms⁻¹) = 10^{-5}; A_z (m²s⁻¹) = 10^{-4}; U(ms⁻¹)=10^{-2}	10^{-1}	A_z = 10^{-3}; 10^{-2}	10^{-1}	w_s = 10^{-4}; A_z = 10^{-4}; 10^{-2}	10^{-1}	A_z = 10^{-3}; 10^{-2}	10^{-1}
10^2	1	0.1	0.316	0.032	10	1	3.16	0.32
10^3	3.16	0.32	1	0.1	31.62	3.16	10	1

The sediment concentration is computed from eq. 13 by use of scaling parameters (14)—(16). These scaling parameters are calculated with the physical constants listed in Table I, except that $w_s = 10^{-4}$ m s⁻¹ is used instead of 10^{-5} m s⁻¹. The vertical sections within 10 km are calculated almost every hour after discharge and approach a steady state after ~ 10 hr. The vertical section shown in Fig. 9 is at 11.1 hr. after the start of the discharge. The horizontal eddy diffusivity used is 10^2 m²s⁻¹ and is larger by an order of

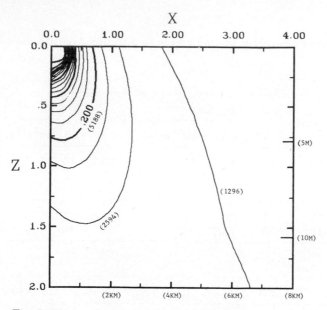

Fig. 9. The vertical distribution of waste sediment concentration along the central section near a point-source 11.5 hr. after the start of discharge. Dimensional values (S in $\mu g\,l^{-1}$) are for physical constants listed in Table I.

magnitude than the value calculated by eq. 6 for a length scale of 10 km. Therefore, a realistic concentration may be about three times the one shown in Fig. 9 except very near the source, considering the scaling parameters S_2 and L_{2x} given by eqs. 14 and 15, respectively.

For the concentration a long time after the start of the discharge and in the domain far from the source, different scaling is more convenient because the characteristic time and horizontal length scales are much larger than those defined by relations (14)—(16). Such scaling is based on the settling velocity and the vertical diffusivity. The characteristic concentration and time are respectively given by:

$$S_3 \ = \ Rw_s A_z \, (2A_x A_y)^{-1/2} \qquad \text{and} \qquad T_3 \ = \ A_x w_s^{-2} \tag{18}$$

The characteristic distances in the three directions x, y and z are respectively given by:

$$L_{3x} \ = \ (A_x A_z)^{1/2} w_s^{-1}, \qquad L_{3y} \ = \ (A_x A_z)^{1/2} w_s^{-1}$$
$$\text{and} \qquad L_{3z} \ = \ A_z w_s^{-1} \tag{19}$$

Since the settling velocity is one of the scaling parameters, ω equals unity whereas μ is now defined by:

$$\mu \ = \ U(A_x/A_y)^{1/2} w_s^{-1} \tag{20}$$

The scaling parameters for the probable range of physical parameters in the

TABLE III

Scaling parameters for the distant region from a point-source for Lagrangian diffusion

Time and space scaling:

A_z (m² s⁻¹) w_s (ms⁻¹)	T_3 (s)		L_{3z} (m)		L_x (km) $(A_x = 10^2\,\text{m}^2\,\text{s}^{-1})$		$(A_x = 10^3\,\text{m}^2\,\text{s}^{-1})$	
	10^{-5}	10^{-4}	10^{-5}	10^{-4}	10^{-5}	10^{-4}	10^{-5}	10^{-4}
10^{-4}	10^{6}	10^{4}	10	1	10	1	31.6	3.16
10^{-3}	10^{7}	10^{5}	10^{2}	10	31.6	3.16	10^{2}	10

Characteristic concentration, $S_3\,(\mu g\,l^{-1})$:

A_z (m² s⁻¹) w_s (ms⁻¹)	$(A_x = A_y = 10^2\,\text{m}^2\,\text{s}^{-1})$		$(A_y = 10^3\,\text{m}^2\,\text{s}^{-1} = 10A_x)$	
	10^{-5}	10^{-4}	10^{-5}	10^{-4}
10^{-4}	8,183	81,828	2,588	25,879
10^{-3}	818	8,183	258	2,588

Parameter, μ (non-dimensional):

A_z (m² s⁻¹) w_s (ms⁻¹)	$(A_x = 10^2\,\text{m}^2\,\text{s}^{-1})$ $(U = 0.1\,\text{m s}^{-1})$		$(U = 1\,\text{m s}^{-1})$	
	10^{-5}	10^{-4}	10^{-5}	10^{-4}
10^{-4}	10	1	100	10
10^{-3}	31.6	3.16	316.2	31.6

D.O.M.E.S. area are listed in Table III. For the physical parameters listed in Table I, these are:

$$T_3 = 10^7\,\text{s}; \quad L_{3z} = 100\,\text{m}; \quad L_{3x} = L_{3z} = 31.6\,\text{km};$$

$$S_3 = 810\,\mu g\,l^{-1}; \quad \text{and} \quad \mu = 63.2$$

As an example, vertical distributions of S far from the source are plotted in Fig. 10A and B for $y = 0$ and $y = 63$ km, respectively. The dimensional

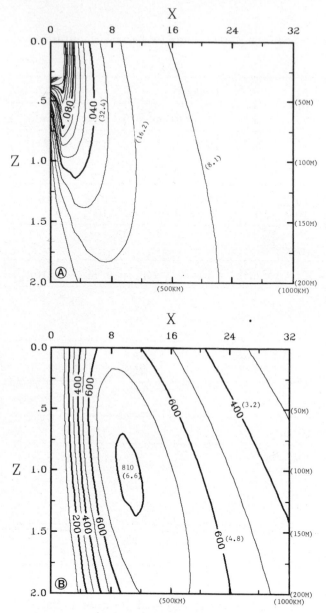

Fig. 10. The vertical distribution of waste sediment concentration: (A) along the central section far from a point-source 185 days after the start of discharge; and (B) along the section at a distance of 73 km from the central section. Dimensional values (S in $\mu g\,l^{-1}$) are for physical constants listed in Table I except that $U = 0.032\,\mathrm{m\,s}^{-1}$.

quantities used are listed in Table I, except that U is taken as $0.032\,\mathrm{m\,s^{-1}}$ instead of $0.2\,\mathrm{m\,s^{-1}}$, considering that the mean current has a low speed if averaged over a long period and large area. These figures represent the concentration at 185 days after commencement of the discharge. Calculation of the distributions at different times indicates that the distributions within 500 km from the source reach a steady state about four months after the start of the discharge (Ichiye and Carnes, 1976a, b, 1977). Fig. 10B shows the high concentrations core at $\sim 100\,\mathrm{m}$ from the surface. This is due to the condition that the sediment is diffused laterally by A_y while sinking and being advected.

There are no data available for the vertical distributions either near the source or far from the source from the monitoring of the pilot-mining operations. However, the time series of the horizontal distributions of suspended sediment are presented in Chapter 15 and thus, comparison of these data with the distribution expressed by eqs. 12 with the near-source parameterizations (13)—(15) is in progress. Also, computer printouts of non-dimensional vertical sections similar to Figs. 8 and 9 are available to interested parties for different values of t, ω and μ (Ichiye and Carnes, 1976a, b).

EULERIAN DIFFUSION: STEADY-STATE DIFFUSION

When mining is constantly operated from a ship at a fixed position or moving in a limited area within a broad, constant ocean current, the sediment distribution reaches steady state after several hours as discussed previously. Under such conditions, the concentration of the sediment, S, can be determined from a solution of the Fickian diffusion equation which describes the concentration at fixed points, i.e. in the Eulerian way. The solution for a point-source can be obtained for the Fickian equation. However, it is more useful to determine the solution for a line-source, because it is more simple in the analysis than the solution for a point-source. Also, the solution is applicable to the actual situation, when it is applied to the upper mixed layer, because it is reported that most of the waste sediment discharged from a mining ship reached at least 30 m at $\sim 700\,\mathrm{m}$ from the ship during the pilot test (Chapter 15).

The Fickian diffusion equation for the concentration, S, is given by:

$$u\,\frac{\partial S}{\partial x} + w_s\,\frac{\partial S}{\partial z} = \frac{\partial}{\partial y}\left(K_y\,\frac{\partial S}{\partial y}\right) + \frac{\partial}{\partial z}\left(K_z\,\frac{\partial S}{\partial z}\right) \tag{21}$$

where u is the current speed which is assumed constant, and K_y and K_z are the Fickian (Eulerian) eddy diffusivities in the y- and z-directions, respectively. The longitudinal eddy diffusivity K_x is neglected for the sake of analytical simplicity, because its effect can be parametrically included in the values of u.

A solution of eq. 21 with constant values of u, K_y and K_z and for a line-source is expressed by:

$$S = S_1(\pi x)^{-1/2} \exp(-\alpha y^2/x + \beta z) \sum_{m=0}^{\infty} S_m(z) \exp(-k_m x) \tag{22}$$

This solution corresponds to a line-source at $x = 0$ where the discharge rate per unit depth is proportional to $\exp(-\alpha z)$. The non-dimensional parameters α and β in solution (22) are respectively defined by:

$$\alpha = h^2 u^2 (4K_y K_z)^{-1} \quad \text{and} \quad \beta = w_s h(2K_z)^{-1} \tag{23}$$

The function $S_m(z)$ and k_m are eigenfunctions and eigenvalues, respectively, and satisfy the equation:

$$d^2 S_m/dz^2 + (k_m^2 - \beta^2)S_m = 0 \tag{24}$$

The scaling horizontal distance L_1 and concentration S_1 are given by:

$$L_1 = uh^2 K_z^{-1} \tag{25}$$

$$S_1 = R\gamma(K_z/K_y)^{1/2} [uh^2(1 - e^{-\gamma})]^{-1} \tag{26}$$

To solve eq. 24, two boundary conditions are necessary. The boundary condition at the surface is simply no flux of the sediment there. Two kinds of boundary conditions are possible at the bottom of the mixed layer. Since the vertical diffusivity in the thermocline is extremely low due to high static stability, these two conditions can be expressed in terms of the flux or the vertical gradient of S. If there is upwelling comparable to the settling velocity below the mixed layer as with Ekman pumping, then there is no flux of the sediment at this level (B_γ). On the other hand, when there is no strong upwelling, the gradient of S vanishes at the bottom of the mixed layer (B_f).

The concentration distribution depends on β, γ, and the boundary conditions. Condition B_f seems to occur more commonly than the condition B_γ, thus, distributions at $y = 0$ (center line) with condition B_f are plotted in Fig. 11A, B and C for $\gamma = 2$, 5 and 10, respectively, in all cases for $\beta = 0.1$. The physical constants listed in Table I yield $\beta = 0.125$. The characteristic concentration S_1 is 668, 1472 and 2928 $\mu g l^{-1}$ for $\gamma = 2$, 5 and 10, respectively. Parameter γ is the depth decay of the discharge rate. As expected, a large value of γ produces a distribution similar to the Lagrangian diffusion from a point-source discussed on p. 499.

Distributions for $\gamma = 5$ and 10 indicate that the concentration near the lower boundary increases as the distance from the source increases, suggesting the accumulation near the bottom of the mixed layer. This tendency becomes more conspicuous for larger β and for the condition B_γ if γ is the same. Larger β corresponds to a higher settling velocity against the upward

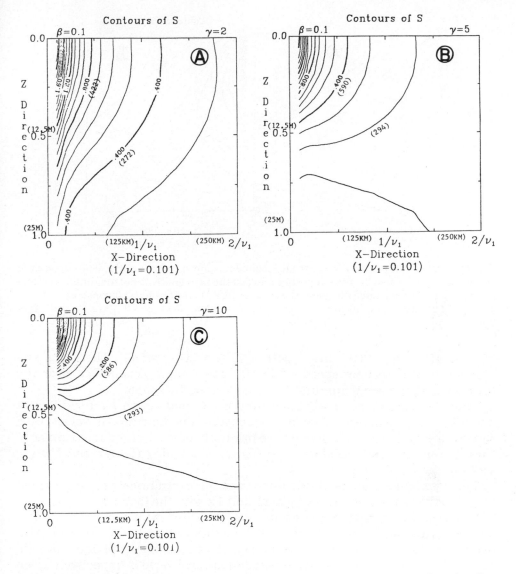

Fig. 11. Vertical distributions of waste sediment concentration at the central section from a line-source for the zero gradient bottom boundary condition. Dimensional values are shown within parentheses (S in $\mu g\,l^{-1}$) for physical constants listed in Table I.

eddy diffusion. Fig. 12 shows the distributions under conditions B_f and B_γ for $\beta = 5$ and $\gamma = 2$. Under the condition B_f, the concentration curves show a sinking tendency as the distance from the source increases for larger β, as expected. The condition B_γ shows a conspicuous accumulation near the bottom. The value of $\beta = 5$ corresponds to $w_s = 4 \cdot 10^{-4}\,\mathrm{m\,s^{-1}}$ with $K_z =$

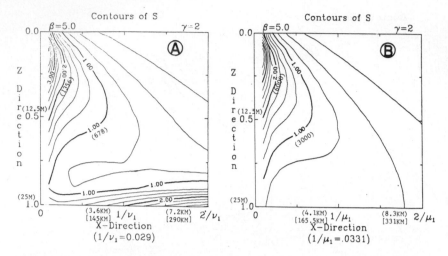

Fig. 12. Vertical distributions of waste sediment concentration at the central sections from a line-source for: (A) the non-flux and (B) the zero gradient bottom boundary conditions. Dimensional values in parentheses (S in $\mu g\,l^{-1}$) are for physical constants listed in Table I. The distances in *brackets* are for $h = 50\,m$, $w_s = 5 \cdot 10^{-5}\,m\,s^{-1}$, and $A_z = 2.5 \cdot 10^{-4}\,m^2\,s^{-1}$.

$10^{-3}\,m^2\,s^{-1}$ or $w_s = 10^{-5}\,m\,s^{-1}$ with $K_z = 2.5 \cdot 10^{-5}\,m^2\,s^{-1}$. Therefore, fine sediments with settling speed of $\sim 10^{-5}\,m\,s^{-1}$ may accumulate near the lower boundary several hundred kilometers from the source.

Table IV lists characteristic concentration S_1 and length L_1 for different dimensional parameters in order to convert non-dimensional values of S shown in Figs. 11 and 12. Computer printouts of vertical sections similar to Figs. 10 and 11 are available for different α, β and γ (Ichiye and Carnes, 1976b).

The distribution on sections other than the central one can be obtained by multiplying those shown in Figs. 11 and 12 with the factor exp $(- \alpha y^2/x)$. This factor shows that the contours are elongated with increasing α. The plume width y_b is defined from the condition that the y gradient of S reaches maximum at the visual boundary of a dye or smoke plume (Pasquill, 1974). The non-dimensional value and dimensional value is respectively given by:

$$y_b = (x/2\alpha)^{1/2} \qquad \text{and} \qquad y_b' = (2x'K_y/u)^{1/2} \tag{27}$$

The dimensional value does not depend on K_y nor on the discharge rate. The data of the plume during the pilot-mining operation (Chapter 15) indicate that the plume width increases almost proportional to the square root of the distance from the ship. The width at 10 km from the ship was ~ 1 km. Since the current was reported as $0.25\,m\,s^{-1}$, relation (27) leads to $K_y = 1.25\,m^2\,s^{-1}$. This value is close to $1\,m^2\,s^{-1}$ for the distance scale $l = 10\,km$ determined from Okubo's relation (6).

TABLE IV

Scaling parameters for the Fickian diffusion from a line-source

	L_1 (km)				S_1 ($\mu g\,l^{-1}$)		
	($K_z = 10^{-3}\,m^2\,s^{-1}$)				($u = 0.2\,m\,s^{-1}$, $h = 25\,m$, $R = 10^3\,t\,s^{-1}$)		
h (m)	25	50	75	γ	2	5	10
u ($m\,s^{-1}$)				K_y/K_z			
0.2	125	500	1,125	10^5	677	1,474	2,928
0.5	312.5	1,250	2,812.5	10^6	214	466	926
1.0	625	2,500	5,625				

	α (non-dimensional)				β (non-dimensional)		
	($K_y = 10^2\,m^2\,s^{-1}$, $K_z = 10^{-3}\,m^2\,s^{-1}$)				($K_z = 10^{-3}\,m^2\,s^{-1}$)		
h (m)	25	50	75	h (m)	25	50	75
u ($m\,s^{-1}$)				w_s ($m\,s^{-1}$)			
0.2	62.5	250	562.5	10^{-5}	0.125	0.25	0.375
0.5	390.6	1,562.5	3,515.6	$5 \cdot 10^{-5}$	0.625	1.25	3.75
1.0	1,562.5	6,250	14,062.5				

SPECULATION ON BENEFITS OF DEEP-OCEAN MINING OPERATIONS

It is estimated that a mining ship will discharge $\sim 2 \cdot 10^4\,m^3\,day^{-1}$ of interstitial water at the surface. It is reported that the mining ship will move with a speed of ~ 3 kt. The concentration of three nutrients, nitrate, phosphate and silicate, are shown for the ambient surface water and for the surface discharge in Table V (N.O.A.A., 1976). It is seen that the discharge has a nutrient concentration $3.5 \cdot 10^2$ times in nitrate, 6 times in phosphate and 70 times in silicate of the ambient water. If this water is discharged for 100 days within an eddy with a 50-km radius and is mixed only with the ambient water in the mixed layer 25 m deep, the dilution factor is $\sim 1.02 \cdot 10^{-3}$. Thus the increase of nitrate and silicate reaches 36% and 7%, respectively, as shown in Table V. The effects of the increase of these nutrients by such magnitude can be felt only on the long-term basis and may not be observed during short-period monitoring. If the discharge water is maintained above 10 m, i.e. the draft of a mining ship moving at 3 kt., and is mixed laterally ~ 100 m during one day as observed for the plume in the pilot-mining

TABLE V

Increment of nutrients in the mixed layer due to the surface discharge

	Ambient concentration (μgAt.l^{-1})	Concentration of surface discharge (μgAt. l^{-1})	Increased nutrients within an eddy[1]		Increased nutrients for one-day operation[2]	
			(μgAt. l^{-1})	(%)	(μgAt. l^{-1})	(%)
Nitrate	0.1	35	0.036	36	$5.2 \cdot 10^{-3}$	5.2
Phosphate	0.4	2.4	0.0026	0.6	$3.6 \cdot 10^{-6}$	0.09
Silicate	2	140	0.4	7	$2.1 \cdot 10^{-2}$	1.0

[1] Based on the condition that the bottom water is discharged for 100 days within a gyre of 50 km radius and 25 m depth.
[2] Based on the condition that a mining ship moving at 3 kt. discharges the water above 10 m depth within a width of 100 m for one day.

operation, then the dilution factor is $1.48 \cdot 10^{-4}$. Thus the increments of the nutrients are unobservable because variances in the values of the ambient water nutrients far exceed those values.

As a long-term effect, the direct uprising of the bottom water can be compared to the natural upwelling process. Since the mining ship is planned to sweep the bottom with a mining device 16 m wide, the swath of the bottom area during a one-day operation is $\sim 2.1 \text{ km}^2$ ($= 132 \text{ km} \times 16 \text{ m}$) and 1.4 km^2 ($= 84 \text{ km} \times 16 \text{ m}$) for speeds of the ship of 3 and 2 kt., respectively. Therefore, the uprising of the bottom water corresponds to an upwelling of the bottom water of $1.1 \cdot 10^{-5} \text{ cm s}^{-1}$ ($= 2.10^4 \text{ m}^3/2.1 \text{ km}^2 \times 8640 \text{ s}$) and $1.7 \cdot 10^{-5} \text{ cm s}^{-1}$ for speeds of the ship of 3 and 2 kt., respectively. These values are comparable to the natural upwelling speed $10^{-5} \text{ cm s}^{-1}$ in the interior of the Pacific Ocean as estimated by Munk (1966). Further, the mining operation is more efficient in bringing the nutrients contained in the bottom water to the surface than the natural upwelling, because in the latter process the vertical eddy diffusivity counteracts upwelling in upward transport of the nutrients.

It is necessary to describe the pre-mining background of the primary productivity and nutrient distributions in the mining area to assess the possible impacts of mining operations. Fig. 13 shows the primary productivity in the Eastern Pacific Ocean modified from a chart compiled by Koblentz-Mishke et al. (1970). The chart indicates that the productivity is highest along the coast of North and South America. The highly productive zone extends from the coast of South America to the west along the equator. Many biologists (Koblentz-Mishke et al., 1970; Bunt, 1975) consider the equatorial and coastal highly productive areas to be the result of enrichment

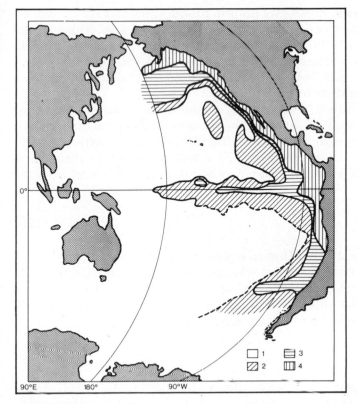

Fig. 13. Distribution of primary production in the Eastern Pacific Ocean based on Koblentz-Mishke et al. (1970). $1 = < 100$; $2 = 100-150$; $3 = 150-150$; and $4 = 250-500$ mg C m^{-2} day^{-1}.

of the upper water by nutrients transported from the bottom by upwelling. However, as discussed on p. 484 the continuous upwelling zone by Ekman pumping is between 5° and 10°N all the year round (Fig. 5A and C) and the upwelling along the equator is not zonally continuous (Ichiye, 1966).

The distribution of primary productivity of Fig. 13 may be compared to the distribution of PO$_4$-P in the Pacific Ocean at the surface and at 100 m (Reid, 1962). The phosphate-rich zone in the Eastern Equatorial Pacific at the surface reflects the advection pattern from the phosphate-rich area off the Peruvian coast as implied by the surface geopotential anomaly. The phosphate distribution at 100 m depth reflects the thermocline depth as well as the horizontal circulation, since phosphates reach maximum beneath the upper mixed layer because of biological consumption near the surface. Therefore, the PO$_4$-rich zone extends to the west between 10° and 15°N from the Central American coast, corresponding to the shallow thermocline zone shown in Fig. 4. However, advection from the coastal area also contributes to form this PO$_4$-rich zone.

The D.O.M.E.S. area is apparently poor in productivity and also in surface nutrients in spite of upwelling in this zone throughout the year. This suggests that the upwelling due to the Ekman pumping does not necessarily bring up the nutrients from the bottom water. This seems to result from the Ekman pumping acting only just below the mixed layer; and thus the bottom nutrients are not brought up to the surface.

The other possible effect of the mining operation is to disturb the stable stratification below the mixed layer and increase the turbulence there caused by dragging the mining equipment. It is difficult to assess this effect particularly on the long-term basis. However, the effect of change in the vertical diffusivity on the nutrient distribution can be estimated by use of a simple diffusion relationship:

$$K_z \frac{\partial S}{\partial z} = F(z) \tag{28}$$

which is based on the Fickian diffusion equation. The function $F(z)$ includes advective terms, the vertical transport term and horizontal diffusion terms. It is assumed that the actual distribution of nutrients $S(z)$ satisfies relation (28). Then if the eddy diffusivity $K_z(z)$ increases n times, the fictitious distribution $S_f(z)$ satisfies the following relation:

$$n K_z \frac{\partial S_f}{\partial z} = F(z) \tag{29}$$

if other conditions remain the same. Then the distribution $S_f(z)$ can be computed by use of this gradient, assuming that the value at the thermocline is unchanged. This procedure is applied to the mean distributions of nitrates, phosphates and silicates averaged over nine pump casts during a 30-hr. period of October 8 and 9, 1975, at D.O.M.E.S. Site B at $11°42'N$ and $138°24'W$. The change of the values in the mixed layer for a 25% increase in vertical diffusivity is listed in Table VI. Further the changes in the vertical distributions of phosphates and nitrates for increases of the diffusivity up to 100% are plotted in Fig. 14A and B, respectively. The mixed layer values show a great increase particularly for nitrates even for a modest increase in K_z. Of course it remains to be seen how long the effect of mining operation

TABLE VI

Initial and increased concentrations ($\mu gAt.l^{-1}$) of nutrients in the mixed layer due to 25% increase of vertical diffusivity in the thermocline

	Nitrate	Phosphate	Silicates
From	0.5	0.5	2.5
To	7.6	0.9	7.4

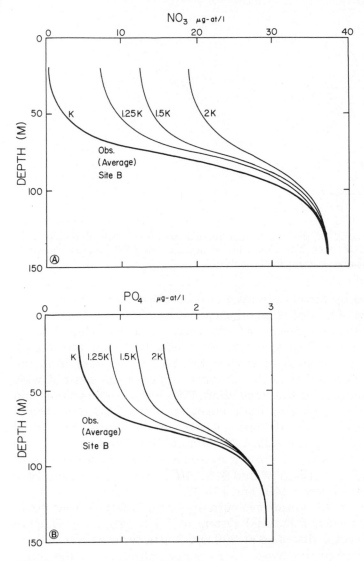

Fig. 14. Increase of nutrients in the thermocline due to increment of the vertical eddy diffusivity calculated for the averaged data at D.O.M.E.S. site *B*: (A) for nitrates; and (B) for phosphates.

lasts, but a substantial increase in the nutrients in the mixed layer is expected.

The increase of nutrients in the mixed layer is small for direct pumping of the bottom and interstitial water, and is uncertain for the stirring effect of the mining operation. However, if these effects are confined within a small, stable gyre, such an increase may produce a significant effect on the bio-productivity. As discussed on p. 477 the eastern tropical Pacific Ocean is

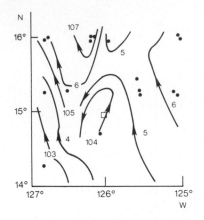

Fig. 15. Dynamic topography of the sea surface relative to 500 dbar in dynamic centimeters from the D.O.M.E.S. data. The *closed circles* indicate the STD stations. The *open rectangle* indicates the moored current meter station.

not simply dominated by zonal currents, but is occupied by several cyclonic and anticyclonic gyres of almost climatological nature.

There is evidence that meanders and gyres have smaller scales of zonal length than those of climatological nature. The surface dynamic topography for the 500-m reference level is plotted in Fig. 15 for the data from the D.O.M.E.S. cruise in September 1975. Although the coverage was not wide enough for general features of the circulation, this chart shows cyclonic and anticyclonic gyres and a large-amplitude meander of the NEC. The second example shows trajectories of drifters released in the North Equatorial Counter Current (NECC) by the N.O.A.A. National Data Buoy Office group (Kerut, 1976). These trajectories show a cyclonic and anticyclonic gyre of ~ 40 km diameter at 6.2°N,154.3°W and 9°N,140°W, respectively; and large meanders of the NECC between 140° and 155°W.

The third example is the subsurface current change inferred from direct current measurements with a moored system at 15°N,126°W by Halpern (1975). The current vector diagrams at 200 and 300 m show a reversal starting in early October from northwest to southeast, although in the upper layer the current continued northward. The buoy station was located within the NEC according to the D.O.M.E.S. chart of Fig. 15. This chart (Fig. 15) also indicates a cyclonic gyre west of the buoy site. Therefore, the reversal of flow direction at 200 and 300 m could be attributed to this cyclonic gyre. However, this may not be the case, since the reversal occurred only at deep layers and not at the upper layers, and it also continued more than four weeks. This leads to a conjecture about baroclinic instability.

A mathematical model for baroclinic instability for a stratified fluid was first developed by Eady (1949), and later elaborated by many meteorologists. Pedlosky (1971) provided a heuristic explanation of the baroclinic instability based on displacement of a parcel. This instability is caused by the vertical shear of the current, modified by the static stability and the stabilizing effect of the Coriolis force. Mathematically, the perturbation velocities are determined from momentum equations and the continuity with a factor $\exp[ik(x-ct)]$, where k is the wave number in the x-direction directed eastward and c is the phase velocity of the perturbation. If c has an imaginary part c_i, the factor kc_i indicates the rate of the exponential growth of the amplitude with time and is called the amplication factor. The boundary conditions at the top and bottom of the layer determine the relationship between the amplification factor and wave length as the frequency equation of the perturbation wave. For a constant Väisälä frequency N between the horizontal boundaries of thickness D, the amplification factor is given by:

$$k_{c_i} = kU\lambda^{-1}[\lambda \coth(-1+\lambda-\lambda^2/4)]^{1/2} \tag{30}$$

where

$$\lambda = NDf^{-1}(k^2 + n^2\pi^2L^{-2})^{1/2} \tag{31}$$

and U and L are the mean horizontal velocity and meridional width of the basic current, respectively, and n is the mode number in the meridional perturbation which is proportional to $\sin(n\pi/L)$.

The square root of eq. 30 is positive only for $0 \leqslant \lambda \leqslant \lambda_c$ (= 2.399). Therefore, there is no instability for wave number $k > \lambda_c f(ND)^{-1}$ (= k_c) or for the narrow basic flow satisfying $L < \pi k_c^{-1}$ (= L_c).

The Väisälä frequency is calculated from STD data at D.O.M.E.S. station OC-42 (15°30'N,127°00'W). The results indicate that the value of N is almost constant below 200 m at $2.0 \cdot 10^{-3}$ s^{-1}. The EASTROPAC data indicate that uniformity of N reaches to almost 550 m. The value of N at OC-42 becomes large peaking at $\sim$ 130 and 110 m. The large density gradient between 100 and 150 m may be interpreted as a horizontal boundary for disturbances. Therefore, the current could change differently below 200 m and above 100 m as observed in early October of 1975 (Halpern, 1975). For $D = 350$ m, corresponding to a layer between 200 and 550 m, $N = 4.5 \cdot 10^{-3}$ s^{-1} and $f = 3.8 \cdot 10^{-5}$ s^{-1}, so that πk_c^{-1} is about 54 km. Since the NEC near the buoy site is much broader than this between 200 and 550 m, there exists baroclinic instability.

The growth rate of the unstable wave is also consistent with observations. For the meridional wave number π/L much smaller than k, eq. 30 leads to the most unstable wave of $\alpha \approx 1.7$. The growth rate k_{c_i} for this is $7.5 \cdot 10^{-6}$ s^{-1} since the mean speed $U \approx 10$ cm s^{-1}. Thus the e-folding time is $\sim$ 15 days and the wave length is 146 km. Halpern's (1975) data show that the change of the current actually started on September 25 and was completed about October 7. Therefore, this time scale seems to agree well with the e-folding time. There is no way to determine wave lengths of the disturbance independently from the data of only one station.

The life time of these gyres after becoming isolated from the mean flow can be inferred fiom the behavior of cyclonic rings observed in the Sargasso Sea in recent years (Richardson, 1976; Rhines, 1977). The life time of the latter is estimated as two years or more, if the gyre does not land on a shelf or is not coalesced with the Gulf Stream. Therefore, if nutrients are added to one such gyre in the D.O.M.E.S. area, it is likely that it will be contained in a small area for a few years.

In the present stage there is no effective way to contain fish within a gyre which is fed nutrients from the mining discharge (Hanson, 1974). However, if the nutrients prove effective in increasing primary productivity assisted by intense solar radiation in the tropical ocean, it might be possible to use floating oyster cloisters or sub-surface clam culture devices as speculated by Hanson (1974).

CONCLUSIONS

The transportation and dispersion of the waste sediments discharged from a mining ship are mathematically modelled. One of the results indicates that the sediments with settling velocity comparable with the upward speed may accumulate just below the mixed layer, if there is such upwelling mainly due to the Ekman pumping. Although two pilot-mining tests in 1978 failed to indicate such a phenomenon (Chapter 15) the evidence is inconclusive because of insufficient vertical sampling and lack of data on vertical velocity. The long-term effects can be monitored only after the full mining operation is started.

The enrichment of nutrients near the surface discharge of the bottom and interstitial water was also not detected by monitoring in the pilot-mining tests as stated in Chapter 15. This is expected because the estimated increase in nutrients in only a few percent at the most, during the duration of the experiments, and is within the range of variances in the natural conditions. This effect can also be assessed on a long-term basis after start of the full-scale operation.

The effect of the mining operation on turbulence in the thermocline is also difficult to determine during the pilot-mining operation. However, it could be done by use of the dye technique applied in the wake of a mining ship. The long-term effect might be determined by use of repeated sampling of temperature and salinity in the mining area.

If the concept of mariculture by enrichment of the surface water is to be applied to the mining operation, it is necessary to monitor generation, movement and change of the meso-scale gyres in the D.O.M.E.S. area. For such purposes, the conventional methods of using research vessels and moored current meters are ineffective and exorbitantly expensive, because it is necessary to obtain data over small intervals both in time and space. The only feasible method is remote sensing with satellites.

There are passive and active methods in satellite remote sensing. The infrared and other spectral band sea-surface imagery and the tracking of drifters (Kerut, 1976) belong to the former. The latter has recently been developed as laser-altimetry of the sea-level anomaly from the geoid (Leitao et al., 1978; Huang, 1979). When this method achieves an accuracy of several centimeters in determining the sea-level anomaly, the geostrophic method can be used to determine the surface current with a sufficient accuracy for gyres of tens of kilometers or more. Then it will be possible to locate a suitable gyre for enrichment of the surface nutrients with the mining operation and then to track the gyre for a few years for mariculture operation.

ACKNOWLEDGEMENTS

Doyle Horne allowed us to use preliminary results from his thesis which has been prepared under guidance of the senior author (T.I.). Dr. Richard Geyer is appreciated for his enthusiasm for our enterprise, without whose prodding this chapter would never have appeared.

REFERENCES

(See also references listed as Selected Bibliography in the Appendices of Chapters 7 and 11 on p. 260 and pp. 376—378, respectively.)

A.H.F. (Allan Hancock Foundation), 1976. Environmental investigations and analyses, Los Angeles—Long Beach Harbors, 1973—1976. Final Rep. to U.S. Army Corps Eng., Los Angeles District. Harbors Environmental Projects, University of Southern California, Los Angeles, Calif., 773 pp.

Alabaster, J. S. (Editor), 1977. Biological monitoring of inland fisheries. In: Proceedings of a Symposium of European Inland Fisheries Advisory Commission. Applied Science Publishers, London, 226 pp.

Albright, L. J. and Wentworth, J. W., 1973. Use of the heterotrophic activity techniques as a measure of eutrophication. Environ. Pollut., 5: 59—72.

Albright, L. J., Wentworth, J. W. and Wilson, E. M., 1972. Techniques for measuring metallic salt effects upon the indigenous heterotrophic microflora of a natural water. Water Res., 6: 1589—1596.

Allen, M. J., 1975. Regional variation in the structure of fish communities. In: Coastal Water Research Project, Annual Report, 1975. U.S. Dep. Commer., Springfield, Va., Natl. Tech. Info. Serv., PB 274467/AS, pp. 99—102.

Allen, M. J. and Voglin, R. M., 1976. Regional and local variation of bottom fish and invertebrate populations. In: Coastal Water Research Project, Annual Report, 1976. U.S. Dep. Commer., Springfield, Va., Natl. Tech. Info. Serv., PB 274461/AS, pp. 217—221.

Allen, M. J. and Voglin, R. M., 1977. Commercial fish catches. In: Coastal Water Research Project, Annual Report, 1977. U.S. Dept. Commer., Springfield, Va., Natl. Tech. Info. Serv., PB 274463/AS, pp. 121—126.

Allen, M. J., Isaacs, J. B. and Voglin, R. M., 1975. Hook-and-line survey of demersal fishes in Santa Monica Bay. Coastal Water Res. Proj., Long Beach, Calif., Tech. Mem. 222, 23 pp.

Allen, M. J., Pecorelli, H. and Word, J., 1976. Marine organisms around outfall pipes in Santa Monica Bay. J. Water Pollut. Control Fed., 48(8): 1881—1893.

Alzieu, C., 1976, Présence de diphénylpolychlorés chez certains poissons de l'Atlantique et de la Méditerranée. Sci. Pêche, 258: 1—11.

Alzieu, C., Michel, P. and Thibaud, Y., 1976. Présence de micropolluants dans les mollusques littoraux. Sci. Pêche, 264: 1—18.

Amiel, A. J. and Navrot, J., 1978. Nearshore sediment pollution in Israel by trace metals derived from sewage effluent. Mar. Pollut. Bull., 9·10—14.

Anderson, A. R. and Mueller, J. A., 1978. Estimate of New York Bight future (2000) contaminant inputs. Rep. to M.E.S.A. New York Bight Project, Stony Brook, N.Y. (unpublished).

Anderson, D. V., Jurek, R. M. and Keith, J. O., 1977. The status of brown pelicans at Anacapa Island in 1975. Calif. Fish Game, 63(1): 4—10.

Anderson, J. J., 1978. Deep ocean mining and the ecology of the tropical North Pacific. Dep. Oceanogr., Univ. Washington, Seattle, Wash., Spec. Rep. No. 83.

Angel, M. V., 1979. Studies on Atlantic halocyprid ostracods: their vertical distributions and community structure in the central gyre region along latitude 30°N from off Africa to Bermuda. Prog. Oceanogr., 8: 1—122.

Angel, M. V., in press. Vertical distribution of the standing crop of macroplankton and micronekton at three stations in the northeast Atlantic. Biol. Oceanogr.

Angelillo, B., De Riu, G. C. and Renga, G., 1973. Studio longitudinale sullo stato di inquinamento delle acque di mare costiere del litorale Napoletano. Fondazione Politecnica per il Mezzogiorno d'Italia, Naples, Quad. No. 73, 19 pp.

Anonymous, 1972. Minamata. Natl. Geogr. 142: 502—527.

Anonymous, 1978. Surveillance program of sedimentation effects of hydraulic dredging. U.S. Army Corps of Engineers, District X, Mobile, Ala.

Armstrong, R. S., 1979. Physical interpretation. In: R. L. Swanson and C. J. Sindermann (Editors), Oxygen Depletion and Associated Benthic Mortalities in New York Bight, 1976, Ch. 6, U.S. Dep. Commer., Washington, D.C., N.O.A.A. Prof. Pap.

Aston, S. R. and Chester, R., 1973. The influence of suspended particles on the precipitation of iron in natural waters. Estuarine Coastal Mar. Sci., 1: 225—231.

Aubert, M., 1975. Le problème du mercure en Méditerranée. Rev. Int. Océanogr. Méd., 37/38: 215—231.

Azam, F., Vaccaro, R. F., Gillespie, P. A., Moussalli, E. I. and Hodson, R. E., 1977. Controlled ecosystem pollution experiment: effect of mercury on enclosed water columns, II. Marine bacterioplankton. Mar. Sci. Commun., 3: 313—329.

Azarovitz, T. R., Byrne, C. S., Silverman, M. J., Freeman, B. L., Smith, W. G., Turner, S. C. Halgren, B. A. and Festa, P. J., 1979. R. L. Swanson and C. J. Sindermann (Editors), Oxygen Depletion and Associated Benthic Mortalities in New York Bight, 1976, Ch. 13. U.S Dep. Commer., Washington, D.C., N.O.A.A. Prof. Pap.

Baker, E. T., Feely, R. A. and Takahashi, K., 1979. Chemical composition, size distribution and particle morphology of suspended particulate matter at DOMES Sites *A, B* and *C*: relationship with local sediment composition. In: J. L. Bischoff and D. Z. Piper (Editors), Marine Geology and Oceanography of the Pacific Manganese Nodule Province, Marine Science 9, Plenum, New York, N.Y., pp. 163—202.

Ballister, A., Cros, L. and Ras, T., 1978. Conenido en mercurio de algunos organismos marinos comerciales des Mediterraneo Catalan. Presented at M.A.B. Congr., Montpellier, Sept. 15, 1977, 27 pp. (preprint).

Baluja, C. J., Franco, J. M. and Murado, M. A., 1973. Contaminacion del medio por plaguicidas organoclorados, VI. Residuos de insecticidas y PCB en especies marinas, significado ecologico y su relacion con la contaminacion litoral nacional. Invest. Pesq., 37: 593—617.

Bancroft, H. H., 1884. History of the Pacific States of North America, Vol. I, 1542—1800, Ch. 13, California, A. L. Bancroft, San Francisco, Calif., 744 pp.

Barnard, E. A., 1973. Comparative biochemistry and physiology of digestion. In: C. L. Prosser (Editor), Comparative Animal Physiology W. B. Saunders, Philadelphia, Pa., pp. 138—139.

Barnard, W. D., 1978. Prediction and control of dredged material dispersion around dredging and open-water pipeline disposal operations. U.S. Army Eng. Waterways Exp. Stn., Vicksburg, Miss., Tech. Rep. DS-78-13.

Barnes, A. T., Quetin, L. B., Childress, J. J. and Pawson, D. L., 1976. Deep-sea macroplanktonic sea cucumbers: suspended sediment feeders captured from deep submergence vehicle. Science, 194: 1083—1085.

Barry, M., 1978. Behavioral response of yellowfin tuna, *Thunnus albacores*, and kawakawa, *Euthynnus affinis*, to turbidity. S.W. Fish. Cent. Honolulu Lab., Nat. Mar. Fish. Serv., Honolulu, Hawaii, Natl. Tech. Info. Serv., Springfield, Va., Rep. No. PB/297106/AS, 31 pp. plus appendices.

Barsdate, R. S., Fenchel, T. and Prentiki, R. T., 1974. Phosphorus cycle of model ecosystems: significance of decomposer food chains and effect of bacterial grazers. Oikos, 25: 239—251.

Bascom, W., 1979a. Life in the bottom: San Diego (SIC, Pedro) and Santa Monica Bays, Palos Verdes and Point Loma Peninsulas. In: Coastal Water Research Project, Annual Report, 1978. U.S. Dep. Commer., Springfield, Va., Natl. Tech. Info. Serv. PB 299830/AS.

Bascom, W., 1979b. Measuring the effects of man's wastes on the ocean. In: C. G. Gunnerson and J. M. Kalbermatten (Editors), Appropriate Technology for Water

Supply and Waste Disposal. American Society of Civil Engineers, New York, N.Y., pp. 249—263.

Bascom, W., Word, J. Q. and Mearns, A. J., 1979. Establishing boundaries for normal, changed, and degraded bottom conditions. In: Coastal Water Research Project Annual Report, 1978. U.S. Dep. Commer., Springfield, Va., Natl. Tech. Info. Serv. PB 299830/AS.

Bassi, D. E. and Basco, D. R., 1974. Field study of an unconfined spoil disposal area of the Gulf intracoastal waterway in Galveston Bay, Texas. Texas A & M University, College Station, Texas, C.O.E. Rep. No. 174, TAMU-SG-74-208.

Batchelor, G. K., 1967. An Introduction to Fluid Mechanics. Cambridge University Press, London, pp. 233—234.

Beardsley, R. C., Boicourt, W. C. and Hansen, D. V., 1976. Physical oceanography of the Middle Atlantic Bight, 1976. In: M. G. Cross (Editor), Middle Atlantic Continental Shelf and the New York Bight, Vol. 2. Am. Soc. Limnol. Oceanogr., Spec. Symp., pp. 20—84.

Bechtel, T. J. and Copeland, B. J., 1970. Fish species diversity indices as indicators of pollution in Galveston Bay, Texas. N.C. State Univ., Contrib. Mar. Sci., 15: 103—132.

Beecher, J., 1915. The history of Los Angeles Harbor. Thesis, University of Southern California, Los Angeles, Calif., M.S., 63 pp.

Beers, J. R., Reeve, M. R. and Grice, G. D., 1977. Controlled ecosystem pollution experiment: effect of mercury on enclosed water columns, IV. Zooplankton population dynamics and production. Mar. Sci. Commun., 3: 355—394.

Belloni, S., Cattaneo, R., Orlando, P. and Pessani, D., 1976. Alcuni considerazioni sul contenuto di metalli pesanti in *Meganychtiphanes norvegica* (Sars, 1957, Crustacea, Euphausiacea) nel Mar Ligure. Boll. Mus. Ist. Biol. Univ. Genova, 44: 113—133.

Bender, M. L., 1971. Does upward diffusion supply the excess manganese in pelagic sediments? J. Geophys. Res., 76: 4212—4215.

Benninger, I. K., Aller, R. C., Cochran, J. K. and Turekian, K. K., 1979. Effects of biological sediment mixing on the ^{210}Pb chronology and trace metal distribution in a Long Island Sound sediment core. Earth Planet. Sci. Lett., 43: 241—259.

Benvegnú, F., Brondi, A. and Ferretti, O., 1974. Criteri di campionamento nella prospezione mineralogica alluvionale. Rend. Soc. Ital. Mineral. Petrol., 30: 165—190.

Bernard, J. L., 1961. Gammaridean Amphipoda from depths of 400 to 6000 metres. Galathea Rep., 5: 23—128.

Bernhard, M., 1976. Manual of methods in aquatic environment research, Part 3. Sampling and analysis of biological material. In: Guidelines for the F.A.O.(G.F.C.M.)—U.N.E.P. Joint Coordinated Project on Pollution in the Mediterranean, F.A.O. Fish. Tech. Pap. 158, 124 pp.

Bernhard, M. and Renzoni, A., 1977. Mercury concentrations in marine organisms from the Mediterranean: Anthropogenic or natural origin. Thalassia Jugosl., 13: 265—300.

Bernhard, M. and Zattera, A., 1975. Major pollutants in the marine environment. In: E. Pearson and E. Frangipane (Editors), Marine Pollution and Marine Waste Disposal. Pergamon, Oxford, pp. 195—300.

Bernhard, M. and Zattera, A., 1979. The role of chemical speciation in the uptake and loss of elements by marine organisms. Proc. Int. Symp. on Interaction Between Water and Living Matter, Odessa, Oct. 6—10, 1975, Vol. II, Nauka, Moscow, pp. 134—143 (in Russian with English summary).

Bernhard, M., Cagnetti, P., Nassogne, A., Peroni, C., Piro, A. and Zattera, A., 1972. Radioecological investigations in the Gulf of Taranto, 2. Preliminary estimations of the receptivity for low level radioactive wastes of a site in the Gulf of Taranto. In: Radioecology Applied to the Protection of Man and His Environment. E.U.R. 4800 *D.F.I.E.* Euratom, Brussels, pp. 347—380.

Bernhard, M., Goldberg, E. D. and Piro, A., 1975. Zinc in seawater — an overview, 1975. In: E. D. Goldberg (Editor), The Nature of Seawater. Dahlem Workshop Rep., Dahlem Konf., Berlin, 33: 43—68.

Bever, K. and Dunn, A., 1976. The energetic role of amino acid and protein metabolism in the kelp bass (Paralabrax clathratus). In: Marine Studies of San Pedro Bay, California, Part 12. Allan Hancock Foundation and Sea Grant Programs, University of Southern California, Los Angeles, Calif., pp. 131—144.

Bewers, J. M. and Yeats, P. A., 1977. Oceanic residence times of trace metals. Nature, (London), 268: 595—598.

Bidleman, T. F. and Olney, C. E., 1974. Chlorinated hydrocarbons in the Sargasso Sea, Atmosphere and surface water. Science, 183: 516—518.

Bisagni, J. J., 1976. Passage of anticyclonic Gulf Stream eddies through Deep Water Dumpsite 106 during 1974 and 1975. N.O.A.A. Dumpsite Evaluation Rep. 76-1, 33 pp.

Bisceglia, A., Fasano, E. and Guida, A., 1973. Inquinamento del mare dovuto al trasporto del petrolio e suoi derivati a mezzo di petroliere. Fondazione Politecnica per il Mezzogiorno d'Italia, Naples, Quad. No. 74, 21 pp.

Bischoff, J. L. and Piper, D. Z., 1976. Chemical composition of marine sediments of DOMES Site C. In: J. L. Bischoff (Editor), Deep Ocean Mining Environmental Study, NE Pacific Nodule Province, Site C, Geology and Geochemistry, U.S. Geol. Surv., Menlo Park, Calif., Open-File Rep. No. 76—548.

Bishop, J. K. B., Edmonds, J. M., Ketten, D. R., Bacon, M. P. and Silker, W. B., 1977. The chemistry, biology, and vertical flux of particulate matter from the upper 400 m of the equatorial Atlantic Ocean. Deep-Sea Res., 24: 511—548.

Bishop, J. K. B., Ketten, D. R. and Edmond, J. M., 1978. The chemistry, biology and vertical flux of particulate matter in the upper 400 m of the Cape Basin in the south-east Atlantic Ocean. Deep-Sea Res., 25: 1121—1162.

Blackburn, M., 1973. Regressions between biological oceanographic measurements in the eastern tropical Pacific and their significance to ecological efficiency. Limnol. Oceanogr., 18: 552—563.

Blackburn, M., 1976. Review of existing information on fishes in the Deep Ocean Mining Environmental Study (DOMES) area of the tropical Pacific. Inst. Mar. Resour., Univ. Calif., La Jolla, Calif., IMR Ref. No. 76-1, 79 pp.

Blazevich, J. N., Gahler, A. R., Vasconcelos, G. J., Rieck, R. H. and Pope, S. V., 1977. Monitoring of trace constituents during PCB recovery dredging operation—Duwamish Waterways. U.S. Environ. Prot. Agency, Region X, Seattle, Wash., Rep. 910/9—077—039.

Borole, D. V., Mohanti, M., Ray, S. B. and Somayajulu, B. L. K., 1979. Dissolved U, Si and major elements in Mahanadi estuary. (Unpublished.)

Bostrom, K., Lysen, L. and Moore, C., 1978. Biological matter as a source of authigenic matter in pelagic sediments. Chem. Geol., 23: 11—20.

Bouchet, P., 1976a. Mise en évidence de stades larvaires planctoniques chez de Gastéropodes Prosobranches des étages bathyals et abyssals. Bull. Mus. Natl. Hist. Nat., 3ème Sér., Zoöl., 277: 947—972.

Bouchet, P., 1976b. Mise en évidence d'une migration des larves veligères entre l'étage abyssal et la surface. C.R. Acad. Sci., Paris, 283: 821—824.

Boudreau, B. P. and Scott, M. R., 1978. A model for the diffusion-controlled growth of deep-sea manganese nodules. Am. J. Sci., 278: 903—929.

Bowen, V. T., Livingston, H. D. and Burke, J. C., 1976. Distribution of transuranium nuclides in sediment and biota of the North Atlantic Ocean. In: Transuranic Nuclides in the Environment. International Atomic Energy Agency, Vienna, pp. 107—120.

Bowman, M. J. and Wunderlich, L. D., 1977. Hydrographic properties, M.E.S.A. New York Bight Atlas, Monogr. 1. New York Sea Grant Institute, Albany, N.Y., 78 pp.

Bragg, D. (Editor), 1978. Gulf coast manganese nodule processing plant location criteria. Ind. Econ. Res. Div., Texas Eng. Exp. Stn., Texas A & M Univ., for U.S. Dep. Commer., N.O.A.A., 169 pp.

Braman, R. S. and Foreback, C. C., 1973. Methylated forms of arsenic in the environment. Science, 182: 1247—49.

Brandsma, M. G. and Divoky, D. J., 1976. Development of models for prediction of short-term fate of dredged material in the estuarine environment, U.S. Army Eng. Waterways Exp. Stn., Environ. Effects Lab., Vicksburg, Miss. Grant No. DACW39-74-C-0075.

Branica, M., Sipos, L., Bubic, S. and Kozar, S., 1976. Electroanalytical determination and characterization of some heavy metals in sea-water. In: Accuracy in Trace Analysis: Sampling, Sample Handling and Analysis. Proc. 7th I.M.R. Symp., 7—11 Oct., 1974, Gaithersburg, Md., Natl. Bar. Stand., Spec. Publ., 422: 917—928.

Brannon, J. M., 1978. Evaluation of dredged material pollution potential. U.S. Army Eng. Waterways Exp. Stn., Vicksburg, Miss., Tech. Rep. DS-78-6.

Brannon, J. M., 1980. Evaluative summary of regulatory criteria for Public Laws 92-500 and 92-532. U.S. Army Eng. Waterways Exp. Stn., Vickburg, Miss.

Brannon, J. M., Engler, R. M., Rose, J. R., Hunt, P. G. and Smith, I., 1976. Selective analytical partitioning of sediments to evaluate potential mobility of chemical constituents during dredging and disposal operations. U.S. Army Eng. Waterways Exp. Stn., Vickburg, Miss., Tech. Rep. D-76-7.

Brewer, G. D., 1976. The effects of waste effluents on fish populations in the Los Angeles Harbor. In: D. Soule and M. Oguri (Editors), Marine Studies of San Pedro Bay, California, Part 12. Allan Hancock Foundation and Sea Grant Program, University of Southern California, Los Angeles, Calif., pp. 165—214.

Bricker, O. P., 1965. Some stability relations in the system $Mn—O_2—H_2O$ at 25° and one atmosphere total pressure. Am. Mineral., 50: 1296—1354.

Briggs, G., 1968. A survey of the Mississippi River—Gulf Outlet Project with emphasis on shoaling problems and factors influencing sediment distribution. U.S. Army Eng. Dist., New Orleans, La.

Brinck, J. W. and Van Wambeke, L., 1974. World resources of mercury. In: Misc. Pap. Proc. Int. Congr. on Mercury, Vol. 1, Barcelona, May 6—10, 1974. Fabrica National de Moneda y Timbre, Madrid, pp. 49—53.

Brinckman, F. E. and Iverson, W. P., 1975. Chemical and bacterial cycling of heavy metals in the estuarine system. In: T. M. Church (Editor), Marine Chemistry in the Coastal Environment, Am. Chem. Soc., Symp. Ser., No. 18, pp. 319—342.

Brock, T. D., 1966. Principals of Microbial Ecology. Prentice-Hall, Englewood Cliffs, N.J., 306 pp.

Brooks, J. M., 1975. Sources, sinks, concentrations and sublethal effects of light aliphatic and aromatic hydrocarbons in the Gulf of Mexico, Ph.D. Dissertation, Texas A & M University, College Station, Texas.

Brooks, J. M., 1976. The flux of light hydrocarbons into the Gulf of Mexico via runoff. In: H. L. Windom and R. A. Duce (Editors), Marine Pollutant Transfer, D.C. Heath, Lexington, Mass., pp. 185—299.

Brooks, J. M., Bernard, B. B. and Sackett, W. M., 1977. Input of low-molecular-weight hydrocarbons from petroleum operations into the Gulf of Mexico. In: D. A. Wolfe (Editor), Proc. Symp. on Fate and Effects of Petroleum Hydrocarbons in Marine Ecosystems and Organisms, Pergamon, New York, N.Y., pp. 373—384.

Brooks, R. R., Presley, B. J. and Kaplan, I. R., 1967. The APDC—MIBK extraction system for the determination of trace elements in saline water by atomic absorption spectrometry. Talanta, 14: 809—816.

Brown, D. A. and Parsons, T. R., 1978. Relationship between cytoplasmic distribution of mercury and toxic effects to zooplankton and chum salmon (*Oncorhynchus keta*) exposed to mercury in a controlled ecosystem. J. Fish. Res. Board Can., 35: 880—884.

Bruland, K. W., Bertine, K., Koide, M. and Goldberg, E. D., 1974. History of metal pollution in Southern California coastal zone. Environ. Sci. Technol., 8: 425—432.

Brun-Cottan, J. C., 1976. Contribution à l'étude de la granulométrie et de la cinétique des particules marines. J. Rech. Océanogr., 1: 41—54.

Buat-Menard, P. and Chesselet, R., 1979. Variable influence of the atmospheric flux on the trace metal chemistry of oceanic suspended matter. Earth Planet. Sci. Lett., 42: 399—411.

Bubic, S., Sipos, L. and Branica, M., 1979. Electroanalytical determination of ionic zinc, cadmium, lead and copper in sea-water—Cruises of RV "Vila Velebita" in the Kvarner region of the Adriatic Sea. 4th Int. Symp. on Chemistry of the Mediterranean, Rovini, May 1976. Thalassia Jugosl., 15.

Bunt, J. S., 1975. Primary productivity of marine ecosystems. In: H. Leith and R. W. Whittaker (Editors), Primary Productivity of the Biosphere. Springer, Berlin, pp. 169—183.

Burks, S. A. and Engler, R. M., 1978. Water quality impacts of aquatic dredged material disposal (laboratory investigations). U.S. Army Eng. Waterways Exp. Stn., Vicksburg, Miss., Tech. Rep. DS-78-4.

Burns, R. E. and Brown, B. A., 1972. Nucleation and mineralogical controls of the composition of manganese nodules. In: D. R. Horne (Editor), Ferromanganese Deposits on the Ocean Floor, National Science Foundation, Washington, D.C., pp. 51—61.

Burns, R. E., Erickson, B., Lavelle, J. W. and Ozturgut, E., 1980. Observations and measurements during the monitoring of deep ocean manganese nodule mining test in the North Pacific, April—May 1978. U.S. Dep. Commer., N.O.A.A. Tech. Mem., E.R.L.—M.E.S.A.-47, 63 pp.

Burton, J. D., 1976. Basic properties and processes in estuarine chemistry. In: J. D. Burton and P. S. Liss (Editors), Estuarine Chemistry, Academic Press, New York, N.Y., pp. 1—36.

Burton, J. D., 1978. The modes of association of trace metals with certain components in the sedimentary cycle. In: E. D. Goldberg (Editor), Biogeochemistry of Estuarine Sediments, UNESCO, Paris, pp. 33—41.

Cabelli, V. J., Levin, M. A., Dufour, A. P. and McCabe, L. J., 1974. Development of criteria for recreational water. In: A. L. H. Gameson (Editor), Discharge of Sewage from Sea Outfalls. Pergamon, London, pp. 63—73.

Cabelli, V. J., Doufor, A. P. and Levin, M. A., 1976. The impact of pollution on marine bathing beaches: an epidemiological study. In: M. G. Gross (Editor), Middle Atlantic Continental Shelf and the New York Bight, Spec. Vol. 2. Am. Soc. Limnol. Oceanogr., pp. 424—432.

Cable, C. C., 1969. Optimum dredging and disposal practices in estuaries. J. Hydraul. Div., Am. Soc. Civ. Eng., Proc. Pap. 6343, 95 (HY1): 103—114.

Callender, E. and Bischoff, J. L., 1976. Recent metalliferous sediment in the Pacific manganese nodule area. In: J. L. Bischoff (Editor) Deep Ocean Mining Environmental Study, NE Pacific Nodule Province, Site *C*, Geology and Geochemistry, U.S. Geol., Surv., Menlo Park, Calif., Open-File Rep. No. 76-548.

Callender, E. and Bowser, C. J., 1976. Freshwater ferromanganese deposits. In: K. H. Wolf (Editor), Handbook of Strata-Bound and Strataform Deposits, Vol. 7, Elsevier, Amsterdam, pp. 341—394.

Cambray, R. S., Jefferies, D. F. and Topping, G., 1979. The atmospheric input of trace elements to the North Sea. Mar. Sci. Commun., 5: 175—194.

Capelli, R., Contardi, V. and Zanicchi, G., 1976. Enquête explorative sur la teneur en métaux lourds (Cd—Co—Cu—Cr—Mn—Ni—Pb) dans les échantillons de moules (*Mytilus galloprovincialis* L.). Journées d'Études sur les Pollutions Marines 25ème Congr. Assem. Plén. Comm. Int. Explor. Sci. Mer, Split, Oct. 1976, pp. 83—88.

Capuzzo, J. M. and Lancaster, B. A., 1980. The effects of pollutants on marine zooplankton at Deepwater Dumpsite *106*—Preliminary Findings. In: B. H. Ketchum, D. R. Kester and P. K. Park (Editors), Ocean Dumping of Industrial Wastes. Plenum, New York, N.Y.

Caracciolo, S., Perna, A. and Di Silvestro, C., 1972. Ricerche sul contenuto in mercurio totale di pesci e di altri prodotti della pesca catturati alla foce del fiume Pescara e nel mar antistante Pescara. Quad. Merceol., 11: 1—11.

Carey, A. G., Pearcy, W. G. and Osterberg, C. L., 1966. Artificial radionuclides in marine organisms in the northeast Pacific Ocean off Oregon. In: Disposal of Radioactive Wastes into Seas, Oceans and Surface Waters. International Atomic Energy Agency, Vienna, pp. 303—319.

Carlisle, Jr., J. G., 1969. Results of a six-year trawl study in an area of heavy waste discharge, Santa Monica Bay, California. Calif. Fish Game, 55: 26—46.

Carls, E. G., 1978. Recreation. M.E.S.A. New York Bight Atlas, Monogr. 19. New York Sea Grant Institute, Albany, N.Y., 32 pp.

Carnes, M., 1975. Calculation of eddy exchange coefficients for heat and salt from BOMEX data. M.S. Thesis, Texas A & M University, College Station, Texas, 78 pp.

Carrada, G. C. and Rigillo-Troncone, M., 1973. Indicazioni bibliografiche per una valutazione della evoluzione chimico-risica e biologica del golfo di Napoli. Fondazione Politecnica per il Mezzogiorno d'Italia, Naples, Quad. No. 70, 29 pp.

Carrada, G. C., Cinelli, F., Rigillo Troncone, M. and Saggiomo, V., 1974. Contributo alla valutazione dei fenomeni di inquinamento nel Golfo di Napoli. Fondazione Politecnica per il Mezzogiorno d'Italia, Naples, Quad. No. 92, 16 pp.

Cassa per il Mezzogiorno, 1975. Progetto speciale per il disinquinamento del Golfo di Napoli, Marzo, 1975—Reticolo fondamentale del sistema depurative dell'area (map).

C.E.Q. (Council on Environmental Quality), 1970. Ocean Dumping. Office of the President, Washington, D.C.

Chamberlain, D. W., 1976. Effects of the Los Angeles Harbor sediment elutriate on the California killifish, *Fundulus parvipinnis* and white croaker, *Genyonemus lineatus*. In: D. Soule and M. Oguri (Editors), Marine Studies of San Pedro Bay, California, Part 11. Allan Hancock Foundation and Sea Grant Program, University of Southern California, Los Angeles, Calif., pp. 33—47.

Chan, A. T., Ozturgut, E., Jacobsen, R. and Anderson, G. C., 1980. Environmental investigation of the effects of deep-sea mining on the phytoplankton in the Tropical Eastern Pacific Ocean, with special attention on primary productivity. Dep. Oceanogr., Univ. Wash., Seattle, Wash., Spec. Rep. No. 92.

Chapman, C. J. and Rice, A. L., 1971. Some direct observations on the ecology and behaviour of the Norway lobster, *Nephrops norvegicus*. Mar. Biol., 10: 321-329.

Chau, A. S. Y. and Sampson, R. C. J., 1975. Electron capture gas chromatography methodology for the quantification of polychlorinated biphenyls: survey and compromises. Environ. Lett., 8: 89—101.

Chekotillo, K. A., 1970. Vertical circulation in the Equatorial Western Pacific. Oceanology, 10: 626—629.

Chew, K. K., 1977. An assessment of the effects of subtidally discharged municipal wastewater effluent on the intertidal biota of several central Puget Sound Beaches. Final Rep. for Municipality of Metropolitan, Seattle, Seattle, Wash., 96 pp.

Childress, J. J., 1977. Physiological approach to the biology of midwater organisms. In: N. R. Andersen and B. J. Zahuranec (Editors), Oceanic Sound Scattering Prediction. Plenum, New York, N.Y., pp. 301—324.

Childress, J. J. and Price, M. H., 1978. Growth rate of the bathypelagic crustacean

Gnathophausia ingens (Mysidacea: Lophogastridae), 1. Dimensional growth and population structure. Mar. Biol., 50: 47—62.

Chung, Y., 1971. Pacific deep and bottom water studies based on temperature, radium, and excess radon measurements. Ph.D. Dissertation, University of California, San Diego, Calif., 239 pp.

Cigna-Rossi, L., Clemente, G. E. and Santarossi, G., 1976. Mercury and selenium in a defined area and in its population. Arch. Environ. Health, 31: 160—165.

Ciusa, W. and Giaccio, M., 1972. Il contenuto di oligoelementi di alcune specie ittiche pescate nell'Adriatico, in relazione alla situazione degli scariche industrial. Boll. Lab. Chim. Prov., 23: 137—145.

Ciusa, W. and Giaccio, M., 1973. Il contenuto di rame, zinco, cadmium, mercurio e piombo di alcune specie ittiche del Mar Tirreno. Quad. Merceol., 12: 137—145.

Ciusa, W., Giaccio, M., Di Donato, F. and Lucianetti, L., 1973. Il contenuto in rame, zinco, cadmio, mercurio e piombo di alcune specie ittiche del Mar Tirreno. Quad. Merceol., 12: 33—48.

Clarke, M. R., 1972. In: G. L. Wood, The Guiness Book of Animal Facts and Feats. Guiness Superlatives, Enfield, 384 pp.

Clarke, M. R. and MacLeod, N., 1976. Cephalopod remains from sperm whales caught off Iceland. J. Mar. Biol. Assoc. U.K., 56: 733—749.

Coffin, D. E., 1963. A method for the determination of free iron in soils and clays. Can. J. Soil Sci., 43: 7—17.

Cohen, M., 1974. Recent surveys of mercury and cadmium in food in the United Kingdom. Proc. Int. Symp. on Problems of the Contamination of Man and His Environment by Mercury and Cadmium. Comm. European Comm., Luxembourg (EUR-5075), pp. 543—570.

Colebrook, J. M., 1978. Continuous plankton records: zooplankton and environment, North-east Atlantic and North Sea 1948—1975. Oceanol. Acta 1: 9—23.

Coleman, J. M. and Gagliano, S. M., 1964. Cyclic sedimentation in the Mississippi River deltaic plain. Trans. Gulf Coast Assoc. Geol. Soc., 14: 67—80.

Collias, E. E. and Lincoln, J. H., 1977. A study of nutrients in the main basin of Puget Sound. Puget Sound Inter. Stud., Munic. Metropol. Seattle, Seattle, Wash. Ref. M 77-2, 151 pp.

Connell, J. H., 1975. Some mechanisms producing structure in natural communities: a model and evidence from field experiments. In: M. L. Cody and J. M. Diamond (Editors), Ecology and Evolution of Communities. Harvard University Press, Cambridge, Mass., 545 pp.

Cotton, F. A. and Wilkinson, G., 1972. Advanced Inorganic Chemistry. Interscience, New York, N.Y., 1145 pp.

Cronin, L. E., 1970. Gross physical and biological effects of overboard spoil disposal in Chesapeake Bay. Natl Resour. Inst., Univ. Md., College Park, Md., Spec. Rep. No. 3.

Csanady, G. T., 1973. Turbulent Diffusion in the Environment. D. Reidel, Boston, Mass., 248 pp.

Csanady, G. T., 1980. An analysis of dump-site diffusion experiments. In: B. H. Ketchum, D. R. Kester and P. K. Park (Editors), Ocean Dumping of Industrial Wastes. Plenum, New York, N.Y.

Cumont, G., Viallex, G., Lelièvre, H. and Bobenrieth, P., 1972. Contamination des poissons de mer par le mercure. Rev. Int. Océanogr. Méd., 28: 95—127.

Cumont, G., Gilles, G., Bernard, F., Briand, M.-B., Stéphan, G., Ramonda, G. and Guillou, G., 1975. Bilan de la contamination des poissons de mer par le mercure á l'occasion d;un contrôle le portant sur 3 années. Ann. Hyg. Lang. Fr., Méd., Nutr., 11: 17—25.

Curtis, W. F., Culbertson, J. K. and Chase, E. B., 1973. Fluvial sediment discharge to the ocean from the conterminous United States. U.S. Geol. Surv., Circ. 670, 17 pp.

Dall'Aglio, M., 1968. The abundance of mercury in 300 natural water samples from the Tuscany and Latium (Central Italy). In: L. A. E. Ahrens (Editor), Origin and Distribution of the Elements. Pergamon, Cambridge, pp. 1065—1081.

Dames & Moore, Inc. and E.I.C. Corp., 1977. Description of manganese nodule processing activities for environmental studies, Vols. 1—11. U.S. Dep. Commer., Washington, D.C., Contract No. 6-35331.

d'Anglejean, B. F. and Smith, E. C., 1973. Distribution, transport and composition of suspended matter in the St. Lawrence estuary. Can. J. Earth Sci., 10: 1380—1396.

Davies, A. G., 1978. Pollution studies with marine plankton, Part II. Heavy metals. Adv. Mar. Biol., 15: 381—508.

Davies, D. K., 1972. Deep sea sediments and their sedimentation, Gulf of Mexico. Am. Assoc. Pet. Geol. Bull., 56: 2212—2239.

Davies, J. M., Gamble, J. C. and Steele, J. H., 1975. Preliminary studies with a large plastic enclosure. In: E. L. Cronin (Editor), Estuarine Research, Vol. 1. Academic Press, New York, N.Y., pp. 251—264.

Dayton, P. K. and Hessler, R. R., 1972.Role of biological disturbance in maintaining diversity in the deep sea. Deep-Sea Res., 18: 199—208.

Defant, A., 1961. Physical Oceanography, Vol. 1. Pergamon, New York, N.Y., 729 pp.

DeGoeij, J. J. M., Guinn, V. P., Young, D. R. and Mearns, A. J., 1974. Neutron activation analysis trace element studies of Dover sole liver and marine sediments. In: Comparative Studies on Food and Environmental Contamination. International Atomic Energy Agency, Vienna, pp. 189—200.

De Lappe, B. W., Risebrough, R. W., Mendola, I. T., Bowen, G. W. and Monod, J. L., 1972. Distribution of polychlorinated biphenyls on the Mediterranean coast of France. In: Journées d'Études sur les Pollutions Marines, Comm. Int. Explor. Sci. Mer. (Athens, Sept. 3—4), pp. 43—45.

De Maio, A. and Moretti, M., 1973. Contributo a un progetto di studio delle correnti del Golfo di Napoli. Fondazione Politecnica per il Mezzogiorno d'Italia, Naples, Quad. No. 71, 29 pp.

Denman, K. L. and Platt, T., 1975. Coherences in the horizontal distributions of phytoplankton and temperature in the upper ocean. Mem. Soc. R. Sci. Liège, 7: 19—30.

Dexter, R. N. and Pavlou, S. P., 1973. Chlorinated hydrocarbons in sediments from southern Greece. Mar. Pollut. Bull., 4: 188—190.

Diaz, H. F., 1979. Atmospheric conditions and comparison with past records. In: R. L. Swanson and C. J. Sindermann (Editors), Oxygen Depletion and Associated Benthic Mortalities in the New York Bight, 1976, Ch. 3. U.S. Dep. Commer., Washington, D.C., N.O.A.a. Prof. Pap.

DiGeorge, III, F. P. and Herbich, J. B., 1978. Laboratory determination of bulking factors for Texas coastal fine-grained materials. Cent. Dredging Stud, Texas A & M Univ., College Station, Texas, CDS Rep. No. 218.

DiSalvo, L. H., Guard, H. E., Hirsch, N. D. and Ng, J., 1977. Assessment and significance of sediment-associated oil and grease in aquatic environments. U.S. Army Eng. Waterways Exp. Stn., Vicksburg, Miss., Tech. Rep. D-77-26.

Doflein, F. and Reichenow, E., 1928. Lehrbuch der Protozoenkunde. G. Fischer, Jena, 528 pp.

Doi, R. and Ui, J., 1975. The distribution of mercury in fish and its form of occurrence. In: P. A. Krenkel (Editor), Heavy Metals in the Aquatic Environment. Pergamon, Oxford, pp. 197—221.

Domenowske, R. S. and Matsuda, R. I., 1969. Sludge disposal and the marine environment. J. Water. Pollut. Control Fed., 41(9): 1613—1624.

Dressel, D. M., Heinle, D. R. and Grote, M. C., 1972. Vital staining to sort dead and live copepods. Chesapeake Sci., 13: 190—193.

Duinker, J. C., 1980. Suspended matter in estuaries: adsorption and desorption processes.

In: E. Olausson and I. Cato (Editors), Chemistry and Biochemistry of Estuaries. Wiley, Chichester, pp. 121—151.

Duinker, J. C. and Nolting, R. F., 1976. Distribution model for particulate trace metals in the Rhine estuary, Southern Bight and Dutch Wadden Sea. Neth. J. Sea Res., 10: 71—102.

Duinker, J. C., van Eck, G. T. M. and Nolting, R. F., 1974. On the behaviour of copper, zinc, iron and manganese in the Dutch Wadden Sea: evidence for mobilization processes. Neth. J. Sea Res., 8: 214—239.

Duncan, A., Scheimer, F. and Klekowski, R. Z., 1974. A preliminary study of feeding rates of bacterial food by adult females of a benthic nematode, *Plectus palustris*, de Man, 1880. Pol. Arch. Hydrobiol., 21: 249—259.

Dunn, B. P. and Young, D. R., 1976. Baseline levels of benzo(a) pyrene in southern California mussels. Mar. Pollut. Bull., 7(12): 231—234.

Durum, W. H. and Haffty, J., 1961. Occurrence of minor elements in water. U.S. Geol. Surv., Circ. 445, 11 pp.

Duursma, E. K. and Marchand, M., 1974. Aspects of organic marine pollution. Oceanogr. Mar. Biol., 12: 315—431.

Duursma, E. K., Marchand, M. and Vas, D., 1974. Chlorinated hydrocarbon residues in biota, sediments and water collected from the Ligurian Sea. In: Activities of the International Laboratory of Marine Radioactivity, Monaco, 1974. Rep. Tech. Doc. IAEA-163, International Atomic Energy Agency, Vienna, pp. 138—150.

Dyer, K. R., 1978. The balance of suspended sediment in the Gironde and Thames estuaries. In: Bj. Kjerfve (Editor), Estuarine Transport Processes. University of South California Press, Columbia, S.C., pp. 135—145.

Eady, E. T., 1949. Long waves and cyclone waves, Tellus, 1: 33—52.

Ebbesmeyer, C. C. and Helseth, J. M., 1975. A study of current properties and mixing using drogue movements observed during summer and winter in central Puget Sound, Washington. Final Rep. for Municipality of Metropolitan Seattle, Seattle, Wash., 81 pp.

Edgington, D. N. and Robbins, J. A., 1976. Records of lead deposition in Lake Michigan sediments since 1800. Environ. Sci. Technol., 10: 266—273.

Edmond, J. M., Boyle, E. D., Brummond, D., Grant, B. and Mislick, T., 1978. Desorption of barium in the plume of the Zaire (Congo) River. Neth. J. Sea Res., 12: 329—337.

Edmonds, J. S. and Francesconi, K. A., 1977. Methylated arsenic from marine fauna. Nature (London), 265: 436.

Edwards, R. L., 1976. Middle Atlantic fisheries: recent changes in populations and outlook. In: M. G. Gross (Editor), Middle Atlantic Continental Shelf and the New York Bight, Spec. Symp. Vol. 2. Am. Soc. Limnol. Oceanogr., pp. 302—311.

Eganhouse, R. P. and Young, D. R., 1976. Mercury in tissues of mussel off southern California. Mar. Pollut. Bull., 7(8): 145—147.

EG & G, 1977a. Measurements of the dispersion of barged waste near 38° 33′ N latitude and 74° 20′ W longitude. Prepared by EG & G Environmental Consultants for E. I. DuPont de Nemours & Co.

EG & G, 1977b. Measurements of the dispersion of barged waste near 38° 50′ N latitude and 72° 15′ W longitude at the "*106*" dump site. Prepared by EG & G Environmental Consultants for E. I. duPont de Nemours & Co.

Eisma, D., 1980. Supply and deposition of suspended matter in the North Sea. In: S. D. Nio, R. T. E. Schüttenhelm and Tj. C. E. van Weering (Editors), Holocene Marine Sedimentation in the North Sea Basin. International Association of Sedimentology (I.A.S.), Spec. Publ. No. 5 (in press).

Eisma, D., Kalf, J. and van der Gaast, S. J., 1978. Suspended matter in the Zaire estuary and the adjacent Atlantic Ocean. Neth. J. Sea Res., 12: 382—406.

Eisma, D., Kalf, J. and Veenhuis, M., 1980. Distribution and particle size of suspended matter in the Southern Bight of the North Sea and the eastern Channel. Neth. J. Sea Res. (in press).

Elder, D. L., 1975. PCB's in N.W. Mediterranean coastal waters. Mar. Pollut. Bull., 7: 63—64.

Elder, D. L. and Villeneuve, J. P., 1977. Polychlorinated biphenyls in the Mediterranean Sea. Mar. Pollut. Bull., 8: 19—22.

Elder, D. L., Villeneuve, G. P., Parsi, P. and Harvey, G. R., 1976. Polychlorinated biphenyls in sea-water, sediments and over-ocean air of the Mediterranean. In: Activities of the International Laboratory of Marine Radioactivity, Monaco, 1976. Rep. Tech. Doc. IAEA-187, International Atomic Energy Agency, Vienna, pp. 136—151.

Elderfield, H., 1976. Manganese fluxes to the oceans. Mar. Chem., 4: 103—132.

El-Sayed, S. Z., Taguchi, S., Franceshini, G., Fryxell, G. and Gates, C., 1977. Phytoplankton and primary productivity studies. Natl. Tech. Info. Serv., Springfield, Va., Rep. No. PB/274699/AS, 197 pp. plus appendices.

Emerson, R. R., 1976a. Bioassay and heavy metal uptake investigations of resuspended sediment on two species of polychaetous annelids. In: D. Soule and M. Oguri (Editors), Marine Studies of San Pedro Bay, California, Part 11. Allan Handock Foundation and Sea Grant Programs, University of Southern California, Los Angeles, Calif., pp. 69—90.

Emerson, R. R., 1976b. Impact of discharges from the main canner, StarKist and the Terminal Island waste treatment plant on the ecology of outer Los Angeles Harbor. In: D. Soule and M. Oguri (Editors), Marine Studies of San Pedro Bay, California, Part 12. Allan Hancock Foundation and Sea Grant Programs, University of Southern California, Los Angeles, Calif., pp. 255—268.

Emery, K. O. and Niino, H., 1963. Sediments of the Gulf of Thailand and adjacent continental shelf. Geol. Soc. Am. Bull., 74: 541—554.

Enright, J. T., 1977. Diurnal vertical migration: adaptive significance and timing, Part 1. Selective advantage: a metabolic model. Limnol. Oceanogr., 22: 856—872.

Eppley, R. W., Sapienza, C. and Renger, E. H., 1978. Gradients in phytoplankton stocks and nutrients off southern California in 1974—76. Estuarine Coastal Mar. Sci., 7: 291—301.

E.Q.A.—M.B.C. (Environmental Quality Analysts—Marine Biological Consultants), 1978. Southern California Edison Company marine monitoring studies. Long Beach, Calif., Generat. Stn., Fin. Rep. 1974—1978, 536 pp.

Establier, R., 1969a. Estudios del contenido en cobre del agua de mar y ostiones (Crassostrea angulata) de las costas de Cadiz. Invest. Pesq., 33: 69—86.

Establier, R., 1969b. Contenido en cobre, hierro, manganeso y cinc de los ostiones (Crassostres angulate) de las costas de Cadiz. Invest. Pesq., 33: 335—343.

Establier, R., 1970a. Contenido en cobre, hierro, manganeso y cinc del atun, Thunnus thynnus (L.) bacoreta, Euthynnus alleteratus (Raf.); bonito, Sarda sarda (Bloch.) y melva, Auxis thazard (Lac.). Invest. Pesq., 34: 171—175.

Establier, R., 1970b. Contenido en cobre, hierro, manganeso y eine de varios organos del atun, Thynnus thynnus (L.) del Golfo de Cadiz. Invest. Pesq., 34: 399—408.

Establier, R., 1972a. Nota sobre el contenido en cobre de los ostiones (Crassostrea angulata) de las costas de Huelva. Invest. Pesq., 36: 293—296.

Establier, R., 1972b. Concentration de mercurio en los tejidos de algunos peces, moluscos y crustaceos del Golfo di Cadiz y caladeros del nordoeste africano. Invest. Pesq., 36: 355—364.

Establier, R., 1973. Nueva aportacion sobre el contenido en mercurio de peces, moluscos y crustaceos del Golfo de Cadiz y caladeros de la costa oeste africana. Invest. Pesq., 37: 107—114.

Eurostaff S.p.A., 1973. Studio sull'inquinamento del Golfo di Napoli, Vol. I: 101 pp.;

Vol. II: 196 pp. Appendix Vol. I (in cooperation with Dagh-Watson Co., England).

Everett, D. E., 1971. Hydrologic and quality characteristics of the lower Mississippi River, La. Dep. Public Works—U.S. Geol. Surv., Baton Rouge, La., 4048 pp.

Fanning, K. A. and Maynard, V. I., 1978. Dissolved boron and nutrients in the mixing plumes of major tropical rivers. Neth. J. Sea Res., 12: 345—354.

F.A.O. (Food and Agriculture Organization of the U.N.), 1938—1975. Yearbook of Fisheries Statistics.

F.A.O. (Food and Agriculture Organization of the U.N.), 1973. Evaluation of mercury, lead, cadmium and the food additives amaranth, diethylpyrocarbonate, and octylgallate. F.A.O. Nutr. Meet., Rep. Ser. No. 51A, 84 pp.

F.A.O. (Food and Agriculture Organization of the U.N.), 1977. Food balance sheets, details, data source F.A.O.—I.C.S. 1972—1974, F.I.P.P., F.A.O., Rome.

Fasham, M. J. R., 1978. The statistical and mathematical analysis of plankton patchiness. Oceanogr. Mar. Biol., Annu. Rev., 16: 43—79.

Favretto, L. and Tunis, F., 1970. Sui molluschi dell'alto Adriatico—Contenuto di metalli del mitilo durante un ciclo annuale. Tech. Ital., 35: 231—233.

Feeley, R. A., 1976. Evidence for aggregate formation in a nepheloid layer and its possible role in the sedimentation of particulate matter. Mar. Geol., 20: M7—M13.

Feeley, R. A., Baker, E. T., Schumacher, J. D., Massoth, G. J. and Landing, W. M., 1979. Processes affecting the distribution and transport of suspended matter in the northeast Gulf of Alaska. Deep-Sea Res., 26A: 445—464.

Fenchel, T., 1969. The ecology of marine microbenthos, IV. Structure and function of the benthic ecosystem, its chemical and physical factors and the microfauna communities with special reference to the ciliated Protozoa. Ophelia, 6: 1—182.

Fenchel, T., 1972. Aspects of decomposer food chains in marine benthos. Verh. Dtsch. Zool. Ges., 65. Jahresversamml. 14: 14—22.

Fenchel, T., 1975. The quantitative importance of the benthic microflora of an arctic tundra panel. Hydrobiologica, 46: 445—464.

Fenchel, T. and Jørgensen, B. B., 1977. Detritus food chains of aquatic ecosystems: the role of bacteria. In: M. Alexander (Editor), Advances in Microbial Ecology, Vol. 1. Plenum, New York, N.Y., pp. 1—58.

Fernandez, M. J. and Franco, J. M., 1976. Presencia de DDT, sus derivados, y benfenilos policlorados (PCB) en sardinas (Sardina pilchardus) y jurel (Trachurus trachurus) del NW de Espana. Inf. Tec. Inst. Invest. Pesq., 39: 3—15.

Fickeisen, D. H. and Schneider, M. J., 1976. Gas bubble disease. Proc. of a Workshop, Richland Wash., Oct. 8—9, 1974. Tech. Info. Cent., Energy Res. Dev., Admin., Richland, Wash.

Figley, W., Pyle, B. and Halgren, B., 1979. Socioeconomic impacts. In: R. I. Swanson and C. J. Sindermann (Editors), Oxygen Depletion and Associated Benthic Mortalities in New York Bight, 1976, Ch. 14. U.S. Dep. Commer., Washington, D.C., N.O.A.A. Prof. Pap.

Figuères, G., Martin, J. M. and Meybeck, M., 1978. Iron behaviour in the Zaire estuary. Neth. J. Sea Res., 12: 329—337.

Fisher, N. S., 1975. Chlorinated hydrocarbon pollutants and photo synthesis of marine phytoplankton; a reassessment. Science, 189: 463—464.

Fisk, H. N., McFarlan, E., Kolb, C. R. and Wilbert, L. J., 1954. Sedimentary framework of the modern Mississippi delta. J. Sediment. Petrol., 24: 76—99.

Fonselius, S. H., 1970. Some trace metal analysis in the Mediterranean, the Red Sea and the Arctic Sea. Bull. Inst. Océanogr. Monaco, 69(1407): 3—15.

Fonselius, S. H. and Koroloff, K. F., 1963. Copper and zinc content of the water in the Ligurian Sea. Bull. Inst. Océanogr. Monaco, 61 (1181): 1—15.

Fowler, S. W., 1977. Trace elements in zooplankton particulate products. Nature (London), 269: 51—53.

Fowler, S. W. and Benayoun, G., 1977. Accumulation and distribution of selenium in mussel and shrimp tissues. Bull. Environ. Contam. Toxicol., 16: 339—346.

Fowler, S. W. and Oregioni, B., 1976. Trace metals in mussels from the N.W. Mediterranean. Mar. Pollut. Bull. 7: 26—29.

Fowler, S. W., Heyraud, M. and La Rosa, J., 1976a. The cycling of mercury in shrimp and mussels. In: Activities of the International Laboratory of Marine Radioactivity, Monaco, 1976 Report. Tech. Doc. IAEA-187, International Atomic Energy Agency, Vienna, pp. 11—20.

Fowler, S. W., Oregioni, B. and La Rosa, J., 1976b. Trace elements in pelagic organisms from the Mediterranean Sea. In: Activities of the International Laboratory of Marine Radioactivity, Monaco, 1976 Report. Tech. Doc. IAEA-187, International Atomic Energy Agency, Vienna, pp. 110—122.

Foxton, P. and Roe, H. S. J., 1974. Observations on the nocturnal feeding of some mesopelagic decapod Crustacea. Mar. Biol., 28: 37—49.

Frache, R., Baffi, F., Dadone, A. and Zanicchi, G., 1976. The determination of heavy metals in the Ligurian Sea I. The determination of Cu, Co, Ni and Cd in surface waters. Mar. Chem., 4: 365—375.

Franco, J. M. and Fernandez, M. J., 1976. Contaminacion por PCB y DDT en el littoral Espanol. In: Mes Redondà III, III Symposio El Agua en la Industria, Madrid, Dec. 1976, pp. 49—67.

Franco-Soler, J. M., 1973. Pesticides organochlorés et PCB dans trois espèces marines sur le littoral espagnol. In: Journées d'Etudes sur les Pollutions Marines, Comm. Int. Explor. Sci. Mer. (Athens, Nov. 3—4, 1972), Monaco, pp. 55—57.

Frascari, F., Frignani, M., Giordani, P., Marabini, F. and Poletti, R., 1976. Distribuzione di metalli pesanti in relazione alle caratteristiche sedimentologiche di campioni di fondo del mare Adriatico tra Ancona e il delta del Po. Ass. Ital. Oceanogr. Limnol., 5 pp. (preprint).

Freeland, G. and Swift, D. J. P., 1978. Surficial sediments. M.E.S.A. New York Bight Atlas, Monogr. 10, New York Sea Grant Institute, Albany, N.Y.

Frenkiel, F. N., 1953. Diffusion from a continuous source. Adv. Appl. Math., 3: 61—107.

Friebertshauser, M. A. and Duxburry, A. C., 1972. A water budget study of Puget Sound and its sub-regions. Limnol. Oceanogr., 17(2): 237—247.

Fukai, R. and Broquet, D., 1965. Distribution of chromium in marine organisms. Bull. Inst. Océanogr. Monaco, 65(1336): 1—19.

Fukai, R. and Huynh-Ngoc, L., 1976a. Trace metals in Mediterranean sea-water. In: Activities of the International Laboratory of Marine Radioactivity, Monaco, 1976 Report. Tech. Doc. IAEA-187, Internatonal Atomic Energy Agency, Vienna, pp. 122—132.

Fukai, R. and Huynh-Ngoc, L., 1976b. Copper, zinc and cadmium in coastal waters of the N.W. Mediterranean. Mar. Pollut. Bull. 7: 9—13.

Fukai, R. and Vas, D., 1967. A differential method of analysis for trivalent and hexavalent chromium in sea-water. J. Oceanogr. Soc. Jpn., 23: 32—38.

Galt, J. A., 1975. Development of a simplified diagnostic model for interpretation of oceanographic data. Natl. Ocean. Atmos. Admin., U.S. Dep. Commer., Boulder, Colo., N.O.A.A. T.R. E.R.L. 339, P.M.E.L. 25.

Galt, J. A. and Watabayashi, G., 1978. Use of a diagnostic circulation model to study shelf circulation. Pac. Mar. Environ. Lab., Natl. Ocean. Atmos. Admin., Seattle, Wash.

Ganther, H. E. and Sunde, M. L., 1974. Effect of tuna fish and selenium on the toxicity of methylmercury: a progress report. J. Food. Sci., 39: 1—5.

Garrels, R. M. and Mackenzie, F. T., 1971. Evolution of Sedimentary Rocks. W. W. Norton, New York, N.Y. 397 pp.

Garside, C. and Malone, T. C., 1978. Monthly carbon and oxygen budgets of the New York Bight Apex. Estuarine Coastal Mar. Sci., 6: 93—104.

G.E.S.A.M.P. (Joint Group of Experts on the Scientific Aspects of Marine Pollution), 1976. Joint Group Experts Sci. Aspects Mar. Pollut., Rep. 6th Session, Food Agric. Org. (F.A.O.), Rome, p. 54.

Giam, C. S., Chan, H. S. and Neff, G. S., 1976. Concentrations and fluxes of phthalates, DDT's and PCB's to the Gulf of Mexico. In: H. L. Windom and R. A. Duce (Editors), Marine Pollutant Transfer, D. C. Heath, Lexington, Mass., pp. 375—386.

Gibbs, R. J., 1973. Mechanisms of trace metal transport in rivers. Science. 180: 71—73.

Gibbs, R. J. (Editor), 1974. Suspended Solids in Water, Vol. 4. Plenum, New York, N.Y., pp. 206 and 218.

Gibbs, R. J., 1977. Transport phases of transition metals in the Amazon and Yukon Rivers. Geol. Soc. Am. Bull., 88: 829—843.

Gillespie, G. C., 1943. Report on a pollution survey of Santa Monica Bay Beaches. Calif. State Dep. Public Health, Berkeley, Calif.

Gillespie, P. A., 1976. Heterotrophic potentials and trophic status of ten New Zealand lakes. N.Z. J. Mar. Fresh Water Res., 10: 91—107.

Gillespie, P. A. and Vaccaro, R. F., 1978. A bacterial bioassay for measuring the copper-chelation capacity of seawater. Limnol. Oceanogr., 23: 543—548.

Gilmartin, M. and Relevante, N., 1975. The concentration of mercury, copper, nickel, cadmium and lead in the Northern Adriatic anchovy (*Engraulis encrasicolus* L.) and sardine (*Sardina pilchardus* Walb.) Fish. Bull. N.M.F.S.—N.O.A.A., 73: 193—202.

Gilmer, R. W., 1972. Free floating mucus webs: a novel feeding adaptation for the open ocean. Science, 176: 1239—1240.

Gocke, G., 1977. Untersuchungen uber die heterotrophe Aktivität in der zentralen Ostsee. Mar. Biol., 40: 87—94.

Goldberg, E. D., 1954. Marine geochemistry, chemical scavengers of the sea. J. Geol., 62: 249—265.

Goldberg, E. D., 1957. Biogeochemistry of trace metals. In: J. W. Hedgepeth (Editor), Treatise on Marine Ecology and Palaeoecology, 1 Ecology, Geol. Soc. Am., New York, N.Y., Mem. 67, pp. 345—358.

Goldberg, E. D., 1971. Man's role in the major sediment cycles. In: D. Dyrssen and D. Jagner (Editors), The Changing Chemistry of The Oceans. Wiley—Interscience, New York, N.Y.,

Goldman, C. R., 1962. A method of studying nutrient limiting factors in situ in water columns isolated by polyethylene films. Limnol. Oceanogr., 7: 99—101.

Gordon, Jr., D. C., 1970. A microscopic study of organic particles in the North Atlantic Ocean. Deep-Sea Res., 17: 175—185.

Gordon, R. B., 1973. Dispersion of dredge spoil dumped in a tidal stream: observations at the New Haven dump site. Yale Rep. to U.S. Army Corps Eng., New Haven, Conn.

Grancini, G., Stievano, B. M., Girardi, G., Guzzi, G. and Pietra, R., 1975. The determination of neutron activation analysis of trace elements in seawater and sediment samples collected in the northern Adriatic Sea. Mem. Biol. Mar. Oceanogr., 5: 77—97.

Grassle, J. F., 1977. Slow recolonisation of deep-sea sediment. Nature (London), 265: 618—619.

Greene, C. S., 1976a. Response and recovery of the benthos at Orange County. In: Coastal Water Research Project, Annual Report, 1976, U.S. Dep. Commer., Springfield, Va. Natl. Tech. Info. Serv., PB 274461/AS, pp. 197—203.

Greene, C. S., 1976b. Responses of benthic infauna to the initiation and termination of sewage discharge. EPA Task Rep. to U.S. Environ. Prot. Agency, Corvallis, Oreg., Grant R801152.

Greig, E. and Wenzloff, R., 1977. Final report on heavy metals and small pelagic fin fish, Euphausid Crustaceans, Apex predators, etc. In: N.O.A.A. Dumpsite Evaluation, Rep. 77-1, pp. 547—564.

Greve, P., 1971. De persistentie van endosulfan in oppervlaktewater. Meded. Rijksfac.

Landbouwwet. Gent, 36: 439—447.

Grice, G. D., Wiebe, P. H. and Hoaglund, E., 1973. Acid-iron waste as a factor affecting the distribution of zooplankton in the New York Bight. Estuarine Coastal Mar. Sci., 1: 45—50.

Grice, G. D., Reeve, M. R., Koeller, P. and Menzel, D. W., 1977. The use of large volume, transparent, enclosed seasurface water columns in the study of stress on plankton ecosystems. Helgol. Wiss. Meeresunters., 30: 118—133.

Grigg, R. W., 1979. Long-term changes in rocky bottom communities off Palos Verdes. Coastal Water Research Project Annual Report 1978. U.S. Dep. Commer., Springfield, Va., Natl. Tech. Info. Serv., PB 299830/AS.

Grigg, R. W. and Kiwala, R. S., 1970. Some ecological effects of discharged wastes on marine life. Calif. Fish Game, 56(3): 145—155.

Grim, R. E., 1968. Clay mineralogy. McGraw-Hill, New York, N.Y., 596 pp.

Grimanis, A. P., Vassilaki-Grimani, M. and Griggs, G. B., 1977. Pollution studies of trace elements in sediments from the Upper Saronikos Gulf, Greece. J. Radioanal. Chem., 37: 761—763.

Grimwood, P. D. and Webb, G. A. M., 1976. Assessment of the radiological protection aspects of disposal of high level waste on the ocean floor. Nat. Radiol. Prot. Board, Harwell, Rep. No. NRPB 48, 117 pp.

Gross, M. G., 1972. Geologic aspects of waste solids and marine waste deposits, New York metropolitan region. Geol. Soc. Am. Bull., 83: 3163—3176.

Gross, M. G. (Editor), 1976a. Middle Atlantic Continental Shelf and the New York Bight, Spec. Symp. Vol. 2, Am. Soc. Limnol. Oceanogr., 441 pp.

Gross, M. G., 1976b. Waste Disposal. M.E.S.A. New York Bight Atlas, Monogr. 26, New York Sea Grant Institute, Albany, N.Y., 32 p.

Gunnerson, C. G., 1955. Discussion of pollution of the Mississippi River near New Orleans. Proc. Am. Soc. Civ. Eng., 80 (Sep. 689): 11—12.

Gunnerson, C. G., 1974. Discharge of sewage from sea outfalls. In: A. L. H. Gameson (Editor), Proceedings Conference on Discharge of Sewage from Sea Outfalls, Pergamon, London, pp. 140—164.

Gunter, G., Mackin, J. G. and Ingle, R. M., 1964. A report to the District Engineer on the effect of disposal of spoil from the Inland Waterway, Chesapeake and Delaware Canal, in upper Chesapeake Bay. U.S. Army Eng. Dist., Philadelphia, Pa.

Haedrich, R. L., 1974. Pelagic capture of the epibenthic rat-tail, Coryphaenoides rupestris. Deep-Sea Res., 21: 977—979.

Haedrich, R. L. and Henderson, N. R., 1974. Pelagic food of Coryphaenoides armatus, a deep benthic rat-tail. Deep-Sea Res., 21: 739—744.

Hall, A. S., Houle, C. R., Teeny, F. M. and Gauglitz, Jr., E. J., 1977. Trace metals— DOMES samples. Natl. Mar. Fish. Serv., Seattle, Wash., 154 pp.

Halpern, D., 1975. Upper ocean circulation studies in the eastern tropical North Pacific during September and October 1975. Offshore Technol. Conf. Pap. No. 2457, pp. 357—369.

Halpern, D., 1979. Observations of the upper ocean currents at DOMES Sites A, B and C in the Tropical Central North Pacific Ocean during 1975 and 1976. In: J. L. Bischoff and D. Z. Piper (Editors), Marine Geology and Oceanography of the Pacific Manganese Nodule Province. Marine Science 9, Plenum, New York, N.Y., pp. 43—82.

Hammon, A., 1977. Port Facilities and Commerce. M.E.S.A. New York Bight atlas, Monogr. 20, New York Sea Grant Institute, Albany, N.Y.

Han, G. and Hansen, D. V., 1978. Steady state diagnostic model of the New York Bight: comparison of model results with observations of time-averaged currents. Atl. Oceanogr. Meteorol. Lab., N.O.A.A., Miami, Fla.

Han, G., Hansen, D. V. and Cantillo, A., 1979. Diagnostic model of water and oxygen transport. In: R. L. Swanson and C. J. Sindermann (Editors), Oxygen Depletion and

Associated Benthic Mortalities in New York Bight, 1976. Ch, 8. U.S. Dep. Commer., N.O.A.A. Prof. Pap.

Hansen, D., 1977. Circulation. M.E.S.A. New York Bight atlas, Monogr. 3. New York Sea Grant Institute, Albany, N.Y., 23 pp.

Hanson, J. A., 1974. Open Sea Mariculture. Dowden, Hutchinson & Ross, Stroudsburg, Pa., pp. 359—374.

Hapgood, W., 1960. Hydrographic observations in the Bay of Naples, January 1957—1958 (station lists). Pubbl. Stn. Zool. Napoli, 31: 337—371.

Harbison, G. R. and Gilmer, R. W., 1976. The feeding rates of the pelagic tunicate *Pegea confederata* and two other salps. Limnol. Oceanogr., 21: 517—528.

Harden Jones, F. R., 1968. Fish Migration. E. Arnold, London, 325 pp.

Harding, Jr., L. W. and Phillips, Jr., J. H., 1978. Polychlorinated biphenyls; transfer from micro particulates to marine phytoplankton and the effects on photosynthesis. Science, 202: 1189—1192.

Harman, R. A., Serwold, J. C. and Sylvester, R. E., 1977. Distribution of subtidal benthic organisms, sediments, and habitats near the West Point outfall. Puget Sound Interim Studies Report, May 1977. Municipality of Metropolitan Seattle, Seattle, Wash.

Harris, J. E., 1977. Characterization of suspended matter in the Gulf of Mexico, II. Particle size analysis of suspended matter from deep water. Deep-Sea Res., 24: 1055—1061.

Harris, J. E., McKee, Th. R., Wilson, Jr., R. C. and Whitehouse, U. G., 1972. Preparation of membrane filter samples for direct examination with an electron microscope. Limnol. Oceanogr., 17: 784—787.

Harrison, W., 1967. Environmental effects of dredging and spoil deposition. Proc. World Dredging Conf., New York, N.Y.

Hartung, R. and Klinger, G. W., 1970. Concentration of DDT by sedimented polluting oils. Environ. Sci. Technol., 4: 407—410.

Harvey, G. and Steinhauser, W. C., 1976. Transport pathways of polychlorinated biphenyls in Atlantic water. J. Mar. Res., 34: 541—575.

Harvey, G., Steinhauer, W. C. and Teal, J. N., 1973. Polychlorinated biphenyls in North Atlantic ocean water. Science, 180: 643—644.

Hattula, M. L., 1974. Some aspects of the recovery of chlorinated residues (DDT-type compounds and PCB) from fish tissue by using different extraction methods. Bull. Environ. Contam. Toxicol., 12: 301—307.

Hayes, S. P., 1979. Benthic current observations at DOMES Sites *A*, *B*, and *C*. In: J. L. Bischoff and D. Z. Piper (Editors), Marine Geology and Oceanography of the Pacific Manganese Nodule Province. Marine Science 9, Plenum, New York, N.Y., pp. 83—112.

Hayes, S. P., 1980. Bottom boundary layer in the eastern tropical Pacific. J. Phys. Ocean. 10: 315—329.

Healy, M. L. and Richards, F. A., 1976. Trace metal baselines in the DOMES study area. Natl. Tech. Info. Serv., Springfield, Va., Rep. No. PB/283243/AS, 7 pp.

Hecker, B., 1977. Baited-traps: benthic baseline DOMES. Lamont—Doherty Geol. Observ., Columbia Univ., Palisades, N.Y., Final Rep. to DOMES Proj., 10 pp.

Hecker, B. and Paul, A. Z., 1979. Abyssal community structure of the benthic infauna of the eastern equatorial Pacific: DOMES Sites *A*, *B*, and *C*. In: J. L. Bischoff and D. Z. Piper (Editors), Marine Geology and Oceanography of the Pacific Manganese Nodule Province, Marine Science 9, Plenum, New York, N.Y., pp. 287—308.

Hein, J. R. and Jones, M. G., 1977. Mineralogy and diagenesis of DOMES Sites *A* and *B*. In: D. Z. Piper (Editor), Deep Ocean Environmental Study: Geology and Geochemistry of DOMES Sites *A*, *B*, and *C*, Equatorial North Pacific, U.S. Geol. Surv., Menlo Park, Calif., Open-File Rep. No. 77-778, 527 pp.

Hein, J. R., Gutmacher, C. and Miller, J., 1976. DOMES Area *C*: General statement about mineralogy, diagenesis and sediment classification. In: J. L. Bischoff (Editor), Deep

Ocean Environmental Study: NE Pacific Nodule Province, Site *C*, geology and geochemistry, U.S. Geol. Surv., Menlo Park, Calif., Open-File Rep. No. 76-548, 275 pp.

Hem, J. D., 1963. Increased oxidation rate of manganese ions in contact with feldspar grains. U.S. Geol. Surv., Prof. Pap., 475-C: C216—C217.

Hendricks, T. J., 1977a. Coastal currents. In: Coastal Water Research Project, Annual Report 1977. U.S. Dep. Commer., Springfield, Va., Natl. Tech. Info. Serv., PB 274463/AS, pp. 53—62.

Hendricks, T. J., 1977b. In situ measurements of initial dilution. In: Coastal Water Research Project, Long Beach, Calif. (U.S. Dep. Commer., Springfield, Va., Natl. Tech. Info. Serv., PB 299830/AS.)

Hendricks, T. J., 1979. Forecasting changes in sediments near municipal wastewater outfalls. In: W. Bascom, 1978 Annual Report, Southern California Coastal Water Research Project. Long Beach, Calif. (U.S. Dep. Commer., Springfield, Va., Natl. Tech. Info. Serv., PB 299830/AS.)

Herbich, J. B., 1975. Coastal and Deep Ocean Dredging. Gulf Publishing Co., Houston, Texas, 622 pp.

Herring, J. R. and Abati, A. L., 1979. Effluent particle dispersion. In: Coastal Water Research Project, Annual Report 1978. Southern California Coastal Water Research Project, El Segundo, Calif., pp. 127—143 (U.S. Dep. Commer., Springfield, Va., Natl. Tech. Info. Serv., PB 299830/AS.)

Hessler, R. R. and Jumars, P. A., 1974. Abyssal community analysis from replicate box cores in the north central Pacific. Deep-Sea Res., 21: 185—210.

Hetherington, J. A., Jefferies, D. F. and Lovett, M. B., 1975. Some investigations into the behaviour of plutonium in the marine environment. In: Impacts of Nuclear Releases into the Aquatic Environment. International Atomic Energy Agency, Vienna, pp. 193—212.

Hirota, J. (Editor), 1977a. Deep Ocean Mining Environmental Study (DOMES) Literature Survey. U.S. Department of Commerce, Washington, D.C., NTIS No. PB-279, 421 pp.

Hirota, J., 1977b. DOMES zooplankton. Natl. Tech. Info. Serv., Springfield, Va., Rep. No. OB/274662/AS, 247 pp.

Hirota, J., 1978. DOMES zooplankton. Hawaii Inst. Mar. Biol.—Dep. Oceanogr., Univ. Hawaii, Manoa, Hawaii (unpublished).

Hirsch, N. D., DiSalvo, L. H. and Peddicord, R., 1978. Effects of dredging and disposal on aquatic organisms. U.S. Army Eng. Waterways Exp. Stn., Vicksburg, Miss., Tech. Rep. DS-78-5.

Hobbie, J. E. and Crawford, C. C., 1969. Respiration corrections for bacterial uptake of dissolved organic compounds in natural waters. Limnol. Oceanogr., 14: 528—532.

Holeman, J. N., 1968. The sediment yield of major rivers of the world. Water Resour. Res., 4: 737—747.

Holle, C. G., 1952. Sedimentation at the mouth of the Mississippi River. In: Proc. 2nd Conf. on Coastal Engineering, University of California Press, Berkeley, Calif., pp. 111—129.

Holliday, B. W., 1978. processes affecting the fate of dredged material. U.S. Army Eng. Waterways Exp. Stn., Vicksburg, Miss., Tech. Rep. DS-78-2.

Holliday, B. W., Johnson, B. H. and Thomas, W. A., 1978. Predicting and monitoring dredged material movement. U.S. Army Eng. Waterways Exp. Stn., Vicksburg, Miss., Rep. DS-78-3.

Hom, W., Risebrough, R. W., Soutar, A. and Young, D. R., 1974. Deposition of DDE and polychlorinated biphenyls in dated sediments of the Santa Barbara Basin. Science, 184: 1197—1199.

Honjo, S., 1978. Sedimentation of materials in the Sargasso Sea at a 5367 m deep station. J. Mar. Res., 36: 469—492.

Honjo, S. and Roman, M. R., 1978. Marine copepod fecal pellets: production, preservation and sedimentation. J. Mar. Res., 36: 45—57.

Honjo, S., Emery, K. O. and Yamamoto, S., 1974. Non-combustible suspended matter in surface waters off eastern Asia. Sedimentology, 21: 555—575.

Horn, M. K. and Adams, J. A. S., 1966. Computer-derived geochemical balances and element abundances. Geochim. Cosmochim. Acta, 30: 279—297.

Horne, D. R., 1980. Variation of surface currents and effects on dispersion in the Tropical Pacific Ocean. M.S. Thesis, Texas A & M University, College Station, Texas, 88 pp.

Horne, D. R., Pelack, M. N. and Horne, B. M., 1973. Metal content of ferromanganese deposits of the oceans. Int. Decade Ocean Explor., Tech. Rep. No. 3, 51 pp.

Horne, D. R., Pelack, M. N. and Horne, B. M., 1974. Physical properties of sedimentary provinces, North Pacific and North Atlantic Ocean. In: A. L. Inderbitzen (Editor), Deep-Sea Sediments, Physical and Mechanical Properties. Plenum, New York, N.Y., pp. 417—439.

Hsueh, Y. and Lee, C.-Y., 1978. A hindcast of barotropic response over the Oregon—Washington continental shelf during the summer of 1972. J. Phys. Oceanogr., 8(5): 799—810.

Hsueh, Y. and Peng, C.-Y., 1977. A diagnostic model of continental shelf circulation. J. Geophys. Res., 83(C6): 3033—3041.

Huang, N. E., 1979. New developments in satellite oceanography and current measurements. Rev. Geophys. Space Phys., 17: 1558—1568.

Hulbert, E. M. and Jones, C. M., 1977. Phytoplankton in the vicinity of Deep Water Dumpsite 106. In: Baseline Report of Environmental Conditions Deep Water Dumpsite 106. N.O.A.A. Dumpsite Evaluation Report 77-1, pp. 219—231.

Humpert, D., 1979. Report to the Pacific Fisheries Management Council on the anchovy fishery for 1978/79. Natl. Mar. Fish. Serv., S.W. Fish. Cent., La Jolla, Calif., 20 pp.

Huston, J. W. and Huston, W. C., 1976. Techniques for reducing turbidity associated with present dredging procedures and operations. U.S. Army Eng. Waterways Exp. Stn., Vicksburg, Miss., Rep. No. D-76-4.

Ichiye, T., 1958. A theory of vertical structure in the ocean. J. Oceanogr. Soc. Jpn., 14: 35—40.

Ichiye, T., 1966. Vertical currents in the Equatorial Pacific Ocean. J. Oceanogr. Soc. Jpn., 22: 274—284.

Ichiye, T. and Carnes, M., 1976a. Radial distribution of waste sediments from a line source. D.O.M.E.S. Model Proj. Rep., III, 167 pp.

Ichiye, T. and Carnes, M., 1976b. Dispersion of waste sediments from a point-source discharge. D.O.M.E.S. Model Proj. Rep., IV, 39 pp.

Ichiye, T. and Carnes, M., 1977. Modeling of sediment dispersion during deep ocean mining operation. 9th Annu. Conf. on Offshore Technology, Houston, Texas, Pap. No. 2778, pp. 421—426.

Ichiye, T., Bassin, N. J. and Harris, J. E., 1972. Diffusivity of suspended matter in the Caribbean Sea. J. Geophys. Res., 77: 6576—6588.

Ichiye, T., Inoue, M. and Carnes, M., 1980. Horizontal diffusion in ocean dumping experiments. In: B. H. Ketchum, D. R. Kester and P. K. Park (Editors), Ocean Dumping of Industrial Wastes. Plenum, New York, N.Y.

I.J.C. (International Joint Commission), 1978. The Ecosystem Approach. Int. Joint Comm., Res. Adv. Board, Great Lakes Regional Office, Windsor, Ont.

Ingham, M. C., Bisagni, J. J. and Mizenko, D., 1977. The general physical oceanography of Deep Water Dumpsite 106. In: Baseline Report of Environmental Conditions at Deep Water Dumpsite 106. N.O.A.A. Dumpsite Evaluation Report 77-1, pp. 29—54.

Insola, A., Romano, A. and Caprio, V., 1973. Inquinanti delle acqua di processo delle lavorazioni di raffineria. Foundazione Politecnica per il Mezzogiorno d'Italia, Naples, Quad. No. 72, 26 pp.

Isaacs, J. D., 1972. Unstructured marine food webs and "pollutant analogues". Fish. Bull., U.S., 70: 1053—1059.

Isaacs, J. D., 1976. Reproductive products in marine food webs. S. Calif. Acad. Sci. Bull.,

75: 220—223.

Isaacs, J. D. and Schwartzlose, R. A., 1975. Active animals of the deep-sea floor. Sci. Am., 233(4): 85—91.

Iseri, K. T. and Langbein, W. B., 1974. Large rivers of the United States. U.S. Geol. Surv., Circ. 686, 10 pp.

Iwata, H., Okamoto, H. and Ohsawa, Y., 1973. Effect of selenium on methyl mercury poisoning. Res. Commun. Pathol. Pharmacol., 5: 673—680.

Jacobs, M. B., Thorndike, E. M. and Ewing, M., 1973. A comparison of suspended particulate matter from nepheloid and clear water. Mar. Geol., 14: 117—128.

Jacobson, S. M. and Boylan, D. B., 1973. Effect of seawater soluble fraction of kerosene on chemotapis in a marine snail, *Nassaruis obsoletris*. Nature (London), 241: 213—215.

Jan, T. -K., Moore, M. D. and Young, D. R., 1977. Metals in seafoods near outfalls. In: Coastal Water Research Project Annual Report, 1977. U.S. Dep. Commer., Springfield, Va., Natl. Tech. Info. Serv., PB 274463/AS.

Jannasch, H. W. and Wirsen, C. O., 1973. Deep-sea microorganisms: in situ responses to nutrient enrichment. Science, 180: 641—643.

Jaschnov, W. A., 1970. Distribution of *Calanus* species in the seas of the Northern Hemisphere. Int. Rev. Ges. Hydrobiol., 55: 197—212.

Jehl, J. R., 1969. A wonderful bird was the pelican. Ocean, 2: 10—19.

Jensen, A. C., 1979. Management of New York Bight Fisheries in an antagonistic environment. N.O.A.A.—M.E.S.A. Tech. Rep., U.S. Dep. Commer., N.O.A.A., Boulder, Colo.

Jensen, S., Renberg, L. and Vaz, R., 1973. Problems in the quantification of PCB in biological material. In: PCB Conference 2. Publ. Natl. Swed. Environ. Prot. Board, 4E: 7—14.

Jerlov, N. G., 1976. Marine Optics. Elsevier, Amsterdam, 232 pp.

Johnson, F. G., 1977. Sublethal biological effects of petroleum hydrocarbon exposures: bacteria, algae and invertebrates. In: D. C. Malines (Editor), Effects of Petroleum on Arctic and Subarctic Marine Environments and Organisms, Vol. II. Academic Press, New York, N.Y., pp. 271—318.

Jokelainen, A., 1967. Cs-137 in some Finnish diets in 1962. In: A. Aberg and F. P. Hungate (Editors), Radioecological Concentration Processes. Pergamon, Oxford, pp. 87—95.

Jones, R. A. and Lee, G. F., 1978. Evaluation of the elutriate text as a method of predicting contaminant release during open water disposal of dredged sediments and environmental impact of open-water dredged material disposal, Vol. I, U.S. Army Eng. Waterways Exp. Stn., Vicksburg, Miss., Tech. Rep. D-78-45.

Jørgensen, C. B., 1966. Biology of Suspension Feeding. Pergamon, Oxford, 357 pp.

Judson, S., 1968. Erosion of the land. Am. Sci., 56: 356—374.

Judson, S. and Ritter, D. F., 1964. Rates of regional denudation in the United States. J. Geophys. Res., 69: 3395—3401.

Jumars, P. A., 1977. Potential environmental impact of deep-sea manganese nodule mining: community analysis and predation. Natl. Tech. Info. Serv., Springfield, Va., Rep. No. PB 28315/AS, 19 pp.

Jumars, P. A. and Hessler, R. R., 1976. Hadal community structure: implications from the Aleutian Trench. J. Mar. Res., 34: 547—560.

Kendall, T. R., 1970. The Pacific Equatorial Countercurrent. International Center for Environmental Research Publ., Laguna Beach, Calif.

Kennedy, G. L., 1975. Paleontologic record of areas adjacent to the Los Angeles—Long Beach Harbors, Los Angeles County, California. In: Marine Studies of San Pedro Bay, California, Part 9. Allan Hancock Foundation and Sea Grant Programs, University of Southern California, Los Angeles, Calif., 119 pp.

Kerut, E. G., 1976. Report on NDBO drifting buoy experiment in the Pacific North Equatorial Countercurrent. N.O.A.A. Data Buoy Office, NSTL Stn., Miss. NDBO-

C62X2A (unpublished).

Kester, D. R., Ahrland, S., Beasley, T. M., Bernhard, M., Branica, M., Campbell, I. D., Eichhorn, G. L., Kraus, K. A., Kreinling, K., Millero, F. J., Nuernberg, H. W., Piro, A., Pytkowicz, R. M., Steffan, L. and Stumm, W., 1975. Chemical speciation in seawater. In: E. D. Goldberg (Editor), The Nature of Seawater, Dahlem Konferenzen. Phys. Chem. Res. Sci. Rep. No. 1, Berlin, pp. 17—42.

Kester, D. R., Hittinger, R. C., Mukherji, R., and HausKnecht, K., 1978. Metal studies at DWD-*106*. Final rep. on N.O.A.A. Grant 04-7-158-44027/44064, 49 pp.

Kester, D. R., Mukherji, R. and Hittinger, R. C., 1979. Summary of recent results based on FY *78* field studies at DWD-*106*. Prelim. Rep., University of Rhode Island, Kingston, R.I. (Unpublished.)

Kester, D. R., Hittinger, R. C. and Mukherji, P., 1980. Effect of acid—iron waste disposal on transition and heavy metals at deepwater dumpsite *106*. In: B. H. Ketchum, D. R. Kester and P. K. Park (Editors), Ocean Dumping of Industrial Wastes. Plenum, New York, N.Y.

Ketchum, B. H. and W. L. Ford, 1948. Preliminary report on acid from waste disposal at sea. Woods Hole Oceanogr. Inst., Woods Hole, Mass.

Kharkar, D. P., Turekian, K. K. and Bertine, K. K., 1968. Stream supply of dissolved Ag, Mo, Sb, Se, Cr, Co, Rb and Cs to the oceans. Geochim. Cosmochim. Acta, 32: 285—298.

Kindle, E. M., 1932. Lacustrine concentrations of manganese. Am. J. Sci., 224: 496—504.

Knauer, G. A. and Martin, J. H., 1973. Seasonal variations of Cd, Cu, Mn, Pb and Zn in water and phytoplankton in Monterey Bay. Limnol. Oceanogr., 18(4): 597—604.

Koblentz-Mishke, O. J., Volkovinsky, V. V. and Kabanova, J. G., 1970. Plankton primary production of the world ocean. In: W. S. Wooster (Editor), Scientific Exploration of the South Pacific. National Academy of Sciences, Washington, D.C., pp. 183—193.

Koditschek, L., 1974. Antimicrobial-resistant coliforms in New York Bight. Mar. Pollut. Bull., 5(5) 71—74.

Koditschek, L., 1976. Antimicrobial-resistant bacteria in the New York Bight. In: M. G. Gross (Editor), Middle Atlantic Continental Shelf and the New York Bight, Spec. Vol. 2. Am. Soc. Limnol. Oceanogr., University of Michigan Press, Ann Arbor, Mich., pp. 383—393.

Koebel, C. and Krueckeberg, D., 1975. Demographic Patterns. M.E.S.A. New York Bight atlas, Monogr. 23, New York Sea Grant Institute, Albany, N.Y., 43 pp.

Koeman, J. H., Peeters, W. H. M. and Koudstaal-Hol, C. H. M., 1973. Mercury—selenium correlations in marine mammals. Nature (London), 245: 385—386.

Koeman, J. H., van de Ven, W. S. M., de Goeij, J. J. M., Tjioe, P. S. and van Haften, J. L., 1975. Mercury and selenium in marine mammals and birds. Sci. Total Environ., 3: 279—287.

Koh, R. C. Y. and Chang, Y. C., 1973. Mathematical Model for Barged Ocean Disposal of Water, U.S.E.P.A., Washington, D.C., Environ. Prot. Agency Ser. EPA 660/2-73-029, Dec. 1973.

Kooyman, G. L. and Andersen, H. T., 1969. Deep diving. In: H. T. Andersen (Editor), The Biology of Marine Mammals. Academic Press, New York, N.Y., pp. 65—94.

Kopp, J. E. and Kroner, R. C., 1967. Trace metals in water of the United States. Fed. Water Pollut. Control Admin., Div. Pollut. Surveillance.

Kosta, L., Byrne, A. R. and Zelenko, V., 1975. Correlation between selenium and mercury in man following exposure to inorganic mercury. Nature (London), 254: 238—239.

Kosta, L., Ravnik, V., Byrne, A. R., Stirn, J., Dermelj, M. and Stegnar, P., 1978. Some trace elements in the waters, marine organisms and sediments of the Adriatic by neutron activation analysis. J. Radioanal. Chem., 44: 317—332.

Kranck, K., 1973. Flocculation of suspended sediment in the sea. Nature (London),

246: 348—350.

Kremer, P., 1978. Dynamic oxygen model of Los Angeles Harbor receiving waters. In: Marine Studies of San Pedro Bay, California, Part 14. Allan Hancock Foundation and Sea Grant Programs, University of Southern California, Los Angeles, Calif., pp. 25—70.

Kremer, P. and Chiang, W.-L., 1976. Quantification of the assimilation capacity of outer Los Angeles Harbor; a dynamic oxygen model. In: Marine Studies of San Pedro Bay, California, Part 12. Allan Hancock Foundation and Sea Grant Programs, University of Southern California, Los Angeles, Calif., pp. 59—112.

Krone, R. B., 1976. Engineering interest in the benthic boundary layer. In: I. N. McCave (Editor), The Benthic Boundary Layer. Plenum, New York, N.Y., pp. 143—156.

Labeyrie, L. D., Livingston, H. D. and Bowen, V. T., 1976. Comparison of the distributions in marine sediments of the fall-out derived nuclides ^{55}Fe and ^{239,240}Pu; a new approach to the chemistry of environmental radionuclides. In: Transuranium Nuclides in the Environment, Proc. Symp. U.S.E.R.D.A.—I.A.E.A., San Francisco, Calif., pp. 121—137.

Lal, D., 1977. The oceanic microcosm of particles. Science, 198: 997—1009.

L.A.R.W.Q.C.B. (Los Angeles Regional Water Quality Control Board), 1969. Review of information pertinent to Los Angeles—Long Beach Harbor and Dominguez Channel. Los Angeles Reg. Water Qual. Control Board, Los Angeles, Calif., 71 pp.

Lasker, R., 1975. Field criteria for survival of anchovy larvae: the relation between inshore chlorophyll maximum layers and successful first feeding. Fish. Bull., 73(3): 453—463.

Latmiral, G., Segre, A. G., Bernabini, M. and Mirabile, L., 1971. Prospezioni sismiche per la riflessione nei Golfo di Napoli e Pozzuoli ed alcuni risultati geologici. Bol. Soc. Geol. Ital., 90: 163—172.

Lavelle, J. W., Ozturgut, E., Baker, E. T. and Swift, S. A., 1978. Discharge and the surface plume measurements during manganese nodule mining tests in the north equatorial Pacific. (Unpublished.)

Lee, R. F. and Anderson, J. W., 1977. Fate and effect of naphthalenes in controlled ecosystem enclosures. Bull. Mar. Sci., 27: 127—134.

Lee, R. F. and Takahashi, M., 1975. Petroleum hydrocarbons in the marine environment. Pap. presented to Workshop No. 65, Int. Counc. Explor. Sea, Aberdeen, Sept. 9—12, 1975.

Leifeste, D. K., 1974. Dissolved-solids discharge to the oceans from the conterminous United States. U.S. Geol. Surv., Circ. 685, 8 pp.

Leitao, C. D., Huang, N. E. and Parra, C. G., 1978. Final report of GEOS-3 ocean current investigation using radar altimeter profiling. NASA TM73280, Houston, Texas, 29 pp.

Lettau, B., Brower, Jr., W. A. and Quayle, R. G., 1976. Marine climatology. M.E.S.A. New York Bight atlas, Monogr. 7. New York Sea Grant Institute, Albany, N.Y., 239 pp.

Levin, M. A., Fischer, J. R. and Cabelli, V. J., 1975. Membrane filter technique for enterococci in marine waters. J. Appl. Microbiol., 30: 66—71.

Li, Yuan-Hui and Lui-Heung Chan, 1979. Desorption of Ba and ^{226}Ra from river-borne sediments in the Hudson estuary. Earth Planet. Sci. Lett., 43: 343—350.

Lie, U., 1967. A quantitative study of benthic infauna in Puget Sound. Fisk Dir. Skr. Hav. Unders., 14(5): 229—556.

Liss, P. S., 1976. Conservative and non-conservative behaviour of dissolved constituents during estuarine mixing. In: J. D. Burton and P. S. Liss (Editors), Estuarine Chemistry, Academic Press, New York, N.Y., pp. 93—130.

Livingstone, D. A., 1963. Chemical composition of rivers and lakes. U.S. Geol. Surv., Prof. Pap. 440-G, 64 pp.

Lockyer, C., 1977. Observations on diving behaviour of the sperm whale, *Physeter catodon.* In: M. V. Angel (Editor), A Voyage of Discovery. Pergamon, Oxford, pp. 591—609.

Longhurst, A. R., 1976. Vertical migration. In: D. H. Cushing and J. J. Walsh (Editors), The Ecology of the Sea. Blackwell, Oxford, pp. 116—137.

Longwell, A. C. and Hughes, J. B., 1980. Cytologic, cytogenic and development states of Atlantic mackerel eggs from sea surface waters of the New York Bight. Prospects for Biological Monitoring with Ichtyoplankton. Rapp. Cons. Int. Explor. Mer. No. 179, Charlottenlund.

Lu, F. C., 1973. Wholesomeness of foodstuffs: the role of WHO. World Health Org. Chron., 27: 245—253.

Lulic, S. and Strohal, P., 1974. The application of neutron activation analysis in studying the marine pollution processes. Rev. Int. Océanogr. Méd., 33: 119—123.

Lund, J. W. G., 1972. Preliminary observations on the use of large experimental tubes in lakes. Verh. Int. Ver. Limnol., 18: 71—77.

L.W.F.C. (Louisiana Wildlife and Fisheries Commission), 1958—1959. 8th Biennial Report. La. Wildlife Fish. Comm., New Orleans, La.

Lynn, D. C. and Bonatti, E., 1965. Mobility of manganese in diagenesis of deep sea sediments. Mar. Geol., 3: 457—474.

Macchi, G., 1966. Chemical environmental factors. In: M. Bernhard (Editor), Annual Report for 1965 of the Laboratory for the Study of Radioactive Contamination, Fiascherino. Euratom, Brussels, Tech. Rep. EUR 3274e, pp. 6—14.

MacIlvaine, J. C. and Ross, D. A., 1979. Sedimentary processes on the continental slope of New England. J. Sediment. Petrol., 49: 563—574.

MacIsaac, J. J., Dugdale, R. C., Huntsman, S. A. and Conway, H. L., 1979. The effect of sewage on uptake of inorganic nitrogen and carbon by natural populations of marine phytoplankton. J. Mar. Res., 37(1): 51—61.

Mackay, N. J., Kazacos, M. N., Williams, R. J. and Leedow, M. I., 1975. Selenium and heavy metals in black marlin. Mar. Pollut. Bull., 6: 57—61.

Mackenthun, K. A., 1961. Impact of pollution upon stream biota in the Wisconsin River. Div. Water Pollut. Control, State Wis., Madison, Wis.

Mahoney, J. B., 1979. Plankton dynamics and nutrient cycling. In: R. L. Swanson and C. J. Sindermann (Editors), Oxygen Depletion and Associated Benthic Mortalities in New York Bight, 1976. Ch. 9, Part 2. U.S. Dep. Commer., Washington, D.C., Natl. Ocean. Atmos. Admin. (N.O.A.A.), Prof. Pap.

Majori, L., Nedocian, G. and Modonutti, G. B., 1976a. Inquinamento da mercurio nell' Alto Adriatico Acqua Aria, 3: 164—172.

Majori, L., Nedoclan, G., Nodonutti, G. B. and Campello, C., 1976b. Pollution par métaux dans la Mer Adriatique du Nord. Note 1: Étude sur les sédiments superficiels du Golfe de Trieste. Preprint, 8 pp; Note II: Étude sur la distribution de quelques éléments en traces dans le *Mytillus galloprovincialis* Lmk. du Golfe de Trieste. Preprint, 8 pp; Note III: Le phénomène d'accumulation des métaux dans le *Mytilus galloprovincialis* Lmk. et son application comme indicateur de pollution. Preprint, 8 pp. XXV Congr. Assem. Plénière de Split, 22—30 Oct. Comité de Lutte Contre les Pollutions Marines. Comm. Int. Explor. Sci. Méd., Monaco.

Majori, L., Nedoclan, G. and Modonutti, G. B., 1976c. L'inquinamento da metalli— acque, sedimenti, miltili e pesci dell'Alto Adriatico. Int. Rep., 1st Ig., Univ. Trieste, pp. 1—103 plus tables and figures.

Malone, T. C., 1977. Plankton systematics and distribution. M.E.S.A. New York Bight atlas, Monogr. 13. New York Sea Grant Institute, Albany, N.Y., 45 pp.

Malone, T. C., Esaias, W. and Falkowski, P., 1979. Water column processes, plankton dynamics and nutrient cycling. In: R. L. Swanson and C. J. Sindermann (Editors), Oxygen Depletion and Associated Benthic Mortalities in New York Bight, 1976, Ch. 9, Part 1. U.S. Dep. Commer., Washington, D.C., N.O.A.A. Prof. Pap.

Marchand, M., Vas, D. and Duursma, E. K., 1975. Résidus de DDT et de polychloro-biphényles (PCB) dans les moules, les sédiments et le plancton de la côte nord-ouest méditerranéenne. In: Iles Journées d'Etudes sur les Pollutions Marines, Comm. Int. Explor. Sci. Méd., Monaco, pp. 171—174.

Marchand, M., Vas, D. and Duursma, E. K., 1976. Levels of PCB's and DDT's in mussels from the N.W. Mediterranean. Mar. Pollut. Bull., 7: 65—69.

Marchelano, R. A. and Ziskowski, J., 1976. Fin rot disease studies in the New York Bight. In: M. G. Gross (Editor), Middle Atlantic Continental Shelf and the New York Bight, Spec. Symp. Vol. 2. Am. Soc. Limnol. Oceanogr., pp. 329—336.

Marshall, N. B. and Merrett, N. R., 1977. The existence of a benthopelagic fauna in the deep-sea. In: M. V. Angel (Editor), A Voyage of Discovery. Pergamon, Oxford, pp. 483—497.

Mart, L., Nürnberg, H. W., Valenta, P. and Stoeppler, M., 1978. Determination of levels of toxic trace metals dissolved in sea-water and inland waters by differential anodic stripping voltametry. Thalassia Juogosl., 13: 171—188.

Martin, D. F., 1970. Marine Chemistry, Vol. 2. Marcel Dekker, New York, N.Y.

Martin, J. M., Jednačak, and Pravdić, V., 1971. The physico-chemical aspects of trace element behaviour in estuarine environments. Thalassia, 7: 619—637.

Martin, J. M., Meybeck, M., Salvadori, F. and Thomas, A., 1976. Pollution chimique des estuaires: état actuel des connaisances. C.N.E.X.O., Rapp. Sci. Technol., 22, 286 pp.

Martin, J. M., Meybeck, M. and Pusset, M., 1978. Uranium behaviour in the Zaire estuary. Neth. J. Sea Res., 12: 338—344.

Mason, B., 1958. Principles of Geochemistry, Wiley, New York, N.Y., 310 pp.

Mathews, T. D., Fredericks, A. D. and Sackett, W. M., 1973. The geochemistry of radio-carbon in the Gulf of Mexico. In: Radioactive Contamination of the Marine Environment. International Atomic Energy Agency, Vienna, pp. 725—734.

Mauchline, J., 1972. The biology of bathypelagic organisms, especially Crustacea. Deep-Sea Res., 19: 753—780.

May, R. M., 1973. Stability and Complexity in Model Ecosystems. Princeton University Press, Princeton, N.J.

Mayer, D. A., Hansen, D. V. and Minton, S. M., 1979. Water movement on the New Jersey shelf. In: R. L. Swanson and C. J. Sindermann (Editors), Oxygen Depletion and Associated Benthic Mortalities in New York Bight, 1976, Ch. 7. U.S. Dep. Commer., Washington, D.C., N.O.A.A. Prof. Pap.

McAnally, Jr., W. H., 1975. Tidal verification and base circulation tests. Los Angeles and Long Beach Harbors model study, Rep. 5. U.S. Army Corps Eng., Tech. Rep. H-75-4.

McCain, B. B., Pierce, K. V., Wellings, S. R. and Miller, B. S., 1977. Hepatomas in marine fish from an urban estuary. Bull. Environ. Contam. Toxicol., 18(1): 1—2.

McConaugha, J. R., 1976a. Toxicity and heavy metals in three species of Crustacea from Los Angeles Harbor sediments. In: Marine Studies of San Pedro Bay, California, Part 11. Allan Hancock Foundation and Sea Grant Programs, University of Southern California, Los Angeles, Calif., pp. 49—67.

McConaugha, J. R., 1976b. Microheterotrophic uptake of organics in seawater. In: Marine Studies of San Pedro Bay, California, Part 12. Allan Hancock Foundation and Sea Grant Programs, University of Southern California, Los Angeles, Calif., pp. 226—232.

McDermott-Ehrlich, D. J., Sherwood, M. J., Heesen, T. C., Young, D. R. and Mearns, A. J., 1977. Chlorinated hydrocarbons in Dover sole, *Microstomus pacificus:* local migrations and fin erosion. Fish. Bull., 75(3): 513—517.

McGowan, J. A., 1974. The nature of oceanic ecosystems. In: C. B. Miller (Editor), The Biology of the Oceanic Pacific. Oregon State University Press, Corvallis, Oreg., pp. 9—28.

McHugh, J. L., 1977. Fisheries and fishery resources of New York Bight. U.S. Dep.

Commer., Washington, D.C., N.O.A.A. Tech. Rep. N.M.F.S. Circ. 401.

McHugh, J. L. and Ginter, J. C., 1978. Fisheries, M.E.S.A. New York Bight Atlas, Monogr. 16, New York Sea Grant Institute, Albany, N.Y.

McLaughlin, D. B. and Elder, J. A., 1976. A conceptual representation of the New York Bight ecosystem. In: M. G. Gross (Editor), Middle Atlantic Continental Shelf and the New York Bight, Spec. Symp., Vol. 2. Am. Soc. Limnol. Oceanogr., pp. 249—259.

McLaughlin, D. B., Elder, J. A., Orlob, G. T., Kibler, D. F. and Evenson, D. E., 1975. A conceptual representation of the New York Bight ecosystem. U.S. Dep. Commer., Boulder, Colo., N.O.A.A. Tech. Mem. ERL MESA-4.

Meade, R. H., 1972. Transport and deposition of sediments in estuaries. Geol. Soc. Am. Mem., 133: 91—120.

Meade, R. H., Sachs, P. L., Manheim, F. T., Hathaway, J. C. and Spencer, D. W., 1975. Sources of suspended matter in waters of the middle Atlantic Bight. J. Sediment. Petrol., 45: 171—188.

Meade, R. H., Nordin, Jr., C. F., Curtis, W. F., Costa Rodrigues, F. M., do Vale, C. M. and Edmond, J. M., 1979. Sediment loads in the Amazon. Nature (London), 278: 161—163.

Means, J. L., Crerar, D. A., Borcsik, M. P. and Duguid, J. O., 1978. Adsorption of Co and selected actinides by Mn and Fe oxides in soils and sediments. Geochim. Cosmochim. Acta, 42: 1763—1773.

Mearns, A. J., 1974. Southern California's inshore demersal fishes: diversity, distribution, and disease as responses to environmental quality. Calif. Coop. Oceanic Fish. Invest. Rep., 17: 141—148.

Mearns, A. J., 1977a. Coastal gradients in sportfish catches. In: Coastal Water Research Project, Annual Report, 1977. U.S. Dep. Commer., Springfield, Va., Natl. Tech. Info., Serv., PB 274463/AS.

Mearns, A. J., 1977b. Abundance of bottomfish off Orange County. In: Coastal Water Research Project, Annual Report, 1977. U.S. Dep. Commer., Springfield, Va., Natl. Tech. Info. Serv., PB 274463/AS.

Mearns, A. J., 1981. Effects of municipal discharges on open coastal ecosystems. In: R. A. Geyer (Editor), Marine Environmental Pollution, Vol. 2, Dumping and Mining, Ch. 1, Elsevier, Amsterdam, pp. 25—66 (this volume).

Mearns, A. J. and Greene, C. S. (Editors), 1974. A comparative trawl survey of three areas of heavy waste discharge. Coastal Water Res. Proj., Tech. Mem. 215, Long Beach, Calif., 76 pp.

Mearns, A. J. and Harris, L., 1975. Age, length, and weight relationships in southern California populations of Dover sole. Coastal Water Res. Proj., Long Beach, Calif., Tech. Mem. 219, 17 pp.

Mearns, A. J. and Sherwood, M. J., 1977. Distribution of neoplasms and other diseases in marine fishes relative to the discharge of wastewater. Ann. N.Y. Acad. Sci., 298: 210—224.

Mearns, A. J. and Young, D. R., 1978. Impact of nearshore development on open coastal resources. In: J. N. Baskin, M. D. Dailey, S. N. Murray and E. Segal (Editors), Proc. Symp. The Urban Environment, Tech. Pap. 1, Southern California, Ocean Studies Consortium, Long Beach, Calif., pp. 23—47.

Mearns, A. J., Hanan, D. A. and Harris, L., 1977. Recovery of kelp forest off Palos Verdes. In: Coastal Water Research Project Annual Report, 1977. U.S. Dep. Commer., Springfield, Va., Natl. Tech. Info. Serv. PB 274463/AS.

Mellino, M., Pinto, A. and Ricciardo, M., 1969. Sull'inquinamento microbico dell'acqua di mare in una zone costiera del Golfo di Napoli. In: L'Igiene Moderna, Anno LXIII: Novembre—Dicembre 1969, Nos. 11—12, pp. 1049—1057.

Menard, H. W., 1964. Marine Geology of the Pacific. McGraw-Hill, New York, N.Y., 271 pp.

Menard, H. W., 1976. Time, chance, and the origin of manganese nodules. Am. Sci., 64: 519—529.

Mendia, L., d'Elia, E. and d'Antonio, G., 1976. Sul problema dell'Inquinamento del Golfo di Napoli. Fondazione Politecnica per il Mezzogiorno d'Italia, Naples, Quad. No. 75, 53 pp.

Menzel, D. W., 1977. Summary of experimental results: controlled ecosystem pollution experiments. Bull. Mar. Sci., 27(1): 142—145.

Menzel, D. W. and Case, J., 1977. Concept and design: controlled ecosystem pollution experiment. Bull. Mar. Sci., 27: 1—7.

M.E.Q. (Marine Environmental Quality), 1971. Report of the Ocean Science Committee of the N.A.S.—N.R.C. Ocean Affair Board. National Academy of Sciences, Washington, D.C., 107 pp.

Merrett, N. R. and Roe, H. S. J., 1974. Patterns and selectivity in the feeding of certain mesopelagic fishes. Mar. Biol., 28: 115—126.

M.E.S.A. (Marine Ecosystem Analysis), 1975. Ocean Dumping in the New York Bight. U.S. Dep. Commer., N.O.A.A., Boulder, Colo. MESA-2, N.O.A.A. TR ERL 321.

M.E.S.A. (Marine Ecosystem Analysis), 1977. Long Island beach pollution: June 1976. U.S. Dep. Commer., N.O.A.A., E.R.L., Boulder, Colo.

M.E.S.A. (Marine Ecosystem Analysis), 1978. MESA New York Bight Project, annual report for fiscal year 1977. U.S. Dep. Commer., N.O.A.A., Boulder, Colo.

Mestres, R., Pagnon, M. and Duboul-Razavet, Ch., 1975. Étude des résidus de pesticides et d'hydrocarbures organochlores dans les sédiments du plateau continental laguedocien en Méditerranée. Trav. Soc. Pharm. Montpellier, 35(2): 181—194.

Mignoit, C., 1968. A study of the physical properties of various very fine sediments and their behaviour under hydrodynamic action. Houille Blanche, 23(7).

Miller, B. S., McCain, B. B., Wingert, R. C., Borton, S. F., Pierce, K. V. and Griggs, D. T., 1977. Ecological and disease studies of demersal fishes in Puget Sound near METRO-operated sewage treatment plants and in the Duwamish River. Puget Sound Interim Studies Report, FRI-UW-7721, Municipality of Metropolitan Seattle, Seattle, Wash., 164 pp.

Miller, G. E., Grant, P. M., Kishore, R., Steinkruger, F. J., Rowland, F. S. and Guinn, V. P., 1972. Mercury concentrations in museum specimens of tuna and swordfish. Science, 175: 1121—1122.

Milliman, J. D., Summerhayes, C. P. and Barretto, H. T., 1975. Oceanography and suspended matter off the Amazon River, February—March 1973. J. Sediment. Petrol., 45: 189—206.

Mizenko, D. and Chamberlain, J. L., in press. Gulf Stream anticyclonic eddies and shelf water at Deep Water Dumpsite 106 during 1977. N.O.A.A. Dumpsite Evaluation Rep. 79-1, Rockville, Md.

Moody, C. L., 1967. Gulf of Mexico Distributive Province. Am. Assoc. Pet. Geol. Bull., 51: 179—199.

Moore, P. G., 1978. Inorganic particulate suspensions in the sea and their effects on marine animals. Oceanogr. Mar. Biol. Annu. Rev., 15: 225—363.

Moore, W. J., 1962. Physical Chemistry. Prentice-Hall, Englewood Cliffs, N.J., 3rd ed., 844 pp.

Morel, F. and Morgan, J. J., 1968. A numerical method for solution of chemical equilibria in aqueous systems. California Institute of Technology, Pasadena, Calif. (Also in: J. J. Morgan and W. Stumm, 1970, Aquatic Chemistry, Wiley—Interscience, New York, N.Y., pp. 292 ff.)

Morey-Gaines, G., 1978 . Microcosm enrichment studies of tuna cannery waste. In: D. Soule and M. Oguri (Editors), Marine Studies of San Pedro Bay, California, Part 14. Allan Hancock Foundation and Sea Grant Programs, University of Southern California, Los Angeles, Calif., pp. 71—113.

Morgan, C. L. and Lovorn, F. T., 1979. At-sea discharge of ocean mining process rejects—A preferable Alternative to Land Disposal. Proc. Conf. Offshore Technology Conference, May 1—3, 1979.

Morgan, J. J. and Stumm, W., 1970. Aquatic Chemistry. Wiley—Interscience, New York, N.Y., 582 pp.

Morgan, K. C. and Kalff, J., 1972. Bacterial dynamics of two high-arctic lakes. Freshwater biol., 2: 217—228.

Mueller, J. A. and Anderson, A. R., 1978. Industrial wastes. M.E.S.A. New York Bight atlas, Monogr. 30. New York Sea Grant Institute, Albany, N.Y., 39 pp.

Mueller, J. A., Jeris, J. S., Anderson, A. R. and Hughes, C. F., 1976a. Contaminant inputs to the New York Bight. N.O.A.A., Boulder, Colo., Tech. Memo. E.R.L. M.E.S.A.-6.

Mueller, J. A., Anderson, A. R. and Jeris, J. S., 1976b. Contaminants entering the New York Bight: sources, mass loads, significance. In: M. G. Gross (Editor), Middle Atlantic Continental Shelf and the New York Bight, Spec. Symp. Vol. 2. Am. Soc. Limnol. Oceanogr., pp. 162—170.

Munk, W. H., 1966. Abyssal recipes. Deep-Sea Res., 13: 707—730.

Munk, W. H. and Riley, G. A., 1952. Adsorption of nutrients by aquatic plants. J. Mar. Res., 11: 215—240.

Murphy, L. S., Hoar, P. R. and Belastock, R. A., 1980. The effects of industrial wastes on marine phytoplankton. In: B. H. Ketchum, D. R. Kester and P. K. Park (Editors), Ocean Dumping of Industrial Wastes. Plenum, New York, N.Y.

Murray, D. J., Healy, T. W. and Fuerstenau, D. W., 1967. Adsorption of aqueous metal on colloidal hydrous manganese oxide. In: Adsorption from Aqueous Solution. Adv. Chem. Ser., 79: 74—81.

Murray, J. and Irvine, R., 1895. On the manganese oxides and manganese nodules in marine deposits. Trans. R. Soc. Edinburgh, 37: 721—742.

Murray, J. W., 1973. Cobalt adsorption onto hydrous MnO_2. Presented at Annu. Geophys. Union Convent., Washington, D.C.

Murray, J. W., 1974. Surface chemistry of hydrous manganese dioxide. J. Colloid Interface Sci., 46: 357—371.

Murray, J. W., 1975. The interaction of metal ions at the manganese dioxide—solution interface. Geochim. Cosmochim. Acta, 39(4): 505—520.

Myers, E. P. and Gunnerson, C. G., 1976. Hydrocarbons in the ocean. M.E.S.A. Spec. Rep., Boulder, Colo., April 1976. Mar. Admin.—N.O.A.A. Environ. Res. Lab., 42 pp.

N.A.S. (National Academy of Sciences), 1971. Marine Environmental Quality. Natl. Acad. Sci., Washington, D.C., 107 pp.

N.A.S. (National Academy of Sciences), 1973. Water quality criteria, 1972. Rep. Comm. Water Qual. Criteria, Natl. Acad. Sci., Washington, D.C., 593 pp.

N.A.S. (National Academy of Sciences), 1975. Assessing Potential Ocean Pollutants. Natl. Acad. Sci., Washington, D.C., 438 pp.

N.C.H.D. (Nassau County Health Department), 1977. Annual Report for 1976. Nassau County Health Dep., Mineola, N.Y.

Neal, V. T. (Editor), 1977. Identification of representative west coast areas for manganese nodule processing activities. Sch. Oceanogr. Oreg. State Univ., Corvallis, Oreg. for U.S. Dep. Commer., N.O.A.A., under Contract No. T-35416, 189 pp.

Neff, J. W., Foster, R. S. and Slowey, J. F., 1978. Availability of sediment-adsorbed heavy metals to benthos with particular emphasis on deposit-feeding infauna. U.S. Army Eng. Waterways Exp. Stn., Vicksburg, Miss., Tech. Rep. D-78-42.

Neihof, R. H. and Loeb, G. I., 1972. The surface charge of particulate matter in seawater. Limnol. Oceanogr., 17: 7—16.

Nichols, M. M., Faas, R. W. and Thompson, G. S., 1980. A field study of fluid mud dredged material: its physical nature and dispersal. U.S. Army Eng. Waterways Exp.

Stn., Vicksburg, Miss., Tech. Rep. D-78-40.

Niino, H. and Emery, K. O., 1961. Sediments of shallow portions of East China Sea and South China Sea. Geol. Soc. Am. Bull., 72: 731—762.

N.O.A.A. (National Oceanic and Atmospheric Administration), 1975. May 1974 baseline investigation of deepwater Dumpsite *106*. U.S. Dep. Commer., N.O.A.A., Rockville, Md.

N.O.A.A. (National Oceanic and Atmospheric Administration), 1976. Deep Ocean Mining Environmental Study—Phase I, Mar. EcoSystems Anal. Program Off., Natl. Oceanic Atmos. Admin., Boulder, Colo., N.O.A.A. Tech. Mem. E.R.L. M.E.S.A.-15, 178 pp.

N.O.A.A. (National Oceanic and Atmospheric Administration), 1978. Draft environmental impact statement and fishery management plan for the northern anchovy. U.S. Dep. Commer., Natl. Ocean. Atmos. Admin., Washington, D.C.

Noshkin, V. E. and Bowen, V. T., 1975. Concentrations and distributions of long-lived fallout radionuclides in open ocean sediments. In: Radioactive Contamination of the Marine Environment. International Atomic Energy Agency, Vienna, pp. 671—686.

Nowlin, W. D., 1971. Water masses and general circulation of the Gulf of Mexico. Oceanol. Int., 6: 28—33.

Nürnberg, H. W., Mart, L. and Valenta, P., 1977. Concentration of Cd, Pb and Cu in Ligurian and Tyrrhenian coastal waters. In: Proc. 25th Congr., Plenary Assem. Comm. Int. Explor. Sci. Mediterr., Split, 1976.

N.Y.C.D.E.P. (New York City Department of Environmental Protection), 1978. Section *208* Areawide Waste Treatment Management Planning Program, New York, N.Y.

O'Connor, D. J., Thomann, R. V. and Salas, H. J., 1977. Water quality. M.E.S.A. New York Bight atlas, Monogr. 27. New York Sea Grant Institute, Albany, N.Y., 104 pp.

O'Connor, J. S., 1979. A perspective on natural and man-related factors. In: R. L. Swanson and C. J. Sindermann (Editors), Oxygen Depletion and Associated Benthic Mortalities in New York Bight, 1976, Ch. 15. U.S. Dep. Commer., Washington, D.C., N.O.A.A. Prof. Pap.

O'Connor, T. P. and Kester, D. R., 1975. Adsorption of copper and cobalt from fresh and marine systems. Geochim. Cosmochim. Acta, 39: 1531—1543.

O'Connor, T. P. and Park, P. K., 1981. Consequences of industrial waste disposal at the 106-mile Ocean Waste Disposal Site. In: G. S. Mayer, A. Calabrese, F. A. Gross, D. C. Malins, D. W. Menzel, J. S. O'Connor and F. J. Vernberg (Editors), Ecological Stress in the New York Bight: Science and Management. Estuarine Research Foundation, Columbia, S.C. (in press).

Odum, H. T., 1972. An energy circuit language for ecological and social systems. In: B. C. Patten (Editor), Systems Analysis and Simulation in Ecology. Academic Press, New York, N.Y., pp. 140—212.

Odum, W. E., Woodwell, G. M. and Wurster, C. F., 1969. DDT residues absorbed from organic detritus by fiddler crabs. Science, 164: 576—577.

Ohi, G., Nishigaka, S., Seki, H., Tamura, Y., Maki, T., Konni, H., Ochiai, S., Yamada, H., Shimamura, Y., Mizoguchi, I. and Yagyu, H., 1976. Efficacy of selenium in tuna and selenite in modifying methylmercury intoxication. Environ. Res., 12: 49—57.

Ohman, M., Ozturgut, E. and Ozretich, R. J., 1979. A seasonal and spatial summary of oceanographic data from the northeastern tropical Pacific (DOMES region), summer 1975 and winter 1976. Dep. Oceanogr. Univ. Wash., Seattle, Wash., Spec. Rep. No. 91.

Okamura, O., 1970. Studies on macrourid fishes of Japan. Rep. Usa Mar. Biol. Stn. (Usa Rinkai Jikkanjo Kenkyu Hokaku), 17: 1—179.

Okubo, A., 1971. Oceanic diffusion diagrams. Deep-Sea Res., 18: 789—798.

Oliver, J. S., Slattery, P. N., Hulberg, L. W. and Nybakken, J. W., 1977. Patterns of succession in benthic infaunal communities following dredging and dredged material disposal in Monterey Bay. U.S. Army Eng. Waterways Exp. Stn., Vicksburg, Miss.,

Tech. Rep. D-77-27.

O'Neal, G. and Sceva, J., 1971. Effects of dredging on water quality in the Northwest. Environ. Prot. Agency, Off. Water Programs, Region X, Seattle, Wash.

Orr, M. H. and Hess, F. R., 1978. Acoustic monitoring of industrial chemical waste released at Deep Water Dumpsite 106. J. Geophys. Res., 83: 6145—6154.

Orr, M. H., Baxter, L. and Hess, F. R., 1980. Remote acoustic sensing of the particulate phase of industrial chemical wastes and sewage sludge. Woods Hole Oceanogr. Inst., Woods Hole, Mass., WHOI Tech. Rep. 79-38, 153 pp.

Ortner, P. B. and Murphy, L. S., 1977. Biological effects of a dump at Deep Water Dumpsite 106. N.O.A.A. Dumpsite Evaluation Report 79-1.

Overstreet, R. and Rattray, Jr., M., 1969. On the roles of vertical velocity and eddy conductivity in maintaining a thermocline. J. Mar. Res., 27: 172—190.

Ozretich, R. J., 1979. Final report on the experiments and field measurements to assess the stimulation of bacterial growth as an impact of prototype deep-ocean mining. Dep. Oceanogr., Univ. Wash., Seattle, Wash., Spec. Rep. 90.

Ozretich, R. J. and Ozretich, R. A., 1978. Bacteria and sediment interaction. Final Rep. to D.O.M.E.S. Project, Natl. Tech. Info. Serv., Springfield, Va, Rep. No. PB 279630/AS, 22 pp.

Ozturgut, E., Anderson, G. C., Burns, R. E., Lavelle, J. W. and Swift, S. A., 1978. Deep ocean mining of manganese nodules in the north Pacific: Pre-mining environmental conditions and anticipated mining effects. Natl. Tech. Info. Serv., Springfield, Va., N.O.A.A. Tech. Mem. E.R.L. M.E.S.A.-33, Rep. No. PB/293545/OST, 185 pp.

Ozturgut, E., Lavelle, J. W., Steffin, O. and Swift, S. A., 1980. Environmental investigations during manganese nodule mining tests in the North Equatorial Pacific in November 1978. N.O.A.A. Tech. Mem., E.R.L.—M.E.S.A.-48, 50 pp.

Paccagnella, B., Prati, L. and Bigoni, A., 1973. Studio epidemiologico sul mercurio nei pesci e la salute umana in un'isola Italiana del Mediterraneo. Ig. Moderna, 66: 479—503.

Paoletti, A., 1964. Microviventi patogeni nell'ambiente marino. Abstract of Meeting Atti del Convegno Nazionale su Problemi Attuali di Igiene e Medicina Sociale, Naples, Jan. 20—23, 1963.

Paoletti, A., 1968. Organismes prédateurs dans l'autoépuration des eaux de mer essais d'Etude avec bacteries radioactives ou autrement marquées. Rev. Int. Oceanogr. Méd., 10: 229—247.

Paoletti, A., 1970. Facteurs biologiques d'autoépuration des eaux de mer: points clairs et points obscurs d'une question discutée. Rev. Int. Océanogr. Méd., 10—19: 33—68.

Paoletti, A., 1975a. Oceanografia Medica Id Inquinamento. Liguori, Naples. 321 pp.

Paoletti, A., 1975b. Studies of marine waste disposal for the city of Maiori utilizing a continuous monitoring system. E. A. Pearson and E. De Fraga Frangipane (Editors), Marine Pollution and Marine Waste Disposal. Pergamon, Oxford, pp. 441—449.

Paoletti, A., Parrella, A., Melluso, F. and De Fusco, R., 1973. Gli autoanalyzer su piccolo battello per lo studio de diluizione e diffusione degli effluenti urbani in mare. Nuovi Ann. Ig. Microbiol., 24(4): 199—251.

Papadopoulou, C., 1972. The elementary composition of marine invertebrates as a contribution to the sea pollution investigation. "Demokritos" Nuclear Research Center, Chemistry Department, Aghia Paraskevi Attikis, Athens. Int. Rep. 17 pp. (mimeographed).

Papadopoulou, C. and Kanias, G. D., 1976. Trace element distribution in seven mollusc species from Saronikos Gulf. Acta Adriat., 18: 367—378.

Papadopoulou, C., Grimanis, A. P. and Hadjistelios, I., 1973a. Mercury and arsenic in a fish collected in polluted and non-polluted sea-waters. Thalassia Jugosl., 9: 211—218.

Papadopoulou, C., Hadzistelios, J. and Grimanis, A. P., 1973b. Studies on the trace element distribution in different parts of the fish Pagellus erythrinus. Hell. Oceanol.

Limnol., 10: 1—3.

Papakostidis, G., Grimanis, A. P., Zafiropoulos, D., Griggs, G. B. and Hopkins, T. S., 1975. Heavy metals in sediments from the Athens sewage outfall area. Mar. Pollut. Bull., 136—139.

Parizek, J. and Ostadalova, I., 1967. The protective effect of small amounts of selenite in sublimate intoxication. Experientia, 23: 142—145.

Park, P. K. and O'Connor, T. P., 1980. Historical and internation considerations. In: B. H. Ketchum, D. R. Kester and P. K. Park (Editors), Ocean Dumping of Industrial Wastes. Plenum, New York, N.Y.

Parks, G. A., 1967. Aqueous surface chemistry of oxides and complex oxide minerals in equilibrium concepts in natural water systems. Adv. Chem. Ser., 7.

Parsons, T. R., Thomas, W. H., Seibert, D., Beers, J. R., Gillespie, P. and Bowden, C., 1977a. The effect of nutrient enrichment on the plankton community in enclosed water columns. Int. Rev. Ges. Hydrobiol., 62: 565—572.

Parsons, T. R., Reeves, M. R., Holm-Hansen, O., von Brockel, K., Koeller, P. and Takahashi, M., 1977b. The distribution of organic carbon in a marine planktonic food web following nutrient enrichment. J. Exp. Biol. Mar. Ecol., 26: 235—247.

Pasquill, F., 1974. Atmospheric Diffusion. Halsted, New York, N.Y., 2nd ed., 423 pp.

Patterson, C. C., 1965. Contaminated and natural lead environments of man. Arch. Environ. Health, 11: 344—360.

Patterson, C. C. and Settle, D. M., 1976. The reduction of orders of magnitude errors in lead analysis of biological materials and natural waters by evaluating and controlling the extent and sources of industrial lead contamination introduced during sample collecting and analysis. In: P. LaFleur (Editor), Accuracy in Trace Analysis. National Bureau of Standards, Washington, D.C., pp. 321—351.

Paul, A., Thorndike, E. M., Sullivan, L. G., Heezen, B. C. and Gerard, R. D., 1978. Observations of the deep-sea floor from 202 days of time lapse photography. Nature (London), 272: 812—814.

Paul, J. and Meischner, D., 1976. Heavy metal analysis from sediments of the Adriatic Sea. Senckenb. Marit., 8: 91—102.

Pearcy, W. G., 1976. Pelagic capture of abyssobenthic macrourid fish. Deep-Sea Res., 23: 1065—1066.

Pearcy, W. G. and Ambler, J. W., 1974. Food habits of deep-sea macrourid fishes of the Oregon coast. Deep-Sea Res., 21: 745—759.

Pedlosky, J., 1971. Geophysical fluid dynamics. Lect. Appl. Math., 13: 1—60.

Pentreath, R. J., 1978a. [237]Pu experiments with the plaice, Pleuronectes platessa. Mar. Biol., 48: 327—335.

Pentreath, R. J., 1978b. [237]Pu experiments with the thornback ray, Raja clavata. Mar. Biol., 48: 337—342.

Pequegnat, W. E. and Smith, D. D., 1977. Potential impact of deep ocean disposal of dredged material. Proc. 2nd Int. Symp. on Dredging Technology, B.H.R.A.—Texas A & M Univ., College Station, Texas, Pap. F4.

Pérès, J. M., 1978. Vulnérabilité des ecosystèmes méditerranéens à la pollution. Ocean Manage., 3: 205—217.

Perna, A., Di Silvestro, C. and Carracciolo, S., 1972. La presenza di mercurio totale nella carne dei pesci e di altri prodotti della pesca del Mare Adriatico. Nuovo Prog. Vet., 10: 961—964.

Peterson, C. L., Klawe, W. L. and Sharp, G. D., 1973. Mercury in tunas: a review. Fish. Bull., 71: 603—613.

Peterson, L. L., 1974. The propagation of sunlight and the size distribution of suspended particles in a municipally polluted ocean water. Ph.D. Thesis, California Institute of Technology, Pasadena, Calif.

Peterson, R. E., 1977. A study of suspended particulate matter: Arctic Ocean and

548

northern Oregon continental shelf. Thesis, Oregon State University, Corvallis, Oreg., 122 pp.

Petrocelli, S. R., Anderson, J. W., Sackett, W. M., Presley, B. J. and Giam, C. S., 1974. Results of research into the effects of sublethal concentrations of selected compounds on the physiological response of marine and estuarine organisms, I.D.O.E. Meet. on Biological Effects, Sidney, B.C.

Picer, M., Picer, N. and Ahel, M., 1977. Chlorinated insecticides and PCB's residues in fish and mussels of the east coastal waters of middle and the north Adriatic Sea. J. Pestic. Monit.

Pierce, K. V., McCain, B. and Sherwood, M. S., 1977. Histology of liver tissue from Dover sole. In: Coastal Water Research Project, Annual Report, 1977, U.S. Dep. Commer., Springfield, Va., Natl. Tech. Info. Serv., PB 274463/AS, pp. 207—212.

Pierce, R. H., Olney, C. E. and Felbeck, G. T., 1974. p,p'-DDT adsorption to suspended particulate matter in sea water. Geochim. Cosmochim. Acta, 38: 1061—1073.

Pomeroy, L., 1974. The oceanic food web: a changing paradigm. Bioscience, 24: 499—503.

Pore, N. A. and Barrientos, C. S., 1976. Storm surge. M.E.S.A. New York Bight atlas, Monogr. 6. New York Sea Grant Institute, Albany, N.Y., 44 pp.

Port of Long Beach, 1976. SOHIO West Coast to Mid-continent Pipeline Project environmental impact report. Draft, Final and Supplement, Vols. 1—3 (1976), Vols. 4 and 5 (1977). Soils International, Inc., Harbors Environmental Projects, Allan Hancock Foundation, University of Southern California, and SocioEconomic Systems, Inc.

Postma, H., 1967. Sediment transport and sedimentation in the marine environment. In: G. H. Lauff (Editor), Estuaries. Am. Assoc. Adv. Sci., Publ. No. 83, pp. 150—179.

Postma, H., 1980. Sediment transport and sedimentation. In: E. Olausson and I. Cato (Editors), Chemistry and Biochemistry of Estuaries. Wiley, Chichester, pp. 153—186.

Pravdić, V., 1970. Surface charge characterization of sea sediments. Limnol. Oceanogr., 15: 230—233.

Preston, A., 1967. Site evaluation and discharge of aqueous radioactive wastes from civil nuclear power plants in England and Wales. In: Disposal of Radioactive Wastes in the Seas. International Atomic Energy Agency, Vienna, pp. 725—732.

Preston, A., 1975. The radiological consequences of releases from nuclear facilities to the aquatic environments. In: Impacts of Nuclear Releases into the Aquatic Environment. International Atomic Energy Agency, Vienna, pp. 3—23.

Pruter, A. T. and Alverson, D. L., 1972. The Columbia River Estuary and Adjacent Ocean Waters. University of Washington Press, Seattle, Wash., 868 pp.

Raybaud, H., 1972. Les biocides organochlorés et les détergents dans le milieu marin. Thesis, University of Aix-Marseille II, Marseille, 64 pp.

Redfield, A. C. and Walford, L. A., 1951. A Study of the disposal of chemical waste at sea. Natl. Acad. Sci—Natl. Res. Counc., Washington, D.C., Publ. 201.

Reid, Jr., J. L., 1962. On the circulation, phosphate-phosphorus content and zooplankton volume in the upper part of the Pacific Ocean. Limnol. Oceanogr., 7: 287—306.

Reish, D. J., 1954. Ecological study of lower San Gabriel River, California, with special reference to pollution. Calif. Dep. Fish Game, 42(1): 51—61.

Reish, D. J., 1971. Effect of pollution abatement in Los Angeles Harbor. Mar. Pollut. Bull., 2(5): 71—74.

Relevante, N. and Gilmartin, M., 1975. DDT, related compounds and PCB, in tissues of 19 species of northern Adriatic commercial fishes. Invest. Pesq., 39: 391—507.

Renfro, W. C. and Oregioni, 1974. Metal concentrations in nearshore Mediterranean sediments. In: Activities of the International Laboratory of Marine Radioactivity, Monaco. Rep. Tech. Doc., Int. At. Energy Agency, Vienna.

Renzoni, A., 1977. A case of mercury abatement along the Tuscan coast. In: Workshop on Marine Pollution, 25th Congr. of Comm. Int. Explor. Sci. Mediterr., Split, Oct.

1976.

Renzoni, A. and Baldi, F., 1975. Osservazioni sulla distribuzione di mercurio nella fauna del Mar Ligure e del Mar Tirreno. Acqua Aria, 8: 507—602.

Renzoni, A., Bacci, E. and Falciai, L., 1973. Mercury concentration in the water, sediments and fauna of an area of the Tyrrhenian coast. Rev. Int. Océanogr. Méd., 36/37: 17—45.

Rhines, P., 1977. The dynamics of unsteady currents. In: A. E. Maxwell (Editor), The Sea, Vol. 6. Wiley—Interscience, New York, N.Y., pp. 189—318.

Rhoads, D. C., 1973. The influence of deposit-feeding benthos on water turbidity and nutrient recycling. Am. J. Sci. 273: 1—22.

Rice, A. L. and Chapman, C. J., 1971. Observations on the burrows and burrowing behaviour of two mud-dwelling decapod crustaceans, Nephrops norvegicus and Goneplax rhomboides. Mar. Biol., 10: 330—342.

Richards, A. F., 1977. Geotechnical testing of U.S. Geological Survey cores collected from the pelagic nodule area, central northeast Pacific Ocean. In: D. Z. Piper (Editor), Deep Ocean Environmental Study: Geology and Geochemistry of DOMES Sites A, B, and C, Equatorial North Pacific. U.S. Geol. Surv., Open-File Rep. No. 77-778.

Richardson, P., 1976. Gulf Stream rings. Oceanus, 19: 65—68.

Riley, G. A., 1963. Organic aggregates in seawater and the dynamics of deep-sea clam determined by ^{228}Ra chronology. Proc. Nat. Acad. Sci., U.S.A., 72: 2829—2832.

Riley, G. A., 1970. Particulate organic matter in sea water. Adv. Mar. Biol., 8: 1—118.

Riley, J. P. and Chester, R., 1971. Introduction to Marine Chemistry. Academic Press, London, 465 pp.

Rimerman, R. A., Buhler, D. R. and Whanger, P. D., 1977. Metabolic interactions of selenium with heavy metals. In: S. D. Lee (Editor), Biochemical Effects of Environmental Pollutants. Ann Arbor Science, Ann Arbor, Mich., pp. 377—396.

Riolfatti, M., 1977. Ulteriori indagini epidemiologiche sulle concentrazioni di mercurio nel pesce alimentare e nel sangue e capelli umani. Ig. Moderna, 70: 169—185.

Risebrough, R. W., Huggett, R. J., Griffin, J. J. and Goldberg, E. D., 1968. Pesticides: Transatlantic movements in the northeast trades. Science, 159: 1233—1236.

Ritchie, D. W., 1970. Fish. In: Gross Physical and Biological Effects of Overboard Spoil Disposal in Upper Chesapeake Bay. Univ. Md., College Park, Md., Nat. Resour. Inst., Spec. Rep. No. 3.

Robertson, D. E. and Carpenter, R., 1976. Activation analysis. In: E. D. Goldberg (Editor), Strategies for Marine Pollution Monitoring. Wiley, New York, N.Y., pp. 93—156.

Robertson, D. E. and Rancitelli, L. H., 1973. Trace element concentrations in seawater collected in the DOMES area. In: The Environmental Impact of Deep-sea Mining Progress Report. Environ. Res. Lab., Natl. Oceanic Atmos. Admin., Boulder, Colo., N.O.A.A. Tech. Rep. No. ERL 290-OD 11.

Robertson, D. E., Rancitelli, L. A., Langford, J. C. and Perkins, R. W., 1972. Battelle Northwest contribution to the IDOE base-line study. In: Workshop on Base-line Studies of Pollutants in Marine Environment. Brookhaven Natl. Lab., May 24—26, pp. 231—274.

Robinson, K. S. and Porath, H., 1974. Current measurements in the outer Los Angeles Harbor. In: D. Soule and M. Oguri (Editors), Marine Studies of San Pedro Bay, California, Part 6. Allan Hancock Foundation and Sea Grant Programs, University of Southern California, Los Angeles, Calif.

Robinson, M. K. and Bauer, R. A., 1971. Atlas of monthly mean sea surface and subsurface temperature and depth of the top of the thermocline. North Pacific Ocean. Fleet Numerical Weather Control, Monterey, Calif.

Rodolfo, K. S., 1969. Suspended sediments in surface Andaman Sea waters off the Irrawaddy delta, northeastern Indian Ocean. Abstr. Annu. Meet. Atlantic City, N.J., pp. 190—191.

Roels, O. A., Amos, A. F., Anderson, O. R., Garside, C., Haines, K. C., Malone, T. C., Paul, A. Z. and Rice, G. E., 1973. The environmental impact of deep-sea mining, progress report. Natl. Tech. Inf. Serv., Springfield, Va., N.O.A.A. Tech. Rep. E.R.L. 290-OD 11, Rep. No. COM7450489/5.

Roger, C., 1975. Rythmes nutritionnels et organisation trophique d'une population de crustaces pélagiques (Euphausiacea). Mar. Biol., 32: 365—378.

Ropes, J. W., Merrill, A. S., Murawski, S. A., Chang, S. and MacKenzie, Jr., C. L., 1979. Field survey assessments, impact on clams and scallops. In: R. L. Swanson and C. J. Sindermann (Editors), Oxygen Depletion and Associated Benthic Mortalities in New York Bight, 1976, Ch. 11, Part 1, U.S. Dep. Commer., Washington, D.C., N.O.A.A. Prof. Pap.

Rosenbauer, R. J. and Bischoff, J. L., 1978. Chemical changes in seawater caused by resuspension of deep-sea sediments from DOMES Sites A, B, and C. Mar. Min., 1(4): 283—304.

Roth, I. and Hornung, H., 1977. Heavy metal concentration in water, sediments and fish from the Mediterranean coastal area, Israel. Environ. Sci. Technol., 11: 265—269.

Ryan, W. B. T. and Heezen, B. C., 1976. Smothering of deep-sea benthic communities from natural disasters. Natl. Tech. Info. Serv., Springfield, Va., Rep. No. PB 279527/AS, 132 pp.

Saila, S. B., Pratt, S. D. and Polgour, T. T., 1972. Dredge spoil disposal in Rhode Island Sound. Univ. of Rhode Island, Mar. Exp. Stn., Kingston, R.I., Mar. Tech. Rep., No. 2.

Sanders, H. L., 1968. Marine benthic diversity: a comparative study. Am. Nat., 102: 243—282.

Sartor, J. D. and Boyd, G. B., 1972. Water pollution aspects of street surface contaminants. Environ. Prot. Technol. Ser., EPA-R2-72-081. Nov. 1972, 236 pp.

Sawyer, T. K., MacLean, S. A., Bodammer, J. E. and Harke, B. A., 1980. Gross and microscopical observations on gills of rock crabs (Cancer irroratus) and lobsters (Homerus americanus) from nearshore waters of the eastern United States. Texas A & M Univ., Sea Grant Publ., College Station, Texas.

S.C.C.W.R.P. (Southern California Coastal Water Research Project), 1973. The ecology of the Southern California Bight: implications for water quality management. In: Southern California Coastal Water Research Project, U.S. Dep. Commer., Springfield, Va., Natl. Tech. Info. Serv., PB 274462/AS, p. TR104.

S.C.C.W.R.P. (Southern California Coastal Water Research Project), 1975. Environmental effects of the disposal of municipal waste waters in open coastal waters. Rep. submitted to Natl. Comm. Water Quality, Aug., 1975. Southern California Coastal Water Research Project, Long Beach, Calif., 71 pp.

Schafer, H. A., 1977. Characteristics of municipal wastewater discharges, 1976. In: Coastal Water Research Project, Annual Report, 1977. U.S. Dep. Commer., Springfield, Va., Natl. Tech. Inf. Serv., PB 274463/AS, pp. 19—23.

Schafer, H. A. and Bascom, W., 1976. Sludge in Santa Monica Bay. In: Coastal Water Research Project, Annual Report, 1976. U.S. Dep. Commer., Springfield, Va., Natl. Tech. Info. Serv., PB 274461/AS, pp. 77—82.

Schelenz, R. and Diehl, J. F., 1973. Anwendung der Neutronenaktivierungsanalyse zur Quecksilberbestimmung in Lebensmitteln. Z. Anal. Chem., 265: 93—97.

Schell, W. R. and Nevissi, A., 1977. Heavy metals from waste disposal in central Puget Sound. Environ. Sci. Technol., 11(9): 887—893.

Schubel, J. R., Carter, H. H., Wilson, R. E., Wise, W. M., Heaton, M. G. and Gross, M. G., 1978. Field investigations of the nature degree and extent of turbidity generated by open-water pipeline disposal operations. U.S. Army Eng. Waterways Exp. Stn., Vicksburg, Miss., Tech. Rep. D-78-30.

Schultz, C. D., Crear, D., Perason, J. E., Rivers, J. B. and Hylin, J. W., 1976. Total and organic mercury in the Pacific blue marlin. Bull. Environ. Contam. Toxicol., 15: 230—234.

Schwartz, J. R., Yayanos, A. A. and Colwell, R. R., 1976. Metabolic activities of the intestinal microflora of a deep-sea invertebrate. Appl. Environ. Microbiol., 31: 46—48.

Scura, E. D. and McClure, V. E., 1975. Chlorinated hydrocarbons in sea-water: analytical method and levels in the northeastern Pacific. Mar. Chem., 3: 337—346.

Segar, D. A., 1976. A review of the impact of dredged material in New York Bight Apex, with an emphasis on chemical processes. Rep. M.E.S.A. New York Bight Project, Stony Brook, N.Y. (unpublished).

Segar, D. A. and Berberian, G. A., 1976. Oxygen depletion in the New York Bight. In: M. G. Gross (Editor), Middle Atlantic Continental Shelf and the New York Bight, Spec. Symp. Vol. 2, Am. Soc. Limnol. Oceanogr., pp. 220—239.

Segar, D. A., Collins, J. D. and Riley, J. P., 1971.The distribution of the major and some minor elements in marine animals, Part II. Molluscs. J. Mar. Biol. Assoc. U.K., 51: 131—136.

Selli, R., Frignani, M., Rossi, C. M. and Viviani, R., 1973. The mercury content in the sediments of the Adriatic and the Tyrrhenian. Bull. Geol. Soc. Greece, 10: 177—179.

Selli, R., Frignani, M. and Giordani, P., 1977. Inquinamento dei sedimenti marini. Acc. Naz. Lincei (Roma) Atti Convegni Lincei, 31: 169—192.

Sheldon, R. W. and Parsons, T. R., 1967. A practical manual on the use of the Coulter counter in marine science. Coulter Electronics, Inc., Toronto, Ont., 66 pp.

Sheldon, R. W. and Sutcliffe, W. H., 1978. Generation times of 3 h for Sargasso Sea microplankton determined by ATP analysis. Limnol. Oceanogr., 23: 1051—1055.

Sherman, K., Busch, D. and Bearse, D., 1977. Deep Water Dumpsite 106: zooplankton studies. In: Baseline Report of Environmental Conditions at Deep Water Dumpsite 106. N.O.A.A. Dumpsite Evaluation Report 77-1, pp. 233—308.

Sherwood, M. J., 1979. The fin erosion syndrome. In: Coastal Water Research, Annual Report, 1978. Coastal Water Res. Proj., U.S. Dep. Commer., Springfield, Va., Natl. Tech. Info. Serv., PB 299830/AS, pp. 203—231.

Sherwood, M. J. and Mearns, A. J., 1977. Environmental significance of fin erosion in southern California demersal fishes. Ann. N.Y. Acad. Sci., 298: 177—189.

Shokes, R. F., 1976. Rate-dependent distributions of lead-210 and interstitial sulfate in sediments of the Mississippi River Delta. Dep. Oceanogr., Texas A & M Univ., College Station, Texas, Tech. Rep. 76-1-T. 122 pp.

Sholkovitz, E. R., 1976. Flocculation of dissolved organic and inorganic matter during the mixing of river water and seawater. Geochim. Cosmochim. Acta, 40: 831—845.

Sholkovitz, E. R., 1978. The flocculation of dissolved Fe, Mn, Al, Cu, Ni, Co and Cd during estuarine mixing. Earth Planet. Sci. Lett., 41: 77—86.

Silver, S., Schottel, J. and Weiss, A., 1976. Bacterial resistance to toxic metals determined by extrachromosomal R factors. In: J. M. Sharpley and A. M. Kaplan (Editors), Proc. 3rd Int. Symp. on Biodegradation. Applied Science Publishers. Essex, pp. 919—936.

Simpson, D. C., Devine, M. F., Warsh, C. E., Meyer, E. R., O'Connor, T. P. and Park, P. K., 1978. Ocean dumping research and monitoring: strategies and tools. J. Ocean Eng., OE-3(4): 165—171.

Simpson, F., Wallin, C. S., Crawford, H. M. and Springer, P. C., 1977. Geotechnical properties of NE Pacific sediments from DOMES Sites A, B, and C. In: D. Z. Piper (Editor), Deep Ocean Environmental Study: Geology and Geochemistry of DOMES Sites A, B, and C, Equatorial North Pacific, U.S. Geol. Surv., Open-File Rep. No. 77-778.

Sindermann, C. J., 1979. Pollution-associated diseases and abnormalities of fish and shellfish: a review. Fish. Bull., 76(4): 717—749.

Sindermann, C. J. and Swanson, R. L., 1979. Historical and regional perspective. In:

R. L. Swanson and C. J. Sindermann (Editors), Oxygen Depletion and Associated Benthic Mortalities in New York Bight, 1976, Ch. 1. U.S. Dep. Commer., Washington, D.C., N.O.A.A. Prof. Pap.

Sly, P. G., 1977. A report on studies of the effects of dredging and disposal in the Great Lakes with emphasis on Canadian waters. Can. Cent. Inland Waters, Sci. Ser. 77.

Smayda, T. J., 1969. Some measurements on the sinking rate of fecal pellets. Limnol. Oceanogr., 14: 621—625.

Smith, G. B., 1974. Some effects of sewage discharge to the marine environment. Ph.D. Dissertation, Scripps Institution of Oceanography, University of California, San Diego, Calif.

Smith, K. L. and Hessler, R. R., 1974. Respiration of benthopelagic fishes, in situ measurements at 1230 meters. Science, 184: 72—73.

Smith, R. W. and Greene, C. S., 1976. Biological communities near submarine outfall. J. Water Pollut. Control Fed., 48(8): 1894—1912.

Snedecor, G. W., 1962. Statistical Methods. Iowa University, Iowa City, Iowa, 534 pp.

Sokolova, M. N., 1965. The uneven distribution of food groupings of the deep water uneven sedimentation. Oceanology, 5: 85—92

Sorokin, Y. I., 1973. Microbiological aspects of the productivity of coral reefs. In: O. A. Jones and R. Endean (Editors), Biology and Geology of Coral Reefs, Vol. 2. Biology 1. Academic Press, New York, N.Y., pp. 17—45.

Sorokin, Y. I., 1978. Microbial production in the coral-reef community. Arch. Hydrobiol., 83(3): 281—323.

Soule, D. and Oguri, M., 1972. Circulation patterns in Los Angeles—Long Beach Harbor drogue study atlas and data report. In: Marine Studies of San Pedro Bay, California, Part 1. Allan Hancock Foundation and Sea Grant Programs, University of Southern California, Los Angeles, Calif.

Soule, D. and Oguri, M., 1976. Bioenhancement studies of the receiving waters in outer Los Angeles Harbor. In: Marine Studies of San Pedro Bay, California, Part 12. Allan Hancock Foundation and Sea Grant Programs. University of Southern California, Los Angeles, Calif.

Soule, D. and Oguri, M. (Editors), 1978. Biological investigations. In: Marine Studies of San Pedro Bay, California, Part 14. Allan Hancock Foundation and Sea Grant Programs, University of Southern California, Los Angeles, Calif.

Soule, D. and Oguri, M. (Editors), 1979a. Ecological changes in outer Los Angeles—Long Beach Harbors following initiation of secondary waste treatment and cessation of fish cannery waste effluent. In: Marine Studies of San Pedro Bay, California, Part 16. Allan Hancock Foundation and Sea Grant Programs, University of Southern California, Los Angeles, Calif.

Soule, D. and Oguri, M. (Editors), 1979b. The marine environment in Los Angeles and Long Beach Harbors during 1978. In: Marine Studies of San Pedro Bay, California, Part 16. Allan Hancock Foundation and Sea Grant Programs, University of Southern California, Los Angeles, Calif.

Soule, D., Oguri, M. and Soule, J., 1978. Urban and fish-processing wastes in the marine environment; bioenhancement studies at Terminal Island, California. Bull. Calif. Water Pollut. Control Assoc. 15(1): 58—63.

Southard, J. B. and Stanley, D. J., 1976. Shelf-break processes and sedimentation. In: D. J. Stanley and D. J. P. Swift (Editors), Marine Sediment Transport and Environmental Management, Ch. 16. Wiley, New York, N.Y., pp. 351—377.

Spencer, D. W., Brewer, P. G., Fleer, A., Honjo, S., Krishnaswani, S. and Nozaici, Y., 1978. Chemical fluxes from a sediment trap experiment in the deep Sargasso Sea. J. Mar. Res., 36: 493—523.

Squires, D. F., 1981. The Bight of the Big Apple. Natl. Ocean. Atmos. Admin. (N.O.A.A.), Spec. Rep., U.S. Dep. Commer., Washington, D.C. (in press).

Stancher, B. and Chimenti, M., 1970. Sui molluschi dell'alto Adriatico—Contenuto di metalli in alcuni Gasteropodi e Lamellibranchi. Rass. Chim., 3: 78—83.

Starr, R. B. and Steimle, F. W., 1979. Temporal development of physical characteristics. In: R. L. Swanson and C. J. Sindermann (Editors), Oxygen Depletion and Associated Benthic Mortalities in New York Bight, 1976, Ch. 2. U.S. Dep. Commer., Washington, D.C., N.O.A.A. Prof. Pap.

Stauffer, G., 1979. Estimate of the spawning biomass of the northern anchovy central subpopulation for the 1979—80 fishing season. Natl. Mar. Fish. Serv., S.W. Fish. Cent., La Jolla, Calif., 20 pp.

Steele, J. H. (Editor), 1978. Spatial Pattern in Plankton Communities. Plenum, New York, N.Y., 470 pp.

Steele, R. L., 1977. Effects of certain petroleum products on reproduction and growth of zygotes and juvenile stages of the alga *Fucus edentatus* De la Pyl (Phaeophyccae: Fucales). In: D. A. Wolfe (Editor), Proceedings of Symposium on Fate and Effects of Petroleum Hydrocarbons in Marine Ecosystems and Organisms. Pergamon, New York, N.Y., pp. 138—142.

Steemann Nielsen, E., 1975. Marine Photosynthesis. Elsevier, Amsterdam, 142 pp.

Steemann Nielsen, E. and Wium-Andersen, S., 1971. The influence of copper on photosynthesis and growth in diatoms. Physiol. Plant., 24: 480—484.

Stefanile, M. and Vergine, F., 1973. Il Golfo di Napoli. Sagep, Genoa, 130 pp.

Stefanini, S., 1971. Distribuzione di Li, Na, K, Sr, Cr, Mn, Fe, Ni, Cu, Zn e Pl nel sedimenti superficiali dell'Adriatico Settentrionale tra Venezia e Trieste. Mem. Mus. Trientino Sci. Nat., 28: 173—213.

Steffin, O., Burns, R. E., Erickson, B. H., Lavelle, J. W. and Ozturgut, E., 1979. Environmental monitoring of a deep-ocean mining system test. 11th Annu. Conf. on Offshore Technology, Houston, Texas, April 30—May 3, 1979, O.T.C. Paper No. 3450, 6 pp.

Steimle, Jr., F. W. and Radosh, D. J., 1979. Effects of the benthic invertebrate community. In: R. L. Swanson and C. J. Sindermann (Editors), Oxygen Depletion and Associated Benthic Mortalities in New York Bight, 1976, Ch. 12. U.S. Dep. Commer., Washington, D.C., N.O.A.A., Prof. Pap.

Stephen-Hassard, Q. D. (Editor), 1978. The feasibility of and potential impact of manganese nodule processing in Hawaii, Dep. Plann. Econ. Dev., State Hawaii, Honolulu, Hawaii, Feb. 1978, 58 pp. and 11 appendices.

Stern, E. M. and Stickle, W. B., 1978. Effects of turbidity and suspended material in aquatic environments, literature review. U.S. Army Eng. Waterways Exp. Stn., Vicksburg, Miss., Tech. Rep. D-78-21.

Stilling, B. R., Lagally, A., Bauersteld, P. and Soares, J., 1974. Effect of cystine, selenium and fish protein on the toxicity and metabolism of methyl mercury in rats. Toxicol. Appl. Pharmacol., 30: 243—254.

Stirn, J., Avcin, A., Cenclj, J., Dorer, M., Gomiscek, S., Kveder, S. Malej, A., Meischner, D., Nozina, I., Paul, J. and Tusnik, P., 1974. Pollution problems of the Adriatic Sea—An interdisciplinary approach. Rev. Int. Océanogr. Méd., 35/36: 21—78.

Stober, Q. J., Dinnel, P. A., Wert, M. A. and Nakatani, R. E., 1977. Toxicity of West Point effluent to marine indicator organisms. Municipality of Metropolitan Seattle, Seattle, Wash., Final Biannu. Rep.

Stoeppler, M., Bernhard, M., Schulte, E. and Backhaus, F., 1979. Mercury in marine organisms from central and western Mediterranean. Sci. Total. Environ., 13: 209—223.

Stoewsand, G. S., Bache, C. A. and Lisk, D. J., 1974. Dietary selenium protection of methylmercury intoxication of Japanese quail. Bull. Environ. Contam. Toxicol., 11: 152—156.

Strickland, J. D. H. and Terhune, L. D. B., 1961. The Study of in situ marine photosynthesis using a large plastic bag. Limnol. Oceanogr., 6: 93—96.

Strohal, P. and Dzajo, M., 1975. Concentration of mercury in North Adriatic biota.

Thalassia Jugosl., 11: 221—229.

Suess, E., 1978. Distinction between natural and anthropogenic materials in sediments. In: E. D. Goldberg (Editor), Biogeochemistry of Estuarine Sediments. UNESCO, Paris, pp. 224—237.

Sullivan, B. K. and Hancock, D., 1977. Zooplankton and dredging: research perspectives from a critical review. Water Res. Bull., 13(3): 461—468.

Sumino, K., Yamamoto, R. and Kitamura, S., 1977. A role of selenium against methylmercury toxicity. Nature (London), 268: 73—74.

Summers, A. D., 1972. Mercury resistance in a plasmid-bearing strain of *Escherichia coli*. J. Bacteriol., 112: 1228—1236.

Sunda, W. G. and Gillespie, P. A., 1979. The response of a marine bacterium to cupric ion and its use to estimate cupric ion activity in sea water. J. Mar. Res., 37(4): 761—777.

Sunda, W. G. and Guillard, R. R., 1976. The relationship between cupric ion activity and the toxicity of copper to phytoplankton. J. Mar. Res., 34: 511—529.

Sustar, J. F. and Ecker, R. M., 1972. Monitoring dredge disposal of San Francisco Bay. Proc. Conf. on Offshore Technology. Houston, Texas, Pap. No. 1600.

Swanson, R. L., 1976. Tides. M.E.S.A. New York Bight atlas, Monogr. 4. New York Sea Grant Institute, Albany, N.Y., 34 pp.

Swanson, R. L., 1977. Status of ocean dumping research in the New York Bight. J. Waterways Div., Proc. Am. Soc. Civ. Eng., 103(WW1): 9—24.

Swanson, R. L., 1979. Pollution of the coastal ocean of the New York Bight. La Recherche, Paris.

Swanson, R. L. and Sindermann, C. J., 1979 (Editors), 1979. Oxygen Depletion and Associated Benthic Mortalities in New York Bight, 1976. U.S. Dep. Commer., Washington, D.C., N.O.A.A. Prof. Pap.

Swanson, R. L., Stanford, H. M., O'Connor, J. S., Chanesman, S., Parker, C. A., Eisen, P. A. and Mayer, G. F., 1978. June 1976 pollution of Long Island ocean beaches. J. Environ. Eng. Div., Proc. Am. Soc. Civ. Eng., 104(EE6): 1067—1083.

Swanson, R. L., Sindermann, C. J. and Han, G., 1979. Oxygen depletion and the future: an evaluation. In: R. L. Swanson and C. J. Sindermann (Editors), Oxygen Depletion and Associated Benthic Mortalities in New York Bight, 1976. U.S. Dep. Commer., Washington, D.C., N.O.A.A. Prof. Pap.

Swift, D. J. P., Freeland, G. L., Gadd, P. E., Han, G., Lavelle, J. W. and Stubblefield, W. L., 1976. In: M. G. Gross (Editor), Middle Atlantic Continental Shelf and the New York Bight, Spec. Symp. Vol. 2. Am. Soc. Limnol. Oceanogr., pp. 69—82.

Takahashi, F. T. and Kittredge, J. S., 1973. Sublethal effects of the water soluble component of oil: chemical communication in the marine environment. In: D. G. Ahearn and S. P. Meyers (Editors), The Microbial Degradation of Oil Pollutants. Center Wetland Resources, Louisiana State University, Baton Rouge, La., Publ. No. LSU-SG-73-01, pp. 259—264.

Takahashi, M., Thomas, W. H., Siebert, D. L. R., Beers, J., Koeller, P. and Parsons, T. R., 1975. The replication of biological events in enclosed water columns. Arch. Hydrobiol., 76: 5—23.

Taylor, W. R., 1964. Abundance of chemical elements in the continental crust: a new table. Geochim. Cosmochim. Acta, 28: 1273—1285.

Thibaud, Y., 1971. Teneur en mercure dans quelques poissons de consommation courante. Sci. Pêche, 209: 1—10.

Thibaud, Y., 1973. Teneur en mercure dans les moules du littoral français. Sci. Pêche, 221: 1—15.

Thibaud, Y. and Duguy, R., 1973. Teneur en mercure chez les cétacés des côtes de France. In: Comité des Mammifères Marins I.C.E.S., C.M. 1973/n:2. I.C.E.S., Copenhagen.

Thom, R. M., Chew, K. K. and Word, J. Q., 1979. Abundance, biomass and trophic structure of the subtidal infaunal communities of the eastern side of Central Puget Sound, Rep. to Municipality of Metropolitan Seattle, Wash., May, 1979, 43 pp.

Thomas, J. P., Phoel, W. C., Steimle, F. W., O'Reilly, J. E. and Evans, C. A., 1976. Seabed oxygen consumption—New York Bight Apex. Am. Soc. Limnol. Oceanogr., Spec. Symp.., 2, pp. 354—369.

Thomas, J. P., O'Reilly, J. E., Draxler, A., Babinchak, J. A., Robertson, C. N., Phoel, C., Waldhauer, R., Evans, C. A., Matte, A., Cohn, M., Nitkowski, M. and Dudley, S., 1979. In: R. L. Swanson and C. J. Sindermann (Editors), Oxygen Depletion and Associated Benthic Mortalities in New York Bight, 1976, Ch. 10. U.S. Dep. Commer., Washington, D.C., N.O.A.A. Prof. Pap.

Thomas, W. H. and Seibert, D. L. R., 1977. Effects of copper on the dominance and the delivery of algae: controlled ecosystem pollution experiment. Bull. Mar. Sci., 27: 23—33.

Thomas, W. H., Seibert, D. L. R. and Dodson A. N., 1974. Phytoplankton enrichment experiments and bioassay in natural coastal seawater and in sewage outfall receiving waters off southern California. Estuarine Coastal Mar. Sci., 2: 191—206.

Thomas, W. H., Seibert, D. L. R. and Takahashi, M., 1977a. Controlled ecosystem pollution experiment: effect of mercury on enclosed water columns, III. Phytoplankton population dynamics and production. Mar. Sci. Commun., 3: 331—354.

Thomas, W. H., Holm-Hansen, O., Seibert, D. L. R., Azam, F., Hodson, R. and Takahashi, M., 1977b. Effects of copper on phytoplankton standing crop and productivity: controlled ecosystem pollution experiment. Bull. Mar. Sci., 27: 34—43.

Timoney, J. F., Port, J., Giles, J. and Spanier, J., 1978. Heavy metal and antibiotic resistance in the bacterial flora of sediments of New York Bight. J. Appl. Environ. Microbiol., 36: 465—472.

Tingle, A. G. and Dieterle, D. A., 1977. A numerical oil trajectory model. Brookhaven Natl. Lab., BNL 50649, Upton, N.Y.

Tingle, A. G., Dieterle, D. A. and Walsh, J. J., 1979. Perturbation analysis of the New York Bight, Brookhaven National Laboratory, Upton, N.Y., Contract No. EY-76-C-01-0016.

Torre, J. A. and Masso, C., 1975. El contenido en mercurio de los moluscos de la ria de Pontevedra, como medida de su grado de contaminacion en dicho metal. Bol. Inst. Español Oceanogr. Madrid, 191: 1—10.

Toth, J. R., 1977. Determination of mercury in manganese nodules by cold vapor atomic absorption spectrometry. Anal. Chim. Acta, 92: 409—412.

Trefry, J. H., 1977. The transport of heavy metals by the Mississippi River and their fate in the Gulf of Mexico. Ph.D. Thesis, Texas A & M University, College Station, Texas, 223 pp.

Trefry, J. H. and Presley, B. J., 1976a. Heavy metals in sediments from San Antonio Bay and the northwest Gulf of Mexico. Environ. Geol., 1: 283—294.

Trefry, J. H. and Presley, B. J., 1976b. Heavy metal transport from the Mississippi River to the Gulf of Mexico. In: H. L. Windom and R. A. Duce (Editors), Marine Pollutant Transfer. D. C. Heath, Lexington, Mass., pp. 39—76.

Tsuchiya, M., 1968. Upper waters of the intertropical Pacific ocean. John Hopkins Oceanogr. Stud., No. 4, 50 pp.

Tsuchiya, M., 1974. Variation of the surface geostrophic flow in the eastern intertropical Pacific Ocean. Fish. Bull., 72(4): 1075—1086.

Turekian, K. K., 1969. The oceans, streams, and atmosphere. In: K. H. Wedepohl (Editor), Handbook of Geochemistry, Vol. I. Springer, Berlin, pp. 297—323.

Turekian, K. K., 1977. The fate of metals in the oceans. Geochim. Cosmochim. Acta, 41: 1139—1144.

Turekian, K. K. and Wedepohl, K. H., 1961. Distribution of the elements in some major units of the earth's crust. Bull. Geol. Soc. Am., 72: 175—191.

Turekian, K. K., Cochran, J. K., Kharkar, D. P., Cerrato, R. M., Vaisnys, J. R., Sanders, H. L., Grassle, J. F. and Allen, J. A., 1975. Slow growth rate of a deep-sea clam determined by ^{228}Ra chronology. Proc. Natl. Acad. Sci. U.S.A., 72: 2829—2832.

U.N.O.E.T.O. (United Nations Oceans Economics and Technology Office), 1979. Manganese Nodules: Dimensions and Perspectives. D. Reidel, Dordrecht, 194 pp.

U.S.A.C.E. (U.S. Army Corps of Engineers), 1950—1975. Stages and discharges of the Mississippi River and tributaries and other watersheds in the New Orleans District. U.S. Army Corps Eng., New Orleans, La.

U.S.A.C.E. (U.S. Army Corps of Engineers), 1975. Mississippi River—Baton Rouge to the Gulf: maintenance dredging sediment and water quality assessment. U.S. Army Corps Eng., New Orleans, La., 41 pp. and appendices.

U.S. Congress, 1978. Modification of secondary treatment requirements for discharges into marine waters. House of the U.S. Congress, Subcommittee on Water Resources, 95th Congr., 2nd Sess., Hearing May 24 and 25, 1978, Rep. 95-54. U.S. Government Printing Office, Washington, D.C., pp. 133—218.

U.S.D.C. (U.S. Department of Commerce), 1979. Report to the Congress on Ocean Dumping Research, January—December 1978. U.S. Dep. Commer., Washington, D.C.

U.S.E.P.A. (U.S. Environmental Protection Agency), 1973. Ocean dumping criteria. Fed. Reg., 38: 12872—12877.

U.S.E.P.A. (U.S. Environmental Protection Agency), 1976. National Interim Primary Drinking Water Regulations. U.S. Environ. Prot. Agency, Off. Water Supply, EPA-570/9-76-003, 159 pp.

U.S.G.S. (U.S. Geological Survey), 1972. Water resources data for Louisiana. U.S. Geol. Survey, Water Resour. Div., Baton Rouge, La., 226 pp.

U.S.G.S. (U.S. Geological Survey), 1973. Water resources data for Louisiana. U.S. Geol. Surv., Water Resour. Div., Baton Rouge, La., 315 pp.

U.S.G.S. (U.S. Geological Survey), 1974. Water resources data for Louisiana. U.S. Geol. Surv., Water Resour. Div., Baton Rouge, La., 473 pp.

U.S.G.S. (U.S. Geological Survey), 1975. Water resources data for Louisiana. U.S. Geol. Surv., Water Resour. Div., Baton Rouge, La., 816 pp.

Vaccaro, R. F. and Dennett, M. R., 1977. The environmental response of marine bacteria to waste disposal activities at Deep Water Dumpsite 106. In: NOAA Dumpsite Evaluation Report 79-1.

Vaccaro, R. F. and Dennett, M. R., 1980. The environmental response of marine bacteria to waste disposal activities at Deep Water Dumpsite 106. In: B. H. Ketchum, D. R. Kester and P. K. Park (Editors), The Ocean Dumping of Industrial Wastes. Plenum, New York, N.Y.

Vaccaro, R. F., Grice, G. D., Rowe, G. T. and Wiebe, P. H., 1972. Acid—iron waste disposal and the summer distribution of standing crops in the New York Bight. Water Res., 6(3): 231256.

Vaccaro, R. F., Azam, F. and Hodson, R. E., 1977. Response of natural marine bacterial populations to copper: controlled ecosystem pollution experiment. Bull. Mar. Sci., 27: 17—22.

van Bennekom, A. J. and Jager, J. E., 1978. Dissolved aluminium in the Zaire River plume. Neth. J. Sea. Res., 12: 358—367.

van der Weijden, C. H., Arnoldus, M. J. H. L. and Meurs, C. J., 1977. Desorption of metals from suspended material in the Rhine estuary. Neth. J. Sea Res., 11: 210—225.

van Olphen, H., 1963. Clay Colloid Chemistry. Interscience, New York, N.Y., 301 pp.

Verber, J. L., 1976. Safe shellfish from the sea. In: M. G. Gross (Editor), Middle Atlantic Continental Shelf and the New York Bight, Spec. Symp. Vol. 2, Am. Soc. Limnol. Oceanogr., pp. 433—441.

Vettorazzi, G., 1975. Toxicological decisions and recommendations resulting from the safety assessment of pesticide residues in food. Crit. Rev. Toxicol., 4: 125—183.

Viviani, R., Borgatti, A. R., Cansellieri, D., Crisetig, G. and Cortesi, P., 1969. Residui di DDT e di suoi metaboliti nei tessuti di clupeiformi Adriatici. Atti. Soc. Ital. Sci. Vet., 23: 1—4.

Viviani, R., Crisetig, G., Petruzzi, V. and Cortesi, P., 1973a. Residui di pesticidi clorurati e di bifenili policlorurati nei clupeiformi adriatici. In: Proc. 5th Int. Colloq. Med. Oceanogr., Messina, Oct. 4—7 1971. Bonanzinga, Messina, pp. 607—621.

Viviani, R., Rossi, C. M., Frignani, M. and Rabbi, E., 1973b. Recherche sur la présence de mercure dans les sédiments de la Mer Adriatique du Nord en face du delta du Po. Journées d'Etudes Pollutions Marines (Comm. Int. l'Explor. Sci. Mer., Athens, Nov. 1972), Comm. Int. Explor. Sci. Mer., Monaco (preprint).

Viviani, R., Crisetig, G., Cortesi, P. and Carpene, E., 1974. Résidue de polychlorobi-phényles (PCB) et des pesticides chlorés dans les poissons et les oiseaux du delta du Po. Rev. Océanogr. Méd., 35/36: 79—90.

Vucetic, T., Vernberg, W. B. and Anderson, G., 1974. Long-term annual fluctuations of mercury in the zooplankton of the east central Adriatic. Rev. Int. Océanogr. Méd., 33: 75—82.

Wahlgren, M. A., Alberts, J. J., Nelson, D. M. and Orlandini, K. A., 1976. Study of the behaviou of transuranics and possible chemical homologues in Lake Michigan water and biota. In: Transuranium Nuclides in the Environment. International Atomic Energy Agency, Vienna, pp. 9—24.

Wallen, I. E., 1951. The direct effect of turbidity on fishes. Bull. Okla. A & M Coll., 48(2): 1—27.

Walsh, F. and Mitchell, R., 1973. Inhibition of bacterial chemoreception by hydro-carbons. In: D. G. Ahern and S. P. Meyers (Editors), The Microbial Degradation of Oil Pollutants. Center Wetlands Resources, Louisiana State University, Baton Rouge, La., Publ. No. LSU-SG-73-01, pp. 275—278.

Walsh, J. J., Falkowski, P. G., Stoddard, A., Tingle, A., Dieterle, D., Han, G. and Esaias, W. E., 1980. Climatology, phytoplankton species succession, and oxygen depletion within the New York Bight. Brookhaven National Laboratory, Upton, N.Y.

Waterman, T., Nunnemacher, R. F., Chace, F. A. and Clarke, G. L., 1939. Diurnal vertical migration of deep-sea plankton. Biol. Bull. Woods Hole, 76: 256—279.

Wavre, M. and Brinkhurst, R. O., 1971. Interaction between some tubificid oligochaetes and bacteria found in the sediments of Toronto Harbor, Ontario. J. Fish. Res. Board Can. 28: 335—341.

Wechsler, B. A. and Cogley, D., 1977. Laboratory study related to predicting the tur-bidity-generated potential of sediments to the dredged. U.S. Army Eng. Waterways Exp. Stn., Vicksburg, Miss., Tech. Rep. D-77-14.

Wedepohl, K. H., 1970a. Environmental influences on the chemical composition of shales and clays. Phys. Chem. Earth, 8: 310—333.

Wedepohl, K. H., 1970b. Geochemische Daten von sedimentären Karbonaten und Kar-bonategesteinen in ihrem faziellen und petrogenetischen Aussagewert. Verh. Geol. Bundesanst., pp. 692—705.

Weiss, H. V., Koide, M. and Goldberg, E. D., 1971. Mercury in a Greenland ice sheet: evidence of recent input by man. Science, 174: 690—694.

Welling, C. G., 1972. Some environmental factors associated with deep ocean mining. 8th Annu. Mar. Technol. Soc. Meet., Sept. 11—13, 1972.

Welling, C. G., 1979. The future outlook for the nodule industry. In: Manganese Nodules: Dimensions and Perspectives. United Nations Ocean Economics and Tech-nology Office, D. Reidel, Dordrecht, pp. 139—148.

W.H.O. (World Health Organisation), 1973. Les oligo-éléments en nutrition humaine. World Health Org. Tech. Rep., Ser., No. 532, 70 pp.

558

W.H.O. (World Health Organisation), 1976a. Mercury. Environmental Health Criteria 1. World Health Organisation, Geneva, 132 pp.

W.H.O. (World Health Organisation), 1976b. Polychlorinated biphenyls and terphenyls. Environmental Health Criteria 2. World Health Organisation, Geneva, 85 pp.

W.H.O. (World Health Organisation), 1977. Lead. Environmental Health Criteria 3. World Health Organisation, Geneva, 160 pp.

Wiebe, P. H., Boyd, S. H. and Winget, C., 1976. Particulate matter sinking to the deep-sea floor at 2000 m in the Tongue of the Ocean, Bahamas, with a description of a new sedimentation trap. J. Mar. Res., 34: 341—354.

Wiebe, P. H., Madin, L. P., Haury, L. R., Harbison, G. R. and Philbin, L. M., 1978. Diel vertical migration by Salpa aspera: Potential for large-scale particulate organic matter transport to the deep-sea. Mar. Biol., 53: 249—253.

Williams, R. B. and Murdock, M. B., 1973. Effects of continuous low-level gamma radiation on sessile marine invertebrates. In: Radioactive Contamination of the Marine Environment, International Atomic Energy Agency, Vienna, pp. 551—563.

Windom, H. L., 1975. Eolian contributions to marine sediments. J. Sediment. Petrol., 45: 520—529.

Woodhead, D. S., 1970. The assessment of the radiation dose to developing fish embryos due to the accumulation of radioactivity by the egg. Radiat. Res., 43: 582—586.

Woodwell, G. M., Wurster, C. F. and Isaacson, P. A., 1967. DDT residues in an East Coast estuary: a case of biological concentration of a persistent insecticide. Science, 156: 821—824.

Word, J. Q., 1978. An evaluation of benthic invertebrate sampling devices for investigating feeding habits of fish. In: C. A. Simenstad and S. J. Lipovsky (Editors), Fish Food Habit Studies. Washington Sea Grant, WSG-WO 77-2, University of Washington, Seattle, Wash., pp. 43—56.

Word, J. Q., 1979. The infaunal trophic index. In: Coastal Water Research Project, Annual Report, 1978. U.S. Dep. Commer., Springfield, Va., Natl. Tech. Info. Serv., PB 299830/AS, pp. 19—39.

Word, J. Q. and Mearns, A. J., 1979. The 60-meter control survey. In: Coastal Water Research Project, Annual Report, 1978. U.S. Dep. Commer., Springfield, Va., Natl. Tech. Info. Serv., PB 299830/AS, pp. 41—56.

Word, J. Q., Myers, B. L. and Mearns, A. J., 1977. Animals that are indicators of marine pollution. In: Coastal Water Research Project, Annual Report, 1977. U.S. Dep. Commer., Springfield, Va., Natl. Tech. Info. Serv., PB 274463/AS, pp. 199—206.

Wright, R. T. and Hobbie, J. E., 1966. Use of glucose and acetate by bacteria and algae in aquatic ecosystems. Ecology, 47: 447—464.

Wright, T. D., 1978. Aquatic Dredged Material Disposal Impacts. U.S. Army Eng. Waterways Exp. Stn., Vicksburg, Miss., Tech. Rep. DS-78-1.

Wright, T. D., Hamil, B. M., Draft, K. J., Leddy, D. G. and Nordeng, S. C., 1975. Analysis report—Keweenaw Waterway maintenance dredging. Vol. I, Analysis Report, pp. 1—141; Vol. II, Data Base, pp. 142—350; Vol. III, Symap Figures G and C: pp. 351—491; and Vol. IV, Symap Figure B: pp. 492—631. Rep. to St. Paul District, U.S. Army Corps of Engineers, St. Paul., Minn., under Contract No. DACW-37-74-C-0149.

Wurster, C. F., 1968. DDT reduces photosynthesis by marine phytoplankton. Science, 159: 1474—1475.

Wyrtki, K., 1966. Oceanography of the Eastern Equatorial Pacific Ocean. Oceanogr. Mar. Biol. Annu. Rev., 4: 33—68.

Wyrtki, K. and Meyers, G., 1975. The trade wind field over the Pacific Ocean, Part I. Hawaii Inst. Geophys., Univ. Hawaii, Honolulu, Hawaii, HIG-75-1.

Yentsch, D. S., 1977. Plankton production. M.E.S.A. New York Bight atlas, Monogr. 12. New York Sea Grant Institute, Albany, N.Y., 25 pp.

Young, D. R., 1979. Priority pollutants in municipal waste waters. In: Coastal Water Research Project, Annual Report, 1978. U.S. Dep. Commer., Springfield, Va., Natl. Tech. Info. Serv., PB 299830/AS, pp. 103—112.

Young, D. R. and Heesen, T. C., 1978. DDT, PCB, and chlorinated benzenes in the marine ecosystem off southern California. In: R. L. Jolley, H. Gorchev and D. H. Hamilton, Jr. (Editors), Water chlorination: Environmental Impact and Health Effects, Vol. 2. Ann Arbor Science, Ann Arbor, Mich., pp. 267—290.

Young, D. R. and Jan, T. K., 1975. Trace metals in nearshore seawater. In: Coastal Water Research Project Annual Report, 1975, U.S. Dep. Commer., Springfield, Va., Natl. Tech. Info. PB 274467/AS, pp. 143—146.

Young, D. R. and Mearns, A. J., 1979. Pollutant flow through marine food webs. In: Coastal Water Research Project, Annual Report, 1978. U.S. Dep. Commer., Springfield, Va., Natl. Tech. Info. Serv., PB 299830/AS, pp. 185—202.

Young, D. R., Jan, T. K. and Heesen, T. C., 1978. Cycling of trace metal and chlorinated hydrocarbon wastes in the Southern California Bight. In: M. L. Wiley (Editor), Estuarine Interactions. Academic Press. New York, N.Y., pp. 481—496.

Young, D. R., Heesen, T. C., Esra, G. N. and Howard, E. B., 1979. DDE-contaminated fish off Los Angeles are suspected cause in deaths of captive marine birds. Bull. Environ. Contam. Toxicol., 21: 584—590.

Zabawa, C. F., 1978. Microstructure of agglomerated suspended sediments in northern Chesapeake Bay estuary. Science, 202: 49—51.

Zafiropoulos, D. and Grimanis, A. P., 1977. Trace elements in *Acartia clausi* from Elefsis Bay of the upper Saronikos Gulf, Greece. Mar. Pollut. Bull., 8: 79—81.

Zaripov, B. R. and Rzheplinskiy, D. G., 1977. Long term seasonal water circulation of the northeastern Atlantic and of the Norwegian, Greenland and North Seas (diagnostic calculations). Oceanology, 17: 520—524.

ZoBell, C. E. and Feltham, C. B., 1937. Bacteria as food for certain marine invertebrates. J. Mar. Res., 1: 312—327.

INDEX*

abalone, 55, 60
Acartia hamata, 460
acid volatile sulfides, 58
Acrocalanus inermis, 460
acute effects of waste discharges, 352
Adriatic Sea, 1, 14, 145, 147, 154, 177
—, north, 165
adsorption chemistry, 421
Aegean Sea, 147
aerosols, 95
African coast, northwestern, 161
air-borne pollutants, 100
air pollution, 94
albacore, 463
aldrin, 177, 191
Alepacephalus agassize, 345
algae, 53
algal spores, 457
aluminum, 189, 198, 203, 208, 450, 462
aluminum/iron, 462
Amazon River, 16, 199, 291
ambient air temperature, 442
Ambrose Light Tower, 383
American Cyanamid, 396
—, waste, 384, 385, 396, 399
ammonia, 30—32, 189, 259, 469
—, levels
Amphiodia urticia, 63
amphipods, 472
Anacapri, 99
anaerobic areas, 119
anaerobic conditions, 126, 453
anchovy (*Engraulis moirax*), 36, 156, 175, 272
anemones (*Medtridium senile*), 35
anion-exchange chromatography, 196
anisole, 383
Anisomysis sp., 460
Anodonia sp., 164
anoxia and fish kill, 359
anoxic bottom waters, 353
Antarctic Ocean, 298
anthropods, 36
Antilles Current, 383
Anzio, 149
Aphanopus carbo, 300
aquaculture study, San Diego, CA, 18

---

*Italicized page numbers refer to illustrations.

aquatic organisms, 259
Armadia bioculata (polychaete), 363
Arno River, 168
arsenic, 150, 166, 171, 188, 189, 294
artificial island, wildlife habitat, 251
Atchafalaya Bay, 257
Atchafalaya River, 193
Atlantic Ocean, 16, 158, 163, 285, 402
—, coast, 179
atmospheric input, trace metals, North
    Sea, 293
atomic absorption spectrophotometer,
    196
authigenic deposits, 429

bacteria, 269, 282, 395, 399, 452, 469
—, assays, 399
—, biomass, 219, 452
—, conversion, 469
—, decomposition, 469
—, growth, 452, 453, 469
—, ocean, 399
—, populations, 399
—, species, 399
—, tolerance development, 222
Baltic Sea, 289
Barcelona, 177, 178
barium, 189
baseline data, 383
baseline values, 8
bathypelagic zone, 300
Bay of Marseilles, 175
Bay of Naples, 67, 77, 136
—, currents in, 76
—, description of, 73
—, hydrocarbons, 118, 126
—, salinity, 87
Benguela Current, 306
benthic, bioassay procedures, 259
—, biomass, 58, 60
—, communities, 37, 303, 472
—, diatoms, 53
—, environment, 35
—, fauna, 261, 472
—, feeding fish, 59
—, fish, 56, 59, 179
—, groups, 305
—, infauna, 36, 41, 59, 60
—, invertebrates, 246, 247, 334
—, macrofauna, 473
—, movement, 30
—, organisms, 259, 471